MACHINE LEARNING FOR
BUSINESS ANALYTICS

MACHINE LEARNING FOR BUSINESS ANALYTICS

Concepts, Techniques, and Applications with JMP Pro®

Second Edition

GALIT SHMUELI
National Tsing Hua University
Taipei, Taiwan

PETER C. BRUCE
statistics.com
Arlington, USA

MIA L. STEPHENS
JMP Statistical Discovery LLC
Cary, USA

MURALIDHARA ANANDAMURTHY
SAS Institute Inc
Mumbai, India

NITIN R. PATEL
Cytel, Inc.
Cambridge, USA

Library of Congress Cataloging-in-Publication Data Applied for:

Hardback: 9781119903833

Cover Design: Wiley
Cover Image: © AdobeLibrary/Adobe Stock Photos

Set in 10/12pt TimesLTStd by Straive, Chennai, India

SKY10045409_040523

To our families

Boaz and Noa
Liz, Lisa, and Allison
Michael, Jade Ann, and Audrey L
Seetha and Ananda
Tehmi, Arjun, and in memory of Aneesh

CONTENTS

PART II   DATA EXPLORATION AND DIMENSION REDUCTION

3   Data Visualization   59

PART IV   PREDICTION AND CLASSIFICATION METHODS

6   Multiple Linear Regression   147

7 k-Nearest Neighbors (k-NN)   175

PART VII   FORECASTING TIME SERIES

17 Handling Time Series   409

18 Regression-Based Forecasting   423

FOREWORD

When I began my career back in the last century, most corporate computing took place on mainframe computers, data was scarce, organizations were far more hierarchical, and managerial decision-making was often driven by the loudest person in the room or the "golden gut" of an experienced executive. By contrast, today's business world features a wide variety of digitally connected professionals who interact with their customers and their colleagues through software applications (many of them web- and cloud-based), remarkably powerful personal computers, and always-connected smart phones. Data is everywhere, though useful data is often still elusive. And more and more of the systems that companies and individuals rely upon are utilizing techniques from machine learning to deliver data-driven insights, make predictions, and drive decision making.

For the past decade, I have been teaching courses in machine learning and predictive analytics to business students at the University of San Francisco. My students have a wide variety of academic backgrounds and professional interests. My goal is to prepare them for careers in this rapidly evolving, digitally enabled, and increasingly data- and algorithmically-driven business world.

I was fortunate enough to find *Data Mining for Business Analytics: Concepts, Techniques, and Applications with JMP Pro* several years ago. This book provides a clear roadmap to the fundamentals of machine learning as well as a number of pathways to explore a variety of specific machine learning methods for business analytics including prediction, classification, and clustering. In addition, this textbook and the JMP Pro software combine to provide a great platform for interactive learning. The textbook utilizes the JMP Pro software to illustrate machine learning fundamentals, exploratory data analysis methods, and data visualization concepts, and a broad range of supervised and unsupervised machine learning methods. The book also provides exercises that also enable you to utilize JMP Pro to learn and master machine learning techniques.

I was very excited when I learned that the next edition of this book was ready to be released. Now entitled *Machine Learning for Business Analytics: Concepts, Techniques, and Applications with JMP Pro*, this 2nd edition is based on the most recent version of the JMP Pro software, and both the text and the software have been significantly expanded and updated. This new edition includes all the first edition material (supervised learning, unsupervised methods, visualization, and time series), as well as a number of new topics: recommendation systems, text mining, ethical issues in data science, deep learning, and interventions and reinforcement learning.

Along with the JMP Pro software, this book will provide you with a foundation of knowledge about machine learning. Its lessons and insights will serve you well in today's dynamic and data-intensive business world. Welcome aboard!

VIJAY MEHROTRA
University of San Francisco

PREFACE

This textbook first appeared in early 2007 and has been used by numerous students and practitioners and in many courses, including our own experience teaching this material both online and in person for more than 15 years. The first edition, based on the Excel add-in Analytic Solver Data Mining (previously XLMiner), was followed by two more Analytic Solver editions, a JMP Pro® edition, two R editions, a Python edition, a RapidMiner edition, and now this second JMP Pro edition, with its companion website, www.jmp.com/dataminingbook. JMP Pro is a desktop statistical package from JMP Statistical Discovery that runs natively on Mac and Windows machines.[1]

The first JMP Pro edition was the first edition to fully integrate JMP Pro. As in the previous JMP edition, the focus in this new edition is on machine learning concepts and how to implement the associated algorithms in JMP Pro. All examples, special topics boxes, instructions, and exercises presented in this book are based on JMP Pro 17, the professional version of JMP, which has a rich array of built-in tools for interactive data visualization, analysis, and modeling.[2]

For this new JMP Pro edition, a new co-author, Muralidhara Anandamurthy, comes on board bringing extensive experience in analytics and data science at Genpact, Target, and Danske, and as a member of the JMP Academic Team.

The new edition provides significant updates both in terms of JMP Pro and in terms of new topics and content. In addition to updating software routines and outputs that have changed or become available since the first edition, this edition also incorporates updates and new material based on feedback from instructors teaching MBA, MS, undergraduate, diploma, and executive courses, and from their students. Importantly, this edition includes several new topics:

- A new chapter on *Responsible Data Science* (Chapter 21) covering topics of fairness, transparency, model cards and datasheets, legal considerations, and more, with an illustrative example.
- A dedicated section on *deep learning* in Chapter 11.
- A new chapter on recommendations, covering association rules and collaborative filtering (Chapter 15).
- A new chapter on Text Mining covering main approaches to the analysis of text data (Chapter 20).
- The *Performance Evaluation* exposition in Chapter 5 was expanded to include further metrics (precision and recall, F1).

---

[1] JMP Statistical Discovery LLC, 100 SAS Campus Drive Cary, NC 27513.
[2] See https://www.jmp.com/pro

- A new chapter on *Generating, Comparing, and Combining Multiple Models* (Chapter 13) that covers ensembles and AutoML.
- A new chapter dedicated to *Interventions and User Feedback* (Chapter 14) that covers A/B tests, uplift modeling, and reinforcement learning.
- A new case (Loan Approval) that touches on regulatory and ethical issues.

A note about the book's title: The first two editions of the book used the title *Data Mining for Business Intelligence*. Business intelligence today refers mainly to reporting and data visualization ("what is happening now"), while business analytics has taken over the "advanced analytics," which include predictive analytics and data mining. Later editions were therefore renamed *Data Mining for Business Analytics*. However, the recent AI transformation has made the term *machine learning* more popularly associated with the methods in this textbook. In this new edition, we therefore use the updated terms *Machine Learning* and *Business Analytics*.

Since the appearance of the first JMP Pro edition, the landscape of the courses using the textbook has greatly expanded: whereas initially the book was used mainly in semester-long elective MBA-level courses, it is now used in a variety of courses in business analytics degrees and certificate programs, ranging from undergraduate programs to postgraduate and executive education programs. Courses in such programs also vary in their duration and coverage. In many cases, this textbook is used across multiple courses. The book is designed to continue supporting the general "predictive analytics" or "data mining" course as well as supporting a set of courses in dedicated business analytics programs.

A general "business analytics," "predictive analytics," or "data mining" course, common in MBA and undergraduate programs as a one-semester elective, would cover Parts I–III, and choose a subset of methods from Parts IV and V. Instructors can choose to use cases as team assignments, class discussions, or projects. For a two-semester course, Part VII might be considered, and we recommend introducing the new Part VIII (Data Analytics).

For a set of courses in a dedicated business analytics program, here are a few courses that have been using our book:

Predictive Analytics—Supervised Learning: In a dedicated business analytics program, the topic of predictive analytics is typically instructed across a set of courses. The first course would cover Parts I–III, and instructors typically choose a subset of methods from Part IV according to the course length. We recommend including "Part VIII: Data Analytics."

Predictive Analytics—Unsupervised Learning: This course introduces data exploration and visualization, dimension reduction, mining relationships, and clustering (Parts II and VI). If this course follows the Predictive Analytics: Supervised Learning course, then it is useful to examine examples and approaches that integrate unsupervised and supervised learning, such as the new part on "Data Analytics."

Forecasting Analytics: A dedicated course on time series forecasting would rely on Part VII.

Advanced Analytics: A course that integrates the learnings from predictive analytics (supervised and unsupervised learning) can focus on Part VIII: Data Analytics, where social network analytics and text mining are introduced, and responsible data science is discussed. Such a course might also include Chapter 13, Generating, Comparing, and Combining Multiple Models from Part IV, as well as Part V, which covers experiments, uplift, and reinforcement learning. Some instructors choose to use the cases (Chapter 22) in such a course.

In all courses, we strongly recommend including a project component, where data are either collected by students according to their interest or provided by the instructor (e.g., from the many machine learning competition datasets available). From our experience and other instructors' experience, such projects enhance the learning and provide students with an excellent opportunity to understand the strengths of machine learning and the challenges that arise in the process.

GALIT SHMUELI, PETER BRUCE, MIA STEPHENS, MURALIDHARA ANANDAMURTHY, AND NITIN PATEL
2022

ACKNOWLEDGMENTS

We thank the many people who assisted us in improving the book from its inception as *Data Mining for Business Intelligence* in 2006 (using XLMiner, now Analytic Solver), its reincarnation as *Data Mining for Business Analytics*, and now *Machine Learning for Business Analytics*, including translations in Chinese and Korean and versions supporting Analytic Solver Data Mining, R, Python, RapidMiner, and JMP.

Anthony Babinec, who has been using earlier editions of this book for years in his data mining courses at Statistics.com, provided us with detailed and expert corrections. Dan Toy and John Elder IV greeted our project with early enthusiasm and provided detailed and useful comments on initial drafts. Ravi Bapna, who used an early draft in a data mining course at the Indian School of Business, and later at University of Minnesota, has provided invaluable comments and helpful suggestions since the book's start.

Many of the instructors, teaching assistants, and students using earlier editions of the book have contributed invaluable feedback both directly and indirectly, through fruitful discussions, learning journeys, and interesting data mining projects that have helped shape and improve the book. These include MBA students from the University of Maryland, MIT, the Indian School of Business, National Tsing Hua University, and Statistics.com. Instructors from many universities and teaching programs, too numerous to list, have supported and helped improve the book since its inception.

Kuber Deokar, instructional operations supervisor at Statistics.com, has been unstinting in his assistance, support, and detailed attention. We also thank Anuja Kulkarni, Poonam Tribhuwan, and Shweta Jadhav, assistant teachers. Valerie Troiano has shepherded many instructors and students through the Statistics.com courses that have helped nurture the development of these books.

Colleagues and family members have been providing ongoing feedback and assistance with this book project. Vijay Kamble at UIC and Travis Greene at NTHU have provided valuable help with the section on reinforcement learning. Boaz Shmueli and Raquelle Azran gave detailed editorial comments and suggestions on the first two editions; Bruce McCullough and Adam Hughes did the same for the first edition. Noa Shmueli provided careful proofs of the third edition. Ran Shenberger offered design tips. Ken Strasma, founder of the microtargeting firm HaystaqDNA and director of targeting for the 2004 Kerry campaign and the 2008 Obama campaign, provided the scenario and data for the section on uplift modeling.

Marietta Tretter at Texas A&M shared comments and thoughts on the time series chapters, and Stephen Few and Ben Shneiderman provided feedback and suggestions on the data visualization chapter and overall design tips.

Susan Palocsay and Margret Bjarnadottir have provided suggestions and feedback on numerous occasions. We also thank Catherine Plaisant at the University of Maryland's Human–Computer Interaction Lab, who helped out in a major way by contributing exercises and illustrations to the data visualization chapter. Gregory Piatetsky-Shapiro, founder of KDNuggets.com, was generous with his time and counsel in the early years of this project.

We thank colleagues at the Sloan School of Management at MIT for their support during the formative stage of this book—Dimitris Bertsimas, James Orlin, Robert Freund, Roy Welsch, Gordon Kaufmann, and Gabriel Bitran. As teaching assistants for the data mining course at Sloan, Adam Mersereau gave detailed comments on the notes and cases that were the genesis of this book, Romy Shioda helped with the preparation of several cases and exercises used here, and Mahesh Kumar helped with the material on clustering.

Colleagues at the University of Maryland's Smith School of Business: Shrivardhan Lele, Wolfgang Jank, and Paul Zantek provided practical advice and comments. We thank Robert Windle and University of Maryland MBA students Timothy Roach, Pablo Macouzet, and Nathan Birckhead for invaluable datasets. We also thank MBA students Rob Whitener and Daniel Curtis for the heatmap and map charts.

Anand Bodapati provided both data and advice. Jake Hofman from Microsoft Research and Sharad Borle assisted with data access. Suresh Ankolekar and Mayank Shah helped develop several cases and provided valuable pedagogical comments. Vinni Bhandari helped write the Charles Book Club case.

We would like to thank Marvin Zelen, L. J. Wei, and Cyrus Mehta at Harvard, as well as Anil Gore at Pune University, for thought-provoking discussions on the relationship between statistics and data mining. Our thanks to Richard Larson of the Engineering Systems Division, MIT, for sparking many stimulating ideas on the role of data mining in modeling complex systems. Over two decades ago, they helped us develop a balanced philosophical perspective on the emerging field of machine learning.

We thank the folks at Wiley for this successful journey of nearly two decades. Steve Quigley at Wiley showed confidence in this book from the beginning, helped us navigate through the publishing process with great speed, and together with Curt Hinrichs's encouragement and support helped make this JMP Pro® edition possible. Jon Gurstelle, Kathleen Pagliaro, Allison McGinniss, Sari Friedman, and Katrina Maceda at Wiley, and Shikha Pahuja from Thomson Digital, were all helpful and responsive as we finalized the first JMP Pro edition. Brett Kurzman has taken over the reins and is now shepherding the project. Becky Cowan, Sarah Lemore, and Kavya Ramu greatly assisted us in pushing ahead and finalizing this new JMP Pro edition. We are also especially grateful to Amy Hendrickson, who assisted with typesetting and making this book beautiful.

Finally, we'd like to thank the reviewers of the first JMP Pro edition for their feedback and suggestions, and members of the JMP Documentation, Education and Development teams, for their support, patience, and responsiveness to our endless questions and requests. We thank L. Allison Jones-Farmer, Maria Weese, Ian Cox, Di Michelson, Marie Gaudard, Curt Hinrichs, Rob Carver, Jim Grayson, Brady Brady, Jian Cao, Elizabeth Claassen, Peng Liu, Chris Gotwalt, Russ Wolfinger, and Fang Chen. Most important, we thank John Sall, whose innovation, inspiration, and continued dedication to providing accessible and user-friendly desktop statistical software made JMP, and this book, possible.

PART I

PRELIMINARIES

1

INTRODUCTION

1.1 WHAT IS BUSINESS ANALYTICS?

Business analytics (BA) is the practice and art of bringing quantitative data to bear on decision-making. The term means different things to different organizations.

Consider the role of analytics in helping newspapers survive the transition to a digital world. One tabloid newspaper with a working-class readership in Britain had launched a web version of the paper, and did tests on its home page to determine which images produced more hits: cats, dogs, or monkeys. This simple application, for this company, was considered analytics. By contrast, the *Washington Post* has a highly influential audience that is of interest to big defense contractors: it is perhaps the only newspaper where you routinely see advertisements for aircraft carriers. In the digital environment, the *Post* can track readers by time of day, location, and user subscription information. In this fashion the display of the aircraft carrier advertisement in the online paper may be focused on a very small group of individuals—say, the members of the House and Senate Armed Services Committees who will be voting on the Pentagon's budget.

Business analytics, or more generically, *analytics*, includes a range of data analysis methods.

Many powerful applications involve little more than counting, rule checking, and basic arithmetic. For some organizations, this is what is meant by analytics.

The next level of business analytics, now termed *business intelligence* (BI), refers to the use of data visualization and reporting for becoming aware and understanding "what happened and what is happening." This is done by use of charts, tables, and dashboards to display, examine, and explore data. Business intelligence, which earlier consisted mainly of generating static reports, has evolved into more user-friendly and effective tools and practices, such as creating interactive dashboards that allow the user not only to access real-time data, but also to directly interact with it. Effective dashboards are those that tie directly to company data, and give managers a tool to see quickly what might not readily be apparent in a large complex database. One such tool for industrial operations managers

Machine Learning for Business Analytics: Concepts, Techniques, and Applications with JMP Pro®,
Second Edition. Galit Shmueli, Peter C. Bruce, Mia L. Stephens, Muralidhara Anandamurthy, and Nitin R. Patel.
© 2023 John Wiley & Sons, Inc. Published 2023 by John Wiley & Sons, Inc.

displays customer orders in one two-dimensional display using color and bubble size as added variables. The resulting 2 by 2 matrix shows customer name, type of product, size of order, and length of time to produce.

Business analytics now typically includes BI as well as sophisticated data analysis methods, such as statistical models and machine learning algorithms used for exploring data, quantifying and explaining relationships between measurements, and predicting new records. Methods like regression models are used to describe and quantify "on average" relationships (e.g., between advertising and sales), to predict new records (e.g., whether a new patient will react positively to a medication), and to forecast future values (e.g., next week's web traffic).

Readers familiar with the earlier edition of this book might have noticed that the book title changed from *Data Mining for Business Analytics* to *Machine Learning for Business Analytics*. The change reflects the more recent term BA, which overtook the earlier term BI to denote advanced analytics. Today, BI is used to refer to data visualization and reporting. The change from *data mining* to *machine learning* reflects today's common use of *machine learning* to refer to algorithms that learn from data. This book uses primarily the term *machine learning*.

WHO USES PREDICTIVE ANALYTICS?

The widespread adoption of predictive analytics, coupled with the accelerating availability of data, has increased organizations' capabilities throughout the economy. A few examples:

Credit scoring: One long-established use of predictive modeling techniques for business prediction is credit scoring. A credit score is not some arbitrary judgement of creditworthiness; it is based mainly on a predictive model that uses prior data to predict repayment behavior.

Future purchases: A more recent (and controversial) example is Target's use of predictive modeling to classify sales prospects as "pregnant" or "not-pregnant." Those classified as pregnant could then be sent sales promotions at an early stage of pregnancy, giving Target a head start on a significant purchase stream.

Tax evasion: The US Internal Revenue Service found it was 25 times more likely to find tax evasion when enforcement activity was based on predictive models, allowing agents to focus on the most likely tax cheats (Siegel, 2013).

The business analytics toolkit also includes statistical experiments, the most common of which is known to marketers as A/B testing. These are often used for pricing decisions:

- Orbitz, the travel site, has found that it could price hotel options higher for Mac users than Windows users.
- Staples online store found that it could charge more for staplers if a customer lived far from a Staples store.

Beware the organizational setting where analytics is a solution in search of a problem: a manager, knowing that business analytics and machine learning are hot areas, decides that her organization must deploy them too, to capture that hidden value that must be lurking somewhere. Successful use of analytics and machine learning requires both an understanding of the business context where value is to be captured and an understanding of exactly what the machine learning methods do.

1.2 WHAT IS MACHINE LEARNING?

In this book, *machine learning* or *data mining* refers to business analytics methods that go beyond counts, descriptive techniques, reporting, and methods based on business rules. While we do introduce data visualization, which is commonly the first step into more advanced analytics, the book focuses mostly on the more advanced data analytics tools. Specifically, it includes statistical and machine learning methods that inform decision-making, often in automated fashion. Prediction is typically an important component, often at the individual level. Rather than "what is the relationship between advertising and sales?" we might be interested in "what specific advertisement, or recommended product, should be shown to a given online shopper at this moment?" Or we might be interested in clustering customers into different "personas" that receive different marketing treatment, then assigning each new prospect to one of these personas.

The era of Big Data has accelerated the use of machine learning. Machine learning methods, with their power and automaticity, have the ability to cope with huge amounts of data and extract value.

1.3 MACHINE LEARNING, AI, AND RELATED TERMS

The field of analytics is growing rapidly, both in terms of the breadth of applications, and in terms of the number of organizations using advanced analytics. As a result, there is considerable overlap and inconsistency in terms of definitions. Terms have also changed over time.

The older term *data mining* means different things to different people. To the general public, it may have a general, somewhat hazy and pejorative meaning of digging through vast stores of (often personal) data in search of something interesting. *Data mining*, as it refers to analytic techniques, has largely been superseded by the term *machine learning*.

Other terms that organizations use are *predictive analytics*, *predictive modeling*, and most recently *machine learning* and *artificial intelligence (AI)*.

Many practitioners, particularly those from the IT and computer science communities, use the term AI to refer to all the methods discussed in this book. AI originally referred to the general capability of a machine to act like a human, and, in its earlier days, existed mainly in the realm of science fiction and the unrealized ambitions of computer scientists. More recently, it has come to encompass the methods of statistical and machine learning discussed in this book, as the primary enablers of that grand vision, and sometimes the term is used loosely to mean the same thing as *machine learning*. More broadly, it includes generative capabilities such as the creation of images, audio, and video.

Statistical Modeling vs. Machine Learning

A variety of techniques for exploring data and building models have been around for a long time in the world of statistics: linear regression, logistic regression, discriminant analysis, and principal components analysis, for example. But the core tenets of classical statistics—computing is difficult and data are scarce—do not apply in machine learning applications where both data and computing power are plentiful.

This is what gives rise to Daryl Pregibon's description of "data mining" (in the sense of machine learning) as "statistics at scale and speed" (Pregibon, 1999). Another major difference between the fields of statistics and machine learning is the focus in statistics on inference from a sample to the population regarding an "average effect"—for example, "a \$1 price increase will reduce average demand by 2 boxes." In contrast, the focus in machine learning is on predicting individual records—"the predicted demand for person i given a \$1 price increase is 1 box, while for person j it is 3 boxes." The emphasis that classical statistics places on inference (determining whether a pattern or interesting result might have happened by chance in our sample) is absent from machine learning. Note also that the term *inference* is often used in the machine learning community to refer to the process of using a model to make predictions for new data, also called *scoring*, in contrast to its meaning in the statistical community.

In comparison to statistics, machine learning deals with large datasets in an open-ended fashion, making it impossible to put the strict limits around the question being addressed that classical statistical inference would require. As a result, the general approach to machine learning is vulnerable to the danger of *overfitting*, where a model is fit so closely to the available sample of data that it describes not merely structural characteristics of the data, but random peculiarities as well. In engineering terms, the model is fitting the noise, not just the signal.

In this book, we use the term *machine learning algorithms* to refer to methods that learn directly from data, especially local data, often in layered or iterative fashion. In contrast, we use *statistical models* to refer to methods that apply global structure to the data that can be written as a simple mathematical equation. A simple example is a linear regression model (statistical) vs. a k-nearest neighbors algorithm (machine learning). A given record would be treated by linear regression in accord with an overall linear equation that applies to *all* the records. In k-nearest neighbors, that record would be classified in accord with the values of a small number of nearby records.

1.4   BIG DATA

Machine learning and Big Data go hand in hand. *Big Data* is a relative term—data today are big by reference to the past and to the methods and devices available to deal with them. The challenge Big Data presents is often characterized by the four V's—volume, velocity, variety, and veracity. *Volume* refers to the amount of data. *Velocity* refers to the flow rate—the speed at which it is being generated and changed. *Variety* refers to the different types of data being generated (time stamps, location, numbers, text, images, etc.). *Veracity* refers to the fact that data is being generated by organic distributed processes (e.g., millions of people signing up for services or free downloads) and not subject to the controls or quality checks that apply to data collected for a study.

Most large organizations face both the challenge and the opportunity of Big Data because most routine data processes now generate data that can be stored and, possibly, analyzed.

The scale can be visualized by comparing the data in a traditional statistical analysis on the large size (e.g., 15 variables and 5000 records) to the Walmart database. If you consider the traditional statistical study to be the size of a period at the end of a sentence, then the Walmart database is the size of a football field. And that probably does not include other data associated with Walmart—social media data, for example, which comes in the form of unstructured text.

If the analytical challenge is substantial, so can be the reward:

- OKCupid, the dating site, uses statistical models with their data to predict what forms of message content are most likely to produce a response.
- Telenor, a Norwegian mobile phone service company, was able to reduce subscriber turnover 37% by using models to predict which customers were most likely to leave and then lavishing attention on them.
- Allstate, the insurance company, tripled the accuracy of predicting injury liability in auto claims by incorporating more information about vehicle type.

The examples above are from Eric Siegel's *Predictive Analytics* (2013, Wiley).

Some extremely valuable tasks were not even feasible before the era of Big Data. Consider web searches, the technology on which Google was built. In early days, a search for "Ricky Ricardo Little Red Riding Hood" would have yielded various links to the "I Love Lucy" show, other links to Ricardo's career as a band leader, and links to the children's story of Little Red Riding Hood. Only once the Google database had accumulated sufficient data (including records of what users clicked on) would the search yield, in the top position, links to the specific *I Love Lucy* episode in which Ricky enacts, in a comic mixture of Spanish and English, Little Red Riding Hood for his infant son.

1.5 DATA SCIENCE

The ubiquity, size, value, and importance of Big Data has given rise to a new profession: the *data scientist*. *Data science* is a mix of skills in the areas of statistics, machine learning, math, programming, business, and IT. The term itself is thus broader than the other concepts we discussed above, and it is a rare individual who combines deep skills in all the constituent areas. Harlan Harris, in *Analyzing the Analyzers* (with Sean Murphy and Marck Vaisman, O'Reilly 2013) describes the skill sets of most data scientists as resembling a "T"—deep in one area (the vertical bar of the T), and shallower in other areas (the top of the T).

At a large data science conference session (Strata-Hadoop World, October 2014) most attendees felt that programming was an essential skill, though there was a sizable minority who felt otherwise. And, although Big Data is the motivating power behind the growth of data science, most data scientists do not actually spend most of their time working with terabyte-size or larger data.

Data of the terabyte or larger size would be involved at the deployment stage of a model. There are manifold challenges at that stage, most of them IT and programming issues related to data handling and tying together different components of a system. Much work must precede that phase. It is that earlier piloting and prototyping phase on which this book focuses—developing the statistical and machine learning models that will eventually be plugged into a deployed system. What methods do you use with what sorts of data and problems? How do the methods work? What are their requirements, their strengths, their weaknesses? How do you assess their performances?

1.6 WHY ARE THERE SO MANY DIFFERENT METHODS?

As can be seen in this book or any other resource on machine learning, there are many different methods for prediction and classification. You might ask yourself why they coexist and whether some are better than others. The answer is that each method has advantages and disadvantages. The usefulness of a method can depend on factors such as the size of the dataset, the types of patterns that exist in the data, whether the data meet some underlying assumptions of the method, how noisy the data are, and the particular goal of the analysis. A small illustration is shown in Figure 1.1, where the goal is to find a combination of *household income level* and *household lot size* that separate owners (blue markers) from non-owners (orange markers) of riding mowers. The first method (top panel) looks only for horizontal lines to separate owners from non-owners, whereas the second method (bottom panel) looks for a single diagonal line.

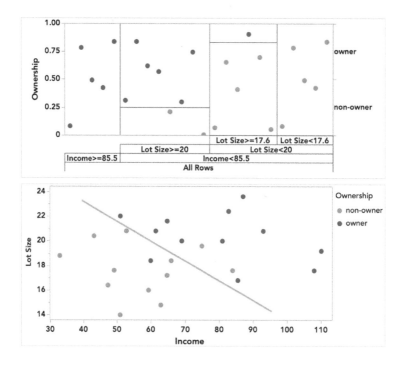

FIGURE 1.1   Two methods for separating owners from non-owners

Different methods can lead to different results, and their performances can vary. It is therefore customary in machine learning to apply several different methods and perhaps their combination and select the one that appears most useful for the goal at hand.

1.7 TERMINOLOGY AND NOTATION

Because of the hybrid origins of data science, its practitioners often use multiple terms to refer to the same thing. For example, in the machine learning (artificial intelligence) field, the variable being predicted is the *output variable* or *target variable*. A categorical output variable is often called a *label*. To a statistician or social scientist, the variable being predicted is the *dependent variable* or the *response*. Here is a summary of terms used:

Algorithm   Refers to a specific procedure used to implement a particular machine learning technique: classification tree, discriminant analysis, and the like.

Attribute   see **Predictor**.

Binary variable   A variable that takes on two discrete values (e.g., fraud and non-fraud transactions); also called *binominal variable*.

Case   see **Observation**.

Confidence   A performance measure in association rules of the type "IF A and B are purchased, THEN C is also purchased." Confidence is the conditional probability that C will be purchased IF A and B are purchased.

Dependent variable   see **Response**.

Estimation   see **Prediction**.

Feature   see **Predictor**. The term feature is also used in the context of selecting variables (feature selection) or generating new variables (feature generation) through some mechanism. More broadly, this process is called feature engineering.

Holdout sample   (or **Holdout set**) A sample of data not used in fitting a model, used to assess the performance of that model; this book uses the term *validation set* or, if one is used in the problem, *test set* instead of *holdout sample*.

Inference   In statistics, the process of accounting for chance variation when making estimates or drawing conclusions based on samples; in machine learning, the term often refers to the process of using a model to make predictions for new data (see Score).

Input variable   see **Predictor**.

Label   A nominal (categorical) attribute being predicted in supervised learning.

Model   An algorithm as applied to a dataset, complete with its settings (many of the algorithms have parameters that the user can adjust).

Nominal variable   A variable that takes on one of several fixed values, for example, a flight could be on-time, delayed, or canceled; also called *categorical variable* or *polynominal variable*.

Numerical variable   A variable that takes on numerical (integer or real) values.

Observation   The unit of analysis on which the measurements are taken (a customer, a transaction, etc.); also called *instance*, *sample*, *example*, *case*, *record*, *pattern*, or *row*. (Each row typically represents a record; each column, a variable. Note that the use of the term "sample" here is different from its usual meaning in statistics, where it refers to all the data sampled from a larger data source, not simply one record.)

Outcome variable   see **Response**.

Output variable   see **Response**.

$P\ (A|B)$ The conditional probability of event A occurring given that event B has occurred. Read as "the probability that A will occur given that B has occurred."

Profile   The set of measurements on an observation (e.g., the height, weight, and age of a person).

Positive class   The class of interest in a binary outcome variable (e.g., purchasers in the outcome variable *purchase/no purchase*); the positive class need not be favorable.

Prediction   The prediction of the value of a continuous output variable; also called *estimation*.

Predictor   A variable, usually denoted by X, used as an input into a predictive model; also called a *feature*, *input variable*, *independent variable*, or, from a database perspective, a *field*.

Record   see **Observation**.

Response   A variable, usually denoted by Y, which is the variable being predicted in supervised learning; also called *dependent variable*, *output variable*, *target variable*, or *outcome variable*.

Score   A predicted value or class. *Scoring new data* means to use a model developed with training data to predict output values in new data.

Success class   see **Positive class**

Supervised learning   The process of providing an algorithm (logistic regression, classification tree, etc.) with records in which an output variable of interest is known, and the algorithm "learns" how to predict this value with new records where the output is unknown.

Target   see **Response**.

Test data   (or **Test set**) The portion of the data used only at the end of the model building and selection process to assess how well the final model might perform on additional data.

Training data   (or **Training set**) The portion of data used to fit a model.

Unsupervised learning   An analysis in which one attempts to learn something about the data other than predicting an output value of interest (e.g., whether it falls into clusters).

Validation data   (or **Validation set**) The portion of the data used to assess how well the model fits, to adjust some models, and to select the best model from among those that have been tried.

Variable   Any measurement on the records, including both the input (X) variables and the output (Y) variable.

1.8   ROAD MAPS TO THIS BOOK

The book covers many of the widely used predictive and classification methods as well as other machine learning tools. Figure 1.2 outlines machine learning from a process perspective and where the topics in this book fit in. Chapter numbers are indicated beside the topic. Table 1.1 provides a different perspective: it organizes supervised and unsupervised machine learning procedures according to the type and structure of the data.

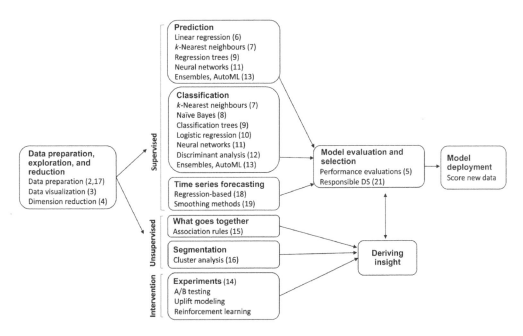

FIGURE 1.2   **Machine learning from a process perspective. Numbers in parentheses indicate chapter numbers**

TABLE 1.1   **Organization of machine learning methods in this book, According to the nature of the data**[a]

| | Supervised | | Unsupervised |
	Continuous response	Categorical response	No response
Continuous predictors	Linear regression (6)	Logistic regression (10)	Principal components (4)
	Neural nets (11)	Neural nets (11)	Cluster analysis (16)
	k-Nearest neighbors (7)	Discriminant analysis (12)	Collaborative filtering (15)
	Ensembles (13)	k-Nearest neighbors (7)	
		Ensembles (13)	
Categorical predictors	Linear regression (6)	Neural nets (11)	Association rules (15)
	Neural nets (11)	Classification trees (9)	Collaborative filtering (15)
	Regression trees (9)	Logistic regression (10)	
	Ensembles (13)	Naive Bayes (8)	
		Ensembles (13)	

[a]Numbers in parentheses indicate the chapter number.

Order of Topics

The book is divided into nine parts: Part I (Chapters 1 and 2) gives a general overview of machine learning and its components. Part II (Chapters 3 and 4) focuses on the early stages of data exploration and dimension reduction.

Part III (Chapter 5) discusses performance evaluation. Although it contains only one chapter, we discuss a variety of topics, from predictive performance metrics to misclassification costs. The principles covered in this part are crucial for the proper evaluation and comparison of supervised learning methods.

Part IV includes eight chapters (Chapters 6–13), covering a variety of popular supervised learning methods (for classification and/or prediction). Within this part, the topics are generally organized according to the level of sophistication of the algorithms, their popularity, and ease of understanding. The final chapter introduces ensembles and combinations of methods.

Part V (Chapter 14) introduces the notions of experiments, intervention, and user feedback. This single chapter starts with A/B testing, then its use in uplift modeling, and finally expands into reinforcement learning, explaining the basic ideas and formulations that utilize user feedback for learning best treatment assignments.

Part VI focuses on unsupervised learning of relationships. It presents association rules and collaborative filtering (Chapter 15) and cluster analysis (Chapter 16).

Part VII includes three chapters (Chapters 17–19), with the focus on forecasting time series. The first chapter covers general issues related to handling and understanding time series. The next two chapters present two popular forecasting approaches: regression-based forecasting and smoothing methods.

Part VIII presents the data analytics topic called text mining (Chapter 20). This method applies machine learning methods to text data. The final chapter on responsible data science (Chapter 21) introduces key issues to consider for when carrying out a machine learning project in a responsible way. Finally, Part IX includes a set of cases.

Although the topics in the book can be covered in the order of the chapters, each chapter stands alone. We advise, however, to read Parts I–III before proceeding to chapters in Parts IV–VI. Similarly, Chapter 17 should precede other chapters in Part VII.

USING JMP AND JMP Pro

To facilitate the learning experience, this book uses JMP Pro (Figure 1.3), a desktop statistical package for Mac OS and Windows operating systems from JMP Statistical Discovery LLC (see jmp.com/system for complete system requirements).

JMP comes in two primary flavors: JMP (the standard version) and JMP Pro (the professional version). The standard version of JMP is dynamic and interactive, and it offers a variety of built-in tools for graphing and analyzing data. JMP has extensive tools for data visualization and data preparation, along with statistical and machine learning techniques for classification, prediction, and forecasting. It offers a variety of supervised machine learning tools, including neural nets, classification and regression trees, linear and logistic regression, and discriminant analysis. It also offers unsupervised algorithms, such as principal component analysis, k-means clustering, and hierarchical clustering.

JMP Pro has all of the functionality of JMP, but adds advanced tools for predictive modeling, including K-Nearest neighbor, penalized regression, naive Bayes, text analytics, association analysis, uplift modeling, advanced trees, and additional options for creating neural networks and ensemble models. Importantly, it also provides a number of modeling utilities for preparing data for modeling and includes built-in tools for data partitioning and model validation, model comparison, model selection, and generating model deployment code.

For these reasons, we use JMP Pro throughout this book. JMP Pro is required for the built-in tools for model cross-validation and comparison that are not available in the standard version of JMP. However, the standard version of JMP can be used for creating many of the graphs, summaries, analyses, and models presented in this book.

While we provide JMP instructions throughout this book, there are many resources available for new users. For tips on getting started with JMP, go to jmp.com/gettingstarted. An in-depth introduction to JMP, *Discovering JMP*, is available online or in JMP (under *Help > JMP Documentation Library*). Additional resources on specific topics, along with short videos, can be found in the JMP Learning Library at jmp.com/learn. For additional details on the features available only in JMP Pro, see jmp.com/pro.

JMP Pro is available through academic licenses at most colleges and universities and through site licenses in many organizations—see your software IT administrator for availability and download information. For academic licensing information, or to request an evaluation of JMP Pro, write to *academic@jmp.com*. If you do not qualify for an academic license, an evaluation copy of JMP Pro can be requested at jmp.com/proeval.

JMP runs natively on both the Windows and Mac OS. On Windows, the JMP Home window (shown top, Figure 1.3) and a Tip of the Day window appear when you open JMP. On Mac, a JMP Starter window (Figure 1.4) appears instead of the JMP Home window. These windows can be accessed from the *View* and *Windows* menus in JMP. Use the *File > New* menu to open a new data table (bottom Figure 1.3), or *File > Open* to open an existing JMP data table or data in another file type.

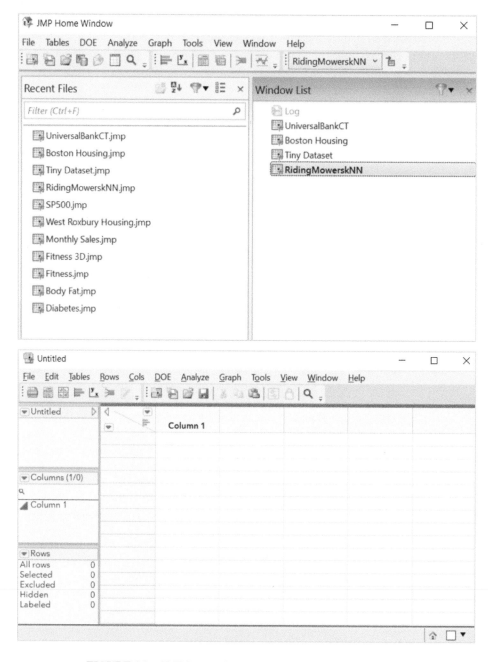

**FIGURE 1.3   JMP home window (top) and data table (bottom)
(on a Windows PC)**

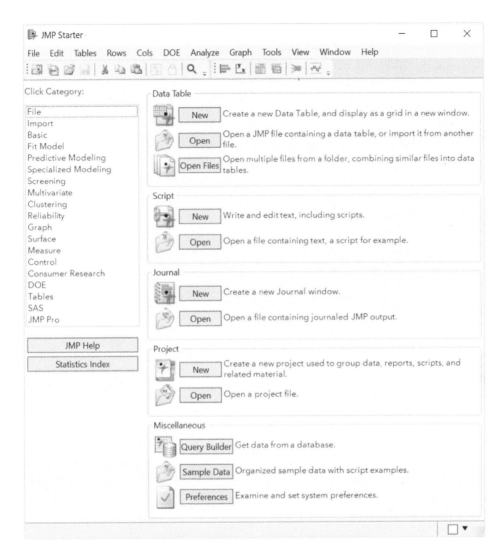

FIGURE 1.4   JMP starter window (on a Windows PC)

2

OVERVIEW OF THE MACHINE LEARNING PROCESS

In this chapter, we give an overview of the steps involved in machine learning (ML), starting from a clear goal definition and ending with model deployment. The general steps are shown schematically in Figure 2.1. We also discuss issues related to data collection, cleaning, and preprocessing. We explain the notion of data partitioning, where methods are trained on a set of training data and then model performance is evaluated on a separate set of validation data, and how this practice helps avoid overfitting. Finally, we illustrate the steps of model building by applying them to data.

Define purpose	Obtain data	Explore and clean data	Determine ML task	Choose ML methods	Apply methods, select final model	Evaluate performance	Deploy

FIGURE 2.1   Schematic of the machine learning process

2.1   INTRODUCTION

In Chapter 1, we saw some very general definitions of machine learning. In this chapter, we introduce a variety of machine learning methods. The core of this book focuses on what has come to be called *predictive analytics*, the tasks of classification and prediction as well as pattern discovery, that have become key elements of a *business analytics* function in most large firms. These terms are described next.

Machine Learning for Business Analytics: Concepts, Techniques, and Applications with JMP Pro®,
Second Edition. Galit Shmueli, Peter C. Bruce, Mia L. Stephens, Muralidhara Anandamurthy, and Nitin R. Patel.
© 2023 John Wiley & Sons, Inc. Published 2023 by John Wiley & Sons, Inc.

2.2   CORE IDEAS IN MACHINE LEARNING

Classification

Classification is perhaps the most basic form of predictive analytics. The recipient of an offer can respond or not respond. An applicant for a loan can repay on time, repay late, or declare bankruptcy. A credit card transaction can be normal or fraudulent. A packet of data traveling on a network can be benign or threatening. A bus in a fleet can be available for service or unavailable. The victim of an illness can be recovered, still be ill, or be deceased.

A common task in machine learning is to examine data where the classification is unknown or will occur in the future, with the goal of predicting what that classification is or will be. Similar data where the classification is known are used to develop rules, which are then applied to the data with the unknown classification.

Prediction

Prediction is similar to classification, except that we are trying to predict the value of a numerical variable (e.g., amount of purchase) rather than a class (e.g., purchaser or nonpurchaser). Of course, in classification we are trying to predict a class, but the term *prediction* in this book refers to the prediction of the value of a continuous numerical variable.

> Sometimes in the machine learning literature, the terms *estimation* and *regression* are used to refer to the prediction of the value of a continuous variable, and *prediction* may be used for both continuous and categorical data.

Association Rules and Recommendation Systems

Large databases of customer transactions lend themselves naturally to the analysis of associations among items purchased, or "what goes with what." *Association rules*, or *affinity analysis*, is designed to find such general associations patterns among items in large databases. The rules can then be used in a variety of ways. For example, grocery stores can use such information for product placement. They can use the rules for weekly promotional offers or for bundling products. Association rules derived from a hospital database on patients' symptoms during consecutive hospitalizations can help find "which symptom is followed by what other symptom" and help predict future symptoms for returning patients.

Online recommendation systems, such as those used on Amazon and Netflix, use *Collaborative Filtering*, a method that uses individual users' preferences and tastes given their historic purchase, rating, browsing, or any other measurable behavior indicative of preference, as well as other users' histories. In contrast to association rules that generate rules general to an entire population, collaborative filtering generates "what goes with what" at the individual user level. Hence, collaborative filtering is used in many recommendation systems that aim to deliver personalized recommendations to users with a wide range of preferences. (Note that these topics are not covered in this book.)

Predictive Analytics

Classification, prediction, and, to some extent, association rules and collaborative filtering constitute the analytical methods employed in *predictive analytics*. However, the term predictive analytics is sometimes used to also include data pattern identification methods such as clustering.

Data Reduction and Dimension Reduction

The performance of some machine learning algorithms is often improved when the number of variables is limited, and when large numbers of records can be grouped into homogeneous groups. For example, rather than dealing with thousands of product types, an analyst might wish to group them into a smaller number of groups and build separate models for each group. Or a marketer might want to classify customers into different "personas" and must first group customers into homogeneous groups to define the personas. This process of consolidating a large number of records (or cases) into a smaller set is termed *data reduction*. Methods for reducing the number of cases are often called *clustering*.

Reducing the number of variables is typically called *dimension reduction*. Dimension reduction is a common initial step before deploying supervised learning methods, intended to improve predictive power, manageability, and interpretability.

Data Exploration and Visualization

One of the earliest stages of engaging with data is exploring it. Exploration is aimed at understanding the global landscape of the data and for detecting unusual values. Exploration is used for data cleaning and manipulation as well as for visual discovery and "hypothesis generation."

Methods for exploring data include looking at various data aggregations and summaries, both numerically and graphically. This includes looking at each variable separately as well as looking at relationships between variables. The purpose is to discover patterns and exceptions. Exploration by creating charts and dashboards is called *data visualization* or *visual analytics*. For numerical variables, we use histograms and boxplots to learn about the distribution of their values, to detect outliers (extreme observations), and to find other information that is relevant to the analysis task. Similarly, for categorical variables, we use bar charts. We can also look at scatterplots of pairs of numerical variables to learn about possible relationships, the type of relationship, and, again, to detect outliers. Visualization can be greatly enhanced by adding features such as color, zooming, interactive navigation, and data filtering.

Supervised and Unsupervised Learning

A fundamental distinction among machine learning techniques is between supervised and unsupervised methods. *Supervised learning algorithms* are those used in classification and prediction. We must have data available in which the value of the outcome of interest (e.g., purchase or no purchase) is known. Such data are also called "labelled data" or the "outcome," since they contain the outcome value (label value) for each record. The use of the term "label" reflects the fact that the outcome of interest for a record may often be a characterization applied by a human; a document may be labeled as relevant, or an object in an X-ray may be labeled as malignant. These *training data* are the data from which the classification or prediction algorithm "learns," or is "trained," about the relationship between predictor variables and the outcome variable. Once the algorithm has learned from the training data, it is then applied to another sample of data (the *validation data*) where the outcome is known, to see how well it does in comparison to other models. If many different models are being tried out, it is prudent to save a third sample, which also includes known outcomes (the *test data*) to use with the model finally selected to predict how well it will do. The model can then be used to classify or predict the outcome of interest in new cases where the outcome is unknown.

Simple linear regression is an example of a supervised learning algorithm (although rarely called that in the introductory statistics course where you probably first encountered it). The Y variable is the (known) outcome variable, and the X variable is a predictor variable. A regression line is drawn to minimize the sum of squared deviations between the actual Y values and the values predicted by this line. The regression line can now be used to predict Y values for new values of X for which we do not know the Y value.

Unsupervised learning algorithms are those used where there is no outcome variable to predict or classify. Hence, there is no "learning" from cases where such an outcome variable is known. Association rules, dimension reduction methods, and clustering techniques are all unsupervised learning methods.

Supervised and unsupervised methods are sometimes used in conjunction. For example, unsupervised clustering methods are used to separate loan applicants into several risk-level groups. Then, supervised algorithms are applied separately to each risk-level group for predicting propensity of loan default.

SUPERVISED LEARNING REQUIRES GOOD SUPERVISION

In some cases, the value of the outcome variable (the "label") is known because it is an inherent component of the data. Web logs will show whether a person clicked on a link or didn't click. Bank records will show whether a loan was paid on time or not. In other cases, the value of the known target must be supplied by a human labeling process to accumulate enough data to train a model. E-mail must be labeled as spam or legitimate, documents in legal discovery must be labeled as relevant or irrelevant. In either case, the machine learning algorithm can be led astray if the quality of the supervision is poor.

Gene Weingarten reported in the January 5, 2014, *Washington Post* magazine how the strange phrase "defiantly recommend" is making its way into English via autocorrection. "Defiantly" is closer to the common misspelling *definatly* than *definitely*, so Google, in the early days, offered it as a correction when users typed the misspelled word "definatly" on google.com. In the ideal supervised learning model, humans guide the autocorrection process by rejecting *defiantly* and substituting *definitely*. Google's algorithm would then learn that this is the best first-choice correction of "definatly." The problem was that too many people were lazy, just accepting the first correction that Google presented. All these acceptances then cemented "defiantly" as the proper correction.

2.3   THE STEPS IN A MACHINE LEARNING PROJECT

This book focuses on understanding and using machine learning algorithms (steps 4–7 below). However, some of the most serious errors in analytics projects result from a poor understanding of the problem—an understanding that must be developed before we get into the details of algorithms to be used. Here is a list of steps to be taken in a typical machine learning effort:

1. *Develop an understanding of the purpose of the machine learning project*: What is the problem? How will the stakeholder use the results? Who will be affected by the results? Will the analysis be a one-shot effort or an ongoing procedure?

2. *Obtain the data to be used in the analysis*: This often involves random sampling from a large database to capture records to be used in an analysis. How well this sample reflects the records of interest affects the ability of the machine learning results to generalize to records outcode of this sample. It may also involve pulling together data from different databases or sources. The databases could be internal (e.g., past purchases made by customers) or external (credit ratings). While machine learning deals with very large databases, usually the analysis to be done requires only thousands or tens of thousands of records.

3. *Explore, clean, and preprocess the data*: This step involves verifying that the data are in reasonable condition. How should missing data be handled? Are the values in a reasonable range, given what you would expect for each variable? Are there obvious outliers? The data are reviewed graphically: for example, a matrix of scatterplots showing the relationship of each variable with every other variable. We also need to ensure consistency in the definitions of fields, units of measurement, time periods, and so on. In this step, new variables are also typically created from existing ones. For example, "duration" can be computed from start and end dates.

4. *Reduce the data dimension, if necessary*: Dimension reduction can involve operations such as eliminating unneeded variables, transforming variables (e.g., turning "money spent" into "spent > \$100" vs. "spent ≤ \$100"), and creating new variables (e.g., a variable that records whether at least one of several products was purchased). Make sure that you know what each variable means and whether it is sensible to include it in the model.

5. *Determine the machine learning task* (classification, prediction, clustering, etc.): This involves translating the general question or problem of step 1 into a more specific machine learning question.

6. *Partition the data* (for supervised tasks): If the task is supervised (classification or prediction), partition the dataset into three parts: training, validation, and test datasets.

7. *Choose the machine learning techniques to be used* (regression, neural nets, hierarchical clustering, etc.).

8. *Use algorithms to perform the task*: This is typically an iterative process—trying multiple variants and often using multiple variants of the same algorithm (choosing different variables or parameter settings within the algorithm). Where appropriate, feedback from the algorithm's performance on validation data is used to refine the parameter settings.

9. *Interpret the results of the algorithms*: This involves making a choice as to the best algorithm to deploy, and where possible, testing the final choice on the test data to get an idea as to how well it will perform. (Recall that each algorithm may also be tested on the validation data for tuning purposes; in this way, the validation data become a part of the fitting process and are likely to underestimate the error in the deployment of the model that is finally chosen.)

10. *Deploy the model*: This step involves integrating the model into operational systems and running it on real records to produce decisions or actions. For example, the model might be applied to a purchased list of possible customers, and the action might be "include in the mailing if the predicted amount of purchase is >$10." A key step here is "scoring" the new records, or using the chosen model to predict the target value for each new record.

The foregoing steps encompass the steps in SEMMA, a methodology developed by SAS Institute:

Sample: Take a sample from the dataset; partition into training, validation, and test datasets.

Explore: Examine the dataset statistically and graphically.

Modify: Transform the variables and impute missing values.

Model: Fit predictive models (e.g., regression tree, collaborative filtering).

Assess: Compare models using a validation dataset.

IBM SPSS Modeler (previously SPSS-Clementine) has a similar methodology, termed CRISP-DM (CRoss-Industry Standard Process for Data Mining). All these frameworks include the same main steps involved in predictive modeling.

2.4 PRELIMINARY STEPS

Organization of Data

Datasets are nearly always constructed and displayed so that variables are in columns and records are in rows. In the example shown in Section 2.6 (home value in West Roxbury, Boston, in 2014), 14 variables are recorded for over 5000 homes. The JMP data table (WestRoxburyHousing.jmp) is organized so that each row represents a home—the first home's assessed value was $344,200, its tax was $4430, its size was 9965 ft^2, it was built in 1880, and so on. In supervised learning situations, one of these variables will be the outcome variable, typically listed at the end or the beginning (in this case, it is TOTAL VALUE, in the first column).

Sampling from a Database

Quite often, we want to perform our machine learning on less than the total number of records that are available. Machine learning algorithms will have varying limitations on what they can handle in terms of the numbers of records and variables, limitations that may be specific to computing power and capacity as well as software limitations. Even within those limits, many algorithms will execute faster with smaller samples.

Accurate models can often be built with as few as several hundred or thousand records (as discussed later). Hence, we may want to sample a subset of records for model building.

Oversampling Rare Events in Classification Tasks

If the event we are interested in classifying is rare, such as customers purchasing a product in response to a mailing, or fraudulent credit card transactions, sampling a random subset of records may yield so few events (e.g., purchases) that we have little information on them. We would end up with lots of data on nonpurchasers and non fraudulent transactions but little on which to base a model that distinguishes purchasers from nonpurchasers or fraudulent from nonfraudulent. In such cases we would want our sampling procedure to overweight the rare class (purchasers or frauds) relative to the majority class (nonpurchasers, nonfrauds) so that our sample would end up with a healthy complement of purchasers or frauds.

Assuring an adequate number of responders or "success" cases to train the model is just part of the picture. A more important factor is the costs of *misclassification*. That is, the cost of incorrectly classifying outcomes. Whenever the response rate is extremely low, we are likely to attach more importance to identifying a responder than to identifying a nonresponder. In direct-response advertising (whether by traditional mail, email, web, or mobile advertising), we may encounter only one or two responders for every hundred records—the value of finding such a customer far outweighs the costs of reaching him or her. In trying to identify fraudulent transactions, or customers unlikely to repay debt, the costs of failing to find the fraud or the nonpaying customer are likely to exceed the cost of more detailed review of a legitimate transaction or customer.

If the costs of failing to locate responders were comparable to the costs of misidentifying responders as nonresponders, our models would usually achieve highest overall accuracy if they identified everyone (or almost everyone, if it is easy to identify a few responders without catching many nonresponders) as a nonresponder. In such a case, the misclassification rate is very low—equal to the rate of responders—but the model is of no value.

More generally, we want to train our model with the asymmetric costs in mind so that the algorithm will catch the more valuable responders, probably at the cost of "catching" and misclassifying more nonresponders as responders than would be the case if we assume equal costs. This subject is discussed in detail in Chapter 5.

Preprocessing and Cleaning the Data

Types of Variables   There are several ways of classifying variables. Variables can be numerical or text (character/string). They can be continuous (able to assume any real numerical value, usually in a given range), integer (taking only integer values), or categorical (assuming one of a limited number of values), or date-time (able to assume data and time values). Categorical variables can be either coded as numerical ($1, 2, 3$) or text (payments current, payments not current, bankrupt). Categorical variables can also be unordered (called *nominal variables*) with categories such as North America, Europe, and Asia; or they can be ordered (called *ordinal variables*) with categories such as high value, low value, and nil value. Date-time variables can have values denoting either date or time, or both.

In JMP, *modeling types* are applied to distinguish the type of variable and are used in many platforms to determine the appropriate graphs and analyses. Colored icons are used to indicate the modeling types applied (see Figure 2.2). Continuous variables are coded with the *continuous* modeling type (represented by blue triangles), variables with unordered categories are given the *nominal* modeling type (represented by red bars), and variables that represent ordered categories are given the *ordinal* modeling type (represented by green bars). *Unstructured text* includes the actual verbatim in multiple languages and they are represented by horizontal white broken lines.

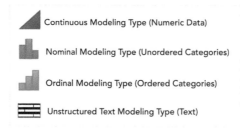

FIGURE 2.2   Four JMP modeling types

Working with Continuous Variables   Continuous variables can be handled by most machine learning routines. In JMP, all supervised routines can take continuous predictor variables. The machine learning grew out of problems with categorical outcomes; the roots of statistics lie in the analysis of continuous variables. Sometimes it is desirable to convert continuous variables to categorical variables. This is often done in the case of outcome variables, where the numerical variable is mapped to a decision (e.g., credit scores above a certain level mean "grant credit," a medical test result above a certain level means "start treatment"). This type of conversion can be easily done in JMP.

Handling Categorical Variables   Categorical variables can also be handled by most routines but often require special handling. If the categorical variable is ordered (age group, degree of creditworthiness, etc.), we can sometimes code the categories numerically (1, 2, 3, …) and treat the variable as if it were a continuous variable. The smaller the number of categories, and the less they represent equal increments of value, the more problematic this approach becomes, but it often works well enough.

Nominal categorical variables, however, often cannot be used as is. In many cases, they must be decomposed into a series of binary variables called *indicator variables* or *dummy variables*. For example, a single categorical variable that can have possible values of "student," "unemployed," "employed," or "retired" could be split into four separate indicator variables:

Student—Yes/No
Unemployed—Yes/No
Employed—Yes/No
Retired—Yes/No

In some algorithms (such as linear and logistic regression), only three of the indicator variables need to be used; if the values of three are known, the fourth is also known. For example, given that these four values are the only possible ones, we can know that if a person is neither student, unemployed, nor employed, he or she must be retired. In fact, due to this redundant information, linear regression and logistic regression algorithms will produce incorrect results or fail if you use all four indicator variables. Yet, in other algorithms (e.g., *k*-nearest neighbors), we must include all four indicator variables. This is called one-hot encoding.

JMP has a utility to convert categorical variables to binary indicator variables (*Make Indicator Columns*, from the *Cols > Utilities* menu). But, for the most part, JMP automatically codes categorical variables behind the scenes if needed, thereby not requiring the user to manually create indicator variables.

CHANGING MODELING TYPES IN JMP

Modeling types in JMP should be specified prior to analysis. To change a modeling type, right-click on a column in the *Columns Panel* of the data table and select *Column Info*. Then, select the correct Modeling Type. Note that for continuous data you may also need to change the *Data Type* to *Numeric*.

Figure 2.3 shows the *Columns Panel*, along with the general layout of JMP data tables. The dataset is *Companies*, which is available from JMP Sample Data Folder (under the *Help* menu). This figure was borrowed from the one-page guide, *JMP Data Tables*, found at jmp.com/learn.

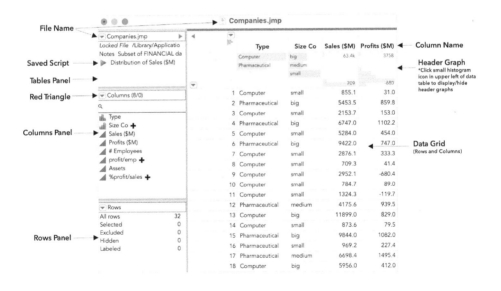

FIGURE 2.3 Example dataset, showing the columns panel (on the left) and the general layout of JMP data tables

Variable Selection   More is not necessarily better when it comes to selecting variables for a model. Other things being equal, parsimony, or compactness, is a desirable feature in a model. For one thing, the more variables we include, the greater the number of records we will need to assess relationships among the variables. Fifteen records may suffice to give us a rough idea of the relationship between Y and a single predictor variable X. If we now want information about the relationship between Y and 15 predictor variables $X_1 \ldots X_{15}$, then 15 records will not be enough (each estimated relationship would have an average of only one record's worth of information, making the estimate very unreliable). In addition, models based on many variables are often less robust, as they require the collection of more variables in the future, are subject to more data quality and availability issues, and require more data cleaning and preprocessing. The process of variable selection is also called *feature selection*.

How Many Variables and How Much Data?   Statisticians give us procedures to learn with some precision how many records we would need to achieve a given degree of reliability with a given dataset and a given model. These are called "power calculations" and are intended to assure that an average population effect will be estimated with sufficient precision from a sample. Data miners' needs are usually different, since the focus is not on identifying an average effect but rather predicting individual records. This purpose typically requires larger samples than those used for statistical inference. A good rule of thumb is to have 10 records for every predictor variable. Another, used by Delmaster and Hancock (2001, p. 68) for classification procedures, is to have at least $6 \times m \times p$ records, where m is the number of outcome classes and p is the number of variables.

In general, compactness or parsimony is a desirable feature in a machine learning model. Even when we start with a small number of variables, we often end up with many more after creating new variables (e.g., converting a categorical variable into a set of dummy variables). Data visualization and dimension reduction methods help reduce the number of variables so that redundancies are avoided.

Even when we have an ample supply of data, there are good reasons to pay close attention to the variables that are included in a model. Someone with domain knowledge (i.e., knowledge of the business process and the data) should be consulted, as knowledge of what the variables represent is typically critical for building a good model and avoiding errors and legal violations.

For example, suppose that we're trying to predict the total purchase amount spent by customers, and we have a few predictor columns that are coded $X_1, X_2, X_3, \ldots$, where we don't know what those codes mean. We might find that X_1 is an excellent predictor of the total amount spent. However, if we discover that X_1 is the amount spent on shipping, calculated as a percentage of the purchase amount, then obviously a model that uses shipping amount cannot be used to predict purchase amount, since the shipping amount is not known until the transaction is completed. This is known as *target leakage*. Another example is if we are trying to predict loan default at the time a customer applies for a loan. If our dataset includes only information on approved loan applications, we will not have information about what distinguishes defaulters from nondefaulters among denied applicants. A model based on approved loans alone can therefore not be used to predict defaulting behavior at the time of loan application, but rather only once a loan is approved. Finally, in certain applications (e.g., credit scoring), variables such as gender and race are legally prohibited.

Outliers   The more data we are dealing with, the greater the chance of encountering erroneous values resulting from measurement error, data-entry error, or the like. If the erroneous value is in the same range as the rest of the data, it may be harmless. If it is well outside the range of the rest of the data (e.g., a misplaced decimal), it might have a substantial effect on some of the machine learning procedures we plan to use.

Values that lie far away from the bulk of the data are called *outliers*. The term *far away* is deliberately left vague because what is or is not called an outlier is basically an arbitrary decision. Analysts use rules of thumb such as "anything over three standard deviations away from the mean is an outlier," but no statistical rule can tell us whether such an outlier is the result of an error. In this statistical sense, an outlier is not necessarily an invalid data point; it is just a distant one.

The purpose of identifying outliers is usually to call attention to values that need further review. We might come up with an explanation looking at the data—in the case of a misplaced decimal, this is likely. We might have no explanation, but know that the value is wrong—a temperature of 178°F for a sick person. We might conclude that the value is within the realm of possibility and leave it alone. Or, it might be that the outliers are what we are looking for—unusual financial transactions or travel patterns. All these are judgments best made by someone with *domain knowledge*, knowledge of the particular application being considered: direct mail, mortgage finance, and so on, as opposed to technical knowledge of statistical or machine learning procedures. Statistical procedures can do little beyond identifying the record as something that needs review.

If manual review is feasible, some outliers may be identified and corrected. In any case, if the number of records with outliers is very small, they might be treated as missing data. How do we inspect for outliers? One technique is to plot all of the variables and review the graphs for very large or very small values in each graph. Another option is to examine the minimum and maximum values of each column using the JMP *Columns Viewer* (from the *Cols* menu). For a more automated approach that considers each record as a unit, automated procedures could be used to identify outliers across many variables. Several procedures for analyzing both univariate and multivariate outliers are available in the JMP *Explore Outliers* utility (from the *Analyze > Screening* menu).

Missing Values   Typically some records will contain missing values. In JMP, missing values for continuous variables are denoted with a "." in the cell. For categorical (nominal or ordinal) variables, the cell is empty. In some modeling routines, such as linear regression, records with missing values are omitted. If the number of records with missing values is small, this might not be a problem. However, if we have a large number of variables, a small proportion of missing values can affect a lot of records. Even with only 30 variables, if only 5% of the values are missing (spread randomly and independently among cases and variables), almost 80% of the records would have to be omitted from the analysis. (The chance that a given record would escape having a missing value is $0.95^{30} = 0.215$.)

An alternative to omitting records with missing values is to replace the missing value with an imputed value, based on the other values for that variable across all records. For example, if among 30 variables, household income is missing for a particular record, we might substitute the mean household income across all records. Doing so does not, of course, add any information about how household income affects the outcome variable. It merely allows us to proceed with the analysis and not lose the information contained in this record for the other 29 variables. Note that using such a technique will understate the variability in a dataset. However, we can assess variability and the performance of our machine learning technique, using the validation data, and therefore this need not present a major problem. More sophisticated procedures do exist—for example, using regression, based on other variables, to fill in the missing values. These methods have been elaborated mainly for analysis of medical and scientific studies, where each patient or subject record comes at great expense. In machine learning, where data are typically plentiful, simpler methods usually suffice.

Some datasets contain variables that have a very large number of missing values. In other words, a measurement is missing for a large number of records. In that case, dropping records with missing values will lead to a large loss of data. Imputing the missing values might also be useless, as the imputations are based on a small number of existing records. An alternative is to examine the importance of the predictor. If it is not very crucial, it can be dropped. If it is important, perhaps a proxy variable with fewer missing values can be used instead. When such a predictor is deemed central, the best solution is to invest in obtaining the missing data.

Significant time may be required to deal with missing data, as not all situations are susceptible to automated solutions. In a messy dataset, for example, a "0" might mean two things: (1) the value is missing or (2) the value is actually zero. In the credit industry, a "0" in the "past due" variable might mean a customer who is fully paid up, or a customer with no credit history at all—two very different situations. Likewise, in many situations a "999" is used to denote missing values. Human judgment may be required for individual cases or to determine a special rule to deal with the situation. *Note*: It is important to distinguish between missing values and zeros. JMP will include zeros in calculations and analyses but will omit missing values from analyses. If a variable contains codes for missing values, the *Missing Value Code* column property for the variable can be used to indicate this.

In JMP, the *Explore Missing Values* utility (from the *Analyze > Screening* menu) can be used to both explore and impute missing values, and the *Missing Data Pattern* utility, from the *Tables* menu, can be used to explore whether there are any patterns among the missing values. In JMP modeling platforms, *Informative Missing* coding is available to address missing values.

Normalizing (Standardizing) and Rescaling Data   Some algorithms require that the data be normalized before the algorithm can be implemented effectively. To normalize a variable, we subtract the mean from each value and then divide by the standard deviation. This operation is also sometimes called *standardizing*. In effect, we are expressing each value as the "number of standard deviations away from the mean," also called a *z-score*.

Normalizing is one way to bring all variables to the same scale. Another popular approach is rescaling each variable to a [0,1] scale. This is done by subtracting the minimum value and then dividing by the range. Subtracting the minimum shifts the variable origin to zero. Dividing by the range shrinks or expands the data to the range [0,1].

To consider why normalizing or scaling to [0,1] might be necessary, consider the case of clustering. Clustering typically involves calculating a distance measure that reflects how far each record is from a cluster center or from other records. With multiple variables, different units will be used: days, dollars, counts, and so on. If the dollars are in the thousands and everything else is in the tens, the dollar variable will come to dominate the distance measure. Moreover, changing units from (say) days to hours or months can alter the outcome completely.

Some machine learning software, including JMP, provide an option to normalize the data in those algorithms where it may be required. In some JMP platforms it is an option rather than an automatic feature of such algorithms, since there are situations where we want each variable to contribute to the distance measure in proportion to its original scale. Variables in JMP can also be standardized using the *Formula Editor*, or by using the dynamic transformation feature from the data table or from any analysis column selection list.

STANDARDIZING DATA IN JMP

Standardizing data prior to analysis or modeling in JMP is typically not required—it is generally done automatically or is provided as a default option in JMP analysis platforms (where needed). However, standardization can easily be done using the dynamic transformation feature in JMP. From the data table, right-click on the column header for the continuous variable and select *New Formula Column > Distributional > Standardize*. From an analysis dialog or control panel, right-click on a variable in the column selection list and select *Distributional > Standardize* to create a temporary variable.

2.5   PREDICTIVE POWER AND OVERFITTING

In supervised learning, a key question presents itself: how well will our prediction or classification model perform when we apply it to new data? We are particularly interested in comparing the performance of various models so that we can choose the one we think will do the best when it is implemented in practice. A key concept is to make sure that our chosen model generalizes beyond the dataset that we have at hand. To ensure generalization, we use the concept of *data partitioning* and try to avoid *overfitting*. These two important concepts are described next.

Overfitting

The more variables we include in a model, the greater the risk of overfitting the particular data used for modeling. What is overfitting?

In Table 2.1, we show hypothetical data about advertising expenditures in one time period and sales in a subsequent time period (a scatterplot of the data is shown in Figure 2.4). We could connect up these points with a smooth but complicated function, one that interpolates all these data points perfectly and leaves no error (residuals). This can be seen in

TABLE 2.1   Hypothetical Data on Advertising Expenditures and Subsequent Sales

Expenditure	Revenue
239	514
364	789
602	550
644	1386
770	1394
789	1440
911	1354

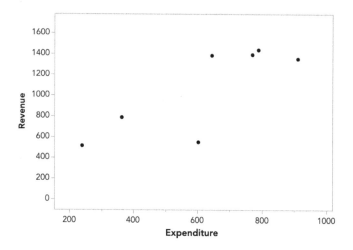

FIGURE 2.4   Scatterplot for expenditure and revenue data

Figure 2.5. However, we can see that such a curve is unlikely to be accurate, or even useful, in predicting future sales on the basis of advertising expenditures (e.g., it is hard to believe that increasing expenditures from $400 to $500 will actually decrease revenue).

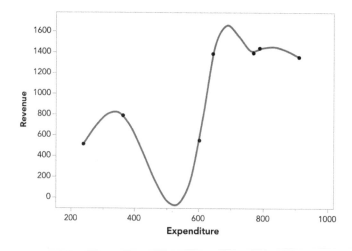

FIGURE 2.5   Overfitting: This function fits the data with no errors

A basic purpose of building a model is to represent relationships among variables in such a way that this representation will do a good job of predicting future outcome values on the basis of future predictor values. Of course, we want the model to do a good job of describing the data we have, but we are more interested in its performance with future data.

In the example above, a simple straight line might do a better job than the complex function in terms of predicting future sales on the basis of advertising. Instead, we devised a complex function that fits the data perfectly, and in doing so, we overreached. We ended up modeling some variation in the data that was nothing more than chance variation. We mislabeled the noise in the data as if it were a signal.

Similarly, we can add predictors to a model to sharpen its performance with the data at hand. Consider a database of 100 individuals, half of whom have contributed to a charitable cause. Information about income, family size, and zip code might do a fair job of predicting whether or not someone is a contributor. If we keep adding more predictors, we can improve the performance of the model with the data at hand and reduce the misclassification error to a negligible level. However, this low error rate is misleading because it probably includes spurious effects, which are specific to the 100 individuals but not beyond that sample.

For example, one of the variables might be height. We have no basis in theory to suppose that tall people might contribute more or less to charity, but if there are several tall people in our sample and they just happened to contribute heavily to charity, our model might include a term for height—the taller you are, the more you will contribute. Of course, when the model is applied to additional data, it is likely that this will not turn out to be a good predictor.

If the dataset is not much larger than the number of predictor variables, it is very likely that a spurious relationship like this will creep into the model. Continuing with our charity example, with a small sample just a few of whom are tall, whatever the contribution level of tall people may be, the algorithm is tempted to attribute it to their being tall. If the dataset is very large relative to the number of predictors, this is less likely to occur. In such a case each predictor must help predict the outcome for a large number of cases, so the job it does is much less dependent on just a few cases, which might be flukes.

Somewhat surprisingly, even if we know for a fact that a higher degree curve is the appropriate model, if the model-fitting dataset is not large enough, a lower degree function (that is not as likely to fit the noise) may perform better. Overfitting can also result from the application of many different models, from which the best performing model is selected.

Creation and Use of Data Partitions

At first glance, we might think it best to choose the model that did the best job of classifying or predicting the outcome variable of interest with the data at hand. However, when we use the same data both to develop the model and to assess its performance, we introduce an "optimism" bias. This is because when we choose the model that works best with the data, this model's superior performance comes from two sources:

- A superior model
- Chance aspects of the data that happen to match the chosen model better than they match other models

The latter is a particularly serious problem with techniques (such as trees and neural nets) that do not impose linear or other structure on the data, and thus end up overfitting it.

To address the overfitting problem, we simply divide (partition) our data and develop our model using only one of the partitions. After we have a model, we try it out on another partition and see how it performs, which we can measure in several ways. In a classification model, we can count the proportion of held-back records that were misclassified. In a prediction model, we can measure the residuals (prediction errors) between the predicted values and the actual values. This evaluation approach in effect mimics the deployment scenario where our model is applied to data that it hasn't "seen."

We typically deal with two or three partitions: a training set, a validation set, and sometimes an additional test set. Partitioning the data into training, validation, and test sets is done either randomly according to predetermined proportions or by specifying which records go into which partition according to some relevant variable (e.g., in time series forecasting, the data are partitioned according to their chronological order). In most cases, the partitioning should be done randomly to avoid getting a biased partition. It is also possible (although cumbersome) to divide the data into more than three partitions by successive partitioning (e.g., divide the initial data into three partitions, then take one of those partitions and partition it further).

Training Partition   The training partition, typically the largest partition, contains the data used to build the various models we are examining. The same training partition is generally used to develop multiple models.

Validation Partition   The validation partition is used to assess the predictive performance of each model so that you can compare models and choose the best one. In some algorithms (e.g., classification and regression trees, k-nearest neighbors), the validation partition may be used in automated fashion to tune parameters and improve the model.

Test Partition   The test partition (sometimes called the *holdout* or *evaluation partition*) is used to assess the performance of the chosen model with new data.

Why have both a validation and a test partition? When we use the validation data to assess multiple models and then choose the model that performs best with the validation data, we again encounter another (lesser) facet of the overfitting problem—chance aspects of the validation data that happen to match the chosen model better than they match other models. In other words, by using the validation data to choose one of several models, the performance of the chosen model on the validation data will be overly optimistic.

The random features of the validation data that enhance the apparent performance of the chosen model will probably not be present in new data to which the model is applied. Therefore, we might have overestimated the accuracy of our model. The more models we test, the more likely it is that one of them will be particularly effective in modeling the noise in the validation data. Applying the model to the test data, which it has not seen before, will provide an unbiased estimate of how well the model will perform with new data. Figure 2.6

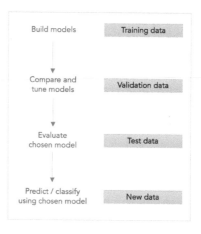

FIGURE 2.6   Three data partitions and their role in the machine learning process

shows the three data partitions and their use in the machine learning process. When we are concerned mainly with finding the best model and less with exactly how well it will do, we might use only training and validation partitions.

Note that with many algorithms, such as classification and regression trees, the model is built using the training data, but the validation set is used to find the best model on the training data. So the validation data are an integral part of fitting the initial model in JMP. For a truly independent assessment of the model, the third test partition should be used.

k-fold Cross-Validation   When the number of records in our sample is small, data partitioning might not be advisable as each partition will contain too few records for model building and performance evaluation. Furthermore, some machine learning methods are sensitive to small changes in the training data, so that a different partitioning can lead to different results. An alternative to data partitioning is k-fold cross-validation, which is especially useful with small samples.

k-fold cross-validation, is a procedure that starts with partitioning the data into "folds," or non-overlapping subsamples. Often we choose $k = 5$ folds, meaning that the data are randomly partitioned into five equal parts, where each fold has 20% of the records. A model is then fit k times. Each time, one of the folds is used as the validation set and the remaining $k - 1$ folds serve as the training set. The result is that each fold is used once as the validation set, thereby producing predictions for every record in the dataset. We can then combine the model's predictions on each of the k validation sets in order to evaluate the overall performance of the model.

PARTITIONING DATA AND CROSS-VALIDATION IN JMP Pro

The JMP Pro *Make Validation Column* which is under *Analyze > Predictive Modeling* has options for partitioning a dataset. This utility creates a new variable, *Validation*, that contains the values "Training" (stored within JMP as "0"), "Validation" (1) and "Test" (2), according to the allocation rates or counts specified. This new variable specifies how the data are allocated to the partitions.

k-fold cross-validation is an option in many modeling platforms in JMP and JMP Pro. For these platforms, you can either specify k-fold cross-validation in the model control panel or in the platform launch window. For other platforms, you can specify k-fold cross-validation through a validation column that contains more than three levels. The *Make Validation Column* utility can be used to create a k-fold cross-validation column.

The *Make Validation Column* utility and built-in model validation are only available in JMP Pro—these features are not available in the standard version of JMP. However, a validation column can be created manually in JMP using the *Column Info > Initialize Data > Random, Indicator* function. And, cross-validation can be done manually for most modeling methods using the *Hide and Exclude* option from the *Rows* menu.

Throughout this book, we will be using JMP Pro for its automatic validation functionality and other model-building features. Many of the modeling techniques, however, are native in the standard version of JMP.

2.6   BUILDING A PREDICTIVE MODEL WITH JMP Pro

Let us go through the steps typical to many machine learning tasks using a familiar procedure: multiple linear regression. This will help you understand the overall process before we begin tackling new algorithms. We illustrate the procedure using JMP Pro.

Predicting Home Values in a Boston Neighborhood

The Internet has revolutionized the real estate industry. Realtors now list houses and their prices on the web, and estimates of house and condominium prices have become widely available, even for units not on the market. In 2014, Zillow (www.zillow.com), a popular online real estate information site in the United States,[1] purchased its major rival, Trulia. By 2015, Zillow had become the dominant platform for checking house prices and, as such, the dominant online advertising venue for realtors. By 2021, another competitor, Redfin, had eclipsed Zillow in market capitalization, largely via its strategy of employing its own agents directly on a salaried basis. What used to be a comfortable 6% commission structure for independent realtors, affording them a handsome surplus (and an oversupply of realtors) was being rapidly eroded by an increasing need to pay for advertising on Zillow and by competition from Redfin. (This, in fact, was the original key to Zillow's business model—redirecting the 6% commission away from realtors and to itself.)

Zillow gets much of the data for its "Zestimates" of home values directly from publicly available city housing data, used to estimate property values for tax assessment. A competitor seeking to get into the market would likely take the same approach. So might realtors seeking to develop an alternative to Zillow.

A simple approach would be a naive, model-less method—just use the assessed values as determined by the city. Those values, however, do not necessarily include all properties, and they might not include changes warranted by remodeling, additions, and the like. Moreover, the assessment methods used by cities may not be transparent or always reflect true market values. However, the city property data can be used as a starting point to build a model, to which additional data (e.g., as collected by large realtors) can be added later.

Let us look at how Boston property assessment data, available from the City of Boston, might be used to predict home values. The data in WestRoxburyHousing.jmp includes data for single family owner-occupied homes in West Roxbury, neighborhood southwest of Boston, MA, in 2014. The data include values for various predictor variables and for a target—assessed home value ("total value"). The dataset includes 5802 homes. A sample of the data[2] is shown in Table 2.2. This dataset has 14 variables, and a description of each variable[3] is given in Table 2.3.

Each row in the data represents a home. For example, the first home was assessed at a total value of \$344.2 thousand (TOTAL VALUE). Its tax bill was \$4330. It has a lot size of 9965 square feet (ft^2), was built in year 1880, has two floors, six rooms, and so on.

---

[1] "Zestimates may not be as right as you'd like" *Washington Post*, February 7, 2015, p. T10, by K. Harney.

[2] The data are a slightly cleaned version of the Property Assessment FY2014 data at https://data.boston.gov/dataset/property-assessment (accessed December 2017).

[3] The full data dictionary provided by the City of Boston is available at https://data.boston.gov/dataset/property-assessment; we have modified a few variable names.

TABLE 2.2 First 10 Records in the West Roxbury home values dataset

Total Value	Tax	Lot Sqft	YR Built	Gross Area	Living Area	Floors	Rooms	Bed Rooms	Full Bath	Half Bath	Kit Chen	Fire Place	Remodel
344.2	4330	9965	1880	2436	1352	2	6	3	1	1	1	0	None
412.6	5190	6590	1945	3108	1976	2	10	4	2	1	1	0	Recent
330.1	4152	7500	1890	2294	1371	2	8	4	1	1	1	0	None
498.6	6272	13773	1957	5032	2608	1	9	5	1	1	1	1	None
331.5	4170	5000	1910	2370	1438	2	7	3	2	0	1	0	None
337.4	4244	5142	1950	2124	1060	1	6	3	1	0	1	1	Old
359.4	4521	5000	1954	3220	1916	2	7	3	1	1	1	0	None
320.4	4030	10000	1950	2208	1200	1	6	3	1	0	1	0	None
333.5	4195	6835	1958	2582	1092	1	5	3	1	0	1	1	Recent
409.4	5150	5093	1900	4818	2992	2	8	4	2	0	1	0	None

TABLE 2.3   Description of Variables in West Roxbury (Boston) Home Value Dataset

TOTAL VALUE	Total assessed value for property, in thousands of USD
TAX	Tax bill amount based on total assessed value multiplied by the tax rate, in USD
LOT SQ FT	Total lot size of parcel in square feet
YR BUILT	Year property was built
GROSS AREA	Gross floor area
LIVING AREA	Total living area for residential properties (ft^2)
FLOORS	Number of floors
ROOMS	Total number of rooms
BEDROOMS	Total number of bedrooms
FULL BATH	Total number of full baths
HALF BATH	Total number of half baths
KITCHEN	Total number of kitchens
FIREPLACE	Total number of fireplaces
REMODEL	When house was remodeled (Recent/Old/None)

Modeling Process

We now describe in detail the various model stages using the West Roxbury home values example.

1. *Determine the purpose*: Let's assume that the purpose of our machine learning project is to predict the value of homes in West Roxbury.

2. *Obtain the data*: We will use the 2014 West Roxbury Housing data. The dataset in question is small enough that we do not need to sample from it—we can use it in its entirety.

3. *Explore, clean, and preprocess the data*: Let us look first at the description of the variables, also known as the "data dictionary," to be sure that we understand them all. These descriptions are available as column notes in the data table (hold your mouse over the variable names in the *Columns Panel* to display) and in Table 2.2. The variable names and descriptions in this dataset all seem fairly straightforward, but this is not always the case. Often, variable names are cryptic and their descriptions are unclear or missing.

   It is useful to pause and think about what the variables mean and whether they should be included in the model. Consider the variable TAX. At first glance, we consider that the tax on a home is usually a function of its assessed value, so there is some circularity in the model—we want to predict a home's value using TAX as a predictor, yet TAX itself is determined by a home's value. TAX might be a very good predictor of home value in a numerical sense, but would it be useful if we wanted to apply our model to homes whose assessed value might not be known? For this reason, we will exclude TAX from the analysis.

   It is also useful to check for outliers that might be errors. For example, suppose that the column FLOORS (number of floors) looked like the one in Table 2.4, after sorting the data in descending order based on floors. We can tell right away that the 15 is in error—it is unlikely that a home has 15 floors. All other values are between 1 and 2. Probably, the decimal was misplaced and the value should be 1.5.

TABLE 2.4   Outlier in West Roxbury data

Floors	Rooms
15	8
2	10
1.5	6
1	6

Last, we may create dummy variables for categorical variables. Here we have one categorical variable: REMODEL, which has three categories. Recall that, in JMP, creating dummy variables for categorical variables is not required. So we can use REMODEL as is, without using dummy coding.

4. *Reduce the data dimension*: Our dataset has been prepared for presentation with fairly low dimension—it has only 12 potential predictor variables (not including TAX), and the single categorical variable considered has only three categories. If we had many more variables, at this stage we might want to apply a dimension reduction technique such as condensing multiple categories into a smaller number, or applying principal components analysis to consolidate multiple similar numerical variables (e.g., LIVING AREA, ROOMS, BEDROOMS, BATH, HALF BATH) into a smaller number of variables.

5. *Determine the machine learning task*: In this case, as noted, the specific task is to predict the value of TOTAL VALUE using the predictor variables. For simplicity, we exclude BLDG TYPE, ROOF TYPE, and EXT FIN from the analysis, because they have many categories. We therefore use all the numerical variables (except TAX) and the dummies created for the remaining categorical variables.

6. *Partition the data (for supervised tasks):* Our task is to predict the house value and then assess how well that prediction performs. We will partition the data into a training set to build the model and a validation set to see how well the model does when applied to new data. We need to specify the proportion of records used in each partition. This technique is part of the "supervised learning" process in classification and prediction problems. These are problems in which we know the class or value of the outcome variable for some data, and we want to use those data in developing a model that can then be applied to other data where that value is unknown.

In JMP Pro, select *Predictive Modeling > Make Validation Column* from the *Analyze* menu. Figure 2.7 shows the initial dialog box (top panel) and the specification dialog (bottom panel). The user can choose between several partitioning options, depending on the type of partitioning needed:

• The partitioning can be done purely at random (*Purely Random*).
• The partitioning can be based on a selected stratifying variable (or multiple stratifying variables) that governs the division into training and validation partitions (*Stratified Random*).
• The partitioning can be based on time periods (for time series).

In this case, we divide the data randomly into two equal partitions, training and validation (50–50 split). This produces a new column in the data table, named *Validation*, which specifies the partition each record is assigned to. The training partition is used to build the model, and the validation partition is used to see how well

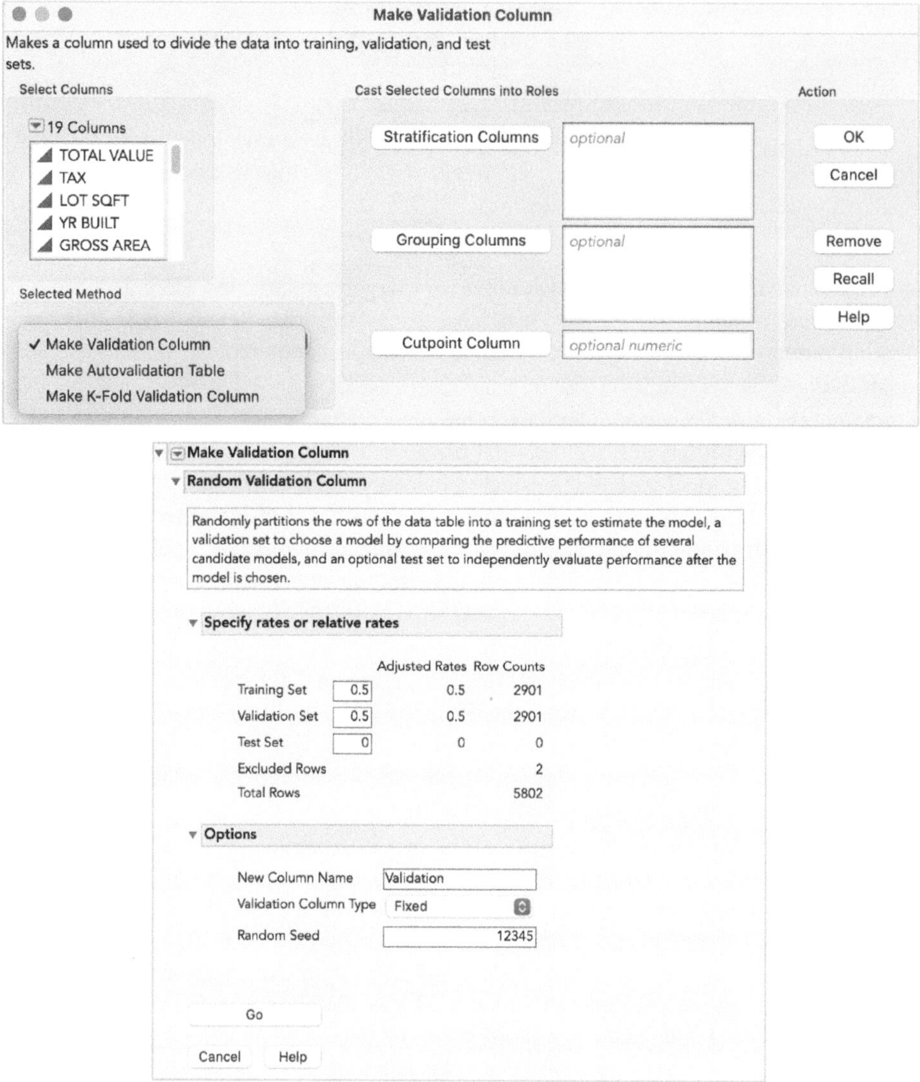

FIGURE 2.7   **Partitioning the data using the JMP Pro** *Make Validation Column* **utility. In this example, a random partition of 50% training, 50% validation, and 0% test is used. Also, note a random seed of 12345 is used**

the model does when applied to new data. Although not used in our example, a test partition might also be used.

When partitioning data, a random seed can be set to duplicate the same random partition later should we need to. Here, we use the Validation column previously created and saved in the data table (this is the last column).

7. *Choose the technique*: In this case, it is multiple linear regression. Having divided the data into training and validation partitions, we can use JMP (or JMP Pro) to build a multiple linear regression model with the training data. We want to

predict the value of a house in West Roxbury on the basis of all the other values (except TAX).

8. *Use the algorithm to perform the task*: In JMP, we select *Fit Model* platform from the *Analyze* menu. The Fit Model dialog is shown in Figure 2.8. The variable TOTAL VALUE is selected as the *Y* (dependent) variable. All the other variables, except TAX and Validation, are selected as model effects (input or predictor variables). We use the partition column, *Validation*, to fit a model on the training data (we put this in the *Validation* field). JMP produces standard regression output and provides both training and validation statistics. For now, we defer discussion of the output, along with other available options within the analysis window. (For more information, see Chapter 6, Multiple Linear Regression, or the JMP Documentation Library under the *Help* menu in JMP.)

FIGURE 2.8   Using the JMP Pro Fit Model platform for multiple linear regression

Here, we focus on the model predictions and the prediction error (the residuals). These can be saved to the data table using the *red triangle* in the analysis window, and selecting the options from the *Save Columns* menu (as shown in Figure 2.9).

FIGURE 2.9   Use the red triangle in the fit least squares analysis window to save the prediction formula and residuals to the data table

Figure 2.10 shows the actual values for the first ten records (observations that are in the validation set are selected), along with the predicted values and the residual values (the prediction errors). Note that the predicted values are often called the *fitted values*, since they are for the records to which the model was fit.

	TOTAL VALUE	TAX	LOT SQFT	YR BUILT	GROSS AREA	LIVING AREA	FLOO RS	ROO MS	BED ROOMS	FULL BATH	HALF BATH	KITC HEN	FIRE PLACE	REMO DEL	Validation	Predicted TOTAL VALUE	Residual TOTAL VALUE
1	344.2	4330	9965	1880	2436	1352	2	6	3	1	1	1	0	None	Validation	383.94	-39.74
2	412.6	5190	6590	1945	3108	1976	2	10	4	2	1	1	0	Recent	Validation	461.42	-48.82
3	330.1	4152	7500	1890	2294	1371	2	8	4	1	1	1	0	None	Validation	360.46	-30.36
4	498.6	6272	13773	1957	5032	2608	1	9	5	1	1	1	1	None	Validation	548.54	-49.94
5	331.5	4170	5000	1910	2370	1438	2	7	3	2	0	1	0	None	Validation	347.44	-15.94
6	337.4	4244	5142	1950	2124	1060	1	6	3	1	0	1	1	Old	Training	286.52	50.88
7	359.4	4521	5000	1954	3220	1916	2	7	3	1	1	1	0	None	Validation	402.34	-42.94
8	320.4	4030	10000	1950	2208	1200	1	6	3	1	0	1	0	None	Training	316.04	4.36
9	333.5	4195	6835	1958	2582	1092	1	5	3	1	0	1	1	Recent	Training	339.45	-5.95
10	409.4	5150	5093	1900	4818	2992	2	8	4	2	0	1	0	None	Validation	503.30	-93.90
11	313	3937	5000	1960	2624	1485	1.5	6	3	2	0	1	1	None	Validation	360.05	-47.05
12	344.5	4333	6768	1958	2844	1460	1.5	6	3	2	0	1	1	None	Validation	381.08	-36.58
13	315.5	3968	5000	1889	2196	1290	2	6	3	1	0	1	0	None	Training	312.26	3.24
14	575	7233	12288	2004	4616	2378	2	9	4	2	1	1	1	None	Validation	578.04	-3.04
15	326.2	4103	5000	1954	2536	1272	1.5	6	3	1	1	1	1	None	Training	345.03	-18.83
16	298.2	3751	5000	1940	2129	864	1	7	3	2	0	1	0	None	Training	271.52	26.68
17	313.1	3938	6949	1880	2612	1438	1.5	7	3	1	1	1	0	Old	Training	351.00	-37.90
18	344.9	4338	10000	1950	2099	1445	1	7	3	1	1	1	1	None	Validation	363.30	-18.40
19	330.7	4160	5000	1910	2408	1470	2	7	3	1	0	1	0	None	Training	331.23	-0.53
20	348	4377	9001	1875	2840	1632	2	7	3	1	0	1	0	None	Validation	384.67	-36.67

FIGURE 2.10   Predictions for the first twenty records in the West Roxbury housing data, made by saving the prediction formula to the data table. Records in the validation set are highlighted

The prediction errors for the training and validation data can be compared using standard JMP platforms, such as *Analyze > Distribution* and *Analyze > Tabulate*. For example, in Figure 2.11, we see the result of using *Tabulate* to summarize the errors for the training and validations sets. Many different statistical summaries are available in *Tabulate*—here, we've selected the *Mean* (the average error), the *Min*, and the *Max*.

▼ ⬇ Tabulate

	Residual TOTAL VALUE		
Validation	Mean	Min	Max
Training	-0.000	-276.1	232.06
Validation	0.420	-209.0	290.74

FIGURE 2.11   Summary statistics for error rates for training (top) and validation (bottom) data produced using *Analyze > Tabulate* (error figures are in thousands of $)

We see that the average error is quite small relative to the units of TOTAL VALUE, indicating that, in general, predictions average about right—our predictions are "unbiased." Of course, this simply means that the positive and negative errors balance out. It tells us nothing about how large these errors are. The *Min* and *Max* (see Figure 2.11) show the range of prediction errors, but there are several other possible measures of prediction error, many of which will be discussed in Chapter 5.

Two measures of prediction error are provided by default in the *Fit Least Squares* analysis window when a validation column is used (see Figure 2.12): *RSquare* and *RASE*. *RSquare*, which is based on the sum of the squared errors, is a measure of the percent of variation in the output variable explained by the inputs. RASE (or RMSE) is a far more useful measure of prediction error. RASE takes the square root of the average squared errors, so it gives an idea of the typical error (whether positive or negative) in the same scale as that used for the original data.

▼ Crossvalidation			
Source	RSquare	RASE	Freq
Training Set	0.8031	43.018	2821
Validation Set	0.8220	42.701	2981

FIGURE 2.12   Crossvalidation statistics from the JMP Pro analysis window for the regression model

In this example, the RASE for the validation data ($42.7 thousand), which the model is seeing for the first time in making these predictions, is close to the RASE for the training data ($43.02 thousand), which were used in training the model.

9. *Interpret the results*: At this stage, we would typically try other prediction algorithms (e.g., regression trees) and see how they do error-wise. We might also try different "settings" on the various models (e.g., we could use the *Stepwise* option in multiple linear regression to chose a reduced set of variables that might perform better with the validation data, or we might try a different algorithm, such as a *regression tree*). For each model, we would save the prediction formula to the data table, and then use the *Model Comparison* platform in JMP Pro to compare validation statistics for the different models.

An additional measure of prediction error, *AAE*, or *average absolute error* is provided in the *Model Comparison* platform (under *Analyze > Predictive Modeling*). This is sometimes reported as *MAD (mean absolute deviation)* or *MAE (mean absolute error)*. (Note that additional measures can also be computed manually using the JMP *Formula Editor*.)

After choosing the best model (typically, the model with the lowest error on the validation data while also recognizing that "simpler is better"), we use that model to predict the output variable using fresh test data. These steps are covered in more detail in the analysis of cases.

10. *Deploy the model*: After the best model is chosen, it is applied to new data to predict TOTAL VALUE for homes where this value is unknown. This was, of course, the overall purpose. Predicting the output value for new records is called *scoring*. For predictive tasks, scoring produces predicted numerical values. For classification tasks, scoring produces classes and/or propensities.

In JMP, we can score new records using the models that we developed. To do this, we simply add new rows to the data table and enter values for the predictors into the appropriate cells. When the prediction formula for a model is saved to the data table, predicted values are automatically computed for these new rows added. Figure 2.13 shows an example of the JMP data table with two homes scored using our saved linear

	TOTAL VALUE	TAX	LOT SQFT	YR BUILT	GROSS AREA	LIVING AREA	FLOO RS	ROO MS	BED ROOMS	FULL BATH	HALF BATH	KITC HEN	FIRE PLACE	REMO DEL	Validation	Pred Formula TOTAL VALUE
5788	393.7	4952	10150	1950	3521	1538	1	5	2	1	1	1	1	None	Training	413.69
5789	472.1	5939	7650	1932	3894	2293	2	10	2	1	1	1	2	None	Validation	506.13
5790	542.6	6825	8280	1928	3700	2102	2	8	3	2	1	1	1	Recent	Validation	518.08
5791	449	5648	7610	1938	3386	1847	2	7	3	1	1	1	1	None	Validation	444.49
5792	564	7095	5000	1950	4108	2421	2	7	4	1	2	1	1	Recent	Training	517.65
5793	408.5	5138	6362	1950	2847	1596	2	7	3	1	1	1	1	None	Validation	404.96
5794	399	5019	6350	1950	2361	1662	2	7	3	1	1	1	1	None	Training	392.81
5795	371.2	4669	6116	1950	2522	1440	2	8	3	1	1	1	1	None	Training	385.55
5796	386.8	4865	6350	1950	2316	1326	2	7	3	2	1	1	1	None	Validation	393.38
5797	413.414	5200	9150	1950	2324	1326	2	7	3	1	1	1	1	None	Validation	399.21
5798	404.8	5092	6762	1938	2594	1714	2	9	3	2	1	1	1	Recent	Validation	452.02
5799	407.9	5131	9408	1950	2414	1333	2	6	3	1	1	1	1	None	Training	403.73
5800	406.5	5113	7198	1987	2480	1674	2	7	3	1	1	1	1	None	Training	408.46
5801	308.7	3883	6890	1946	2000	1000	1	5	2	1	0	1	0	None	Validation	271.83
5802	447.6	5630	7406	1950	2510	1600	2	7	3	1	1	1	1	None	Validation	403.67
5803	•	•	6000	2000	2500	2000	1	6	3	2	0	1	0	None	Validation	354.90
5804	•	•	8000	2010	3000	2200	2	8	4	2	1	1	1	None	Validation	479.80

FIGURE 2.13   Two records scored using the saved linear regression model

regression model. Note that all the required predictor columns are present, and the output column is absent.

Additional information for these predicted values, such as confidence intervals, can be added by using the *Save Columns* option from the top red triangle in the *Fit Least Squares* analysis window for the model. Data in a separate JMP data table, with the same predictor variables, can also be scored by copying the saved prediction formula into a column in the data table via the *Formula Editor*.

Alternatively, in JMP Pro we can publish models to the *Formula Depot*, which can create scoring code for use in an environment outside of JMP. For example, you can generate scoring code in C, Python, JavaScript, SAS, or SQL.

2.7   USING JMP Pro FOR MACHINE LEARNING

An important aspect of this process to note is that the heavy-duty analysis does not necessarily require a huge number of records. The dataset to be analyzed might have millions of records, or be part of a regular ongoing flow, and JMP can generally handle this. However, in applying multiple linear regression or applying a classification tree, the use of a sample of 20,000 is likely to yield as accurate an answer as that obtained when using the entire dataset. The principle involved is the same as the principle behind polling: if sampled judiciously, 2000 voters can give an estimate of the entire population's opinion within one or two percentage points. (See "How Many Variables and How Much Data" in Section 2.4 for further discussion.)

Therefore, in most cases, the number of records required in each partition (training, validation, and test) can be accommodated by JMP. Of course, we need to get those records into JMP, and for this purpose JMP provides the ability to import data in a number of different formats, to connect to external databases using the *Query Builder*, and to randomly sample records from an external database or SAS dataset.

2.8 AUTOMATING MACHINE LEARNING SOLUTIONS

Automating machine learning solutions is often termed "machine learning operations" ("MLOps"), or "AI engineering." In most supervised machine learning applications, the goal is not a static, one-time analysis of a particular dataset. Rather, we want to develop a model that can be used on an ongoing basis to predict or classify new records. Our initial analysis will be in prototype mode, while we explore and define the problem and test different models. We will follow all the steps outlined earlier in this chapter.

At the end of that process, we will typically want our chosen model to be deployed in automated fashion. For example, the US Internal Revenue Service (IRS) receives several hundred million tax returns per year—it does not want to have to pull each tax return out into an Excel sheet or other environment separate from its main database to determine the predicted probability that the return is fraudulent. Rather, it would like that determination be made as part of the normal tax filing environment and process. Pandora or Spotify need to determine "recommendations" for next songs quickly for each of millions of users; there is no time to extract the data for manual analysis.

In practice, this is done by building the chosen algorithm into the computational setting in which the rest of the process or database lies. A tax return is entered directly into the IRS system by a tax preparer, a predictive algorithm is immediately applied to the new data in the IRS system, and a predicted classification is decided by the algorithm. Business rules would then determine what happens with that classification. In the IRS case, the rule might be "if no predicted fraud, continue routine processing; if fraud is predicted, alert an examiner for possible audit."

This flow of the tax return from data entry, into the IRS system, through a predictive algorithm, then back out to a human user is an example of a "data pipeline." The different components of the system communicate with one another via Application Programming Interfaces (API's) that establish locally valid rules for transmitting data and associated communications. An API for a machine learning algorithm would establish the required elements for a predictive algorithm to work—the exact predictor variables, their order, data formats, and so on. It would also establish the requirements for communicating the results of the algorithm. Algorithms to be used in an automated data pipeline will need to be compliant with the rules of the API's where they operate.

Finally, once the computational environment is set and functioning, the machine learning work is not done. The environment in which a model operates is typically dynamic, and predictive models often have a short shelf life—one leading consultant finds they rarely continue to function effectively for more than a year. So, even in a fully deployed state, models must be periodically checked and re-evaluated. Once performance lags, it is time to return to prototype mode and see if a new model can be developed.

As you can imagine, these computational environments are complex, and, as of 2022, large companies are only beginning to construct them. They are often driven to do so by the force of circumstance. In the following, we describe two examples of such environments.

---

[3] This section copyright 2021 Peter Bruce, Galit Shmueli, and Victor Diloreto; used by permission.

Predicting Power Generator Failure[4]

Sira-Kvina, a major hydro-electric power company that supplies 7% of Norway's power, found that generator failures were taking a long time to diagnose, leading to lengthy off-line periods and loss of revenue and customer dissatisfaction. The company was using an engineering checklist approach: imagine a more complex version of the troubleshooting guide that comes with a fancy appliance, or the advice you encounter when trying to fix a software problem. Such an approach, however, could not begin to make use of all the data that comes in from hundreds of sensors producing tens of thousands of readings per day. A team from Elder Research investigated one particularly lengthy and costly outage and, after much data wrangling, visualization, and analysis, identified a set of separate events which, when they happened together, caused the failure. Moreover, generator failures were often preceded by a set of defined temperature spikes in one specific sensor (you can read a summary of the analysis at `www.statistics.com/sira-kvina-hydro-power-the-case-of-the-faulty-generator`). But an after-the-fact detective hunt is inherently unsatisfactory, since the generator down-time has already occurred. What is really needed is *proactive preventive maintenance*, triggered by advance warnings that are predictive of failure. Elder built such a system for Sira-Kvina. Machine learning models that predict generator failure are the key analytical component, but they are embedded in a system that must also handle, in addition to training and fitting a model, the following tasks:

- collecting and ingesting data from sensors
- transforming data (especially aggregating and disaggregating granular time series data)
- passing alerts from models to humans and other elements of the system
- providing visualization tools for analysts
- applying case management of anomalies
- adding manual labels applied by analysts, to facilitate further training.

Dozens of tools are involved in this structure. Figure 2.14 illustrates how different layers of tools contribute to this process.

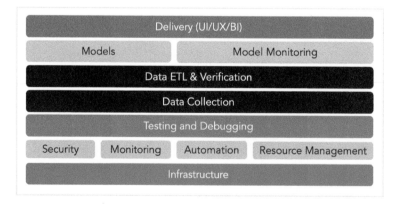

FIGURE 2.14   Layers of tools supporting machine learning automation

[4]Thanks to Victor Diloreto, Chief Technical Officer of Elder Research, Inc., who wrote portions of this segment.

The "Infrastructure" base layer provides base computing capability, memory, and networking, either on-premises or in the cloud. The next layer up supports basic services for the layers above it. "Security" provides user admin capabilities, permission levels, and networking and access rules. "Monitoring" ingests logs, implements needed thresholds, and issues alerts. "Automation" has a more limited meaning than the overall topic of this section: it refers to bringing up, configuring, and tearing down tools and infrastructure as needed. "Resource management" provides the oversight of the various resources used throughout the stack. This area is closely coupled to monitoring: understanding what resources could be coming close to exhaustion is important for maintenance purposes.

"Testing/debugging" is self-explanatory, while the "data collection" layer refers primarily to the data store (data warehouse, data lake, or data lakehouse). In the Sira-Kvina case, there is a raw immutable data store (IDS) where all source data is collected, plus an analytic base table (ABT) that holds derivatives of the main data store needed for analysis. The "data ETL" (Extract, Transform, Load) layer provides the tools to create these derivatives. The verification step within this layer checks that the transformations, aggregations, or any formula used to take a raw piece of data and convert it to something else is doing the job we think it's supposed to do. This verification function is wired into the monitoring capability of the system, which flags any issue with an alert to maintainers of the system. The modeling layer contains the model(s) that make predictions (i.e., the primary focus of this book), and is accompanied by the model-monitoring section, which checks that the model is performing to expectations. Tools in the latter are tied into the lower-level infrastructure monitoring to generate alerts when a model is missing on expectations. Finally, the "delivery" layer is the user window into the system, and can represent simple things like a spreadsheet or a text file, or be as elaborate as an enterprise-wide tool like Tableau or Power BI.

From the above, you can see that Sira-Kvina's predictive models occupy only a small part of the data pipeline through which sensor data could flow and, when needed, trigger alerts to operators about system components that need attention. For a more far-reaching structure, consider Uber.

Uber's Michelangelo

Uber, which provides ride-sharing, meal delivery services, and more, uses real-time predictive models for several purposes including:

- matching the optimal driver for a customer ride request
- predicting customer pickup time
- predicting ride duration
- estimating pricing (including which peak pricing tier applies)
- predicting Uber Eats delivery time

To integrate model-building with deployment, Uber built its own machine learning (ML) structure (MLOps) called Michelangelo. Prior to Michelangelo, in Uber's words, "the impact of ML at Uber was limited to what a few data scientists and engineers could build in a short time frame with mostly open source tools. Specifically, there were no systems in place to build reliable, uniform, and reproducible pipelines for creating and managing training and prediction data at scale."

Uber is a company essentially built on data science; at more traditional companies, the model deployment challenge is exacerbated by the fact that data science and data engineering are often considered separate jobs and disciplines. One major retailer described the problem as follows: "Our highly trained data scientists would build great models, give them to the engineers, who would say 'thank you very much, now we will need to build it again from scratch.'"

Uber's Michelangelo is a development and deployment environment that Uber's data scientists use for a variety of models. It provides integration with the company's data stores and standard tools to build pipelines to ingest data. It has tools to split data into training and validation partitions, and to ensure that new data to be predicted follows the same format.

Some of Uber's goals (e.g., predicting the restaurant prep time component of Uber Eats delivery time) are not extremely time-sensitive, and batch processing can be used. Others (e.g., finding a driver) require nearly instantaneous action ("low-latency" response). Michelangelo supports both batch processing and real-time low-latency processing (Uber calls it "online"). Uber is constantly adding huge amounts of granular customer and driver data to its data store, so Michelangelo must have sufficient compute, data storage and data flow capacity to scale efficiently. It also provides a facility to standardize feature definition and storage, so that data can be shared easily across users. It provides visualization and exploration tools for the prototyping and prep phases. Finally, it collects and makes available (via APIs and via a web user interface) standard information about each model[5]:

- Who trained the model
- Start and end time of the training job
- Full model configuration (features used, hyper-parameter values, etc.)
- Reference to training and test datasets
- Distribution and relative importance of each feature
- Model accuracy metrics
- Standard charts and graphs for each model type (e.g., ROC curve, PR curve, and confusion matrix for a binary classifier)
- Full learned parameters of the model
- Summary statistics for model visualization.

Finally, to ensure that model performance does not degrade over time, Michelangelo has facilities for periodically storing a sample of predictions, comparing predictions to outcomes, and sending signals to operators when the margin exceeds a certain standard.

A variety of tools, both proprietary and open source, can be used to build the MLOps pipeline. In the Sira-Kvina case, the infrastructure layer was built in the cloud using Amazon Web Services Cloud Development Kit (AWS CDK). AWS's Simple Notification Service (SNS) was used to transmit verification and monitoring messages between layers. Uber built Michelangelo with open source tools such as the Cassandra distributed database system and Kafka brokers to stream verification and monitoring messages.

In this book, our focus will be on the prototyping phase—all the steps that go into properly defining the model and developing and selecting a model. You should be aware, though,

---

[5]This list is from the Uber Engineering blog at https://eng.uber.com/michelangelo-machine-learning-platform; see also the discussion of "Model Cards" in Chapter 21.

that most of the actual work of implementing a machine learning solution lies in the automated deployment phase. Much of this work is not in the analytic domain; rather, it lies in the domains of databases and computer engineering, to assure that detailed nuts and bolts of an automated dataflow all work properly. Nonetheless, some familiarity with the requirements and components of automated pipelines will benefit the data scientist, as it will facilitate communication with the engineers who must ultimately deploy their work.

This book's focus is on a comprehensive understanding of the different techniques and algorithms used in machine learning and less on the data management requirements of real-time deployment of machine learning models. JMP's short learning curve makes it ideal for this purpose, and for exploration, prototyping, and piloting of solutions. Many of the methods introduced in this book are available through a point-and-click interface in the standard version of JMP. For more advanced machine learning techniques, JMP Pro is generally required.

2.9   ETHICAL PRACTICE IN MACHINE LEARNING

Prior to the advent of internet-connected devices, the biggest source of Big Data was public interaction on the internet. Social media users, as well as shoppers and searchers on the internet, have made an implicit deal with the big companies that provide these services: users can take advantage of powerful search, shopping and social interaction tools for free, and, in return, the companies get access to user data. Since the first edition of this book was published in 2007, ethical issues in machine learning and data science have received increasing attention, and new rules and laws have emerged. As a business analytics professional or data scientist, you must be aware of the rules and follow them, but you must also think about the implications and use of your work once it goes beyond the prototyping phase. It is now common for businesses, governments, and other organizations to require the explicit incorporation of ethical principles and audits into machine learning and AI applications. Chapter 21, Responsible Data Science, covers this topic in more detail.

MACHINE LEARNING SOFTWARE TOOLS: THE STATE OF THE MARKET

by **Herb Edelstein**

Over time, three things are steadily increasing: the amount of data available for analysis, the power of the computing environment, and the number of organizations taking advantage of the first two changes for a widening variety of analytical and predictive goals. Software products for machine learning must deliver more comprehensive coverage than ever before. The term "data science" has become ubiquitous and covers the spectrum of tasks required for business analytics, starting with data engineering (including data gathering, building databases, and making the data suitable for analytics) and machine learning. Machine learning starts with describing and understanding the data, often utilizing advanced visualization techniques followed by sophisticated model building. After the models are constructed, they must be efficiently deployed so an organization can meet its goals. At the same time, the amount of data collected today presents a great opportunity to tease out subtle relationships that allow precise predictions while unfortunately making it easy to find spurious relationships or statistically significant results that have no practical value. The range of users that must be supported has also increased. Software must support people with diverse backgrounds, ranging from extensive education in machine learning, statistics, and programming to professionals performing such tasks as sales and marketing, risk analysis, fraud detection, reliability prediction, and engineering and scientific analysis. One of the biggest contributors to the hardware environment for machine learning is the spread of high powered GPUs (Graphic Processing Units). While originally intended for accelerating 3-D graphics, meeting the computational requirements for fast 3-D graphics also significantly accelerates machine learning. The algorithms used by the major cloud vendors all take advantage of GPUs. However, not all GPUs have the same interface and therefore not all stand-alone software will necessarily take advantage of the GPUs on your computer. Machine learning has also improved model building. As you read this book, you will see that the various algorithms have a plethora of tuning options to help you build the best model. While many experienced model builders can do a good job of manual tuning, it is impossible to try all the combinations of parameters that control the learning process (called hyperparameters) by hand, so an increasing number of products offer the ability to automatically tune the hyperparameters. The more hyperparameters, the larger the number of combinations to be searched. Autotuning can be particularly useful in building the neural networks used for deep learning.

R and Python

One of the most common open source statistical analysis and machine learning software languages is R. R is the successor to a Bell Labs program called S, which was commercialized as S+. Widely used in the academic community, R has become one of the most popular machine learning tools. New algorithms are frequently available first in R. R contains an enormous collection of machine learning algorithms contained in what are known as "packages," along with an abundance of statistical algorithms, data management tools, and visualization tools. Over 10,000 packages are in the CRAN (Comprehensive R Archive Network) library, and there are numerous packages outside of CRAN. This large number provides a great deal of choice. There are multiple packages that will perform essentially the same task, allowing the user to choose the one they think is most appropriate. However, there is a downside to all these options in that there is no uniformity in the R community. For example, in addition to the base R graphing tools, there are the packages Lattice, ggplot2, Plotly, and others.

Because it is essentially a special purpose programming language, R has enormous flexibility but a steeper learning curve than many of the Graphical User Interface (GUI)-based tools. The most common tool used with R is RStudio, an interactive development environment that facilitates the use of R. RStudio has also created a collection of packages called the "tidyverse" which are designed to enhance the ability of R to manipulate and graph data. These have become widely used in the R community. In particular, R has excelled in creating publication quality visualizations using the ggplot2 package (part of the tidyverse) which is based on Leland Wilkinson's groundbreaking book *The Grammar of Graphics*.

Some GUIs for R try to improve its usability. R Commander is the oldest of these and like R is extensible with add-on packages. Its creator views it as a gateway to writing R code for more complex analyses. BlueSky is a more recent interface that not only is more comprehensive than R Commander but allows the user to create their own dialogs to add R functionality.

Python is a widely used general programming language (developed in the 1980s by Guido van Rossum and named after the British comedy group Monty Python) that has overtaken R as the most popular machine learning package. This is due in part to its speed advantage over R, its data-wrangling capabilities, and its better functionality as a programming language. Furthermore, there is an interface to R that allows you to access R functionality should you need to. Pandas is an open source Python library that provides extensive data-wrangling capabilities. Scikit-Learn is another open source Python library that provides a very comprehensive suite of tools for machine learning that is widely used in the Python community. Python is more often used for deep neural networks than R. While R is still dominating in academia, Python appears to be dominating in commercial applications.

Stand-Alone Machine Learning Software
Many stand-alone software packages have appeared that are intended to simplify the process of creating machine learning models. The number of machine learning packages has grown considerably in response to the increasing demand. IBM offers two machine learning platforms: Watson and SPSS Modeler. Watson is cloud based and offers the same ability to utilize GPUs as the other major cloud providers. SPSS Modeler is aimed at local implementation. However, Watson Studio includes Modeler which can run under Studio. In addition to running on IBM's cloud servers, Watson is also available to run on local computers. SAS is the largest company specializing in statistical software. Recognizing the growing importance of machine learning, SAS added Enterprise Miner, a workflow drag-and-drop software, in 1999 to address this need. More recently, SAS created Viya Machine Learning. SAS Enterprise Miner has multiple client interfaces, whereas Viya's interface is a unified HTML5 interface. While both Enterprise Miner and Viya can operate in the cloud, Viya has a wider range of interfaces to the major cloud providers. It is aimed more directly at the data science and machine learning communities.

JMP Statistical Discovery (pronounced *Jump*) was developed by SAS co-founder John Sall in 1989. JMP is desktop data visualization and analysis software widely used in the science and engineering world, but also competes in commercial and government settings. JMP Pro, the version of JMP designed for predictive analytics and modeling, provides a comprehensive set of machine learning tools. It has a unique GUI that is centered on what you are trying to accomplish. For example, to analyze data you go to the Analyze menu, choose Predictive Modeling, and then select a modeling method (which JMP calls a platform). JMP has its own scripting language, JSL, for tasks that are more complicated than can be accomplished through the GUI, but it also interfaces with R, Python, and MATLAB.

Analytic Solver Data Mining (ASDM) is an add-in for Microsoft Excel that provides access to a suite of machine learning tools as well as data engineering capabilities without the need for programming. It is widely used in MBA programs because of its accessibility to students, but it is also used in industry, especially with its cloud interface to Microsoft's Azure. The original version (XLMiner) was developed by one of the authors of this book (Patel). It provides not only traditional machine learning tools such as tree-based learning and neural nets, but also a variety of ensemble methods which have become central to accurate prediction. Additionally, ASDM provides text mining capabilities.

Drag-and-Drop Workflow Interfaces
Since machine learning solutions are typically deployed in workflow pipelines (data ingestion > model fitting > model assessment > data scoring), some software use a drag-and-drop icon and arrows interface, where icons represent different stages of the process, and arrows connect different stages. While there are a variety of such software, we'll mention a few: IBM Modeler is based on the SPSS Clementine program, which had one of the first workflow-based interfaces. This style of interface links the data engineering, exploration, machine learning, model evaluation, and deployment steps together. Modeler has a wide range of statistical and machine learning algorithms, and provides interfaces to Watson, Python, and R for additional functionality. Orange is a visual interface for Python and Scikit-Learn using wrappers designed to make analysis easier. Developed at the University of Ljubljana in Slovenia, Orange is open source software that uses a workflow interface to provide a greatly improved way to use Python for analysis.

RapidMiner Studio was originally developed in Germany in 2001 (the original name was YALE), and has grown into a widely used platform for machine learning. It has a workflow drag-and-drop interface that supports data import, data cleansing, graphics, a large variety of machine learning tools, and model evaluation. One of the most difficult problems in building good models is feature selection. RapidMiner provides a number of tools including genetic algorithms for feature selection. Its Turbo Prep and Auto Model interfaces offer support for data preparation tasks and auto-tuning hyperparameters for standard machine learning algorithms, respectively. RapidMiner also has integration with R and Python to extend its capabilities.

Cloud Computing
Cloud computing vendors are actively promoting their services for machine learning. These products are oriented more toward application developers than machine learning prototypers and business analysts. A big part of the attraction of machine learning in the cloud is the ability to store and manage enormous amounts of data without requiring the expense and complexity of building an in-house capability. This can also enable a more rapid implementation of large distributed multi-user applications. Many of the non-cloud-based machine learning vendors provide interfaces to cloud storage to enable their tools to work with mammoth amounts of data. For example, RapidMiner offers a product RapidMiner AI Hub (formerly RapidMiner Server) that leverages the standard cloud computing platforms (Amazon AWS, Google Cloud, MS Azure) for automation and deployment of machine learning models. Amazon has several machine learning services. SageMaker is a machine learning platform that provides developers with functionality including acquiring and wrangling data, building models based on its large collection of algorithms, and deploying the resulting

models. Models can be used with data stored in Amazon Web Services (AWS) or exported for use in local environments. There is also an automatic machine learning process that will choose the appropriate algorithm and create the model. AWS also has what it calls Deep Learning AMI's (Amazon Machine Images) that are preconfigured virtual machines with deep learning frameworks such as TensorFlow.

Google is very active in cloud analytics with its Vertex AI. The user can either custom train models or use Vertex AI's AutoML which will automatically build models based on your selected targets. Unfortunately, this is not a transparent process, and there is a limitation in the size of the data—100 GB or 100 million rows of tabular data. Models such as TensorFlow deep learning models can be exported for use in local environments. Microsoft is an active player in cloud analytics with its Azure Machine Learning. Azure ML also supports a workflow interface making it more accessible to the nonprogrammer data scientist. Along with Amazon and Google, Azure Machine Learning also supports an automated machine learning mode. ASDM's cloud version is based on Microsoft Azure. Machine learning plays a central role in enabling many organizations to optimize everything from manufacturing and production to marketing and sales. New storage options and analytical tools promise even greater capabilities. The key is to select technology that's appropriate for an organization's unique goals and constraints. As always, human judgment is the most important component of a machine learning solution. The focus of *Machine Learning for Business Analytics* is on a comprehensive understanding of the different techniques and algorithms used in machine learning, and less on the data management requirements of real-time deployment of machine learning models.

Herb Edelstein is president of Two Crows Consulting (`www.twocrows.com`), a leading data mining consulting firm near Washington, DC. He is an internationally recognized expert in data mining and data warehousing, a widely published author on these topics, and a popular speaker. Herb Edelstein.

PROBLEMS

2.1   Assuming that machine learning techniques are to be used in the following cases, identify whether the task required is supervised or unsupervised learning.

   a. Deciding whether to issue a loan to an applicant based on demographic and financial data (with reference to a database of similar data on prior customers).

   b. In an online bookstore, making recommendations to customers concerning additional items to buy based on the buying patterns in prior transactions.

   c. Identifying a network data packet as dangerous (virus, hacker attack) based on comparison to other packets whose threat status is known.

   d. Identifying segments of similar customers.

   e. Predicting whether a company will go bankrupt based on comparing its financial data to those of similar bankrupt and nonbankrupt firms.

   f. Estimating the repair time required for an aircraft based on a trouble ticket.

   g. Automated sorting of mail by zip code scanning.

   h. Printing of custom discount coupons at the conclusion of a grocery store checkout based on what you just bought and what others have bought previously.

2.2   Describe the difference in roles assumed by the validation partition and the test partition.

2.3   Consider the sample from a database of credit applicants in Table 2.5. Comment on the likelihood that it was sampled randomly, and whether it is likely to be a useful sample.

2.4   Consider the sample from a bank database shown in Table 2.6; it was selected randomly from a larger database to be the training set. *Personal Loan* indicates whether a solicitation for a personal loan was accepted and is the response variable. A campaign is planned for a similar solicitation in the future and the bank is looking for a model that will identify likely responders. Examine the data carefully and indicate what your next step would be.

2.5   Using the concept of overfitting, explain why when a model is fit to training data, zero error with those data is not necessarily good.

2.6   In fitting a model to classify prospects as purchasers or nonpurchasers, a certain company drew the training data from internal data that include demographic and purchase information. Future data to be classified will be lists purchased from other sources, with demographic (but not purchase) data included. It was found that "refund issued" was a useful predictor in the training data. Why is this not an appropriate variable to include in the model?

2.7   A dataset has 1000 records and 50 variables with 5% of the values missing, spread randomly throughout the records and variables. An analyst decides to remove records that have missing values. About how many records would you expect would be removed?

TABLE 2.5 Sample from a database of credit applications

OBS	Chek Acct	Duration	History	New Car	Used Car	Furniture	Radio TV	Educ	Retrain	Amount	Save Acct	Response
1	0	6	4	0	0	0	1	0	0	1169	4	1
8	1	36	2	0	1	0	0	0	0	6948	0	1
16	0	24	2	0	0	0	1	0	0	1282	1	0
24	1	12	4	0	1	0	0	0	0	1804	1	1
32	0	24	2	0	0	1	0	0	0	4020	0	1
40	1	9	2	0	0	0	1	0	0	458	0	1
48	0	6	2	0	1	0	0	0	0	1352	2	1
56	3	6	1	1	0	0	0	0	0	783	4	1
64	1	48	0	0	0	0	0	0	1	14421	0	0
72	3	7	4	0	0	0	1	0	0	730	4	1
80	1	30	2	0	0	1	0	0	0	3832	0	1
88	1	36	2	0	0	0	0	1	0	12612	1	0
96	1	54	0	0	0	0	0	0	1	15945	0	0
104	1	9	4	0	0	1	0	0	0	1919	0	1
112	2	15	2	0	0	0	0	1	0	392	0	1

TABLE 2.6 Sample from a bank database

OBS	Age	Experience	Income	Zip Code	Family	CC Avg	Educ	Mortgage	Personal Loan	Securities Acct
1	25	1	49	91107	4	1.6	1	0	0	1
4	35	9	100	94112	1	2.7	2	0	0	0
5	35	8	45	91330	4	1.0	2	0	0	0
9	35	10	81	90089	3	0.6	2	104	0	0
10	34	9	180	93023	1	8.9	3	0	1	0
12	29	5	45	90277	3	0.1	2	0	0	0
17	38	14	130	95010	4	4.7	3	134	1	0
18	42	18	81	94305	4	2.4	1	0	0	0
21	56	31	25	94015	4	0.9	2	111	0	0
26	43	19	29	94305	3	0.5	1	97	0	0
29	56	30	48	94539	1	2.2	3	0	0	0
30	38	13	119	94104	1	3.3	2	0	1	0
35	31	5	50	94035	4	1.8	3	0	0	0
36	48	24	81	92647	3	0.7	1	0	0	0
37	59	35	121	94720	1	2.9	1	0	0	0
38	51	25	71	95814	1	1.4	3	198	0	0
39	42	18	141	94114	3	5.0	3	0	1	1
41	57	32	84	92672	3	1.6	3	0	0	1

2.8   Normalize the data in Table 2.7, showing calculations. Confirm your results in JMP (create a JMP data table, click on a column header for a variable, right-click, and select *New Formula Column > Distributional*).

TABLE 2.7

Age	Income ($)
25	49,000
56	156,000
65	99,000
32	192,000
41	39,000
49	57,000

2.9   The distance between two records can be measured in several ways. Consider Euclidean distance, measured as the square root of the sum of the squared differences. For the first two records in Table 2.7, it is

$$\sqrt{(25 - 56)^2 + (49,000 - 156,000)^2}.$$

Can normalizing the data change which two records are farthest from each other in terms of Euclidean distance?

2.10   Two models are applied to a dataset that has been partitioned. Model A is considerably more accurate than model B on the training data, but slightly less accurate than model B on the validation data. Which model are you more likely to consider for final deployment?

2.11   The dataset `ToyotaCorolla.jmp` contains data on used cars on sale during the late summer of 2004 in the Netherlands. It has 1436 records containing details on 38 attributes, including Price, Age, Kilometers, HP, and other specifications.

a. Explore the data using the data visualization (e.g., *Graph > Scatterplot Matrix* and *Graph > Graph Builder*) capabilities of JMP. Which of the pairs among the variables seem to be correlated? (Refer to the guides and videos at jmp.com/learn, under *Graphical Displays and Summaries*, for basic information on how to use these platforms.)

b. We plan to analyze the data using various machine learning techniques described in future chapters. Prepare the dataset for machine learning techniques of supervised learning by creating partitions using the JMP Pro *Make Validation Column* utility (from the *Analyze > Predictive Modeling* menu). Use the following partitioning percentages: training (50%), validation (30%), and test (20%), with the random seed 12345. Describe the roles that these partitions will play in modeling. Describe the role of the random seed.

PART II

DATA EXPLORATION AND DIMENSION REDUCTION

3

DATA VISUALIZATION

In this chapter, we describe a set of plots that can be used to explore the multi-dimensional nature of a dataset. We present basic plots (bar charts, line graphs, and scatter plots), distribution plots (boxplots and histograms), and different enhancements that expand the capabilities of these plots to visualize more information. We focus on how the different visualizations and operations can support machine learning tasks, from supervised (prediction, classification, and time series forecasting) to unsupervised tasks, and provide some guidelines on specific visualizations to use with each machine learning task. We also describe the advantages of interactive visualization over static plots. The chapter concludes with a presentation of specialized plots that are suitable for data with special structure (hierarchical and geographical).

Data visualization in *JMP:*   All the methods discussed in this chapter are available in the standard version of JMP.

3.1   INTRODUCTION[1]

The popular saying "a picture is worth a thousand words" refers to the ability to condense diffused verbal information into a compact and quickly understood graphical image. In the case of numbers, data visualization and numerical summarization provide us with both a powerful tool to explore data and an effective way to present results (Few, 2012).

Where do visualization techniques fit into the machine learning as described so far? They are primarily used in the preprocessing portion of the machine learning process. Visualization supports data cleaning by finding incorrect values (e.g., patients whose age is 999 or −1), missing values, duplicate rows, columns with all the same value, and the like.

---

[1] This and subsequent sections in this chapter copyright © 2019 Datastats, LLC and Galit Shmueli. Used by permission.

Machine Learning for Business Analytics: Concepts, Techniques, and Applications with JMP Pro®,
Second Edition. Galit Shmueli, Peter C. Bruce, Mia L. Stephens, Muralidhara Anandamurthy, and Nitin R. Patel.
© 2023 John Wiley & Sons, Inc. Published 2023 by John Wiley & Sons, Inc.

Visualization techniques are also useful for variable derivation and selection: they can help determine which variables to include in the analysis and which might be redundant. They can also help with determining appropriate bin sizes, should binning of numerical variables be needed [e.g., a numerical outcome variable might need to be converted to a binary (nominal modeling type) variable if a yes/no decision is required]. They can also play a role in combining categories as part of the data reduction process. Finally, if the data have yet to be collected and collection is expensive (as with the Pandora project at its outset, see Chapter 7), visualization methods can help determine, using a sample, which variables and metrics are useful.

In this chapter, we focus on the use of graphical presentations for the purpose of *data exploration*, in particular with relation to predictive analytics. Although our focus is not on visualization for the purpose of data reporting, this chapter offers ideas as to the effectiveness of various graphical displays for the purpose of *data presentation*. These offer a wealth of information beyond tabular summaries and basic bar charts, and are currently the most popular form of data presentation in the business environment. For an excellent comprehensive discussion on using graphs to report business data, see Few (2012). In terms of reporting machine learning results graphically, we describe common graphical displays elsewhere in the book, some of which are technique specific (e.g., dendrograms for hierarchical clustering in Chapter 16, and tree charts for classification and regression trees in Chapter 9) while others are more general [e.g., receiver operating characteristic (ROC) curves and lift curves for classification in Chapter 5].

Note: The term "graph" has two meanings in statistics. It can refer, particularly in popular usage, to any of a number of figures to represent data (e.g., line chart, bar plot, histogram, etc.). In a more technical use, it refers to the data structure and visualization in networks. Using the terms "chart" or "plot" for the visualizations we explore in this chapter avoids this confusion.

Data exploration is a mandatory initial step whether or not more formal analysis follows. Graphical exploration can support free-form exploration for the purpose of understanding the data structure, cleaning the data (e.g., identifying unexpected gaps or "illegal" values), identifying outliers, discovering initial patterns (e.g., correlations among variables and surprising clusters), and generating interesting questions. Graphical exploration can also be more focused, geared toward specific questions of interest. In the machine learning context a combination is needed: free-form exploration performed with the purpose of supporting a specific goal.

Graphical exploration can range from generating very basic plots to using operations such as filtering and zooming interactively to explore a set of interconnected plots that include advanced features such as color and multiple panels. This chapter is not meant to be an exhaustive guidebook on visualization techniques, but rather to introduce main principles and features that support data exploration in a machine learning context. We start by describing varying levels of sophistication in terms of visualization and proceed to show the advantages of different features and operations. Our discussion is from the perspective of how visualization supports the subsequent machine learning goal. In particular, we distinguish between supervised and unsupervised learning; within supervised learning, we further distinguish between classification (categorical outcome variable) and prediction (numerical outcome variable).

3.2   DATA EXAMPLES

To illustrate data visualization, we use two datasets used in additional chapters in the book.

Example 1: Boston Housing Data

The Boston Housing data (BostonHousing.jmp) contain information on census tracts in Boston[2] for which several measurements are taken (e.g., crime rate, pupil/teacher ratio). The dataset has 14 variables in all. A description of each variable is given in Table 3.1, and these descriptions have been added as notes in the JMP data table (hold your mouse over the variables in the *Columns Panel* in the data table to view the descriptions). The first 10 records are shown in Figure 3.1. In addition to the original 13 variables, the dataset contains the variable CAT.MEDV, which was created by categorizing median value (MEDV) into two categories, high (1) and low (0).

TABLE 3.1   Description of Variables in the Boston Housing Dataset

CRIM	Crime rate
ZN	Percentage of residential land zoned for lots over 25,000 ft^2
INDUS	Percentage of land occupied by nonretail business
CHAS	Charles River dummy variable (= 1 if tract bounds river; = 0 otherwise)
NOX	Nitric oxide concentration (parts per 10 million)
RM	Average number of rooms per dwelling
AGE	Percentage of owner-occupied units built prior to 1940
DIS	Weighted distances to five Boston employment centers
RAD	Index of accessibility to radial highways
TAX	Full-value property-tax rate per $10,000
PTRATIO	Pupil/teacher ratio by town
LSTAT	% Lower status of the population
MEDV	Median value of owner-occupied homes in $1000s
CAT.MEDV	Binary coding of MEDV (= 1 if MEDV > $30,000; = 0 otherwise)

	CRIM	ZN	INDUS	CHAS	NOX	RM	AGE	DIS	RAD	TAX	PTRATIO	B	LSTAT	MEDV	CAT.MEDV
1	0.00632	18	2.31	0	0.538	6.575	65.2	4.09	1	296	15.3	396.9	4.98	24	0
2	0.02731	0	7.07	0	0.469	6.421	78.9	4.9671	2	242	17.8	396.9	9.14	21.6	0
3	0.02729	0	7.07	0	0.469	7.185	61.1	4.9671	2	242	17.8	392.83	4.03	34.7	1
4	0.03237	0	2.18	0	0.458	6.998	45.8	6.0622	3	222	18.7	394.63	2.94	33.4	1
5	0.06905	0	2.18	0	0.458	7.147	54.2	6.0622	3	222	18.7	396.9	5.33	36.2	1
6	0.02985	0	2.18	0	0.458	6.43	58.7	6.0622	3	222	18.7	394.12	5.21	28.7	0
7	0.08829	12.5	7.87	0	0.524	6.012	66.6	5.5605	5	311	15.2	395.6	12.43	22.9	0
8	0.14455	12.5	7.87	0	0.524	6.172	96.1	5.9505	5	311	15.2	396.9	19.15	27.1	0
9	0.21124	12.5	7.87	0	0.524	5.631	100	6.0821	5	311	15.2	386.63	29.93	16.5	0
10	0.17004	12.5	7.87	0	0.524	6.004	85.9	6.5921	5	311	15.2	386.71	17.1	18.9	0

FIGURE 3.1   First 10 records in the Boston housing dataset

[2]The Boston Housing dataset was originally published by Harrison and Rubinfeld in "Hedonic prices and the demand for clean air." *Journal of Environmental Economics and Management*, vol. 5, pp. 81–102, 1978. A census tract includes 1200–8000 homes.

We consider three possible tasks:

1. A supervised predictive task, where the outcome variable of interest is the median value of a home in the tract (MEDV).
2. A supervised classification task, where the outcome variable of interest is the binary variable CAT.MEDV that indicates whether the median home value is above or below $30,000.
3. An unsupervised task, where the goal is to cluster census tracts.

(MEDV and CAT.MEDV are not used together in any of the three cases).

Example 2: Ridership on Amtrak Trains

Amtrak, a US railway company, routinely collects data on ridership. Here we focus on forecasting future ridership using the series of monthly ridership data between January 1991 and March 2004. The data and their source are described in Chapter 17. Hence our task here is (numerical) time series forecasting. Data are in `Amtrak.jmp`.

3.3 BASIC CHARTS: BAR CHARTS, LINE GRAPHS, AND SCATTER PLOTS

The three most effective basic plots—bar charts, line charts, and scatter plots—are easily created using the JMP *Graph Builder* (from the *Graph* menu). These plots are the commonly used in the current business world, in both data exploration and presentation (unfortunately, pie charts are also popular, although they are usually ineffective visualizations). Basic charts support data exploration by displaying one or more columns of data (variables) at a time. This is useful in the early stages of getting familiar with the data structure, the amount and types of variables, the volume and type of missing values, etc. The initial *Graph Builder* window is shown in Figure 3.2.

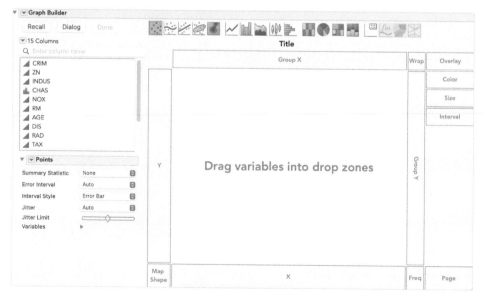

FIGURE 3.2   The JMP *Graph Builder* initial window

The nature of the machine learning task and domain knowledge about the data will affect the use of basic charts in terms of the amount of time and effort allocated to different variables. In supervised learning, there will be more focus on the outcome variable. In scatter plots, the outcome variable is typically associated with the *y*-axis. In unsupervised learning (for the purpose of data reduction or clustering), basic plots that convey relationships (e.g., scatter plots) are preferred.

> ### USING THE JMP GRAPH BUILDER
>
> The JMP *Graph Builder* is a versatile graphing platform for creating basic (and more advanced) graphs for one, two, or many variables in two-dimensional space. Drag variables to zones (e.g., *Y* or *X*), and click the available graph icons at the top to change the graph type. Drag a graph icon onto a graph pane to add an additional graph element, and change display options at the bottom left for selected graph types. To close the control panel and produce the finished graph, click *Done* (see Figure 3.2).

Examples of basic charts created using the *Graph Builder* are shown in Figure 3.3. The top-left panel in Figure 3.3 displays a line chart for the time series of monthly railway passengers on Amtrak. Line graphs are used primarily for showing time series. The choice of the time frame to plot, as well as the temporal scale, should depend on the horizon of the forecasting task and on the nature of the data.

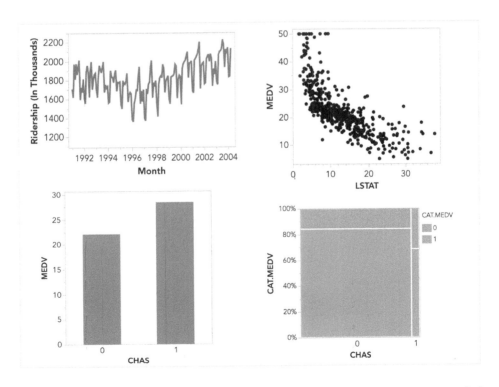

FIGURE 3.3   Basic plots: line chart (top left), scatter plot (top right), bar chart for numerical variable (bottom left), and mosaic plot for categorical variable (bottom right)

Bar charts are useful for comparing a single statistic (e.g., average, count, percentage) across groups. The height of the bar (or length, in a horizontal display) represents the value of the statistic, and different bars correspond to different groups. The bottom-left panel in Figure 3.3 shows a bar chart for a numerical variable (MEDV), which uses the average MEDV on the y-axis. Separate bars are used to denote homes in Boston that are near the Charles River (CHAS = 1) as opposed to those that are not (CHAS = 0), thereby comparing the two categories of CHAS. This supports the second predictive task mentioned earlier: the numerical outcome is on the y-axis and the x-axis is used for a potential categorical predictor.[3] (Note that the x-axis on a bar chart must be used only for categorical variables, since the order of bars in a bar chart should be interchangeable.)

For the classification task, CAT.MEDV is on the y-axis in the bottom-right panel in Figure 3.3, but its aggregation is a percentage (the alternative would be a count). This graph is a form of a stacked bar graph called a *mosaic plot*. The width of the bars relative to the x-axis shows us that the vast majority (nearly 85%) of the tracts do not border the Charles River (the bar for CHAS = 0 is much wider than for CHAS = 1). Note that the labeling of the y-axis can be confusing in this case: the y-axis is simply a percentage of records in each category of CAT.MEDV for CHAS = 0 and CHAS = 1.

The top-right panel in Figure 3.3 displays a scatter plot of MEDV vs. LSTAT. This is an important plot in the prediction task. Note that the output MEDV is again on the y-axis (and LSTAT on the x-axis is a potential predictor). Because both variables in a basic scatter plot must be numerical, the scatter plot cannot be used to display the relation between CAT.MEDV and potential predictors for the classification task (but we can enhance it to do so—see Section 3.4). For unsupervised learning, this particular scatter plot helps us study the association between two numerical variables in terms of information overlap as well as identifying clusters of observations.

All four of these charts highlight global information, such as the overall level of ridership or MEDV, as well as changes over time (line chart), differences between subgroups (bar chart and mosaic plot), and relationships between numerical variables (scatter plot).

Distribution Plots: Boxplots and Histograms

Before moving on to more sophisticated visualizations that enable multidimensional investigation, we note two important plots that are usually not considered "basic charts" but are very useful in statistical and machine learning contexts. The *histogram* and the *boxplot* are two charts that display the entire distribution of a numerical variable. Although averages are very popular and useful summary statistics, there is usually much to be gained by looking at additional statistics such as the median and standard deviation of a variable, and even more so by examining the entire distribution. Whereas bar charts can only use a single aggregation, boxplots and histograms display the entire distribution of a numerical variable.

Boxplots are also effective for comparing subgroups by generating side-by-side boxplots or for looking at distributions over time by creating a series of boxplots.

In JMP, histograms and boxplots for a single numeric variable are generated in the *Graph Builder* (or in *Analyze > Distribution*, which also provides summary statistics).

---

[3] We refer here to a bar chart with vertical bars. The same principles apply if using a bar chart with horizontal lines, except that the x-axis is now associated with the numerical variable and the y-axis with the categorical variable.

Comparative boxplots can be produced using the *Graph Builder* or *Analyze > Fit Y by X* (which also produces summary statistics for the groups).

Distribution plots (histograms and boxplots) are useful in supervised learning for determining potential machine learning methods and variable transformations. For example, skewed numerical variables might warrant transformation (e.g., moving to a logarithmic scale) if used in methods that assume normality (e.g., linear regression, discriminant analysis).

A histogram represents the frequencies of all *x* values with a series of connected bars. For example, in the top-left panel of Figure 3.4, there are over 175 tracts where the median value (MEDV) is between 20 and 25 thousand dollars.

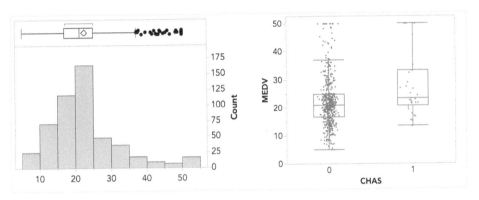

FIGURE 3.4 **Distribution charts for numerical variable MEDV: histogram and boxplot (left), comparative boxplots (right)**

A boxplot represents the variable being plotted on the *y*-axis, although the plot can potentially be turned in a 90° angle so that the boxes are parallel to the *x*-axis (as we see above the histogram in Figure 3.4). In the top-right panel of Figure 3.4 there are two side-by-side boxplots. The box encloses 50% of the data—for example, in the boxplot for CHAS = 1 half of the tracts have median values (MEDV) between 20 and 33 thousand dollars. The horizontal line inside the box represents the median (50th percentile). The top and bottom of the box represent the 75th and 25th percentiles, respectively. Lines extending above and below the box cover the rest of the data range; potential outliers are depicted as points beyond the lines. Boxplots are often arranged in a series with a different plot for each of the various values of a second variable, shown on the *x*-axis.

In the JMP *Distribution* platform, a diamond is displayed in the boxplot, where the center of the diamond is the mean (see the boxplot above the histogram in Figure 3.4). In *Fit Y by X* (not shown), the mean may display as a line or as a diamond (depending on the red triangle option you select). Comparing the mean and the median helps in assessing how skewed the data are. If the data are highly right-skewed, the mean will generally be much larger than the median.

Because histograms and boxplots are geared toward numerical variables, in their basic form they are useful for prediction tasks. Boxplots can also support unsupervised learning by displaying relationships between a numerical variable (*y*-axis) and a categorical variable (*x*-axis). For example, the histogram of MEDV in the left panel in Figure 3.4 reveals a skewed distribution. Transforming the outcome variable to log(MEDV) might improve results of a linear regression prediction model.

The right panel in Figure 3.4 shows side-by-side boxplots comparing the distribution of MEDV for homes that border the Charles River (1) or not (0). We see that not only is the median MEDV for river bounding homes(CHAS = 1) higher than the non-river bounding homes (CHAS = 0), the entire distribution is higher (median, quartiles, min, and max). We also see that all river-bounding homes have MEDV above 10 thousand dollars, unlike non-river-bounding homes. This information is useful for identifying the potential importance of this predictor (CHAS) and for choosing machine learning methods that can capture the nonoverlapping area between the two distributions (e.g., trees).

Boxplots and histograms applied to numerical variables can also provide directions for deriving new variables. For example, they can indicate how to bin a numerical variable (e.g., binning a numerical outcome in order to use a naive Bayes classifier, or in the Boston Housing example, choosing the threshold to convert MEDV to CAT.MEDV).

Finally, side-by-side boxplots are useful in classification tasks for evaluating the potential of numerical predictors. This is done by using the x-axis for the categorical outcome variable and the y-axis for a numerical predictor. An example is shown in Figure 3.5, where we can see the effects of four numerical predictors on CAT.MEDV. These were generated in the *Graph Builder* using the *Column Switcher* (under the red triangle > Redo) to swap out the predictor variables one at a time. The pairs that are most separated (e.g., LSTAT and INDUS) indicate potentially useful predictors.

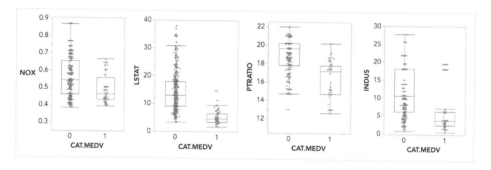

FIGURE 3.5   Side-by-side boxplots for exploring the CAT.MEDV outcome variable by different numerical predictors created in the *Graph Builder* using the *Column Switcher*

The main weakness of basic charts in their basic form (that is, using position in relation to the axes to encode values) is that they only display two variables and therefore cannot reveal high-dimensional information. Each of the basic charts has two dimensions, where each dimension is dedicated to a single variable. In machine learning the data are usually multivariate by nature, and the analytics are designed to capture and measure multivariate information. Visual exploration should therefore incorporate this important aspect.

In the next section, we describe how to extend basic and distribution charts to multidimensional data visualization by adding features, employing manipulations, and incorporating interactivity. We then present several specialized charts geared toward displaying special data structures (Section 3.5).

TOOLS FOR DATA VISUALIZATION IN JMP

The three most popular tools in JMP for creating visualizations are *Distribution*, *Fit Y by X* (from the *Analyze* menu), and the *Graph Builder* (from the *Graph* menu). *Distribution* is used for creating univariate graphs and statistics (bar charts, histograms, frequency distributions, and summary statistics). The graphs and statistics available are based on the types of the variables, and additional options are available under the red triangle for the variable. *Fit Y by X* is used for bivariate graphs and statistics (scatter plots, boxplots and summary statistics, mosaic plots and contingency tables, etc.). Again, the graphs and statistical methods provided are based on the variables selected, and the red triangle offers additional options. The *Graph Builder* is primarily an interactive graphing platform, providing a wide array of graphs for one, two, or many variables.

Heatmaps (Color Maps and Cell Plots): Visualizing Correlations and Missing Values

A *Heatmap* is a graphical display of numerical data where color is used to denote values. In a machine learning context, these charts are especially useful for two purposes: for visualizing correlation tables and for visualizing missing values in the data. In both cases the information is conveyed in a two-dimensional table. A correlation table for p variables has p rows and p columns. A data table contains p columns (variables) and n rows (records). If the number of records is huge, then a subset can be used (although JMP can generally handle a very large number of observations).

Heatmaps are based on the fact that it is much easier and faster to scan the color-coding rather than the values. Note that a heatmap is useful when examining a large number of values, but it is not a replacement for more precise graphical displays, such as bar charts or scatter plots, since color differences cannot be perceived accurately.

Two types of heatmaps in JMP for visualizing correlations and missing values are *color maps* and *cell plots*. An example of a text-colored correlation table and corresponding correlation heatmap (or color map) for the Boston Housing data are shown in Figure 3.6. All of the pairwise correlations between the 13 variables (MEDV and 12 predictors) are given in the correlation table (top). In the color map (bottom), a blue to red color theme has been applied to highlight negative (blue) and positive correlations (red). Note that the y-axis labels are not displayed—the rows correspond to the variables listed across the top of the color map. It is easy to quickly spot the high and low correlations. This output was produced in the *Multivariate* platform (under *Analyze > Multivariate Methods*).

Next we can examine the data table (with n rows and p variables) using a color map to search for missing values. In JMP, a missing value analysis can be conducted using *Tables > Missing Data Pattern*. This option produces a *missing data pattern* data table, which uses a binary coding of the original dataset where 1 denotes a missing value for an variable and 0 otherwise. The pattern indicates which variables, or columns in the data table, are missing values. For example, a pattern of 00000 for five variables indicates that there are no missing

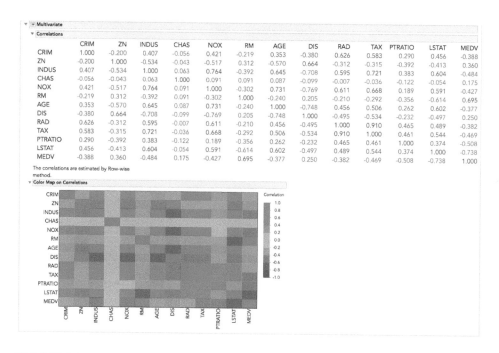

FIGURE 3.6 Color-coded correlation table and heatmap (also called a color map) of correlations with a blue-red color theme. *HINT*: You can right-click on the Color map legend, and select *Gradient* to reverse the color theme

values, while a pattern of 10100 indicates that there are missing values for the first and third variables. The count indicates how many observations are missing values for those particular variables.

For illustration, Figure 3.7 shows a partial missing data table for a dataset with over 40 columns and 539 rows (file Bands Data.jmp from the JMP *Sample Data Folder* under the *Help* menu). The variables were merged from multiple sources, and for each source information was not always available. In Figure 3.7, we see that only 277 rows (see the first row in the missing data pattern table) are not missing values for any the columns and that several rows are missing values for just one or two columns.

The next step is to create a *cell plot*, which highlights the missing values in the data matrix. Figure 3.7 shows the result for our example. The cell plot is created via a shortcut provided as a script in the top-left corner of the missing data pattern table. To create the cell plot, right-click on the script and select *Run Script*.

The cell plot in Figure 3.8 graphically displays missing rows for particular variables as shaded regions, along with rows missing values for multiple variables (shaded regions for multiple variables). The cell plot helps us visualize the amount of "missingness" in the merged data file. Some patterns of missingness easily emerge: variables that are missing values for nearly all observations, as well as rows that are missing many values. Variables with little missingness are also visible. This information can then be used for determining how to handle the missingness (e.g., dropping some variables, dropping some records, imputing values, or via other techniques).

		Count	Number of columns missing	Patterns	Banding?	timestamp
	1	277	0	00000000000000000000000000000000000000000	0	0
	2	1	1	00000000000000000000000000000000100000	0	0
	3	2	1	00000000000000000000000000000010000000	0	0
	4	3	1	00000000000000000000000000010000000000	0	0
	5	36	1	00000000000000000000000001000000000000	0	0
	6	1	3	00000000000000000000000001000000110000	0	0
	7	1	2	00000000000000000000001010000000000	0	0
	8	8	1	00000000000000000000010000000000000	0	0
	9	1	4	00000000000000000000010010000110000	0	0
	10	8	2	00000000000000000000011000000000000	0	0
	11	1	3	00000000000000000000011010000000000	0	0
	12	2	1	00000000000000000001000000000000000	0	0
	13	9	1	00000000000000000010000000000000000	0	0
	14	2	1	00000000000000000100000000000000000	0	0
	15	5	1	00000000000000001000000000000000000	0	0
	16	1	2	00000000000000010000001000000000000	0	0
	17	1	2	00000000000000100000010000000000000	0	0
	18	1	4	00000000000000100100000000000000110	0	0
	19	1	5	00000000000000100100100000000000110	0	0
	20	1	6	00000000000001101000000000000000111	0	0
	21	72	1	00000000000001000000000000000000000	0	0
	22	1	2	00000000000001000000010000000000000	0	0
	23	9	2	00000000000001000000010000000000000	0	0
	24	4	2	00000000000001000000100000000000000	0	0
	25	3	2	00000000000001000100000000000000000	0	0

Missing Data Pattern.jmp
- Source
- Treemap
- Cell Plot
- Cell Plot 2
- Cell Plot 3

Columns (41/0)
- Patterns ✔
- Banding?
- timestamp
- cylinder number
- press
- customer
- job number
- grain screened
- proof on ctd ink
- blade mfg
- paper type
- ink type
- direct steam
- solvent type
- type on cylinder

Rows
All rows	68
Selected	0
Excluded	0
Hidden	0
Labeled	0

FIGURE 3.7 Missing data pattern table. The pattern indicates which columns are missing values. Run the scripts in the top-right corner to produce a treemap or cell plot of missing values

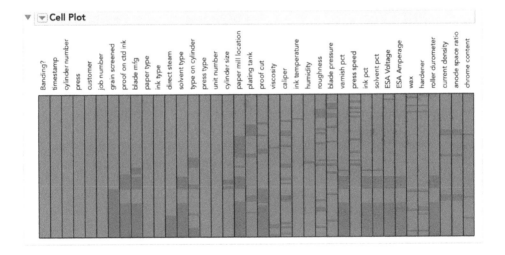

FIGURE 3.8 Missing data pattern cell plot

> Additional options for exploring missing values, including options to impute missing values, are available in JMP from *Analyze > Screening > Explore Missing Values*. The platform provides various options including multivariate imputation and automated data imputation techniques.

3.4   MULTIDIMENSIONAL VISUALIZATION

Basic charts in JMP can convey richer information with features such as color, size, and multiple panels, and by enabling operations such as rescaling, aggregation, interactivity, dynamic linking between graphs and the data table, and dynamic filtering of data based on values of other variables. These additions allow us to look at more than one or two variables at a time. The beauty of these additions is their effectiveness in displaying complex information in an easily understandable way. Effective features are based on understanding how visual perception works (see Few, 2021). The purpose is to make the information more understandable, not just represent the data in higher dimensions (e.g., three-dimensional plots, which are usually ineffective visualizations).

Adding Variables: Color, Size, Shape, Multiple Panels, and Animation

In order to include more variables in a plot, we must consider the types of variables to include. To represent additional categorical information, the best way is to use hue, shape, or multiple panels. For additional numerical information, we can use color intensity or size. Temporal information can be added via animation.

Incorporating additional categorical and/or numerical variables into the basic (and distribution) charts means that we can now use all of them for both prediction and classification tasks. For example, we mentioned earlier that a basic scatter plot cannot be used for studying the relationship between a categorical outcome and predictors (in the context of classification). However, a very effective plot for classification is a scatter plot of two numerical predictors color-coded by the categorical outcome variable. An example is shown in the left panel of Figure 3.9, with color and marker denoting CAT.MEDV (using *Graph Builder*, with CAT.MEDV as both the *overlay* and the *color* variable).

In the context of prediction, color-coding supports the exploration of the conditional relationship between the numerical outcome variable (on the *y*-axis) and a numerical predictor. Color-coded scatter plots then help us assess the need for including interaction terms. For example, is the relationship between MEDV and LSTAT different for homes near vs. away from the river? Simply use the *Graph Builder* to graph MEDV vs. LSTAT, and drag CHAS to the *Overlay* zone. For a more formal exploration of this relationship, click on the *Line of Fit* graph icon. If the resulting lines are not parallel, this is a good indication that an interaction term for LSTAT and CHAS is needed (i.e., the relationship between LSTAT and MEDV depends on the value of CHAS).

Color can also be used to add categorical variables to a bar chart, as long as the number of categories is small. When the number of categories is large, a better alternative is to use *multiple panels*. Creating multiple panels (also called *trellising*) is done by splitting the observations according to a categorical variable and creating a separate plot (of the same type) for each category. An example of trellising is shown in the right-hand panel of

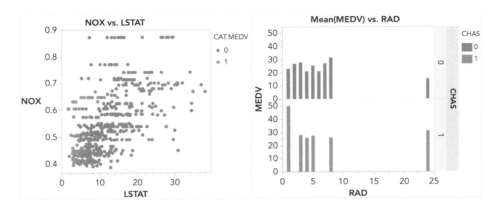

FIGURE 3.9 **Color-coding and trellising by a categorical variable. Left: scatter plot of two numerical predictors, color-coded and marked by the categorical outcome CAT.MEDV. Right: Bar chart of MEDV by two categorical predictors (*CHAS* and *RAD*), using *CHAS* as the *Group Y* variable to create multiple panels**

Figure 3.9, where a bar chart of average MEDV by highway accessibility (RAD) is broken down into two panels by CHAS. In this example, CHAS is used as the *Group Y* variable in the *Graph Builder*. We see that the average MEDV for different highway accessibility levels (RAD) behaves differently for homes near the river (lower panel) compared to homes away from the river (upper panel). This is especially salient for RAD = 1. We also see that there are no near-river homes at RAD levels 2, 6, and 7. Such information might lead us to create an interaction term between RAD and CHAS and to consider combining some of the bins in RAD. All these explorations are useful for prediction and classification.

The *Group* zones creates multiple panels that are aligned either vertically or horizontally. For categorical variables with multiple levels, it is sometimes better to align the panels in multiple rows by columns. To achieve that, use the *Wrap* zone (in the top-right corner of the *Graph Builder*; see Figure 3.2). This will create a *trellis* plot by the variable in the wrap zone, which automatically wraps to efficiently display multiple panels. Figure 3.10 shows an example of a trellis plot that uses wrap. Specifically, we created a scatter plot of NOX vs. LSTAT, color-coded and marked the points by CAT.MEDV, and then trellised the scatter plot by adding RAD into the wrap zone.

Note: When numeric variables are placed in the wrap zone, or in either of the group zones, the variables are automatically binned to create trellis plots. The binning can be changed by right-clicking on the zone and selecting one of the options under *Level*.

A special plot that uses scatter plots with multiple panels is the scatter plot matrix. In this plot, all pairwise scatter plots are shown in a single display. The panels in a scatter plot matrix are organized in a special way, such that each column corresponds to a variable and each row corresponds to an variable; thereby the intersections create all the possible pairwise scatter plots. The scatter plot matrix is useful in unsupervised learning for studying the associations between pairs of numerical variables, detecting outliers, and identifying clusters. For supervised learning, it can be used for examining pairwise relationships (and their nature) between predictors to support variable transformations and variable selection (see the correlation analysis in Chapter 4). For prediction, it can also be used to depict the relationship between the outcome and the numerical predictors.

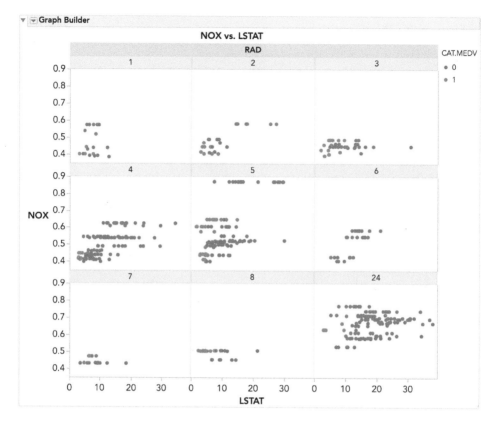

FIGURE 3.10   Scatter plot of *NOX* vs. *LSTAT*, trellised by *RAD*

An example of a scatter plot matrix is shown in Figure 3.11, with MEDV and three predictors. Variable names in each row indicate the *y*-axis variable, and variable names in each column indicate the *x*-axis variable. For example, the plots in the bottom row all have MEDV on the *y*-axis (which allows studying the individual outcome—predictor relations). We can see different types of relationships from the different shapes (e.g., an exponential relationship between MEDV and LSTAT and a highly skewed relationship between CRIM and INDUS) that can indicate potentially useful transformations. Note that the plots above and to the right of the diagonal are mirror images of those below and to the left.

Once hue is used, further categorical variables can be added via shape and multiple panels. However, one must proceed cautiously in adding multiple variables, as the display can become overcluttered and then visual perception is lost.

Adding a numerical variable via size is especially useful in scatter plots (thereby creating "bubble plots"), since, in a scatter plot, points typically represent individual observations. Size variables can be added in the *Graph Builder* and the *Bubble Plot* (under the *Graph* menu). In plots that aggregate across observations (e.g., boxplots, histograms, bar charts), size and hue are not normally incorporated.

Finally, adding a temporal dimension to a chart to show how the information changes over time can be achieved via animation. A famous example is Rosling's animated scatter

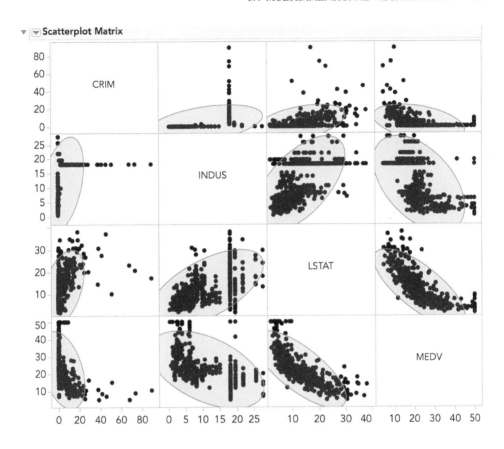

FIGURE 3.11 **Scatter plot matrix for** *MEDV* **and three numerical predictors. The** *density ellipses* **graphically show the nature and strength of correlation among pairs of variables**

plots showing how world demographics changed over the years (www.gapminder.org). In JMP, this form of animation is available with the *Bubble Plot*. Additional animation is available with the *Data Filter* and the *Column Switcher*. However, while animations of this type work for "statistical storytelling," they are usually not very effective for data exploration.

Manipulations: Rescaling, Aggregation and Hierarchies, Zooming, Filtering

Most of the time spent in machine learning projects is spent in data preprocessing. Typically, considerable effort is expended getting all the data in a format that can actually be used in machine learning. Additional time is spent processing the data in ways that improve the performance of the machine learning procedures. The preprocessing step includes variable transformation and derivation of new variables to help models perform more effectively. Transformations include changing the numeric scale of a variable, binning numerical variables, combining categories in categorical variables, and so on. The following manipulations support the preprocessing step as well the choice of adequate machine learning methods. They do so by revealing patterns and their nature.

Rescaling   Changing the scale in a display can enhance the chart and illuminate relationships. For example, in Figure 3.12 we see the effect of changing both axes of the scatter plot
(top) and the *y*-axis of a boxplot (bottom) to logarithmic (log) scale. Whereas the original
charts (left) are hard to understand, the patterns become visible in log scale (right). In the
scatter plots the nature of the relationship between MEDV and CRIM is hard to determine
in the original scale because too many of the points are "crowded" near the *y*-axis. The
rescaling removes this crowding and allows a better view of the linear relationship between
the two log-scaled variables (indicating a log–log relationship). In the boxplots the crowding toward the *x*-axis in the original units does not allow us to compare the two box sizes,
their locations, lower outliers, and most of the distribution information. Rescaling removes
the "crowding to the *x*-axis" effect, thereby allowing a comparison of the two boxplots.
(Double-click on an axis in JMP to change the scale.)

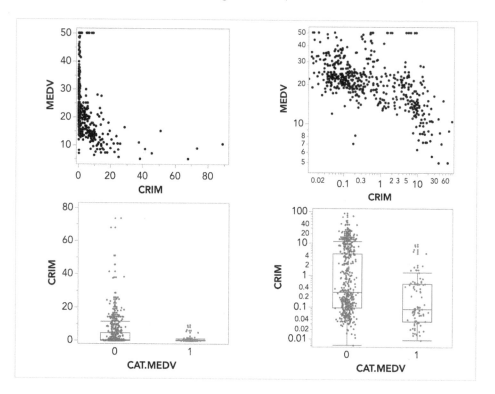

FIGURE 3.12   **Rescaling can enhance plots and reveal patterns: original scale (left), logarithmic scale (right)**

Aggregation and Hierarchies   Another useful manipulation of scaling is changing the
level of aggregation. For a temporal scale, we can aggregate by using different granularity
(e.g., monthly, daily, hourly) or even by a "seasonal" factor of interest such as month-of-
year or day-of-week. A popular aggregation for time series is a moving average, where the
average of neighboring values within a given window of time is plotted. Moving average
plots enhance visualizing a global trend (see Chapter 17).

Non-temporal variables can be aggregated if some meaningful hierarchy exists: geographical (tracts within a zip code in the Boston Housing example), organizational (people

within departments within business units), and so forth. Figure 3.13 illustrates two types of aggregation for the railway ridership time series. The original monthly series is shown in the top-left panel. Seasonal aggregation (by month-of-year) is shown in the top-right panel, where it is easy to see the peak in ridership in August and the dip in February and March. The bottom-right panel shows temporal aggregation, where the series is now displayed in yearly aggregates. This plot reveals the global long-term trend in ridership and the generally increasing trend from 1996 on.

Examining different scales, aggregations, or hierarchies supports both supervised and unsupervised tasks in that it can reveal patterns and relationships at various levels and can suggest new sets of variables with which to work.

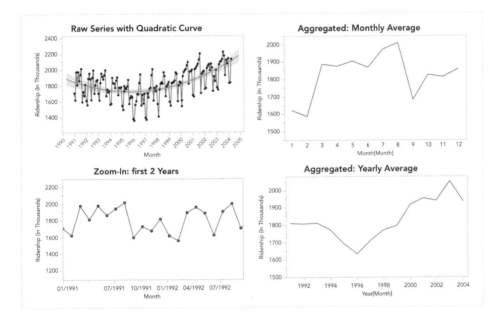

FIGURE 3.13 Time series line charts using different aggregations (right panels), adding curves (top-left panel), and zooming in (bottom-left panel)

Zooming and Panning   The ability to zoom in and out of certain areas of the data on a plot is important for revealing patterns and outliers. Zooming is particularly useful with geographic maps (e.g., produced via the *Graph Builder*). We are often interested in more detail on areas of dense information or of special interest. Panning refers to moving the zoom window to other areas or time frames. An example of zooming is shown in the bottom-left panel of Figure 3.13, where the ridership series is zoomed in to the first two years of the series. Zooming and panning support supervised and unsupervised methods by detecting areas of different behavior, which may lead to creating new interaction terms, new variables, or even separate models for data subsets. In addition, zooming and panning can help us choose between methods that assume global behavior (e.g., regression models) and data-driven methods (e.g., exponential smoothing forecasters and k-nearest neighbors classifiers) and indicate the level of global/local behavior (as manifested by parameters such as k in k-nearest neighbors, the size of a tree, or the smoothing parameters in exponential smoothing).

In JMP, different aggregations and transformations can be applied from the data table, the *Graph Builder*, or in any analysis dialog window. From the data table, right-click on the variable of interest, select *New Formula Column*, and select from the available functions. To create transformed variables from a graphing or analysis platform, right-click on the variable in the dialog, and select the function of interest. This produces a temporary variable that can be used in the current graph or analysis. This transformed variable (and its formula) can then be saved to the data table for future use.

Note that in *Graph Builder* if you use the Wrap zone or a group zone to bin continuous data, you can save the binning as transformed column (right-click on the zone, and select *Levels > Save Transform Column*). This creates a temporary variable, which can also be saved to the data table.

Filtering   Filtering means highlighting or removing some of the observations from the plot (or the analysis). The purpose of filtering is to focus the attention on certain data while eliminating "noise" created by other data. Filtering supports supervised and unsupervised learning in a similar way to zooming and panning: it assists in identifying different or unusual local behavior.

In JMP, the *Data Filter* (from the *Rows* menu) can be used to select, hide, and/or exclude values. Hiding removes observations from graphical displays, while excluding removes observations from analyses and calculations. (*Note:* In analysis platforms, you need

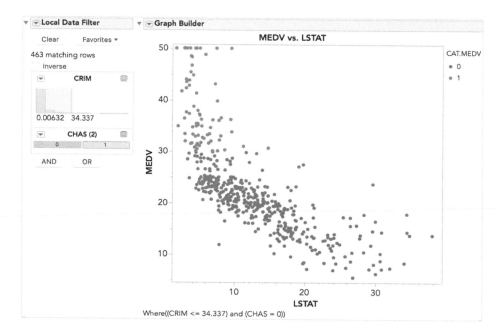

FIGURE 3.14   Using the *Local Data Filter* in the *Graph Builder* to view only data in specific ranges of *CRIM* and *CHAS* values

to re-run the analysis after excluding observations.) The Data Filter is a *global* filter, and it applies to the data table and all output windows. The *Local Data Filter* (available under the *red triangle* menu in any output window, or as a shortcut on the toolbar) selects and includes values of the specified columns for only the current (or active) graphs and analysis. In Figure 3.14, we explore the relationship between MEDV and LSTAT at specified values of two other variables, CRIM and CHAS. The graph dynamically updates as values of the filtering variables are changed.

Reference: Trend Lines and Labels

Trend lines and in-plot labels also help us detect patterns and outliers. Trend lines serve as a reference and allow us to more easily assess the shape of a pattern. Although linearity is easy to visually perceive, more elaborate relationships such as exponential and polynomial trends are harder to assess by eye. Trend lines are useful in line graphs as well as in scatter plots. An example is shown in the top-left panel of Figure 3.13, where a polynomial curve is overlaid on the original line graph (see also Chapter 5).

ADDING TRENDLINES IN THE JMP GRAPH BUILDER

To create a graph with a quadratic curve (as shown in Figure 3.15):

- Drag a continuous variable to the *Y* zone.
- Drag a continuous variable to the *X* zone.
- Click on the *Line of Fit* icon (the third icon above the graph). This fits a trendline to the data (the shaded bands are confidence bands). A series of options for this line of fit will appear on the bottom left, below the list of variables.
- Under the *Line of Fit* outline, change the *Degree* from Linear to Quadratic (the option for a Cubic fit is also available).
- To connect the points, drag the *Line* icon onto the graph.

In Figure 3.15, three graph elements (points, line of fit, and line) have been used—these icons all appear shaded. Options for selected graph elements are provided on the bottom left of the graph. To close the control panel on the left and hide these options, click *Done*.

In displays that are not cluttered or overcrowded, the use of in-plot labels can be useful for better exploration of outliers and clusters. An example is shown in Figure 3.16. The figure shows different utilities on a scatter plot that compares fuel cost with total sales. We might be interested in clustering the data and using clustering algorithms to identify clusters that differ markedly with respect to fuel cost and sales. Figure 3.16 helps us visualize these clusters and their members (e.g., Nevada and Puget are part of a clear cluster with low fuel costs and high sales). For more on clustering, and this example, see Chapter 16. Labels are added as column and row properties in the data table. Within a graph, labels can be repositioned (dragged manually) for optimal spacing and visibility.

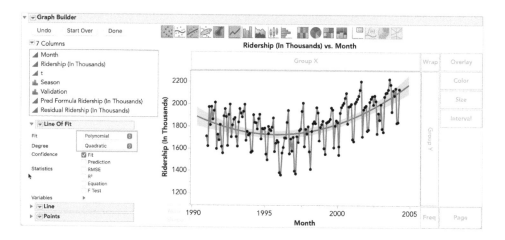

FIGURE 3.15   Adding a trendline in *Graph Builder*

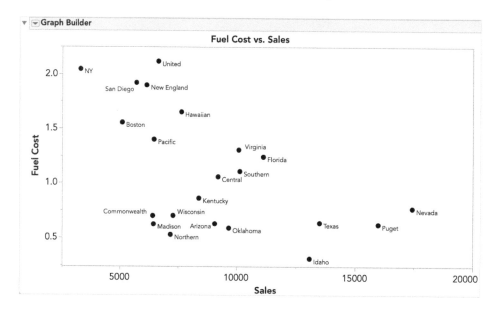

FIGURE 3.16   Scatter plot with labeled points

Points can be labeled in JMP using tags and column/row labels. To display a tag with the plotted values for a point in a graph, hold your mouse over the point—this will display a temporary tag. To pin (retain) this tag on the graph, right-click on the tag and select *Pin*. To add the values for a variable to the tag, right-click on the variable name in the data table and select *Label/Unlabel*. To add permanent variable labels for particular points, select the points and use *Rows > Label/Unlabel*.

Scaling Up to Large Datasets

When the number of observations (rows) is large, plots that display each individual observation (e.g., scatter plots) can become ineffective. Aside from using aggregated charts such as boxplots, some alternatives are as follows:

1. Sampling (by drawing a random sample and using it for plotting). JMP has a sampling utility (under *Tables > Subset*), and sampling can be done directly in the *Graph Builder*.
2. Reducing marker size.
3. Using more transparent marker colors and removing fill.
4. Breaking down the data into subsets (e.g., by creating multiple panels).
5. Using aggregation (e.g., bubble plots where size corresponds to number of observations in a certain range).
6. Using jittering (slightly moving the marker by adding a small amount of noise so that individual markers can be seen).

An example of the advantage of plotting a sample over the large dataset is shown in Chapter 12 (Figure 12.2), where a scatter plot of 5000 records is plotted along with a scatter plot of a sample. The dataset, UniversalBank.jmp, is described in a future chapter. In Figure 3.17, we illustrate techniques for plotting the large datasets described above. In this scatter plot, we use smaller markers, different markers for each class, and more transparent colors (a gray scale was used for printing). We can see that larger areas of the plot are dominated by the gray class (hollow markers). The black class (solid markers) is mainly on the right, while there is a lot of overlap in the top-right area.

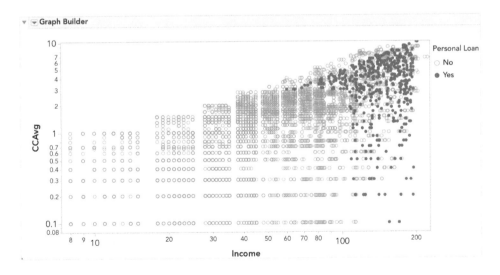

FIGURE 3.17   **Scatter plot of 5000 records with reduced marker size, different markers, and more transparent coloring**

Multivariate Plot: Parallel Coordinates Plot

Another approach toward presenting multidimensional information in a two-dimensional plot is via specialized plots such as the *parallel coordinates plot*. In this plot, a vertical axis is drawn for each variable. Then each observation is represented by drawing a line that connects its values on the different axes, thereby creating a "multivariate profile." An example is shown in Figure 3.18 for the Boston Housing data. In this display, separate panels are used for the two values of CAT.MEDV, in order to compare the profiles of tracts in the two classes (for a classification task). We see that tracts with more expensive homes (right panel) consistently have low CRIM, low LSAT, and high RM compared to tracts with less expensive homes (left panel), which are more mixed on CRIM and LSAT, and have a medium level of RM. This observation gives indication of useful predictors and suggests possible binning for some numerical predictors.

Parallel coordinate plots are also useful in unsupervised tasks. They can reveal clusters, outliers, and information overlap across variables. A useful manipulation is to re-order the columns to better reveal clustering of observations. Parallel coordinates plots are available in JMP under *Graph > Parallel Plot*, and are an option in the *Clustering* platform in JMP.

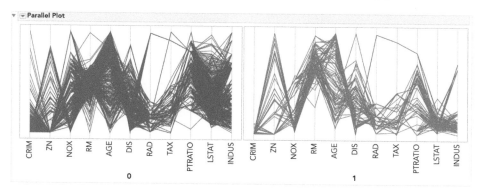

FIGURE 3.18   Parallel coordinates plot for Boston Housing data. Each line is census tract. Each of the variables (shown on the horizontal axis) is scaled to 0–100%. Panels are used to distinguish *CAT.MEDV* (left panel = tracts with median home value below $30,000)

Interactive Visualization

Similar to the interactive nature of the machine learning process, interactivity is key to enhancing our ability to gain information from graphical visualization and is fundamental in JMP. In the words of Stephen Few (2021), an expert in data visualization:

> We can only learn so much when staring at a static visualization such as a printed graph.
> … If we can't interact with the data … we hit the wall.

By interactive visualization, we mean an interface that supports the following principles:

1. Making changes to a chart is *easy, rapid, and reversible*.
2. Multiple concurrent charts and tables can be easily combined and displayed on a single screen.
3. A set of visualizations can be linked so that operations in one display are reflected in the other displays.

JMP is designed, from the ground up, to support all these operations. Let us consider a few examples where we contrast a static plot generator (e.g., Excel) with an interactive visualization interface (i.e., JMP).

Histogram Rebinning Consider the need to bin a numerical variable and using a histogram for that purpose. A static histogram would require replotting for each new binning choice (in Excel it would even require creating the new bins manually). If the user generates multiple plots, then the screen becomes cluttered. If the same plot is recreated, then it is hard to compare to other binning choices. In contrast, an interactive visualization provides an easy way to change bin width interactively, and then the histogram automatically and rapidly replots as the user changes the bin width. The hand tool in JMP (aka the *Grabber* tool) on the toolbar makes this effortless (select the *Grabber*, click on the histogram, and drag up or down to rebin, or drag left or right to rescale.)

Aggregation and Zooming Consider a time series forecasting task, given a long series of data. Temporal aggregation at multiple levels is needed for determining short- and long-term patterns. Zooming and panning are used to identify unusual periods. A static plotting software requires the user to create new variables for each temporal aggregation (e.g., aggregate daily data to obtain weekly aggregates). Zooming and panning in Excel requires manually changing the min and max values on the axis scale of interest (thereby losing the ability to quickly move between different areas without creating multiple charts). In contrast, JMP provides immediate temporal hierarchies that the user can easily switch between (via dynamic column transformations, discussed previously). The cursor, or arrow tool, is also used to quickly zoom, stretch, or reposition axes, thereby allowing direct manipulation and rapid reaction.

Combining Multiple Linked Plots That Fit in a Single Screen In earlier sections, we used charts to illustrate the advantages of visualization, because "a picture is worth a thousand words." The advantages of an interactive visualization are even harder to convey in words. As Ben Shneiderman, a well-known researcher in information visualization and interfaces, puts it:

> A picture is worth a thousand words. An interface is worth a thousand pictures.

In the exploration process, we typically create multiple visualizations. For example, to support a classification task, multiple plots are created of the outcome variable vs. potential categorical and numerical predictors. These can include side-by-side boxplots, color-coded scatter plots with trend lines, multi-panel bar charts, and so on. We might want to detect possible multidimensional relationships (and identify possible outliers) by selecting a certain subset of the data (e.g., a single category of some variable) and locating the observations on the other plots. In a static interface, we would have to manually organize the plots of interest and resize them in order to fit them within a single screen. Moreover a static interface would usually not support inter-plot linkage, and even if so, the entire set of plots would have to be regenerated each time a selection is made.

In contrast, JMP provides an easy way to automatically organize and re-size the set of plots to fit within a screen. All plots are dynamically linked, and plots are dynamically linked to the data table. You can also create dashboards (using the *Dashboard Builder* with related graphs, analyses, and data filters). Multiple observations are selected by drawing a box around the points with the cursor (or using a data filter), and selections in

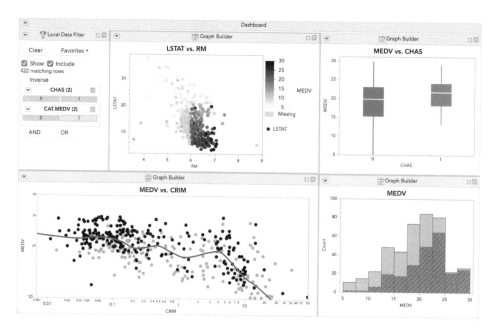

FIGURE 3.19 Multiple inter-linked plots in a single dashboard (a JMP dashboard). All graphs were created using the *Graph Builder*, and these graphs were combined in one window to create a JMP dashboard. The *Local Data Filter* was used to select observations corresponding to *CHAS* = 0. These observations are automatically selected in all of the graphs

one plot are automatically highlighted in the other plots. A simple example is shown in Figure 3.19. For more information on building dashboards in JMP, search for *Dashboards* in the *JMP Help*.

3.5 SPECIALIZED VISUALIZATIONS

In this section, we mention a few specialized visualizations that are able to capture data structures beyond the standard time series and cross-sectional structures—special types of relationships that are usually hard to capture with ordinary plots. In particular, we address hierarchical data, network data, and geographical data—three types of data that are becoming increasingly available.

Visualizing Networked Data

Network analysis techniques were spawned by the explosion of social and product network data. Examples of social networks are networks of sellers and buyers on eBay and networks of people on Facebook. An example of a product network is the network of products on Amazon (linked through the recommendation system). Network data visualization is available in various network-specialized software and in some general-purpose software (although it is only available in JMP through the JMP Scripting Language, or *JSL*).

A network diagram consists of actors and the relations between them. "Nodes" are the actors (e.g., people in a social network or products in a product network) and are represented by circles. "Edges" are the relations between nodes and are represented by lines connecting nodes. For example, in a social network such as Facebook, we can construct a list of users (nodes) and all the pairwise relations (edges) between users who are "Friends." Alternatively, we can define edges as a posting that one user posts on another user's Facebook page. In this setup, we might have more than a single edge between two nodes. Networks can also have nodes of multiple types. A common structure is networks with two types of nodes.

An example of a two-type node network is shown in Figure 3.20, where we see a set of transactions between a network of sellers and buyers on the online auction site www.eBay.com (the data are for auctions selling Swarovski beads, and these auctions took place over a period of several months; see Jank and Yahav, 2010).

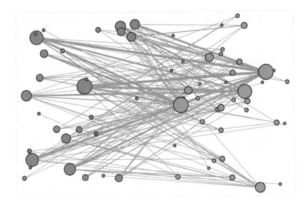

FIGURE 3.20   **Network graph of eBay sellers (left side) and buyers (right side) of Swarovski beads. Circle size represents the node's number of transactions. Line width represents the number of transactions between that pair of seller—buyer. Produced using a** JMP **add-in**

The circles on the left side represent sellers and on the right side buyers. Circle size represents the number of transactions that the node (seller or buyer) was involved in within this network. Line width represents the number of auctions that the bidder—seller pair interacted in (in this case, we use arrows to denote the directional relationship from buyer to seller). We can see that this marketplace is dominated by three or four high-volume sellers. We can also see that many buyers interact with a single seller. The market structures for many individual products could be reviewed quickly in this way. Network providers could use the information, for example, to identify possible partnerships to explore with sellers.

Figure 3.20 was produced using a JMP add-in, the *Transaction Visualizer*. This add-in was developed using the JMP Scripting Language (*JSL*), and it is freely available at the JMP User Community (`community.jmp.com`). For more information on JMP add-ins, search for *Add-ins* in the *JMP Help*.

Visualizing Hierarchical Data: More on Treemaps

We discussed hierarchical data and the exploration of data at different hierarchy levels in the context of plot manipulations. Treemaps (available from the *Graph* menu and within the

Graph Builder) are useful visualizations specialized for exploring large datasets that are hierarchically structured (tree-structured). They allow exploration of various dimensions of the data while maintaining the hierarchical nature of the data. An example is shown in Figure 3.21, which displays a large set of eBay auctions,[4] hierarchically ordered by item category, subcategory, and brand. The levels in the hierarchy of the treemap are visualized as rectangles containing sub-rectangles. Categorical variables can be included in the display by using hue. Numerical variables can be included via rectangle size and color intensity (ordering of the rectangles is sometimes used to reinforce size).

In the example in Figure 3.21, size is used to represent the average closing price (which reflects item value), and color intensity represents the percentage of sellers with negative feedback (a negative seller feedback indicates buyer dissatisfaction in past transactions and is often indicative of fraudulent seller behavior). Consider the task of classifying ongoing auctions in terms of a fraudulent outcome. From the treemap we see that the highest proportion of sellers with negative ratings (dark blue) is concentrated in expensive item auctions (Rolex and Cartier wristwatches).

One example of an interactive online application of treemaps is currently available at www.drasticdata.nl/drastictreemap.htm. One of these examples displays player-level data from the 2014 World Cup, aggregated to team level. The user can choose to explore players and team data. Similar treemaps can be produced in JMP.

Visualizing Geographical Data: Maps

Many data sets used for machine learning now include geographical information. Zip codes are one example of a categorical variable with many categories, where it is not straightforward to create meaningful variables for analysis. Plotting the data on a geographic map can often reveal patterns that are harder to identify otherwise. A map chart uses a geographical map as its background, and then color, hue, and other features to include categorical or numerical variables. Besides specialized mapping software, maps are now becoming part of general-purpose software, and Google Maps provides API's (application programming interfaces) that allow organizations to overlay their data on a Google map.

The ability to plot data on background maps and overlay data on images is readily available in JMP (see www.jmp.com/learn or search for "create background maps" in the JMP Help). Figure 3.22 shows a geographic map (created in the JMP *Graph Builder*), representing fatal and non-fatal airline incidents in the Continental US (from the Aircraft Incidents.jmp dataset in the JMP *Sample Data Folder*). Color coding is used to differentiate fatal and non-fatal incidents. Tools like the JMP *Column Switcher* and *Data Filter*, when used with geographic maps, can bring added dimensions to the visualizations.

Figure 3.23 shows two world map charts comparing countries' "wellbeing" (according to a 2006 Gallup survey) in the top map, to gross domestic product (GDP) in the bottom map.[5] Darker shade means higher value.

---

[4]We thank Sharad Borle for sharing this dataset.

[5]Data from Veenhoven, R., World Database of Happiness, Erasmus University Rotterdam. Available at http://worlddatabaseofhappiness.eur.nl

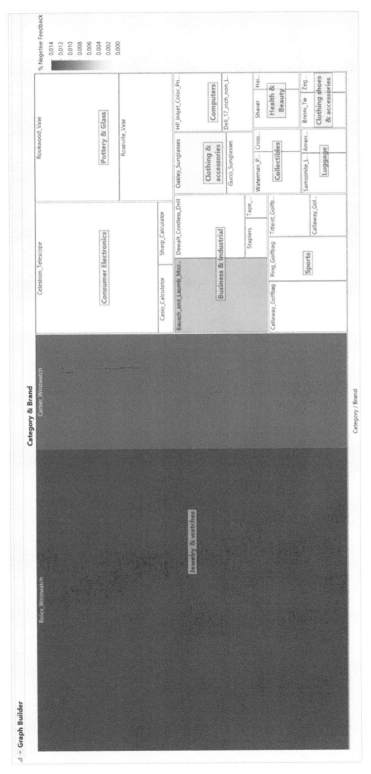

FIGURE 3.21 Treemap showing nearly 11,000 eBay auctions, organized by item category and brand. Rectangle size represents average closing price (reflecting item value). Shade represents % of sellers with negative feedback (darker = higher %)

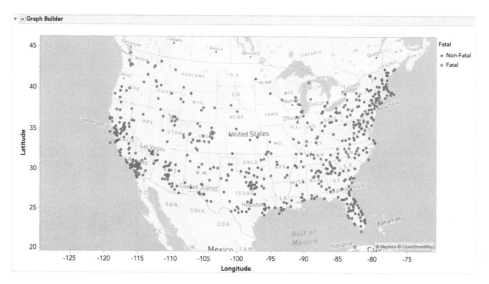

FIGURE 3.22   Aircraft incidents in the USA. Each dot is an incident, colored by fatality level

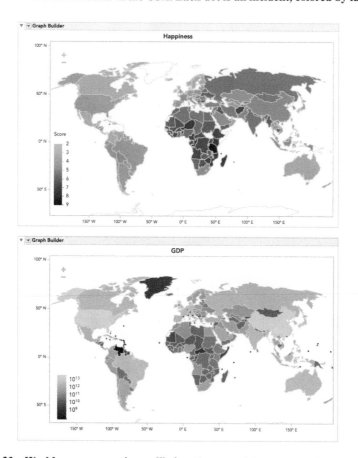

FIGURE 3.23   World maps comparing wellbeing (Top) to GDP (bottom). Shading by Average wellbeing Score (Top) or GDP (bottom) of country. Darker corresponds to higher score. Source: Data from Veenhoven's world database of happiness

3.6   SUMMARY: MAJOR VISUALIZATIONS AND OPERATIONS, ACCORDING TO MACHINE LEARNING GOAL

Prediction

- Plot outcome variable on the numerical axis of boxplots, bar charts, scatter plots.
- Study relation of outcome variable to categorical predictors via side-by-side boxplots, bar charts, and multiple panels.
- Study relation of outcome variable to numerical predictors via scatter plots.
- Use distribution plots (boxplot, histogram) for determining potentially useful transformations of the outcome variable (and/or numerical predictors).
- Examine scatter plots with added color/panels/size to determine the need for interaction terms.
- Use various aggregation levels, zooming, and filtering to determine areas of the data with different behavior and to evaluate the level of global vs. local patterns.

Classification

- Study relation of outcome variable to categorical predictors using bar charts with the outcome variable frequencies/percentages on the numerical axis.
- Study relation of outcome variable to pairs of numerical predictors via color-coded scatter plots (color denotes the outcome).
- Study relation of outcome variable to numerical predictors via side-by-side boxplots: plot boxplots of a numerical variable by outcome. Create similar displays for each numerical predictor. The most dissimilar boxes indicate potentially useful predictors.
- Use color to represent the outcome variable on a parallel coordinate plot.
- Use distribution plots (boxplot, histogram) for determining potentially useful transformations of numerical predictor variables.
- Examine scatter plots of numerical predictor variables with added color/panels for outcome to determine the need for interaction terms.
- Use various aggregation levels, zooming, and filtering to determine areas of the data with different behavior, and to evaluate the level of global vs. local patterns.

Time Series Forecasting

- Create line charts at different temporal aggregations to determine types of patterns.
- Use zooming and panning to examine shorter periods of the series to determine areas of the data with different behavior.
- Use various aggregation levels to identify global and local patterns.
- Identify missing values in the series.
- Overlay trend lines of different types to determine adequate modeling choices.

Unsupervised Learning

- Create scatter plot matrices to identify pairwise relationships and clustering of observations.
- Use heat maps/color maps to examine the correlation table.
- Use various aggregation levels and zooming to determine areas of the data with different behavior.
- Generate a parallel coordinate plot to identify clusters of observations.

PROBLEMS

3.1   Shipments of Household Appliances: Line Graphs. The file `ApplianceShipments.jmp` contains the series of quarterly shipments (in millions of dollars) of US household appliances between 1985 and 1989.

  a. Create a well-formatted time plot of the data using the JMP *Graph Builder*.

  b. Does there appear to be a quarterly pattern? For a closer view of the patterns, try zooming in on different ranges of data on the two axes.

  c. Create one chart with four separate lines, one for each of Q1, Q2, Q3, and Q4. This can be achieved by creating a transformed variable (right-click on Quarter in the *Graph Builder*, and select *Date Time, Quarter*). Then drag this new column to the *Overlay*, *Wrap*, or *Group Y* zone. Does there appear to be a difference between quarters?

  d. Create a line graph of the series at a yearly aggregated level (i.e., the total shipments in each year). Again, you'll need to create a transformed variable in *Graph Builder*.

3.2   Sales of Riding Mowers: Scatter Plots. A company that manufactures riding mowers wants to identify the best sales prospects for an intensive sales campaign. In particular, the manufacturer is interested in classifying households as prospective owners or non-owners on the basis of income (in \$1000s) and lot size (in 1000 ft^2). The marketing expert looked at a random sample of 24 households, given in the file `RidingMowers.jmp`.

  a. Create a scatter plot of lot size vs. income, color-coded by the outcome variable Ownership. Make sure to obtain a well-formatted plot. The result should be similar to Figure 9.2. Describe the potential relationship(s) of ownership to lot size and income.

  b. Explore different methods for saving and sharing your work (data, reports, projects, workflows) in JMP. Search for "Save and Share" in the JMP documentation which is under *Help* or see JMP basics section of the *JMP Learning Library* at jmp.com/learn. Mention two methods.

3.3   Laptop Sales at a London Computer Chain: Bar Charts and Boxplots. The file `LaptopSalesJanuary2008.jmp` contains data for all sales of laptops at a computer chain in London in January 2008. This is a subset of the full dataset that includes data for the entire year.

  a. Create a bar chart, showing the average retail price by store (Store Postcode). Adjust the *y*-axis scaling to magnify differences. Which store has the highest average? Which has the lowest? Use the *Local Data Filter* (under the top *red triangle*) to isolate these two stores.

  b. To better compare retail prices across stores, create side-by-side boxplots of retail price by store. Now compare the prices in the two stores above. Does there seem to be a difference between their price distributions?

3.4   Laptop Sales at a London Computer Chain: Interactive Visualization. The file `LaptopSales.txt` is a comma-separated file with nearly 300,000 rows. ENBIS (the European Network for Business and Industrial Statistics) provided these data as part of a contest organized in the fall of 2009.

Scenario: Imagine that you are a new analyst for a company called Acell (a company selling laptops). You have been provided with data about products and sales. You need to help the company with their business goal of planning a product strategy and pricing policies that will maximize Acell's projected revenues in 2009. Import the data into JMP (for details on importing data into JMP, search for "import text files" in the JMP documentation or at jmp.com/learn). Check to ensure that the data and modeling types in the data table are correct for each of the variables, and answer the following questions.

a. Price Questions:

   i. At what price are the laptops actually selling?

   ii. Does price change with time? *Hint*: Make sure that the date column is recognized as such (check *Column Info*). JMP will then allow dynamic transformations and allow you to plot the data by weekly or monthly aggregates, or even by day of week.

   iii. Are prices consistent over retail outlets?

   iv. How does price change with configuration?

b. Location Questions:

   i. Where are the stores and customers located?

   ii. Which stores are selling the most?

   iii. How far would customers travel to buy a laptop?

   o *Hint 1:* You should be able to aggregate the data, such as by a plot of the sum or average of the prices.

   o *Hint 2:* Use the coordinated highlighting between multiple visualizations in the same page, for example, select a store in one view to see the matching customers in another visualization.

   o *Hint 3:* Explore the use of filters to see differences. Be sure to filter in the zoomed-out view. For example, try to use "store location" as an alternative way to dynamically compare store locations. This might be more useful to spot outlier patterns if there were 50 store locations to compare.

   iv. Try an alternative way of looking at how far customers traveled. Do this by creating a new data column that computes the distance between customer and store. (For information on creating formulas in JMP, search for "Creating Formulas" in the JMP documentation or see jmp.com/learn.)

c. Revenue Questions:

   i. How do the sales volume in each store relate to Acell's revenues?

   ii. How does this depend on the configuration?

d. Configuration Questions:

   i. What are the details of each configuration? How does this relate to price?

   ii. Do all stores sell all configurations?

4

DIMENSION REDUCTION

In this chapter, we describe the important step of dimension reduction. The dimension of a dataset, which is the number of variables, often must be reduced for the machine learning algorithms to operate efficiently. This process is part of the pilot/prototype phase of supervised learning and is done before deploying a model. We present and discuss several dimension reduction approaches: (1) incorporating domain knowledge to remove or combine categories, (2) using data summaries to detect information overlap between variables (and remove or combine redundant variables or categories), (3) using data conversion techniques such as converting categorical variables into numerical variables, and (4) employing automated reduction techniques such as principal component analysis (PCA), where a new set of variables (which are weighted averages of the original variables) is created. These new variables are uncorrelated, and a small subset of them usually contains most of their combined information (hence we can reduce dimension by using only a subset of the new variables). Finally, we mention supervised learning methods such as regression models and classification and regression trees, which can be used for removing redundant variables and for combining "similar" categories of categorical variables.

Dimension reduction in *JMP*:   All the methods discussed in this chapter are available in the standard version of JMP.

4.1   INTRODUCTION

In machine learning, one often encounters situations where there is a large number of variables in the database. Even when the initial number of variables is small, this set quickly expands in the data preparation step, where new derived variables are created (e.g., recoded or transformed variables). In such situations, it is likely that subsets of variables are highly correlated with each other. Including highly correlated variables in a classification or prediction model or including variables that are unrelated to the outcome of interest, can lead to

Machine Learning for Business Analytics: Concepts, Techniques, and Applications with JMP Pro®,
Second Edition. Galit Shmueli, Peter C. Bruce, Mia L. Stephens, Muralidhara Anandamurthy, and Nitin R. Patel.
© 2023 John Wiley & Sons, Inc. Published 2023 by John Wiley & Sons, Inc.

overfitting, and prediction accuracy and reliability can suffer. A large number of variables also poses computational problems for some supervised as well as unsupervised algorithms (aside from questions of correlation). In model deployment, superfluous variables can increase costs due to the collection and processing of these variables.

4.2   CURSE OF DIMENSIONALITY

The *dimensionality* of a model is the number of independent or input variables used by the model. The *curse of dimensionality* is the affliction caused by adding variables to multivariate data models. As variables are added, the data space becomes increasingly sparse, and classification and prediction models fail because the available data are insufficient to provide a useful model across so many variables. An important consideration is the fact that the difficulties posed by adding variables increase exponentially with the addition of each variable. One way to think of this intuitively is to consider the location of an object on a chessboard, which has two dimensions and 64 squares or choices. If you expand the chessboard to a cube, you increase the dimensions by 50%—from two dimensions to three dimensions. However, the location options increase by 800%, to 512 ($8 \times 8 \times 8$). In statistical distance terms, the proliferation of variables means that nothing is close to anything else anymore—too much noise has been added, and patterns and structure are no longer discernible. The problem is particularly acute in Big Data applications, including genomics, where, for example, an analysis might have to deal with values for thousands of different genes. One of the key steps in machine learning therefore is finding ways to reduce dimensionality with minimal sacrifice of information content or accuracy. In the artificial intelligence literature, dimension reduction is often referred to as *factor selection*, *feature selection*, or *feature extraction*.

4.3   PRACTICAL CONSIDERATIONS

Although machine learning prefers automated methods vs. domain knowledge, it is important at the first step of data exploration to make sure that the variables measured are reasonable for the task at hand. The integration of expert knowledge through a discussion with the data provider (or user) will probably lead to better results. Practical considerations include: which variables are most important for the task at hand, and which are most likely to be useless? Which variables are likely to contain much measurement error? Which variables will be available for measurement (and what will it cost to measure them) in the future if the analysis is repeated? Which variables can actually be measured before the outcome occurs? (For example, if we want to predict the closing price of an ongoing online auction, we cannot use the number of bids as a predictor because this will not be known until the auction closes.)

Example 1: House Prices in Boston

We return to the Boston housing example introduced in Chapter 3 (the dataset is `BostonHousing.jmp`). For each census tract (a census tract includes 1200–8000 homes), a number of variables are given, such as the crime rate, the student/teacher ratio, and the median value of a housing unit in the neighborhood. A description of all 14 variables is given in Table 4.1. The first 10 records of the data are shown in Figure 4.1. The first row represents the first census tract, which had an average per capita crime rate of 0.006, 18% of the residential land zoned for lots over 25,000 ft^2, 2.31% of the land devoted to nonretail business, no border on the Charles River, and so on.

TABLE 4.1   Description of Variables in the Boston Housing Dataset

CRIM	Crime rate
ZN	Percentage of residential land zoned for lots over 25,000 ft^2
INDUS	Percentage of land occupied by nonretail business
CHAS	Does tract bound Charles River? (= 1 if tract bounds river, = 0 otherwise)
NOX	Nitric oxide concentration (parts per 10 million)
RM	Average number of rooms per dwelling
AGE	Percentage of owner-occupied units built prior to 1940
DIS	Weighted distances to five Boston employment centers
RAD	Index of accessibility to radial highways
TAX	Full-value property tax rate per $10,000
PTRATIO	Pupil-to-teacher ratio by town
LSTAT	Percentage of lower status of the population
MEDV	Median value of owner-occupied homes in $1000s
CAT.MEDV	Is median value of owner-occupied homes in tract above $30,000 (CAT.MEDV = 1) or not (CAT.MEDV = 0)?

	CRIM	ZN	INDUS	CHAS	NOX	RM	AGE	DIS	RAD	TAX	PTRATIO	LSTAT	MEDV	CAT.MEDV
1	0.00632	18	2.31	0	0.538	6.575	65.2	4.09	1	296	15.3	4.98	24	0
2	0.02731	0	7.07	0	0.469	6.421	78.9	4.9671	2	242	17.8	9.14	21.6	0
3	0.02729	0	7.07	0	0.469	7.185	61.1	4.9671	2	242	17.8	4.03	34.7	1
4	0.03237	0	2.18	0	0.458	6.998	45.8	6.0622	3	222	18.7	2.94	33.4	1
5	0.06905	0	2.18	0	0.458	7.147	54.2	6.0622	3	222	18.7	5.33	36.2	1
6	0.02985	0	2.18	0	0.458	6.43	58.7	6.0622	3	222	18.7	5.21	28.7	0
7	0.08829	12.5	7.87	0	0.524	6.012	66.6	5.5605	5	311	15.2	12.43	22.9	0
8	0.14455	12.5	7.87	0	0.524	6.172	96.1	5.9505	5	311	15.2	19.15	27.1	0
9	0.21124	12.5	7.87	0	0.524	5.631	100	6.0821	5	311	15.2	29.93	16.5	0
10	0.17004	12.5	7.87	0	0.524	6.004	85.9	6.5921	5	311	15.2	17.1	18.9	0

FIGURE 4.1   First 10 records in the Boston Housing data

4.4   DATA SUMMARIES

As we have seen in the chapter on data visualization, an important initial step of data exploration is getting familiar with the data and their characteristics through summaries and graphs. The importance of this step cannot be overstated. The better you understand the data, the better the results from the modeling or mining process.

Numerical summaries and graphs of the data are very helpful for data reduction. The information that they convey can assist in combining categories of a categorical variable, in choosing variables to remove, in assessing the level of information overlap between variables, and more. Before discussing such strategies for reducing the dimension of a dataset, let us consider useful summaries and tools.

Summary Statistics

JMP has several platforms for summarizing data (most are under the *Analyze* menu) and provides a variety of summary statistics for learning about the characteristics of each variable. The most common of these statistics are the sample *average* (or *mean*), the *median*, the *standard deviation*, the *minimum* and *maximum values*, and for categorical data, the *number of levels* and the *counts* for each level. These statistics give us information about the scale and type of values that a variable takes. The minimum and maximum values can be used to detect extreme values that might be errors. The mean and median give a sense of the central values of a variable, and a large difference between the two also indicates skew. The standard deviation gives a sense of how dispersed the data are (relative to the mean). Other statistics, such as *N Missing*, can alert us about missing values.

To quickly summarize all of the variables in a dataset, use the *Columns Viewer* (under the *Cols* menu). Select the variables, then click *Show Summary*. Categorical variables show a value under *N Categories*, while summary statistics are provided for numeric (continuous) variables. If data are missing, the table will contain a column summarizing missing data across the variables (*N Missing*). Checking the *Show Quartiles* box before clicking *Show Summary* provides the statistics in the last four columns shown in Figure 4.2.

Columns	N Categories	Min	Max	Mean	Std Dev	Median	Lower Quartile	Upper Quartile	Interquartile Range
CRIM	.	0.01	88.98	3.61	8.60	0.26	0.08	3.68	3.60
ZN	.	0.00	100.00	11.36	23.32	0.00	0.00	12.50	12.50
INDUS	.	0.46	27.74	11.14	6.86	9.69	5.18	18.10	12.93
CHAS	2	.	.	.	.	.	.	.	.
NOX	.	0.39	0.87	0.55	0.12	0.54	0.45	0.62	0.18
RM	.	3.56	8.78	6.28	0.70	6.21	5.88	6.63	0.74
AGE	.	2.90	100.00	68.57	28.15	77.50	44.85	94.10	49.25
DIS	.	1.13	12.13	3.80	2.11	3.21	2.10	5.21	3.12
RAD	.	1.00	24.00	9.55	8.71	5.00	4.00	24.00	20.00
TAX	.	187.00	711.00	408.2	168.54	330.00	279.00	666.00	387.00
PTRATIO	.	12.60	22.00	18.46	2.16	19.05	17.38	20.20	2.83
LSTAT	.	1.73	37.97	12.65	7.14	11.36	6.93	16.99	10.07
MEDV	.	5.00	50.00	22.53	9.20	21.20	16.95	25.00	8.05
CAT.MEDV	2	.	.	.	.	.	.	.	.

FIGURE 4.2   Summary statistics for the Boston Housing data from the *Columns Viewer*

We immediately see that the different variables have very different ranges of values. We will soon see how variation in scale across variables can distort analyses if not treated properly. Another observation that can be made is that the average of the first variable, CRIM (as well as several others), is much larger than the median, indicating right skew. None of the variables have missing values. There also do not appear to be indications of extreme values that might result from data entry errors (the Min and Max values for all of the variables look reasonable).

For a more detailed look at the distributions and summary statistics for specific variables, select the variables and click the *Distribution* button in the *Columns Viewer*. The *Distribution* platform (also accessible from the *Analyze* menu) provides summary statistics and basic graphical displays. Figure 4.3 shows default output for CRIM and CHAS, from

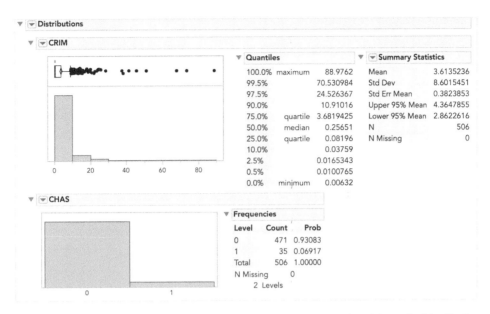

FIGURE 4.3   Basic graphs and summary statistics for *CRIM* and *CHAS* from the *Distribution* platform

the Boston Housing data. Additional summary statistics and graphical options are available under the red triangles (e.g., we used the Stack option from the top red triangle to produce the horizontal layout).

Next we summarize relationships between two or more variables. For numerical variables, we can compute pairwise correlations using the *Analyze > Multivariate Methods > Multivariate* platform (as shown in Chapter 3). This produces a complete matrix of correlations between each pair of variables. Figure 4.4 shows the correlation matrix for a subset of the Boston Housing variables. We see that most correlations are low and that many are negative. Recall also the visual display of a correlation matrix via a color map (see Figure 3.6 in Chapter 3 for the color map corresponding to this correlation table). We will return to the importance of the correlation matrix soon, in the context of correlation analysis.

▼ ☑ Multivariate
　▼ Correlations

	CRIM	ZN	INDUS	CHAS	NOX	RM	AGE	DIS	RAD	TAX	PTRATIO	LSTAT	MEDV
CRIM	1.000	-0.200	0.407	-0.056	0.421	-0.219	0.353	-0.380	0.626	0.583	0.290	0.456	-0.388
ZN	-0.200	1.000	-0.534	-0.043	-0.517	0.312	-0.570	0.664	-0.312	-0.315	-0.392	-0.413	0.360
INDUS	0.407	-0.534	1.000	0.063	0.764	-0.392	0.645	-0.708	0.595	0.721	0.383	0.604	-0.484
CHAS	-0.056	-0.043	0.063	1.000	0.091	0.091	0.087	-0.099	-0.007	-0.036	-0.122	-0.054	0.175
NOX	0.421	-0.517	0.764	0.091	1.000	-0.302	0.731	-0.769	0.611	0.668	0.189	0.591	-0.427
RM	-0.219	0.312	-0.392	0.091	-0.302	1.000	-0.240	0.205	-0.210	-0.292	-0.356	-0.614	0.695
AGE	0.353	-0.570	0.645	0.087	0.731	-0.240	1.000	-0.748	0.456	0.506	0.262	0.602	-0.377
DIS	-0.380	0.664	-0.708	-0.099	-0.769	0.205	-0.748	1.000	-0.495	-0.534	-0.232	-0.497	0.250
RAD	0.626	-0.312	0.595	-0.007	0.611	-0.210	0.456	-0.495	1.000	0.910	0.465	0.489	-0.382
TAX	0.583	-0.315	0.721	-0.036	0.668	-0.292	0.506	-0.534	0.910	1.000	0.461	0.544	-0.469
PTRATIO	0.290	-0.392	0.383	-0.122	0.189	-0.356	0.262	-0.232	0.465	0.461	1.000	0.374	-0.508
LSTAT	0.456	-0.413	0.604	-0.054	0.591	-0.614	0.602	-0.497	0.489	0.544	0.374	1.000	-0.738
MEDV	-0.388	0.360	-0.484	0.175	-0.427	0.695	-0.377	0.250	-0.382	-0.469	-0.508	-0.738	1.000

The correlations are estimated by Row-wise method.

FIGURE 4.4   Correlation table for Boston Housing data, generated using the *Multivariate* platform

Tabulating Data (Pivot Tables)

Another very useful tool for summarizing data in JMP is *Tabulate* (under the *Analyze* menu). *Tabulate* provides interactive tables that can combine information from multiple variables and compute a range of summary statistics (count, mean, percentage, etc.). The initial *Tabulate* window is shown in Figure 4.5. Like *Graph Builder*, *Tabulate* has drop zones for dragging and dropping variables in an interactive table, but a more traditional dialog window is also available.

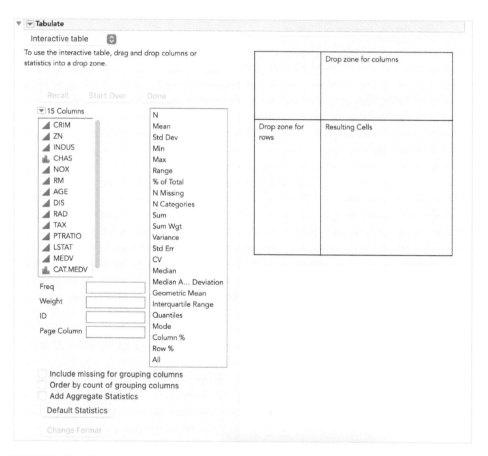

FIGURE 4.5   The initial *Tabulate* window. Drag and drop variables in the drop zones for columns or rows, and select the summary statistics of interest from the list provided

A simple example is the number of census tracts that bound the Charles River vs. those that do not, and the mean MEDV for each group. This is shown in the left-hand panel of Figure 4.6. This tabulation was obtained by dragging MEDV to the drop zone for columns, CHAS to the drop zone for rows, and dragging the statistics Mean and N on top of the existing statistic (the Sum). Clicking the *Done* button (on the left in Figure 4.5) closes the control panel and produces the table. For categorical variables, we obtain a break-down of the records by the combination of categories. For instance, the right-hand panel of Figure 4.6 shows the average of each of the numeric variables for the two levels of CAT.MEDV, along with a count of neighborhoods bordering the Charles River for each

	MEDV	
CHAS	Mean	N
0	22	471
1	28	35

		CAT.MEDV	
		0	1
CRIM	Mean	4.20	0.69
ZN	Mean	7.57	30.44
INDUS	Mean	12.26	5.51
NOX	Mean	0.57	0.49
RM	Mean	6.08	7.29
AGE	Mean	70.97	56.52
DIS	Mean	3.68	4.36
RAD	Mean	10.32	5.69
TAX	Mean	428.80	304.95
PTRATIO	Mean	18.88	16.31
LSTAT	Mean	14.15	5.14
CHAS			
	0	398.00	73.00
	1	24.00	11.00

FIGURE 4.6   **Tabulations for the Boston Housing data, created using** *Tabulate*

category of CAT.MEDV. There are many more possibilities and options for using *Tabulate*. We leave it to the reader to explore these options. For more information, search for *Tabulate* in the *JMP Help*.

In classification tasks, where the goal is to find predictor variables that distinguish between two classes, a good exploratory step is to produce summaries for each class. This can assist in detecting useful predictors that display some separation between the two classes. Data summaries are useful for almost any machine learning task and are therefore an important preliminary step for cleaning and understanding the data before carrying out further analyses.

4.5   CORRELATION ANALYSIS

In datasets with a large number of variables (which are likely to serve as predictors), there is usually much overlap in the information covered by the set of variables. One simple way to find redundancies is to look at a correlation matrix (see the example in Figure 4.4). This shows all the pairwise correlations between variables. Pairs that have a very strong (positive or negative) correlation contain a lot of overlap in information and are good candidates for data reduction by removing one of the variables. Removing variables that are strongly correlated to others is useful for avoiding multicollinearity problems that can arise in various models. (*Multicollinearity* is the presence of two or more predictors sharing the same linear relationship with the outcome variable.)

Correlation analysis is also a good method for detecting duplications of variables in the data. Sometimes the same variable appears accidentally more than once in the dataset (under a different name) because the dataset was merged from multiple sources, the same phenomenon is measured in different units, and so on. Use of correlation table color map, as shown in Chapter 3, can make the task of identifying strong correlations easier.

4.6 REDUCING THE NUMBER OF CATEGORIES IN CATEGORICAL VARIABLES

When a categorical variable (that is, having a nominal or ordinal modeling type) has many categories, and this variable is destined to be a predictor, many machine learning methods will require converting it into many *dummy* or *indicator variables*. In particular, a variable with m categories will be transformed into either m or $m - 1$ dummy variables (depending on the method). Note that JMP does not require converting data into dummy variables per se—it will do this behind the scenes in most modeling and machine learning platforms. Nonetheless, even if we have very few original categorical variables, they can greatly inflate the dimensions of the dataset. One way to handle this is to reduce the number of categories by combining close or similar categories. To combine categories requires incorporating expert knowledge and common sense. Tabulations are useful for this task: we can examine the sizes of the various categories and how the response behaves at each category. Generally, categories that contain very few records are good candidates for combining with other categories. A best practice is to use only the categories that are most relevant to the analysis, recode missing values as "missing," and recode the rest as "other."

In classification tasks (with a categorical output), a tabulation broken down by the output classes can help identify categories that do not separate the classes. Those categories are also candidates for inclusion in the "other" category. An example is shown on the left in Figure 4.7. We can see that the distribution of CAT.MEDV is identical for ZN = 17.5, 90, 95, and 100 (where all neighborhoods have CAT.MEDV = 1). These four categories can then be combined into a single category. Similarly categories ZN = 12.5, 25, 28, 30, and 70

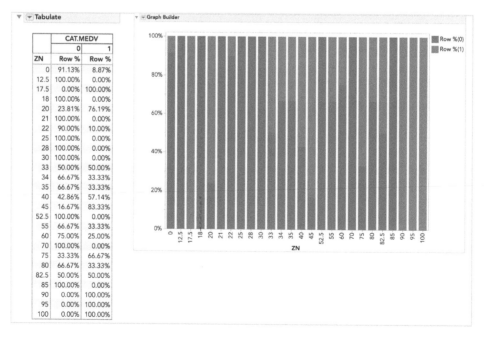

FIGURE 4.7 Distribution of CAT.MEDV (red indicates *CAT.MEDV* = 1) by ZN. Similar bars indicate low separation of classes, suggesting those categories can be combined

(and others where CAT.MEDV = 0) can be combined. Further combination is also possible based on similar bars.

To create the tabulation and stacked bar graph in Figure 4.7:

- Change the modeling type for ZN to Nominal.
- In *Tabulate*, drag CAT.MEDV to the drop zone for columns, drag ZN to the drop zone for rows, and drag Row % to the results area.
- Select *Make into Data Table* from the top red triangle.
- In *Graph Builder*, drag ZN to the X zone, drag the Row % columns for both classes to the Y zone (at once), and click on the bar icon (above the graph). Under Bar Style (on the bottom left), select Stacked.

In a time series context where we might have a categorical variable denoting season (such as month or hour of day) that will serve as a predictor, reducing categories can be done by examining the time series plot and identifying similar periods. For example, the time plot in Figure 4.8 shows the quarterly revenues of Toys "R" Us between 1992 and 1995. Only quarter four periods appear different, and therefore we can combine quarters one to three into a single category.

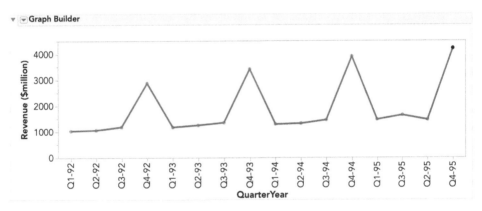

FIGURE 4.8   Quarterly revenues of Toys "R" Us, 1992–1995

Combining categories in JMP is done using *Recode* (right click on the column header and select Recode). We can use Recode, for example, to combine zones 12.5, 25, 28, 30, and 70 into a new group, "zone1," as shown in Figure 4.9. The variable ZN is stored as numeric data in JMP, so first we select *Convert to Character* from the red triangle (or select *Yes* in the popup box to convert). Selecting *Formula Column* will result in the data being recoded in a new column in the data table, with a stored formula for recoding.

If additional data are added to the data table, these data will be recoded as well. Note that additional options for data cleanup for categorical data, such as grouping similar values, trimming white space, and addressing capitalization issues, are available under the red triangle in the *Recode* window (these are not available for numeric data).

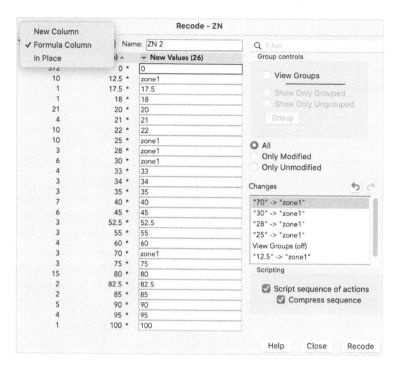

FIGURE 4.9   Combining categories using *Recode*

4.7 CONVERTING A CATEGORICAL VARIABLE TO A CONTINUOUS VARIABLE

Sometimes the categories in a categorical variable represent intervals. Common examples are age group or income bracket. If the interval values are known (e.g., category 2 is the age interval 20–30), we can replace the categorical value ("2" in the example) with the mid-interval value (here "25"). The resulting numeric values can be treated as a single continuous variable, thus simplifying analyses.

4.8 PRINCIPAL COMPONENT ANALYSIS

Principal component analysis (PCA) is a useful method for dimension reduction, especially when the number of variables is large. PCA is especially valuable when we have subsets of measurements that are measured on the same scale and are highly correlated. In that case it provides a few variables (often as few as three) that are weighted linear combinations of the original variables and that retain the majority of the information of the full original set of variables. PCA is intended for use with quantitative variables. For categorical variables, other methods, such as correspondence analysis, are more suitable. (*Multiple Correspondence Analysis* is available in JMP from the *Analyze > Multivariate Methods* platform.)

Example 2: Breakfast Cereals

Data were collected on the nutritional information and consumer rating of 77 breakfast cereals (data are in `Cereals.jmp`).[1] The consumer rating is a rating of cereal "healthiness" for consumer information (not a rating by consumers). For each cereal the data include 13 numerical variables, and we are interested in reducing this dimension. For each cereal the information is based on a bowl of cereal rather than a serving size, since most people simply fill a cereal bowl (resulting in constant volume, but not weight). A snapshot of these data is given in Table 4.2, and the description of the different variables is given in Table 4.3.

TABLE 4.2 Sample from the 77 breakfast cereals dataset

Cereal name	mfr	type	calories	protein	fat	sodium	fiber	carbo	sugars	potass	vitamins
100% Bran	N	C	70	4	1	130	10	5	6	280	25
100% Natural Bran	Q	C	120	3	5	15	2	8	8	135	0
All-Bran	K	C	70	4	1	260	9	7	5	320	25
All-Bran with Extra Fiber	K	C	50	4	0	140	14	8	0	330	25
Almond Delight	R	C	110	2	2	200	1	14	8		25
Apple Cinnamon Cheerios	G	C	110	2	2	180	1.5	10.5	10	70	25
Apple Jacks	K	C	110	2	0	125	1	11	14	30	25
Basic 4	G	C	130	3	2	210	2	18	8	100	25
Bran Chex	R	C	90	2	1	200	4	15	6	125	25
Bran Flakes	P	C	90	3	0	210	5	13	5	190	25
Cap'n'Crunch	Q	C	120	1	2	220	0	12	12	35	25
Cheerios	G	C	110	6	2	290	2	17	1	105	25
Cinnamon Toast Crunch	G	C	120	1	3	210	0	13	9	45	25
Clusters	G	C	110	3	2	140	2	13	7	105	25
Cocoa Puffs	G	C	110	1	1	180	0	12	13	55	25
Corn Chex	R	C	110	2	0	280	0	22	3	25	25
Corn Flakes	K	C	100	2	0	290	1	21	2	35	25
Corn Pops	K	C	110	1	0	90	1	13	12	20	25
Count Chocula	G	C	110	1	1	180	0	12	13	65	25
Cracklin' Oat Bran	K	C	110	3	3	140	4	10	7	160	25

We focus first on two variables: *calories* and consumer *rating*. These are given in Table 4.4 for all cereals. The average calories across the 77 cereals is 106.88 with standard deviation of 19.48. The average consumer rating is 42.67 with a standard deviation of 14.05. This is shown in Figure 4.10. The two variables are strongly correlated, with a negative correlation of −0.69. (In JMP, use *Analyze > Multivariate* to compute the correlation. For descriptive statistics, select *Simple Statistics > Univariate Simple Statistics* from the top red triangle.)

Roughly speaking, 69% of the total variation in both variables is actually "co-variation," or variation in one variable that is duplicated by similar variation in the other variable. Can we use this fact to reduce the number of variables while making maximum use of their unique contributions to the overall variation? Since there is redundancy in the information that the two variables contain, it might be possible to reduce the two variables to a single

---

[1] The data are available at `https://dasl.datadescription.com/datafile/cereals/`

TABLE 4.3   Description of the Variables in the Breakfast Cereal Dataset

Variable	Description
mfr	Manufacturer of cereal (American Home Food Products, General Mills, Kellogg, etc.)
type	Cold or hot
calories	Calories per serving
protein	Grams of protein
fat	Grams of fat
sodium	Milligrams of sodium
fiber	Grams of dietary fiber
carbo	Grams of complex carbohydrates
sugars	Grams of sugars
potass	Milligrams of potassium
vitamins	Vitamins and minerals: 0, 25, or 100, indicating the typical percentage of FDA recommended
shelf	Display shelf (1, 2, or 3, counting from the floor)
weight	Weight in ounces of one serving
cups	Number of cups in one serving
rating	Rating of the cereal calculated by Consumer Reports

variable without losing too much information. The idea in PCA is to find a linear combination of the two variables that contains most, even if not all, of the information so that this new variable can replace the two original variables. Information here is in the sense of variability: what can explain the most variability *among* the 77 cereals?

The total variability here is the sum of the variances[2] of the two attributes, which in this case is $19.48^2 + 14.05^2 = 576.87$. This means that *calories* accounts for 66% = $19.48^2/567.87$ of the total variability and *rating* accounts for the remaining 34%. If we drop one of the variables for the sake of dimension reduction, we lose at least 34% of the total variability. Can we redistribute the total variability between two new variables in a more polarized way? If so, it might be possible to keep only the one new variable that (hopefully) accounts for a large portion of the total variation.

Figure 4.11 shows a scatterplot of *rating* vs. *calories*. The line z_1 is the direction in which the variability of the points is largest. It is the line that captures the most variation in the data if we decide to reduce the dimensionality of the data from two to one. Among all possible lines, it is the line for which, if we project the points in the dataset orthogonally to get a set of 77 (one-dimensional) values, the variance of the z_1 values will be maximum. This is called the *first principal component*. It is also the line that minimizes the sum-of-squared perpendicular distances from the line. The z_2-axis is chosen to be perpendicular to the z_1-axis. In the case of two variables, there is only one line that is perpendicular to z_1, and it has the second largest variability, but its information is uncorrelated with z_1. This is called the *second principal component*. In general, when we have more than two variables, once we find the direction z_1 with the largest variability, we search among all the orthogonal directions to z_1 for the one with the next-highest variability. That is z_2. The idea is then to find the coordinates of these lines and to see how they redistribute the variability.

---

[2]The variance is the squared standard deviation.

TABLE 4.4   Cereal Calories and Ratings

Cereal	calories	rating	Cereal	calories	rating
100% Bran	70	68.40297	Just Right Fruit & Nut	140	36.471512
100% Natural Bran	120	33.98368	Kix	110	39.241114
All-Bran	70	59.42551	Life	100	45.328074
All-Bran with Extra Fiber	50	93.70491	Lucky Charms	110	26.734515
Almond Delight	110	34.38484	Maypo	100	54.850917
Apple Cinnamon Cheerios	110	29.50954	Muesli Raisins, Dates & Almonds	150	37.136863
Apple Jacks	110	33.17409	Muesli Raisins, Peaches & Pecans	150	34.139765
Basic 4	130	37.03856	Mueslix Crispy Blend	160	30.313351
Bran Chex	90	49.12025	Multi-Grain Cheerios	100	40.105965
Bran Flakes	90	53.31381	Nut&Honey Crunch	120	29.924285
Cap'n'Crunch	120	18.04285	Nutri-Grain Almond-Raisin	140	40.69232
Cheerios	110	50.765	Nutri-grain Wheat	90	59.642837
Cinnamon Toast Crunch	120	19.82357	Oatmeal Raisin Crisp	130	30.450843
Clusters	110	40.40021	Post Nat. Raisin Bran	120	37.840594
Cocoa Puffs	110	22.73645	Product 19	100	41.50354
Corn Chex	110	41.44502	Puffed Rice	50	60.756112
Corn Flakes	100	45.86332	Puffed Wheat	50	63.005645
Corn Pops	110	35.78279	Quaker Oat Squares	100	49.511874
Count Chocula	110	22.39651	Quaker Oatmeal	100	50.828392
Cracklin' Oat Bran	110	40.44877	Raisin Bran	120	39.259197
Cream of Wheat (Quick)	100	64.53382	Raisin Nut Bran	100	39.7034
Crispix	110	46.89564	Raisin Squares	90	55.333142
Crispy Wheat & Raisins	100	36.1762	Rice Chex	110	41.998933
Double Chex	100	44.33086	Rice Krispies	110	40.560159
Froot Loops	110	32.20758	Shredded Wheat	80	68.235885
Frosted Flakes	110	31.43597	Shredded Wheat 'n' Bran	90	74.472949
Frosted Mini-Wheats	100	58.34514	Shredded Wheat spoon size	90	72.801787
Fruit & Fibre Dates, Walnuts & Oats	120	40.91705	Smacks	110	31.230054
Fruitful Bran	120	41.01549	Special K	110	53.131324
Fruity Pebbles	110	28.02577	Strawberry Fruit Wheats	90	59.363993
Golden Crisp	100	35.25244	Total Corn Flakes	110	38.839746
Golden Grahams	110	23.80404	Total Raisin Bran	140	28.592785
Grape Nuts Flakes	100	52.0769	Total Whole Grain	100	46.658844
Grape-Nuts	110	53.37101	Triples	110	39.106174
Great Grains Pecan	120	45.81172	Trix	110	27.753301
Honey Graham Ohs	120	21.87129	Wheat Chex	100	49.787445
Honey Nut Cheerios	110	31.07222	Wheaties	100	51.592193
Honey-comb	110	28.74241	Wheaties Honey Gold	110	36.187559
Just Right Crunchy Nuggets	110	36.52368			

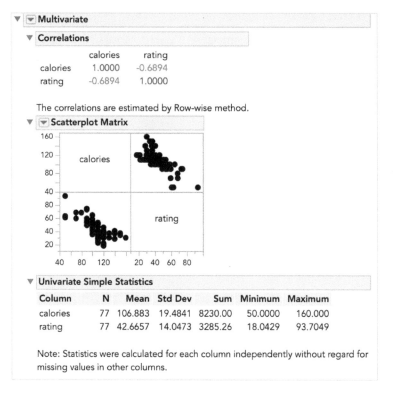

FIGURE 4.10   Correlation analysis, scatterplot matrix, and descriptive statistics for *calories* and *rating*

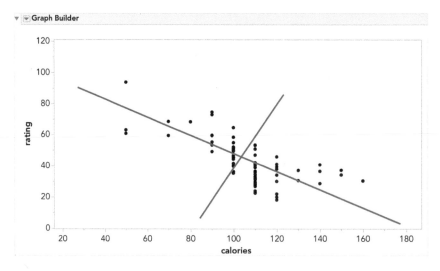

FIGURE 4.11   Scatterplot of *rating* vs. *calories* for 77 breakfast cereals, with the two principal component directions, created using *Graph Builder*

Figure 4.12 (top) shows the initial dialog box and options for running the PCA on these two variables (using *Analyze > Multivariate Methods > Principal Components*, and changing the *Standardize* option to *Unscaled*—more on this later).

Looking at the reallocated variances for the new derived variables (listed under *Eigenvalues* at the bottom of Figure 4.12), we see that z_1 accounts for 86.3% of the total variability and that z_2 accounts for the remaining 13.7%. Therefore, if we drop the second component, we still maintain 86.3% of the total variability in the data.

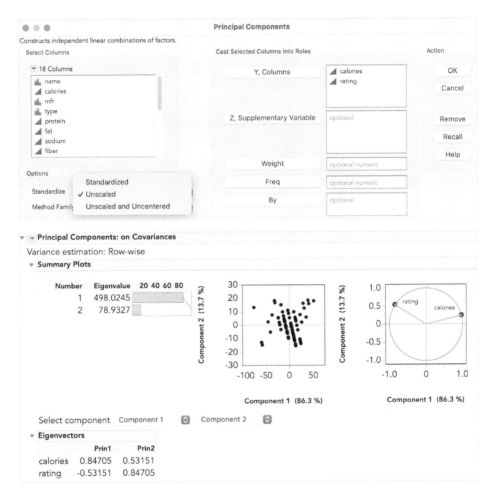

FIGURE 4.12   Initial launch dialog and output from principal component analysis of *calories* and *rating*, using unscaled data. *Eigenvalues* and *eigenvectors* are displayed using red triangle options

Eigenvectors are the weights that are used to project the original data onto the two new directions. The eigenvectors for the cereal data are shown in Figure 4.12. The weights for z_1 (which is labeled *Prin1*) are given by 0.847 for *calories* and −0.532 for *rating*. For z_2, they are given by 0.532 and 0.847, respectively. These weights are used to compute principal component scores for each record, which are the projected values of *calories* and *rating* onto the new axes (after subtracting the means for each variable). Figure 4.13 shows

	name	calories	rating	Prin1	Prin2
1	100%_Bran	70	68.4030	-44.9215	2.1972
2	100%_Natural_Bran	120	33.9837	15.7253	-0.3824
3	All-Bran	70	59.4255	-40.1499	-5.4072
4	All-Bran_with_Extra_Fiber	50	93.7049	-75.3108	12.9991
5	Almond_Delight	110	34.3848	7.0415	-5.3577

FIGURE 4.13   **Principal component scores from PCA of *calories* and *rating* for the first five cereals**

the scores for the two dimensions. The first column is the projection onto z_1 using the weights $(0.847, -0.532)$. The second column is the projection onto z_2 using the weights $(0.532, 0.847)$. For example, the first score for the 100% Bran cereal (with 70 calories and a rating of 68.4) is $(0.847)(70 - 106.88) + (-0.532)(68.4 - 42.67) = -44.92$.

Note that the means of the new variables z_1 and z_2 are zero because we've subtracted the mean of each variable. The variances of z_1 and z_2 are 498.02 and 78.93, respectively, and the sum of these variances var(z_1) + var(z_2) is equal to the sum of the variances of the original variables, var(*calories*) + var(*rating*). So, the first principal component, z_1, accounts for 86% of the total variance. Since it captures most of the variability in the data, it seems reasonable to use one variable, the first principal score, to represent the two variables in the original data. Next, we generalize these ideas to more than two variables.

Graphical and numeric summaries of the principal component analysis are provided under *Summary Plots* (this is the default output). The bar chart displays the percentage of variation accounted for by each principal component (also shown in the *Eigenvalues* table). The *Score Plot* (top, center in Figure 4.12) helps interpret the data relative to the principal components—points that are close together are similar. The *Loadings Plot* is for interpreting the relationships among the variables and their "loadings," or projections, onto the principal components.

Principal Components

Let us formalize the procedure described above so that it can easily be generalized to $p > 2$ variables. Denote by $X_1, X_2, \ldots, X_p$ the original p variables. In PCA we are looking for a set of new variables $Z_1, Z_2, \ldots, Z_p$ that are weighted averages of the original variables (after subtracting their mean):

$$Z_i = a_{i,1}(X_1 - \bar{X}_1) + a_{i,2}(X_2 - \bar{X}_2) + \cdots + a_{i,p}(X_p - \bar{X}_p), \quad i = 1, \ldots, p$$

where each pair of Z's has correlation $= 0$. We then order the resulting Z's by their variance, with Z_1 having the largest variance and Z_p having the smallest variance. JMP computes the weights $a_{i,j}$ which are then used in computing the principal component scores.

A further advantage of the principal components compared to the original data is that they are uncorrelated (correlation coefficient $= 0$). Because we construct regression models using these principal components as independent variables, we will not encounter problems of multicollinearity.

Let us return to the breakfast cereal dataset with all 15 variables and apply PCA to the 13 numerical variables. In the Principal Components initial launch dialog, select *Unscaled* as the *Standardize* option. The resulting output is shown in Figure 4.14.

▼ Principal Components: on Covariances

Variance estimation: Pairwise

▼ Eigenvectors

	Prin1	Prin2	Prin3	Prin4	Prin5	Prin6	Prin7
calories	0.07629		0.61056	-0.61793	0.45563	0.11148	-0.11231
protein					0.05494	0.07855	-0.27998
fat			0.91598	-0.02603	-0.01894	-0.19240	-0.34032
sodium	0.98041	0.13452	-0.14069				
fiber		0.03042				0.26427	-0.08320
carbo	0.01551	-0.01732	0.01227	0.02268	0.36294	-0.44395	0.73170
sugars	0.00407		0.09946	-0.11648	-0.30415	0.70344	0.48650
potass	-0.12850	0.98741	0.03598	-0.04251		-0.04820	
vitamins	0.10134		0.70796	0.69787			
shelf						-0.08087	-0.08151
weight							
cups							
rating	-0.07681	0.07217	-0.30688	0.33775	0.74942	0.41590	-0.07587

▼ Eigenvalues

Number	Eigenvalue	Percent	20 40 60 80	Cum Percent
1	7207.777	55.001		55.001
2	4961.273	37.859		92.860
3	498.364	3.803		96.663
4	357.087	2.725		99.388
5	72.305	0.552		99.939
6	3.853	0.029		99.969
7	2.760	0.021		99.990
8	0.621	0.005		99.995
9	0.436	0.003		99.998
10	0.186	0.001		99.999
11	0.059	0.000		100.000
12	0.033	0.000		100.000
13	0.0037536	0.000		100.000

FIGURE 4.14   PCA output using all 13 numerical variables in the breakfast cereals dataset. The eigenvectors table gives results for the first seven principal components

Note that the first three components account for more than 96% of the total variation associated with all 13 of the original variables. This suggests that we can capture most of the variability in the data with fewer than 25% of the number of original variables. In fact, the first two principal components alone capture 92.9% of the total variation. However, these results are influenced by the scales of the variables, as we describe next.

Standardizing the Data

A further use of PCA is to understand the structure of the data. This is done by examining the weights (the eigenvectors) to see how the original variables contribute to the different principal components. In our example it is clear that the first principal component is dominated by the sodium content of the cereal: it has the highest (in this case, positive) weight. This means that the first principal component is in fact measuring how much sodium is in the cereal. Similarly, the second principal component seems to be measuring the amount of potassium. Since both these variables are measured in milligrams, whereas the other nutrients are measured in grams, the scale is obviously leading to this result. The variances of potassium and sodium are much larger than the variances of the other variables, and thus the total variance is dominated by these two variances. A solution is to standardize the data before performing the PCA. Standardizing a variable is accomplished by subtracting the mean and dividing by its standard deviation. The effect of this standardization is to give all variables equal importance in terms of the variability.

When should we standardize the data like this? It depends on the nature of the data. When the units of measurement are common for the variables (e.g., dollars), and when their scale reflects their importance (sales of jet fuel, sales of heating oil), it is probably best not to standardize (i.e., not to rescale the data so that they have unit variance). If the variables are measured in quite differing units so that it is unclear how to compare the variability of different variables (e.g., dollars for some, parts per million for others), or if for variables measured in the same units, scale does not reflect importance (earnings per share, gross revenues), then it is generally advisable to standardize. In this way the changes in units of measurement do not change the principal components' weights. In the rare situations where we can give relative weights to variables, we multiply the standardized variables by these weights before doing the PCA.

JMP's default in PCA is to standardize the variables. So far we have calculated principal components using unstandardized variables, by selecting the option *Unscaled* under *Standardize* in the launch dialog.

Returning to the breakfast cereal data, we conduct PCA on the standardized data (using the default *Standardize* option) due to the different scales of the 13 variables. The output is shown in Figure 4.15. Now we find that we need seven principal components to account for more than 90% of the total variability. The first two principal components account for only 52% of the total variability, and thus reducing the number of variables to 2 would mean losing a lot of information. Examining the weights, we see that the first principal component measures the balance between two quantities: (1) fiber, potassium, protein, and rating (large positive weights) vs. (2) calories and cups (large negative weights). High scores on principal component 1 mean that the cereal is high in fiber, protein, and potassium and low in calories and the amount per bowl. Unsurprisingly, this type of cereal is associated with a high consumer rating. The second principal component is most affected by the weight of a serving, and the third principal component by the carbohydrate content. We can continue labeling the next principal components in a similar fashion to learn about the structure of the data.

The *Score Plot*, which displays by default when conducting a PCA in JMP, shows a scatter plot of the first two principal component scores for the different breakfast cereals (see Figure 4.16). *Score Plots* for additional principal components are available from the *red triangle*. The records can be labeled if the dataset is not too large. We can see that as we move from right (bran cereals) to left, the cereals are less "healthy" in the sense of high calories, low protein and fiber, and so on. Also, moving from bottom to top, we get heavier cereals (moving from puffed rice to raisin bran). These plots are especially useful if interesting clusterings of records can be found. For instance, we see here that children's cereals are close together on the middle-left part of the plot.

Principal Components: on Correlations

Variance estimation: Pairwise

Eigenvectors

	Prin1	Prin2	Prin3	Prin4	Prin5	Prin6	Prin7
calories	-0.32162	0.36184	0.13241	0.30788	0.08827	-0.21135	0.05469
rating	0.45181	-0.22962	0.18280	0.06200	0.03166	-0.16934	0.04143
protein	0.30330	0.16124	0.26246	0.43397	0.15066	0.15246	0.28307
fat	-0.05605	0.34157	-0.21055	0.38199	0.44654	0.39820	-0.02192
sodium	-0.20152	0.12592	0.37879	-0.15608	-0.32916	0.68307	-0.24368
fiber	0.44003	0.21473	0.07972	-0.10160	-0.24714	0.06170	0.11727
carbo	-0.17448	-0.18662	0.56296	0.19721	0.12406	-0.25809	-0.28654
sugars	-0.24883	0.34541	-0.34513	-0.10790	-0.27827	-0.19973	0.21222
potass	0.38616	0.32584	0.08448	0.03316	-0.16532	0.02412	0.13936
vitamins	-0.13871	0.16729	0.38360	-0.52274	0.21840	0.03482	0.34750
shelf	0.13604	0.27492	0.01801	-0.43535	0.59519	-0.12070	-0.16091
weight	-0.07548	0.43538	0.27569	0.10476	-0.27012	-0.38473	0.02476
cups	-0.27954	-0.24141	0.13905	0.09021	0.06792	0.05981	0.74008

Eigenvalues

Number	Eigenvalue	Percent	20 40 60 80	Cum Percent
1	3.604260	27.725		27.725
2	3.166122	24.355		52.080
3	1.872251	14.402		66.482
4	1.091073	8.393		74.875
5	0.968190	7.448		82.322
6	0.720951	5.546		87.868
7	0.662285	5.094		92.963
8	0.437280	3.364		96.326
9	0.299182	2.301		98.628
10	0.091702	0.705		99.333
11	0.058118	0.447		99.780
12	0.024129	0.186		99.966
13	0.004456	0.034		100.000

FIGURE 4.15 PCA output using all 13 numerical variables (standardized) in the breakfast cereals dataset. The eigenvectors table gives results for the first seven principal components

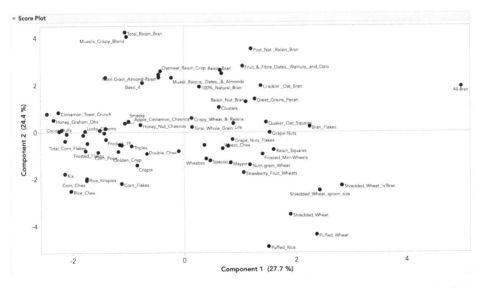

FIGURE 4.16 Score scatter plot of the second vs. first principal components scores for PCA based on correlations

> Reminder: To label records in JMP, right-click on the variable(s) in the data table and select *Label/Unlabel*. When you move your mouse over these points in a graphical display, the labels will appear. To permanently label these points, select the rows in the data table and select *Label/Unlabel* (or select the points directly in the graph, right-click, and select *Row Label*).

Using Principal Components for Classification and Prediction

When the goal of the data reduction is to have a smaller set of variables that will serve as predictors, we can proceed as follows: partition the data into training and holdout sets (*hide* and *exclude* the holdout data). Apply PCA to the predictors using the training data. Use the PCA output to determine the number of principal components to be retained, and save the principal components to the data table (*red triangle > Save Principal Components*). This creates new formula columns in the data table, with the weights for each principal component computed using only the training data. Principal scores for each record are automatically calculated for the using these formulas. Use the holdout data to fit a model, using the principal component columns as the predictors rather than the original variables.

One disadvantage of using a subset of principal components as predictors in a supervised task is that we might lose predictive information that is nonlinear (e.g., a quadratic effect of a predictor on the outcome or an interaction between predictors). This is because PCA produces linear transformations, thereby capturing only linear relationships between the original variables.

4.9 DIMENSION REDUCTION USING REGRESSION MODELS

In this chapter, we discussed methods for reducing the number of columns using summary statistics, plots, and principal component analysis. All these are considered exploratory methods. Some of them completely ignore the outcome variable (e.g., PCA), whereas in other methods we informally try to incorporate the relationship between the predictors and the outcome variable (e.g., combining similar categories, in terms of their behavior with Y). Another approach to reducing the number of predictors, which directly considers the predictive or classification task, is by fitting a regression model. For prediction, a linear regression model is used (see Chapter 6), and for classification, a logistic regression model is used (see Chapter 10). In both cases, we can employ subset selection procedures that algorithmically choose a subset of variables among the larger set (see details in the relevant chapters).

Fitted regression models can be used to further combine similar categories: categories that have coefficients that are not statistically significant (i.e., have a high p-value) can be combined with the reference category because their distinction from the reference category appears to have no significant effect on the output variable. Moreover, categories that have similar coefficient values (and the same sign) can often be combined because their effect on the output variable is similar. See the example in Chapter 10 on predicting delayed flights for an illustration of how regression models can be used for dimension reduction.

4.10 DIMENSION REDUCTION USING CLASSIFICATION AND REGRESSION TREES

Another method for reducing the number of columns and for combining categories of a categorical variable is by applying classification and regression trees (see Chapter 9). Classification trees are used for classification tasks and regression trees for prediction tasks. In both cases the algorithm creates binary splits on the predictors that best classify/predict the outcome (e.g., above/below age 30). Although we defer the detailed discussion to Chapter 9, we note here that the resulting tree diagram can be used for determining the important predictors (using *Column Contributions*). Predictors (numerical or categorical) that do not appear in the tree can be removed. Similarly, categories that do not appear in the tree can be combined.

PROBLEMS

4.1 **Breakfast Cereals.** Use the data for the breakfast cereals example in Section 4.8 (Cereals.jmp) to explore and summarize the data as follows:

    a. Which variables are continuous/numerical? Which are ordinal? Which are nominal?

    b. Calculate the following summary statistics: mean, median, min, max, and standard deviation for each of the continuous variables, and the count for each categorical variable. This can be done using *Cols > Columns Viewer*.

        i. Is there any evidence of extreme values?

        ii. Which, if any, of the variables is missing values?

    c. Use *Analyze > Distribution* to plot a histogram for each of the continuous variables and create summary statistics. Based on the histograms and summary statistics, answer the following questions:

        i. Which variables have the largest variability?

        ii. Which variables seem skewed?

        iii. Are there any values that seem extreme?

    d. Use the *Graph Builder* to plot a side-by-side boxplot comparing the calories in hot vs. cold cereals. What does this plot show us?

    e. Use the *Graph Builder* to plot a side-by-side boxplot of consumer rating as a function of the shelf height (the variable *shelf*). If we were to predict consumer rating from shelf height, does it appear that we need to keep all three categories of shelf height?

    f. Compute the correlation table, and generate a scatterplot matrix for the continuous variables (use *Analyze > Multivariate Methods > Multivariate*).

        i. Which pair of variables is most strongly correlated?

        ii. How can we reduce the number of variables based on these correlations?

        iii. How would the correlations change if we standardized the data first?

    g. Consider the first principal component (the first column on the left under *Eigenvectors*) in Figure 4.14. Describe briefly what this principal component represents.

4.2 **Chemical Features of Wine.** Figure 4.17 shows the PCA output for analyses conducted on standardized (left) and unscaled data (right). Each variable represent a chemical characteristics of wine, and each case is a different wine.

    a. The data are in the file Wine.jmp. Consider the variances in the columns labeled *Eigenvalue* for the principal component analysis conducted using unscaled data. Why is the variance for the first principal component so much larger than the others?

    b. Comment on the use of standardized vs. unscaled variables. Would the results change dramatically if PCA (in this example) were conducted on the standardized variables instead?

Principal Components: on Correlations						Principal Components: on Covariances					
Variance estimation: Row-wise						Variance estimation: Row-wise					
Eigenvalues						Eigenvalues					
Number	Eigenvalue	Percent	20 40 60 80		Cum Percent	Number	Eigenvalue	Percent	20 40 60 80		Cum Percent
1	4.705850	36.199			36.199	1	99201.79	99.809			99.809
2	2.496974	19.207			55.406	2	172.54	0.174			99.983
3	1.446072	11.124			66.530	3	9.44	0.009			99.992
4	0.918974	7.069			73.599	4	4.99	0.005			99.997
5	0.853228	6.563			80.162	5	1.23	0.001			99.998
6	0.641657	4.936			85.098	6	0.84	0.001			99.999
7	0.551028	4.239			89.337	7	0.28	0.000			100.000
8	0.348497	2.681			92.018	8	0.15	0.000			100.000
9	0.288880	2.222			94.240	9	0.11	0.000			100.000
10	0.250902	1.930			96.170	10	0.0717026	0.000			100.000
11	0.225789	1.737			97.907	11	0.037576	0.000			100.000
12	0.168770	1.298			99.205	12	0.0210724	0.000			100.000
13	0.103378	0.795			100.000	13	0.0082037	0.000			100.000

FIGURE 4.17   Principal components of standardized (left) and unscaled (right) wine data

4.3   **University Rankings.** The dataset on American college and university rankings (`Colleges.jmp`) contains information on 1302 American colleges and universities offering an undergraduate program. For each university, there are 17 measurements that include continuous measurements (e.g., tuition and graduation rate) and categorical measurements (e.g., location by state and whether it is a private or a public school).

a. Make sure the variables are coded correctly in JMP (*Nominal*, *Ordinal*, or *Continuous*), and then use the *Columns Viewer* to summarize the data. Are there any missing values? How many Nominal columns are there?

b. Conduct a principal component analysis on the data and comment on the results. Recall that, by default, JMP will conduct the analysis on standardized variables. Is this necessary? Do the data need to be standardized in this case? Discuss key considerations in this decision. Discuss what characterizes the key principal components.

4.4   **Sales of Toyota Corolla Cars.** The file `ToyotaCorolla.jmp` contains data on used cars (Toyota Corollas) on sale during late summer of 2004 in the Netherlands. The data table has 1436 records containing details on 38 variables, including *Price*, *Age*, *Kilometers*, *HP*, and some categorical and dummy-coded variables. The ultimate goal will be to predict the price of a used Toyota Corolla based on its specifications. Although special coding of categorical variables in JMP is generally not required, in this exercise we explore how to create dummy variables (and why).

a. Identify an outlier record, possibly caused due to a data entry error. This can be done by analyzing the distribution of variables or by using *Explore outliers* which is under *Analyze > Screening*. Filter out this observation for the subsequent analysis.

b. Identify the categorical variables.

c. Which variables have been dummy coded?

d. Consider the variable *Fuel_Type*. How many binary dummy variables are required to capture the information for this variable?

e. Use the *Make Indicator Columns* option under *Cols > Utilities* to convert the *Fuel_Type* into dummy variables, and then change the *Modeling Types* for these dummy variables to *Continuous*. Explain in words the values in the derived binary dummies.

f. Produce correlations, a scatterplot matrix, and descriptive statistics for the variables. Comment on the relationships among the variables

PART III

PERFORMANCE EVALUATION

5

EVALUATING PREDICTIVE PERFORMANCE

In this chapter, we discuss how the predictive performance of machine learning methods can be assessed. This is a critical step in any analytics project. We point out the danger of overfitting to the training data, and the need for testing model performance on data that were not used in the training step. We discuss popular performance metrics. For prediction, metrics include average absolute error (*AAE*) and root mean squared error (RMSE) (based on the validation data or test data). For classification tasks, metrics based on the classification matrix include overall accuracy, sensitivity and specificity, and metrics that account for misclassification costs. We also show the relation between the choice of threshold value and method performance, and present the receiver operating characteristic (ROC) curve, which is a popular chart for assessing method performance at different threshold values. When the goal is to accurately classify the most interesting or important cases, called *ranking*, rather than accurately classify the entire sample (e.g., the 10% of customers most likely to respond to an offer, or the 5% of claims most likely to be fraudulent), lift curves are used to assess performance. We also discuss the need for oversampling rare classes and how to adjust performance metrics for the oversampling. Finally, we mention the usefulness of comparing metrics based on the validation data to metrics based on the training data for the purpose of detecting overfitting. While some differences are expected, extreme differences can be indicative of overfitting.

Using JMP: Most of the performance metrics introduced in this chapter are available in the standard version of JMP. However, to evaluate performance based on metrics computed on validation data, JMP Pro is generally required.

Machine Learning for Business Analytics: Concepts, Techniques, and Applications with JMP Pro®,
Second Edition. Galit Shmueli, Peter C. Bruce, Mia L. Stephens, Muralidhara Anandamurthy, and Nitin R. Patel.
© 2023 John Wiley & Sons, Inc. Published 2023 by John Wiley & Sons, Inc.

5.1 INTRODUCTION

In supervised learning, we are interested in predicting the outcome variable for new records. The outcome variable acts as a "supervisor" guiding the machine learning process. Three main types of outcomes of interest are:

Predicted numerical value—when the outcome variable is numerical (e.g., house price)

Predicted class membership—when the outcome variable is categorical (e.g., buyer/nonbuyer)

Propensity—the probability of class membership, when the outcome variable is categorical (e.g., the propensity to default).

Prediction methods are used for generating numerical predictions, while classification methods ("classifiers") are used for generating propensities, and using a threshold value on the propensities, we generate predicted class memberships.

A subtle distinction to keep in mind is the two distinct predictive uses of classifiers: one use, *classification*, is aimed at predicting class membership for new records. The other, *ranking*, is detecting, among a set of new records, the ones most likely to belong to a class of interest.

In this chapter, we examine the approach for judging the usefulness of a prediction method used for generating numerical predictions (Section 5.2), a classifier used for classification (Section 5.3), and a classifier used for ranking (Section 5.4). In Section 5.5, we'll look at evaluating performance under the scenario of oversampling.

5.2 EVALUATING PREDICTIVE PERFORMANCE

First, let us emphasize that predictive accuracy is not the same as goodness-of-fit. Classical measures of performance are aimed at finding a model that fits well to the data on which the model was trained. In predictive modeling, we are interested in models that have high predictive accuracy when applied to *new* records. Measures such as R^2 and standard error of estimate are common metrics in classical regression modeling, and residual analysis is used to gauge goodness-of-fit in that situation. However, these measures do not tell us much about the ability of the model to predict new records.

For assessing prediction performance, several measures are used. In all cases the measures are based on the validation set, which serves as a more objective ground than the training set to assess predictive accuracy. This is because records in the validation set are more similar to the future records to be predicted, in the sense that they are not used to select predictors or to estimate the model coefficients. Models are trained on the training data, applied to the validation set, and measures of accuracy then use the prediction errors on that validation set.

Naive Benchmark: The Average

The benchmark criterion in prediction is using the average outcome (thereby ignoring all predictor information). In other words, the prediction for a new record is simply the average outcome of the records in the training set ($\bar{y}$). This is sometimes called a *naive benchmark*.

Depending on the business context, occasionally it might make sense to use an alternative benchmark value (e.g., when predicting this week's sales, we may use last week's sales as a benchmark). A good predictive model should outperform the benchmark criterion in terms of predictive accuracy.

Prediction Accuracy Measures

The *prediction error*, or *residual*, for record i is defined as the difference between its actual y value and its predicted y value: $e_i = y_i - \hat{y}_i$. There are a few popular numerical measures of predictive accuracy, calculated on the validation data:

- *MAE, MAD*, or *AAE* (mean absolute error/deviation, or average absolute error) $= 1/n \sum_{i=1}^{n} |e_i|$. This gives the magnitude of the average error.
- *Average error* $= 1/n \sum_{i=1}^{n} e_i$. This measure is similar to *MAE* except that it retains the sign of the errors, so that negative errors cancel out positive errors of the same magnitude. It therefore gives an indication of whether the predictions are on average over- or under-predicting the response.
- *MPE* (mean percentage error) $= 100 \times \frac{1}{n} \sum_{i=1}^{n} e_i/y_i$. This gives the percentage score of how predictions deviate from the actual values (on average), taking into account the direction of the error.
- *MAPE* (mean absolute percentage error) $= 100 \times 1/n \sum_{i=1}^{n} |e_i/y_i|$. This measure gives a percentage score of how predictions deviate (on average) from the actual values.
- *RMSE* (root mean squared error) $= \sqrt{1/n \sum_{i=1}^{n} e_i^2}$. This measure is similar to the estimate of prediction error in linear regression, except that it is computed on the validation data rather than on the training data. It has the same units as the variable predicted.
- Total *SSE* (sum of squared errors) $= \sum_{i=1}^{n} e_i^2$.
- *RASE* (root average squared error) $= \sqrt{SSE/n}$. This measure is similar to RMSE, except it is not adjusted for the number of parameters in the model. This characteristic makes *RASE* useful in comparing models with different numbers of parameters.

JMP Pro reports the following measures of predictive accuracy for the validation data (additional measures are available for the training data, and R^2 is also reported):

- *RASE* (or RMSE in some platforms)
- *AAE* (in the *Model Comparison* platform)

Such measures can be used to compare models and to assess their degree of prediction accuracy. Note that all of these measures are influenced by outliers. To check outlier influence, we can compute median-based measures (and compare to the mean-based measures) or simply plot a histogram or boxplot of the errors. Plotting the distribution of prediction errors is in fact very useful and can highlight more information than the metrics alone.

To illustrate the use of predictive accuracy measures and charts of the distribution of prediction error, consider the error metrics and charts shown in Figures 5.1 and 5.2. These are the result of fitting a certain predictive model to prices of used Toyota Corolla cars. The training set includes 600 cars and the validation set includes 400 cars. The prediction errors (residuals) for the validation set are summarized in the bottom row in Figure 5.1. Residuals

▼ Crossvalidation			
Source	RSquare	RASE	Freq
Training Set	0.8699	1327.7	600
Validation Set	0.8646	1407.0	400

FIGURE 5.1   Prediction error metrics from a model for Toyota car prices. Training (top row) and validation (bottom row)

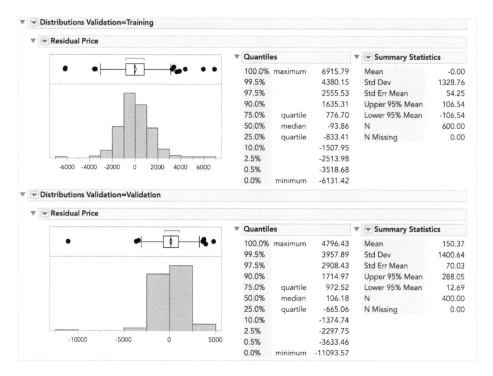

FIGURE 5.2   Histograms, boxplots, and summary statistics of Toyota price prediction errors, for training and validation sets

for this model were saved to the data table. The histogram, boxplot, and summary statistics corresponding to the residuals for the validation set (at the bottom of Figure 5.2) show a small number of outliers, with over half of the errors in the [−900, 800] range. On average, the model overpredicts (the average residual value for the validation set is 150.37).

Comparing Training and Validation Performance

Errors (residuals) that are based on the training set tell us about model fit, whereas errors based on the validation set (called "prediction errors") measure the model's ability to predict new data (predictive performance). We expect training errors to be smaller than the validation errors (because the model was fit using the training set), and the more complex the model, the greater is the likelihood that it will *overfit* the training data (indicated by a greater difference between the training and validation errors). In an extreme case of overfitting, the training errors would be zero (perfect fit of the model to the training data), and the validation errors would be nonzero and nonnegligible. For this reason, it is important to compare the error plots and metrics (RMSE, average error, etc.) of the training and validation sets. Figure 5.1 illustrates this comparison: the RASE for the training set is lower than for the validation set, while the RSquare is much higher.

5.3 JUDGING CLASSIFIER PERFORMANCE

The need for performance measures arises from the wide choice of classifiers and predictive methods. Not only do we have several different methods, but even within a single method there are usually many options that can lead to completely different results. A simple example is the choice of predictors used within a particular predictive algorithm. Before we study these various algorithms in detail and face decisions on how to set these options, we need to know how we will measure success.

A natural criterion for judging the performance of a classifier is the probability of making a *misclassification error*. Misclassification means that the observation belongs to one class, but the model classifies it as a member of a different class. A classifier that makes no errors would be perfect, but we do not expect to be able to construct such classifiers in the real world due to "noise" and to not having all the information needed to classify cases precisely. Is there a minimal probability of misclassification that we should require of a classifier?

Benchmark: The Naive Rule

A very simple rule for classifying a record into one of m classes, ignoring all predictor information $(x_1, x_2, \ldots, x_p)$ that we might have, is to classify the record as a member of the majority class. In other words, "classify as belonging to the most prevalent class." The *naive rule* is used mainly as a baseline or benchmark for evaluating the performance of more complicated classifiers. Clearly, a classifier that uses external predictor information (on top of the class membership allocation) should outperform the naive rule.

Similar to using the sample mean ($\bar{y}$) as the naive benchmark in the numerical outcome case, the naive rule for classification relies solely on the y information and excludes any additional predictor information.

Class Separation

If the classes are well separated by the predictor information, even a small dataset will suffice in finding a good classifier, whereas if the classes are not separated at all by the predictors, even a very large dataset will not help. Figure 5.3 illustrates this for a two-class case. The top panel includes a small dataset ($n = 24$ observations) where two predictors (income and lot size) are used for separating owners from non-owners (we thank Dean Wichern for this example, described in Johnson and Wichern, 2002). In this case the predictor information seems useful in that it separates the two classes (owners/non-owners). The bottom panel shows a much larger dataset ($n = 5000$ observations) where the two predictors (income and average credit card spending) do not separate the two classes well in most of the higher ranges (loan acceptors/nonacceptors).

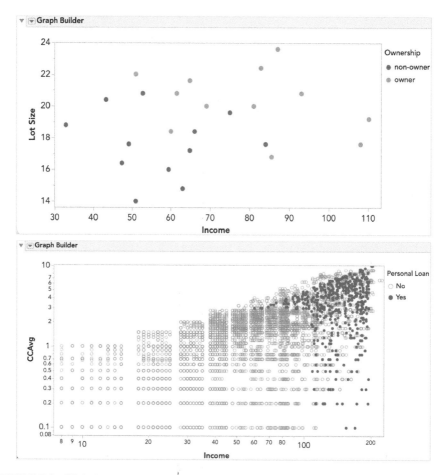

FIGURE 5.3   High (top) and low (bottom) levels of separation between two classes, using two predictors

The Classification (Confusion) Matrix

In practice, most accuracy measures are derived from the *classification matrix*, also called the *confusion matrix*. This matrix summarizes the correct and incorrect classifications that a classifier produced for a certain dataset. Rows and columns of the classification matrix correspond to the true and predicted classes, respectively. Figure 5.4 shows an example of a classification (confusion) matrix for a two-class (0/1) problem resulting from applying

		Predicted Class		
	Count	0	1	Total
Actual Class	0	2689	25	2714
	1	85	201	286
	Total	2774	226	3000

FIGURE 5.4   Classification matrix based on 3000 observations and two classes

a certain classifier to 3000 observations. The two diagonal cells (upper left, lower right) give the number of correct classifications, where the predicted class coincides with the actual class of the observation. The off-diagonal cells give counts of misclassification. The upper right cell gives the number of class 0 members that were misclassified as 1's (in this example, there were 25 such misclassifications). Similarly, the lower-left cell gives the number of class 1 members that were misclassified as 0's (85 such observations).

The classification matrix gives estimates of the true classification and misclassification rates. Of course, these are estimates and they can be incorrect, but if we have a large enough dataset and neither class is very rare, our estimates will be reliable. Sometimes, we may be able to use public data such as US Census data to estimate these proportions. However, in most business settings, we will not know them.

Using the Validation Data

To obtain an honest estimate of future classification error, we use the classification matrix that is computed from the *validation data*. In other words, we first partition the data into training and validation sets by random selection of cases. We then construct a classifier using the training data and then apply it to the validation data. This will yield the predicted classifications for observations in the validation set (see Figure 2.7 in Chapter 2). We then summarize these classifications in a classification matrix. Although we can summarize our results in a classification matrix for training data as well, the resulting classification matrix is not useful for getting an honest estimate of the misclassification rate for new data due to the danger of overfitting.

In addition to examining the validation data classification matrix to assess the classification performance on new data, we compare the training data classification matrix to the validation data classification matrix in order to detect overfitting: although we expect somewhat inferior results on the validation data, a large discrepancy in training and validation performance might be indicative of overfitting.

Accuracy Measures

Different accuracy measures can be derived from the classification matrix. Consider a two-class case with classes C_0 and C_1 (e.g., nonbuyer/buyer). The schematic classification matrix in Table 5.1 uses the notation $n_{i,j}$ to denote the number of cases that are class C_i members, and were classified as C_j members. Of course, if $i \neq j$, these are counts of misclassifications. The total number of observations is $n = n_{0,0} + n_{0,1} + n_{1,0} + n_{1,1}$.

TABLE 5.1   Classification Matrix: Meaning of Each Cell

Actual class	Predicted class	
	C_0	C_1
C_0	$n_{0,0}$ = number of C_0 cases classified correctly	$n_{0,1}$ = number of C_0 cases classified incorrectly as C_1
C_1	$n_{1,0}$ = number of C_1 cases classified incorrectly as C_0	$n_{1,1}$ = number of C_1 cases classified correctly

A main accuracy measure is the *estimated misclassification rate*, also called the *overall error rate*. It is given by

$$\text{err} = \frac{n_{0,1} + n_{1,0}}{n}$$

where n is the total number of cases in the validation dataset. In the example in Figure 5.4, we get err $= (25 + 85)/3000 = 3.67\%$.

We can measure accuracy by looking at the correct classifications—the full half of the cup—instead of the misclassifications. The *overall accuracy* of a classifier is estimated by

$$\text{accuracy} = 1 - \text{err} = \frac{n_{0,0} + n_{1,1}}{n}.$$

In the example, we have $(2689 + 201)/3000 = 96.33\%$.

Propensities and Threshold for Classification

The first step in most classification algorithms is to estimate the probability that a case belongs to each of the classes. These probabilities, also called *propensities*, are typically used either as an interim step for generating predicted class membership (classification), or for rank-ordering the records by their probability of belonging to a class of interest. Let us consider their first use in this section. The second use is discussed in Section 5.4.

If overall classification accuracy (involving all the classes) is of interest, the case can be assigned to the class with the highest probability. In many contexts a single class is of special interest, which is also called the positive class. So, we will focus on the positive class and compare the propensity of belonging to that class to a *threshold (or cutoff)* value set by the analyst. This approach can be used with two classes or more than two classes, though it may make sense in such cases to consolidate classes so that you end up with two: the class of interest and all other classes. If the probability of belonging to the class of interest is above the cutoff, the case is assigned to that class.

The default threshold value in two-class classifiers is 0.5. Thus, if the probability of a record being a class 1 member is greater than 0.5, that record is classified as a 1. Any record with an estimated probability of less than 0.5 would be classified as a 0. It is possible, however, to use a threshold that is either higher or lower than 0.5. A threshold greater than 0.5 will end up classifying fewer records as 1's, whereas a threshold less than 0.5 will end up classifying more records as 1. Typically, the misclassification rate will rise in either case.

Consider the data in Figure 5.5, showing the actual class for 24 records (based on the riding mower example), along with propensities (*Prob[owner]*) and classifications (*Most Likely Ownership*) based on some model. To get the propensities and classifications, a model was fit, and the probability formula was saved to the data table (note that some columns are hidden, and that for illustration the *Actual Class* column was changed from its original name, *Ownership*).

The misclassification rate for the *Most Likely Ownership*, which uses the standard 0.5 as the threshold, is 4/24. This is summarized in a confusion matrix on the left in Figure 5.6. If we instead adopt a threshold of 0.25, we classify more records as owners and the misclassification rate goes up to 5/24 (see column *Predicted Class* at 0.25 in Figure 5.5). If we adopt a threshold of 0.75, fewer records are classified as owners and the misclassification rate again goes up to 6/24, as shown in the last column in Figure 5.5. All of this can be seen in the confusion matrices in Figure 5.6.

	Income	Lot Size	Actual Class	Prob[owner]	Most Likely Ownership	Predicted Class at 0.25	Predicted Class at 0.75
1	60	18.4	owner	0.1746	non-owner	non-owner	non-owner
2	85.5	16.8	owner	0.4333	non-owner	owner	non-owner
3	64.8	21.6	owner	0.8873	owner	owner	owner
4	61.5	20.8	owner	0.7163	owner	owner	non-owner
5	87	23.6	owner	0.9984	owner	owner	owner
6	110.1	19.2	owner	0.9916	owner	owner	owner
7	108	17.6	owner	0.9524	owner	owner	owner
8	82.8	22.4	owner	0.9921	owner	owner	owner
9	69	20	owner	0.7284	owner	owner	non-owner
10	93	20.8	owner	0.9881	owner	owner	owner
11	51	22	owner	0.7148	owner	owner	non-owner
12	81	20	owner	0.9103	owner	owner	owner
13	75	19.6	non-owner	0.7801	owner	owner	owner
14	52.8	20.8	non-owner	0.4904	non-owner	owner	non-owner
15	64.8	17.2	non-owner	0.1018	non-owner	non-owner	non-owner
16	43.2	20.4	non-owner	0.1842	non-owner	non-owner	non-owner
17	84	17.6	non-owner	0.5833	owner	owner	non-owner
18	49.2	17.6	non-owner	0.0287	non-owner	non-owner	non-owner
19	59.4	16	non-owner	0.0192	non-owner	non-owner	non-owner
20	66	18.4	non-owner	0.2915	non-owner	owner	non-owner
21	47.4	16.4	non-owner	0.0076	non-owner	non-owner	non-owner
22	33	18.8	non-owner	0.0154	non-owner	non-owner	non-owner
23	51	14	non-owner	0.0011	non-owner	non-owner	non-owner
24	63	14.8	non-owner	0.0091	non-owner	non-owner	non-owner

FIGURE 5.5   Riding mower example: 24 records with the actual class and probability (propensity) of ownership, as estimated by a classifier

▼ ⊞ Tabulate

	Most Likely Ownership				Predicted Class at 0.25				Predicted Class at 0.75			
	non-owner		owner		non-owner		owner		non-owner		owner	
Actual Class	Row %	N	Row %	N	Row %	N	Row %	N	Row %	N	Row %	N
owner	16.67%	2	83.33%	10	8.33%	1	91.67%	11	41.67%	5	58.33%	7
non-owner	83.33%	10	16.67%	2	66.67%	8	33.33%	4	91.67%	11	8.33%	1

FIGURE 5.6   Classification matrices based on thresholds of 0.5, 0.25, and 0.75 for the riding mower data

Why would we want to use cutoff values different from 0.5 if they increase the misclassification rate? The answer is that it might be more important to classify 1's properly than 0's, and we would tolerate a greater misclassification of the latter. Or the reverse might be true: the costs of misclassification might be asymmetric. We can adjust the threshold value in such a case to classify more records as the high-value class, that is, accept more misclassifications where the misclassification cost is low. Keep in mind that we are doing so after the machine learning model has already been selected—we are not changing that model. It is also possible to incorporate costs into the picture before deriving the model. These subjects are discussed in greater detail below.

CHANGING THE DECISION THRESHOLDS IN JMP Pro

To see how the accuracy or misclassification rates change as a function of the threshold, we can use the *Decision Threshold*. To do this:

1. Fit a classifier model in JMP Pro.
2. Select *Decision Threshold* from the red triangle for the model. This produces a report with an interactive *Fitted Probability* graph, confusion matrices, performance metrics, and other results. See Figure 5.7.
3. Click and enter a new *Probability Threshold* value in the *Fitted Probability* graph, or drag the slider to dynamically change the threshold value.
4. The confusion matrix, Accuracy, and other graphs and metrics will automatically update.

Another, more manual, approach is to save the probability formula for the model to the data table (using the top red triangle for the model), create a new column, and then use the formula editor to create a conditional formula using the desired threshold (see the formula shown in Figure 5.8. This was done to create the predicted class columns in Figure 5.5).

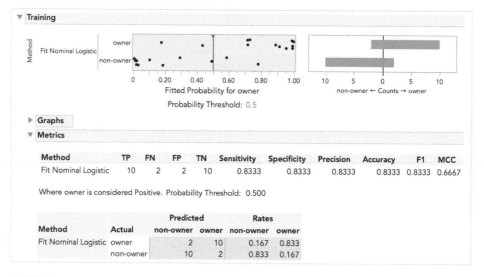

FIGURE 5.7   *Decision Threshold* **report, with fitted probability graph, confusion matrices, and performance metrics**

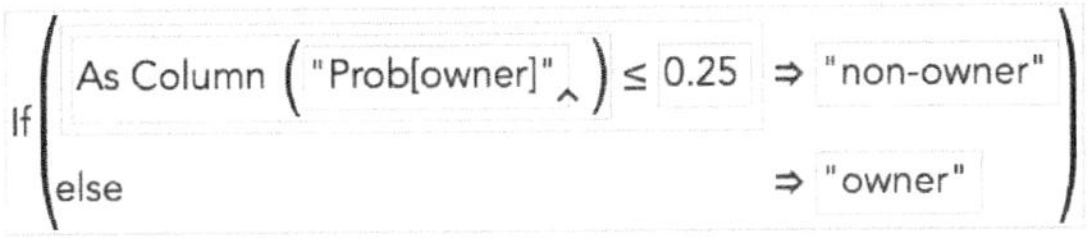

$$\text{If} \left(\begin{array}{ll} \text{As Column} \left(\text{"Prob[owner]"}_\wedge \right) \le 0.25 & \Rightarrow \text{"non-owner"} \\ \text{else} & \Rightarrow \text{"owner"} \end{array} \right)$$

FIGURE 5.8 Using the *Formula Editor* with *Conditional* and *Comparison* functions to change the threshold for classification

---

THRESHOLD VALUES FOR TRIAGE

In some cases it is useful to have two thresholds and allow a "cannot say" option for the classifier. In a two-class situation, this means that for a case we can make one of three predictions: the case belongs to C_0, or the case belongs to C_1, or we cannot make a prediction because there is not enough information to pick C_0 or C_1 confidently. Cases that the classifier cannot classify are subjected to closer scrutiny either by using expert judgment or by enriching the set of predictor variables by gathering additional information that is perhaps more difficult or expensive to obtain. An example is classification of documents found during legal discovery (reciprocal forced document disclosure in a legal proceeding). Under traditional human-review systems, qualified legal personnel are needed to review what might be tens of thousands of documents to determine their relevance to a case. Using a classifier and a triage outcome, documents could be sorted into clearly relevant, clearly not relevant, and the gray area documents requiring human review. This substantially reduces the costs of discovery.

Performance in Unequal Importance of Classes

Suppose that it is more important to predict membership correctly in class 1 than in class 0. An example is predicting the financial status (bankrupt/solvent) of firms. It may be more important to correctly predict a firm that is going to bankrupt than to correctly predict a firm that is going to remain solvent. The classifier is essentially used as a system for detecting or signaling bankruptcy. In such a case, the overall accuracy is not a good measure for evaluating the classifier. Suppose that the important (positive) class is C_1. The following pair of accuracy measures are the most popular:

The **sensitivity** of a classifier is its ability to correctly detect all the important class members. This is measured by $n_{1,1}/(n_{1,0}+n_{1,1})$, the percentage of C_1 positive members that are classified correctly. Sensitivity is also called the *True Positive Rate (TPR)* or *Recall*.

The **specificity** of a classifier is its ability to rule out C_0 (negative class) members correctly. This is measured by $n_{0,0}/(n_{0,0}+n_{0,1})$, the percentage of C_0 members correctly classified. Specificity is also called the *True Negative Rate (TNR)*.

It can be useful to plot these measures vs. the threshold value in order to find a threshold value that balances these measures. Note that graphs for exploring these measures are available in the *Decision Threshold* report.

ROC Curve   A popular method for plotting the sensitivity and specificity of a classifier is through *ROC* (receiver operating characteristic) *curves*. Starting from the lower left the ROC curve plots the pairs {Sensitivity, 1−Specificity} as the threshold value increases from 0 and 1. Better performance is reflected by curves that are closer to the top-left corner.

The comparison curve is the diagonal, which reflects the average performance of a guessing classifier that has no information about the predictors or outcome. This guessing classifier guesses that a proportion α of the records is 1's and therefore assigns each record an equal probability $P(Y = 1) = \alpha$. In this case, on average, a proportion α of the 1's will be correctly classified (Sensitivity = α), and a proportion α of the 0s will be correctly classified (1 − Specificity = α). As we increase the threshold value $1 - \alpha$ from 0 to 1, we get the diagonal line Sensitivity = 1 − Specificity. Note that the naive rule is one point on this diagonal line, where α = proportion of actual 1's.

A common metric to summarize an ROC curve is *area under the curve (AUC)*, which ranges from 1 (perfect discrimination between classes) to 0.5 (no better than random guessing). AUC is commonly used to compare models, especially when there are many models under consideration. It has the advantage of simplicity and can be compared across datasets with different positive class rates. However, the AUC has several critical deficiencies. For example, some areas of the ROC curve are irrelevant (e.g., if extremely low sensitivity is unacceptable). Second, when ROC curves of different models cross, AUC comparisons can be misleading: a larger AUC value might correspond to poorer performance over almost the entire range of threshold values (Hand, 2009). For further issues, see Lobo et al. (2008). We therefore advocate not using AUC as a single performance metric for selecting a model.

ROC curves are available as a red triangle option in JMP. The ROC curve for the riding mower example and its corresponding AUC (called *Area* in JMP) are shown in Figure 5.9. The chart provides separate lines for each class (here red for owner and blue for non-owner).

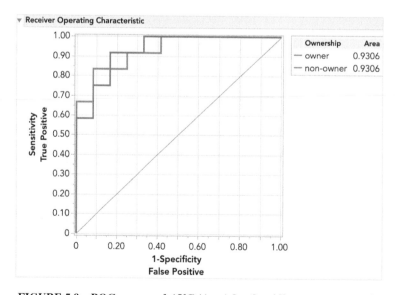

FIGURE 5.9   ROC curve and AUC (*Area*) for the riding mower example

Precision and F1-score   In addition to sensitivity and specificity, there is another performance measure called *precision* that is often considered in tandem with sensitivity (recall):

> The **precision** of a classifier is its ability to correctly detect only the important (positive) class members. This is measured by $n_{1,1}/(n_{1,1} + n_{0,1})$, the percentage of correctly classified positive (C_1) members among all those classified as positive.

For a two-class classifier, there tends to be a tradeoff between precision and recall. While precision is a measure of *exactness* (what percentage of members classified as C_1 actually belong to C_1), recall is a measure of *completeness* (what percentage of C_1 members are classified as such). Depending on the business context, precision may be more important than recall, and vice versa. For example, for credit card fraud detection (positive class = fraudulent), we would want recall close to 1 since we want to find *all* transactions that are truly fraudulent, and we can accept low precision (mistakenly classifying nonfraudulent transactions as fraudulent) if the cost of such misclassification is not too high. The threshold value can be selected based on whether precision or recall is a more important consideration.

A metric called *F1-score* is used to combine precision and sensitivity values into a single measure, giving them equal weights. F1-score is given by

$$F1 = \frac{2 \times \text{precision} \times \text{sensitivity}}{(\text{precision} + \text{sensitivity})}.$$

JMP produces all these metrics as part of *Decision Threshold* report (recall that *Decision Threshold* is a red triangle option for classification models).

COMPUTING RATES: FROM WHOSE POINT OF VIEW?

Sensitivity and specificity measure the performance of a classifier from the point of view of the "classifying agency" (e.g., a company classifying customers or a hospital classifying patients). They answer the question "how well does the classifier segregate the important class members?" It is also possible to measure accuracy from the perspective of the entity that is being predicted (e.g., the customer or the patient), who asks "what is my chance of belonging to the important class?" Then again, this question is usually less relevant in a data mining application. The terms "false-positive rate" and "false-negative rate," which are sometimes used erroneously to describe 1−sensitivity and 1−specificity, are measures of performance from the perspective of the individual entity. If C_1 is the important class, they are defined as:

The *false-positive rate (or false discovery rate (FDR) or fallout)* is the proportion of C_1 predictions that are wrong: $n_{0,1}/(n_{0,1} + n_{1,1})$. Notice that this is a ratio within the column of C_1 predictions (that is, it uses only records that were classified as C_1).

The *false-negative rate (or false omission rate (FOR))* is the proportion of C_0 predictions that are wrong: $n_{1,0}/(n_{0,0} + n_{1,0})$. Notice that this is a ratio within the column of C_0 predictions (that is, it uses only records that were classified as C_0).

Asymmetric Misclassification Costs

Up to this point, we have been using the misclassification rate as the criterion for judging the efficacy of a classifier. However, there are circumstances where this measure is not appropriate. Sometimes the error of misclassifying a case belonging to one class is more serious than for the other class. For example, misclassifying a household as unlikely to respond to a sales offer when it belongs to the class that would respond incurs a greater cost (the opportunity cost of the forgone sale) than the converse error. In the former case, you are missing out on a sale worth perhaps tens or hundreds of dollars. In the latter, you are incurring the costs of mailing a letter to someone who will not purchase. In such a scenario, using the misclassification rate as a criterion can be misleading.

Note that we are assuming that the cost (or benefit) of making correct classifications is zero. At first glance, this may seem incomplete. After all, the benefit (negative cost) of classifying a buyer correctly as a buyer would seem substantial. In other circumstances (e.g., scoring our classification algorithm to fresh data to implement our decisions), it will be appropriate to consider the actual net dollar impact of each possible classification (or misclassification). Here, however, we are attempting to assess the value of a classifier in terms of classification error, so it greatly simplifies matters if we can capture all cost/benefit information in the misclassification cells. So, instead of recording the benefit of classifying a respondent household correctly, we record the cost of failing to classify it as a respondent household. It amounts to the same thing, and our goal becomes the minimization of costs, whether the costs are actual costs or missed benefits (opportunity costs).

Consider the situation where the sales offer is mailed to a random sample of people for the purpose of constructing a good classifier. Suppose that the offer is accepted by 1% of those households. For these data, if a classifier simply classifies every household as a nonresponder, it will have an error rate of only 1%, but the classifier will be useless in practice. A classifier that misclassifies 2% of buying households as nonbuyers and 20% of the nonbuyers as buyers would have a higher error rate but would be better if the profit from a sale is substantially higher than the cost of sending out an offer. In these situations, if we have estimates of the cost of both types of misclassification, we can use the classification matrix to compute the expected cost of misclassification for each case in the validation data. This enables us to compare different classifiers using overall expected costs (or profits) as the criterion.

Suppose that we are considering sending an offer to 1000 more people, 1% of whom respond (1), on average. Naively classifying everyone as a 0 has an error rate of only 1%. Using a machine learning process, suppose that we can produce these classifications:

	Predict class 0	Predict class 1
Actual 0	970	20
Actual 1	2	8

These classifications have an error rate of $100 \times (20 + 2)/1000 = 2.2\%$—higher than the naive rate.

Now suppose that the profit from a 1 is \$10 and the cost of sending the offer is \$1. Classifying everyone as a 0 still has a misclassification rate of only 1% but yields a profit of $0. Using the machine learning process, despite the higher misclassification rate, yields a profit of $60.

The matrix of profit is as follows (nothing is sent to the predicted 0's, so there are no costs or sales in that column):

Profit	Predict class 0	Predict class 1
Actual 0	0	– $20
Actual 1	0	$80

Looked at purely in terms of costs, when everyone is classified as a 0, there are no costs of sending the offer; the only costs are the opportunity costs of failing to make sales to the ten 1's = $100. The cost (actual costs of sending the offer plus the opportunity costs of missed sales) of using the machine learning process, to select people to send the offer to is only $48, as follows:

Costs	Predict class 0	Predict class 1
Actual 0	0	$20
Actual 1	$20	$8

However, this does not improve the actual classifications. A better method is to change the classification rules, hence the misclassification rates), as discussed in the preceding section, to reflect the asymmetric costs.

A popular performance measure that includes costs is the *average misclassification cost*, which measures the average cost of misclassification per classified observation. Denote by q_0 the cost of misclassifying a class 0 observation (as belonging to class 1) and by q_1 the cost of misclassifying a class 1 observation (as belonging to class 0). The average misclassification cost is

$$\frac{q_0 n_{0,1} + q_1 n_{1,0}}{n}.$$

Thus, we are looking for a classifier that minimizes this quantity. This can be computed, for instance, for different threshold values.

It turns out that the optimal parameters are affected by the misclassification costs only through the ratio of these costs. This can be seen if we write the foregoing measure slightly differently:

$$\frac{q_0 n_{0,1} + q_1 n_{1,0}}{n} = \frac{n_{0,1}}{n_{0,0} + n_{0,1}} \frac{n_{0,0} + n_{0,1}}{n} q_0 + \frac{n_{1,0}}{n_{1,0} + n_{1,1}} \frac{n_{1,0} + n_{1,1}}{n} q_1.$$

Minimizing this expression is equivalent to minimizing the same expression divided by a constant. If we divide by q_0, it can be seen clearly that the minimization depends only on q_1 / q_0 and not on their individual values. This is very practical, since in many cases it is difficult to assess the cost associated with misclassifying a 0 member and that associated with misclassifying a 1 member, but estimating the ratio is easier.

This expression is a reasonable estimate of future misclassification cost if the proportions of classes 0 and 1 in the sample data are similar to the proportions of classes 0 and 1 that are expected in the future. If, instead of a random sample, we draw a sample such that one class is oversampled (as described in the next section), the sample proportions of 0's and 1's will be distorted compared to the future or population. We can then correct the average misclassification cost measure for the distorted sample proportions by incorporating estimates of the true proportions (from external data or domain knowledge), denoted by $p(C_0)$ and $p(C_1)$, into the formula

$$\frac{n_{0,1}}{n_{0,0} + n_{0,1}} p(C_0) \, q_0 + \frac{n_{1,0}}{n_{1,0} + n_{1,1}} p(C_1) \, q_1.$$

Using the same logic as above, it can be shown that optimizing this quantity depends on the costs only through their ratio (q_1/q_0) and on the prior probabilities only through their ratio $[p(C_0)/p(C_1)]$. This is why software packages that incorporate costs and prior probabilities might prompt the user for ratios rather than actual costs and probabilities.

ASYMMETRIC MISCLASSIFICATION COSTS IN JMP

In JMP, asymmetric misclassification costs are assigned as a *Column Property* for the outcome variable using the *Profit Matrix*, as shown in Figure 5.10. The *Profit Matrix* assigns weights to predicted outcomes and calculates expected costs and profits for the predicted class when a probability formula for a model is saved to the data table.

To set the *Profit Matrix* from the data table, right-click on the column header, select *Column Properties*, and then *Profit Matrix* (towards the bottom of the list). In addition, the threshold for classification can be set after choosing the desired target level in the *Profit Matrix*.

Alternatively, the *Profit Matrix* can be set from the *Decision Threshold* report.

Figure 5.10 shows the setup of the *Profit Matrix* for the following situation: the profit from a 1 is $10 and the cost of sending the offer is $1.

Generalization to More Than Two Classes

All the comments made above about two-class classifiers extend readily to classification into more than two classes. Let us suppose that we have m classes $C_0, C_1, C_2, \ldots, C_{m-1}$. The classification matrix has m rows and m columns. The misclassification cost associated with the diagonal cells is, of course, always zero. Incorporating prior probabilities of the various classes (where now we have m such numbers) is still done in the same manner.

However, evaluating misclassification costs using the formulas in the previous section becomes much more complicated. For an m-class case, we have $m(m-1)$ types of misclassifications. While the *Profit Matrix* in JMP can accommodate misclassification costs for situations involving more than two classes, it is often difficult to elicit or obtain all the misclassification costs associated with more than two classes.

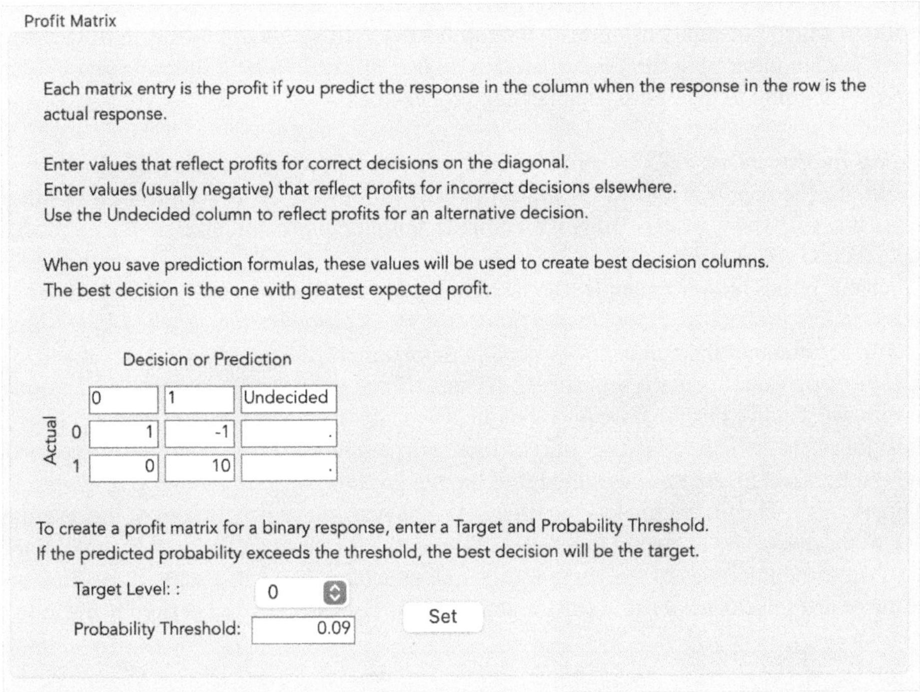

Profit Matrix

Each matrix entry is the profit if you predict the response in the column when the response in the row is the actual response.

Enter values that reflect profits for correct decisions on the diagonal.
Enter values (usually negative) that reflect profits for incorrect decisions elsewhere.
Use the Undecided column to reflect profits for an alternative decision.

When you save prediction formulas, these values will be used to create best decision columns.
The best decision is the one with greatest expected profit.

Decision or Prediction

	0	1	Undecided
Actual 0	1	-1	.
1	0	10	.

To create a profit matrix for a binary response, enter a Target and Probability Threshold.
If the predicted probability exceeds the threshold, the best decision will be the target.

Target Level: :   0   ⬍

Probability Threshold: 0.09   Set

FIGURE 5.10 **Specifying misclassification costs in JMP using the** *Profit Matrix* **column property**

5.4 JUDGING RANKING PERFORMANCE

We now turn to the predictive goal of detecting, among a set of new records, the ones most likely to belong to a class of interest. Recall that this differs from the goal of predicting class membership for each new record.

Lift Curves for Binary Data

Lift curves (also called *lift charts*, gains curves, or gains charts) are available in JMP for models involving categorical outcomes. The lift curve helps us determine how effectively we can "skim the cream" by selecting a relatively small number of records and getting a relatively large portion of the responders. The input required to construct a lift curve is a validation dataset that has been "scored" by appending to each record the propensity that it will belong to a given class.

Let us continue with the case in which a particular class is relatively rare and of much more interest than the other class: tax cheats, debt defaulters, or responders to a mailing. We would like our classification model to sift through the records and sort them according to which ones are most likely to be tax cheats, responders to the mailing, and so on. We can then make more informed decisions. For example, we can decide how many, and which tax returns to examine, looking for tax cheats. The model will give us an estimate of the extent to which we will encounter fewer and fewer noncheaters as we proceed through the sorted data starting with the records most likely to be tax cheats. Or we can use the sorted data to decide to which potential customers a limited-budget mailing should be targeted. In other words, we are describing the case where our goal is to obtain a rank-ordering among the records according to their class membership propensities.

Sorting by Propensity To construct a lift curve, we sort the set of records by propensity in descending order. This is the propensity to belong to the important positive class, say, C_1. Then, in each row, we compute the cumulative number of C_1 members (Actual class = C_1).

For our riding mower example, the table in Figure 5.11 shows the 24 records ordered in descending probability of ownership (under the *Prob[owner]* column). The *Cum Owner* column accumulates the number of owners. In this example, the top 6 rows (portion = 0.25, or 25% of our data) contain 6 of the 12 owners. Since we have 12 owners and 12 non-owners, we would expect to have only 3 of the owners in the top 25% of our data by chance alone (since $12 \times 0.25 = 3$). The ratio of these two numbers, actual owners vs. expected owners, is called *lift*. In this case, the lift at the top 25% of our data is 6/3 = 2.0, as shown in Figure 5.11. The lift curve plots the lift against the percentage (or portion) of the data for each of the classes, as shown in Figure 5.12. The curve for *owner* is the thicker (blue) line. (To change the thickness of the curve for the outcome class in a Lift Curve, right-click on the curve and select *Customize*. Then, select *Level Curves* and enter the desired line width.)

Cols ▾	Income	Lot Size	Actual Class	Prob [owner] (sorted descending)	Cum Owner	Portion	Lift
1	87	23.6	owner	0.9984	1	0.04	2.00
2	82.8	22.4	owner	0.9921	2	0.08	2.00
3	110.1	19.2	owner	0.9916	3	0.13	2.00
4	93	20.8	owner	0.9881	4	0.17	2.00
5	108	17.6	owner	0.9524	5	0.21	2.00
6	81	20	owner	0.9103	6	0.25	2.00
7	64.8	21.6	owner	0.8873	7	0.29	2.00
8	75	19.6	non-owner	0.7801	7	0.33	1.75
9	69	20	owner	0.7284	8	0.38	1.78
10	61.5	20.8	owner	0.7163	9	0.42	1.80
11	51	22	owner	0.7148	10	0.46	1.82
12	84	17.6	non-owner	0.5833	10	0.50	1.67
13	52.8	20.8	non-owner	0.4904	10	0.54	1.54
14	85.5	16.8	owner	0.4333	11	0.58	1.57
15	66	18.4	non-owner	0.2915	11	0.63	1.47
16	43.2	20.4	non-owner	0.1842	11	0.67	1.38
17	60	18.4	owner	0.1746	12	0.71	1.41
18	64.8	17.2	non-owner	0.1018	12	0.75	1.33
19	49.2	17.6	non-owner	0.0287	12	0.79	1.26
20	59.4	16	non-owner	0.0192	12	0.83	1.20
21	33	18.8	non-owner	0.0154	12	0.88	1.14
22	63	14.8	non-owner	0.0091	12	0.92	1.09
23	47.4	16.4	non-owner	0.0076	12	0.96	1.04
24	51	14	non-owner	0.0011	12	1.00	1.00

FIGURE 5.11 **Illustration of lift calculations for the riding mower example ("Prob[owner]" = propensities)**

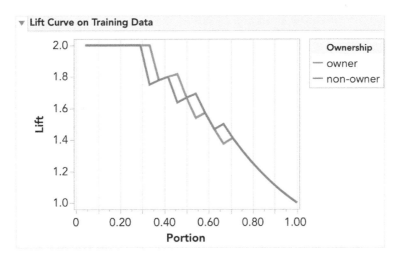

FIGURE 5.12   Lift curve for the riding mower example

Note: The columns Cum Owner, Portion, and Lift shown in Figure 5.11 are computed behind the scenes in JMP to create lift curves within the modeling platforms—you won't see these calculations (they were created here using the *Formula Editor* for illustration).

Interpreting the Lift Curve   What is considered good or bad performance? The ideal ranking performance would place all the owners (or 1's) at the beginning (the actual 1's would have the highest propensities and be at the top of the table) and all the non-owners (the 0's) at the end. In contrast, a useless model would be one that randomly assigns propensities [shuffling the 1's and 0's randomly in the Ownership (or actual class) column]. Such behavior would increase the cumulative number of 1's, on average, by $\frac{\#1's}{n}$ in each row.

If we did not have a model but simply selected cases at random, our lift will hover at around 1.0. As a result, a lift of 1.0 provides a benchmark against which we can evaluate the ranking performance of the model. A more ideal situation is a lift curve that starts with a very high lift for the 1's (in a small portion of the data), and then has a steep downward slope leveling out at a lift of 1.0. In the riding mower example, although our model is not perfect, it seems to perform much better than the random benchmark.

How do we read a lift curve? For a given portion of our data (*x*-axis), the lift value on the *y*-axis tells us how much better we are doing compared to random assignment. For example, looking at Figure 5.12, if we use our model to choose the top 20% of a new sample (portion = 0.20), we would catch twice as many owners than choosing a random sample of the same size. Similarly, if we use the model to choose 50% of a new sample (portion = 0.5), we would catch around 1.7 more owners compared to using a random 50%. (We can see in Figure 5.11 that for portion = 0.2 the lift is 2 and for portion = 0.5 the lift is 1.67.) The lift will vary with the portion of data we choose to act on. A good classifier will give us a high lift when we act on only a small portion of the data.

Beyond Two Classes

In JMP, lift curves can also be used with a multi-class classifier. A lift curve will be provided for each class. Lift curves are not available for prediction models with continuous responses.

Lift Curves Incorporating Costs and Benefits

When the benefits and costs of correct and incorrect classifications are known or can be estimated, the lift curve is still a useful presentation and decision tool (see Figure 5.13). As before, a classifier is needed that assigns to each record a propensity that it belongs to a particular class. The procedure is as follows:

1. Set the misclassification costs using the *Profit Matrix*, as shown in Figure 5.10.
2. Build your model.
3. Save the *Probability Formula* for the model to the data table. This will also calculate the *Profit* for the different classes and the *Expected Profit* for the outcome.
4. Sort the records in descending order of predicted probability of success, where *success* = belonging to the class of interest (right-click on the columns and select *Sort > Descending*).
5. Create a cumulative expected profit column. To do this, right-click on the *Expected Profit* column header, select *New Formula Column > Row > Cumulative Sum*.
6. Create a row number column. To do this, right-click on any column header, select *New Formula Column > Row > Row*.
7. Use the *Graph Builder* to manually create the lift curve, with the *Cumulative Expected Profit* column on the *y*-axis and *Row* on the *x*-axis. Use the Line graph icon (above the graph) to connect the points.
8. Use the *Line* tool on the toolbar to manually add a reference line connecting the first point (row = 1) to the last point (row = N, where *y* = the total net profit).

Note: It is entirely possible for the reference line that incorporates costs and benefits to have a negative slope if the net value for the entire dataset is negative. A hypothetical example is shown in Figure 5.13. If the cost of mailing to a person is \$0.65, the value of a responder is \$25, and the overall response rate is 2%, the expected net value of mailing to a list of 10,000 is $(0.02 \times \$25 \times 10{,}000) - (\$0.65 \times 10{,}000) = \$5000 - \$6500 = -\$1500$. Hence the *y*-value at the far right of the lift curve in Figure 5.13 ($x = 10{,}000$) is -1500, and the slope of the reference line from the origin will be negative. The optimal point will be where the lift curve is at a maximum (i.e., mailing to about 3000 people).

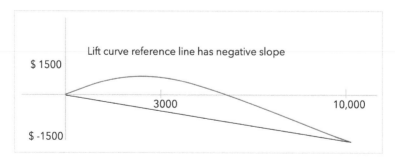

FIGURE 5.13   Hypothetical example, lift curve incorporating costs

5.5   OVERSAMPLING

As we saw briefly in Chapter 2, when classes are present in very unequal proportions, simple random sampling may produce too few of the rare class to yield useful information about what distinguishes them from the dominant class. In such cases stratified sampling is often used to oversample the cases from the more rare class and improve the performance of classifiers.

It is often the case that the more rare events are the more interesting or important ones: responders to a mailing, those who commit fraud, defaulters on debt, and the like. This same stratified sampling procedure is sometimes called *weighted sampling* or *undersampling*, the latter referring to the fact that the more plentiful class is undersampled, relative to the rare class. We will stick to the term *oversampling*.

> In all discussions of *oversampling*, we assume the common situation in which there are two classes, one of much greater interest than the other. Data with more than two classes do not lend themselves to this procedure.

Consider the data in Figure 5.14, where × represents nonresponders and ∘, responders. The two axes correspond to two predictors. The dashed vertical line does the best job of classification under the assumption of equal costs: it results in just one misclassification (one ∘ is misclassified as an ×).

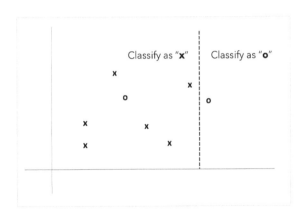

FIGURE 5.14   Classification assuming equal costs of misclassification

If we incorporate more realistic misclassification costs—let us say that failing to catch a ∘ is five times as costly as failing to catch an ×—the costs of misclassification jump to 5. In such a case, a horizontal line as shown in Figure 5.15 does a better job: it results in misclassification costs of just 2.

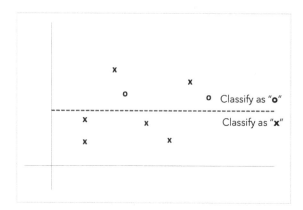

FIGURE 5.15 **Classification assuming unequal costs of misclassification**

Oversampling is one way of incorporating these costs into the training process. In Figure 5.16, we can see that classification algorithms would automatically determine the appropriate classification line if four additional o's were present at each existing o. We can achieve appropriate results either by taking five times as many o's as we would get from simple random sampling (by sampling with replacement if necessary) or by replicating the existing o's fourfold.

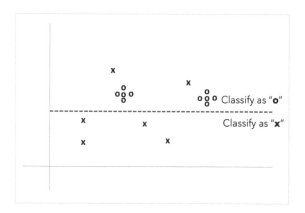

FIGURE 5.16 **Classification using oversampling to account for unequal costs**

Oversampling without replacement in accord with the ratio of costs (the first option above) is the optimal solution, but it may not always be practical. There may not be an adequate number of responders to assure that there will be enough of them to fit a model if they constitute only a small proportion of the total. Also it is often the case that our interest in discovering responders is known to be much greater than our interest in discovering nonresponders, but the exact ratio of costs is difficult to determine. When faced with very low response rates in a classification problem, practitioners often sample equal numbers of responders and nonresponders as a relatively effective and convenient approach. Whatever approach is used, when it comes time to assess and predict model performance, we will need to adjust for the oversampling in one of two ways:

1. Score the model to a validation set that has been selected without oversampling (i.e., via simple random sampling).
2. Score the model to an oversampled validation set, and reweight the results to remove the effects of oversampling.

The first method is more straightforward and easier to implement. We describe how to oversample and how to evaluate performance for each of the two methods.

When classifying data with very low response rates, practitioners typically:

- Train models on data that are 50% responder, 50% nonresponder.
- Validate the models with an unweighted (simple random) sample from the original data.

STRATIFIED SAMPLING AND OVERSAMPLING IN JMP

In JMP, stratified sampling can be accomplished using the *Random, Stratify* option from the *Tables > Subset* platform. Oversampling can be accomplished using the *Stratified Split Balanced* add-in, available from the JMP User Community (`community.jmp.com`, search for *Stratified Data Partitioning* in the *File Exchange*). This add-in allows you to oversample the training data or both the training and the validation data.

Creating an Over-sampled Training Set

How is weighted sampling done? One common procedure, where responders are sufficiently scarce that you will want to use all of them to train the model, follows:

Step 1. The response and nonresponse data are separated into two distinct sets, or *strata*.

Step 2. Records are randomly selected for the training set from each stratum. Typically, one might select half the (scarce) responders for the training set, then an equal number of nonresponders.

Step 3. Remaining responders are put in the validation set.

Step 4. Nonresponders are randomly selected for the validation set in sufficient numbers to maintain the original ratio of responders to nonresponders.

Step 5. If a test set is required, it can be taken randomly from the validation set.

Evaluating Model Performance Using a Nonoversampled Validation Set

Although the oversampled data can be used to train models, they are often not suitable for evaluating model performance, since the number of responders will (of course) be exaggerated. The most straightforward way of gaining an unbiased estimate of model performance is to apply the model to regular data (i.e., data not oversampled). To recap: train the model on oversampled data, but validate it with regular data.

Evaluating Model Performance If Only Oversampled Validation Set Exists

In some cases very low response rates may make it more practical to use oversampled data not only for the training data but also for the validation data. This might happen, for example, if an analyst is given a dataset for exploration and prototyping that is already oversampled to boost the proportion with the rare response of interest (perhaps because it is more convenient to transfer and work with a smaller dataset). In such cases it is still possible to assess how well the model will do with real data, but this requires the oversampled validation set to be reweighted, in order to restore the class of observations that were underrepresented in the sampling process. This adjustment should be made to the classification matrix and to the lift curve in order to derive good accuracy measures. These adjustments are described next.

Adjusting the Confusion Matrix for Oversampling Suppose the response rate in the data as a whole is 2% and that the data were oversampled, yielding a sample in which the response rate is 25 times higher (50% responders). The relationship is as follows:

Responders: 2% of the whole data; 50% of the sample

Nonresponders: 98% of the whole data, 50% of the sample

Each responder in the whole data is worth 25 responders in the sample (50/2). Each nonresponder in the whole data is worth 0.5102 nonresponders in the sample (50/98). We call these values *oversampling weights*.

Assume that the validation classification matrix looks like this:

Classification Matrix, Oversampled Data (Validation)

	Predicted 0	Predicted 1	Total
Actual 0	390	110	500
Actual 1	80	420	500
Total	470	530	1000

At this point, the misclassification rate appears to be $(80 + 110)/1000 = 19\%$, and the model ends up classifying 53% of the records as 1's. However, this reflects the performance on a sample where 50% are responders.

To estimate predictive performance when this model is used to score the original population (with 2% responders), we need to undo the effects of the oversampling. The actual number of responders must be divided by 25, and the actual number of nonresponders divided by 0.5102.

The revised classification matrix is as follows:

Classification Matrix, Reweighted

	Predicted 0	Predicted 1	Total
Actual 0	$390/0.5102 = 764.4$	$110/0.5102 = 215.6$	980
Actual 1	$80/25 = 3.2$	$420/25 = 16.8$	20
Total			1000

The adjusted misclassification rate is $(3.2 + 215.6)/1,000 = 21.9\%$. The model ends up classifying $(215.6 + 16.8)/1000 = 23.24\%$ of the records as 1's, when we assume 2% responders.

In JMP, the misclassification rate can be adjusted using a *Weight* column. This is a column in the data table created using the *Formula Editor* (see Figure 5.17). To create a classification matrix using these weights, use the *Fit Y by X* platform, with the actual class as the *Y, Response*, the most likely class as the *X, Factor*, and the weight column in the *Weight* field.

$$\text{If} \begin{pmatrix} \text{Actual Class} = 0 & \Rightarrow & \dfrac{1}{0.5102} \\ \\ \text{else} & \Rightarrow & \dfrac{1}{25} \end{pmatrix}$$

FIGURE 5.17   **Formula to adjust for oversampling**

Adjusting the Lift Curve for Oversampling   The lift curve is likely to be a more useful measure in low-response situations, where our interest lies not so much in classifying all the records correctly as in finding a model that guides us toward those records most likely to contain the response of interest (under the assumption that scarce resources preclude examining or contacting all the records). Typically, our interest in such a case is in maximizing value or minimizing cost.

The procedure presented earlier for creating a lift curve incorporating costs will be used, with one difference: a new column, *Adjusted Expected Profit*, will be used instead of the *Expected Profit* column. To create this column use the *Formula Editor*, and multiply the expected profit column by a column with the sampling weights (see Figure 5.17).

APPLYING SAMPLING WEIGHTS IN JMP

In JMP, sampling weights can be saved when creating a stratified random sample (using the *Tables > Subset* menu), or a column of sampling weights can be created using the *Formula Editor* (see Figure 5.17).

Sampling weights can be applied when fitting models by using the *Weight* field in the model launch dialogs. The resulting misclassification rates, ROC curves, lift curves, and model statistics will be adjusted based on these weights. Sampling weights can also be applied after the fact (after fitting a model and saving the prediction formula to the data table) to adjust the classification matrix for oversampling.

PROBLEMS

5.1 A machine learning process has been applied to a transaction dataset and has classified 88 records as fraudulent (30 correctly so) and 952 as nonfraudulent (920 correctly so). Construct the classification or confusion matrix and calculate the classification error (overall error) rate.

5.2 Suppose that this process has an adjustable threshold (cutoff) mechanism by which you can alter the proportion of records classified as fraudulent. Describe how moving the threshold up or down would affect the following:

a. The classification error rate for records that are truly fraudulent.

b. The classification error rate for records that are truly nonfraudulent.

5.3 Consider Figure 5.18 in which a lift curve for the transaction data model is applied to new data.

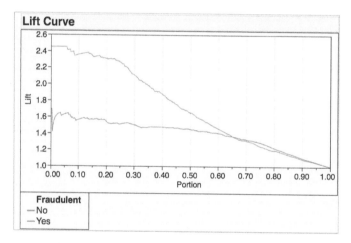

FIGURE 5.18   Lift curve for transaction data

5.4 FiscalNote is a startup founded by a Washington, DC entrepreneur and funded by a Singapore sovereign wealth fund, the Winklevoss twins of Facebook fame, and others. It uses machine learning techniques to predict for its clients whether legislation in the US Congress and in US state legislatures will pass or not. The company reports 94% accuracy (*Washington Post*, November 21, 2014, "Capital Business").

Considering just bills introduced in the US Congress, do a bit of Internet research to learn about numbers of bills introduced and passage rates. Identify the possible types of misclassifications, and comment on the use of overall accuracy as a metric. Include a discussion of other possible metrics and the potential role of propensities.

a. Interpret the meaning of the top curve (for Fraudulent = Yes), at portion = 0.2.

b. Explain how you might use this information in practice.

c. Another analyst comments that you could improve the accuracy of the model by classifying everything as nonfraudulent. If you do that, what is the error rate?

d. Comment on the usefulness, in this situation, of these two metrics of model performance (error rate and lift).

5.5 A large number of insurance records are to be examined to develop a model for predicting fraudulent claims. Of the claims in the historical database, 1% were judged to be fraudulent. A sample is taken to develop a model, and oversampling is used to provide a balanced sample in light of the very low response rate. When applied to this sample ($n = 800$), the model ends up correctly classifying 310 frauds and 270 nonfrauds. It missed 90 frauds and classified 130 records incorrectly as frauds when they were not.

a. Produce the classification matrix for the sample as it stands.

b. Find the adjusted misclassification rate (adjusting for the oversampling).

c. What percentage of new records would you expect to be classified as fraudulent?

5.6 The table below shows a small set of predictive model validation partition results for a classification model, with both actual values and propensities.

Propensity of 1	Actual
0.03	0
0.52	0
0.38	0
0.82	1
0.33	0
0.42	0
0.55	1
0.59	0
0.09	0
0.21	0
0.43	0
0.04	0
0.08	0
0.13	0
0.01	0
0.79	1
0.42	0
0.29	0
0.08	0
0.02	0

a. Calculate error rates, sensitivity, and specificity using threshold values of 0.25, 0.5, and 0.75.

b. Create a lift curve.

PART IV

PREDICTION AND CLASSIFICATION METHODS

6

MULTIPLE LINEAR REGRESSION

In this chapter, we introduce linear regression models for the purpose of prediction. We discuss the differences between fitting and using regression models for the purpose of inference (as in classical statistics) and for prediction. A predictive goal calls for evaluating model performance on a holdout (validation) set and for using predictive metrics. We then raise the challenges of using many predictors and describe variable selection algorithms that are often implemented in linear regression procedures. Finally, we introduce regularized regression techniques, which are particularly useful for datasets with a large number of highly-correlated predictors.

Multiple Linear Regression in JMP: Multiple linear regression models can be fit using the standard version of JMP. However, to compute validation statistics using a validation column or to fit regularized regression models, JMP Pro is required.

6.1 INTRODUCTION

The most popular model for making predictions is the *multiple linear regression model* encountered in most introductory statistics courses and textbooks. This model is used to fit a relationship between a numerical *outcome variable Y* (also called the *response, target, or dependent variable*) and a set of *predictors* $X_1, X_2, \ldots, X_p$ (also referred to as *independent variables, input variables, regressors, effects,* or *covariates*). The assumption is that the following function approximates the relationship between the predictors and outcome variable:

$$Y = \beta_0 + \beta_1 x_1 + \beta_2 x_2 + \cdots + \beta_p x_p + \epsilon, \tag{6.1}$$

where $\beta_0, \ldots, \beta_p$ are *coefficients* and ϵ is the *noise* or *unexplained* part. Data are then used to estimate the coefficients and to quantify the noise. In predictive modeling, the data are also used to evaluate model performance.

Machine Learning for Business Analytics: Concepts, Techniques, and Applications with JMP Pro®, Second Edition. Galit Shmueli, Peter C. Bruce, Mia L. Stephens, Muralidhara Anandamurthy, and Nitin R. Patel. © 2023 John Wiley & Sons, Inc. Published 2023 by John Wiley & Sons, Inc.

Regression modeling means not only estimating the coefficients but also choosing which input variables to include and in what form. For example, a numerical input can be included as-is, or in logarithmic form ($\log(X)$), or in a binned form (e.g., age group). Choosing the right form depends on domain knowledge, data availability, and needed predictive power.

Multiple linear regression is applicable to numerous predictive modeling situations. Examples are predicting customer activity on credit cards from their demographics and historical activity patterns, predicting the time to failure of equipment based on utilization and environmental conditions, predicting expenditures on vacation travel based on historical frequent flyer data, predicting staffing requirements at help desks based on historical data and product and sales information, predicting sales from cross-selling of products from historical information, and predicting the impact of discounts on sales in retail outlets.

6.2   EXPLANATORY VS. PREDICTIVE MODELING

Before introducing the use of linear regression for prediction, we must clarify an important distinction that often escapes those with earlier familiarity with linear regression from courses in statistics. In particular, there are the two popular but different objectives behind fitting a regression model:

1. Explaining or quantifying the average effect of inputs on an output (explanatory or descriptive task, respectively).
2. Predicting the outcome value for new records, given their input values (predictive task).

The classical statistical approach is focused on the first objective. In that scenario, the data are treated as a random sample from a larger population of interest. The regression model estimated from this sample is an attempt to capture the *average* relationship in the larger population. This model is then used in decision making to generate statements such as "a unit increase in service speed (X_1) is associated with an average increase of five points in customer satisfaction (Y), all other factors ($X_2, X_3, \ldots, X_p$) being equal." If X_1 is known to *cause* Y, then such a statement indicates actionable policy changes—this is called explanatory modeling. When the causal structure is unknown, then this model quantifies the degree of *association* between the inputs and output, and the approach is called descriptive modeling.

In predictive analytics, however, the focus is typically on the second goal: predicting new individual records. Here, we are not interested in the coefficients themselves, nor in the "average record," but rather in the predictions that this model can generate for new records. In this scenario, the model is used for micro decision making at the record level. In our previous example we would use the regression model to predict customer satisfaction for each new customer of interest.

Both explanatory and predictive modeling involve using a dataset to fit a model (i.e., to estimate coefficients), checking model validity, assessing its performance, and comparing to other models. However, the modeling steps and performance assessment differ in the two cases, usually leading to different final models. Therefore, the choice of model is closely tied to whether the goal is explanatory or predictive.

In explanatory and descriptive modeling, where the focus is on modeling the average record, we try to fit the best model to the data in an attempt to learn about the underlying relationship in the population. In contrast, in predictive modeling, the goal is to find a regression model that best predicts new individual records. A regression model that fits the existing data too well is not likely to perform well with new data. Hence, we look for a model that has the highest predictive power by evaluating it on a holdout set and using predictive metrics (see Chapter 5).

Let us summarize the main differences in using a linear regression in the two scenarios:

1. A good explanatory model is one that fits the data closely, whereas a good predictive model is one that predicts new cases accurately. Choices of input variables and their form can therefore differ.

2. In explanatory models, the entire dataset is used for estimating the best-fit model, to maximize the amount of information that we have about the hypothesized relationship in the population. When the goal is to predict outcomes of new individual cases, the data are typically split into a training set and a validation set. The training set is used to estimate the model, and the validation set[1] is used to assess this model's predictive performance on new, unobserved data.

3. Performance measures for explanatory models measure how close the data fit the model (how well the model approximates the data) and how strong the average relationship is, whereas in predictive models performance is measured by predictive accuracy (how well the model predicts new individual records).

4. In explanatory models the focus is on the coefficients (β), whereas in predictive models the focus is on the predictions ($\hat{Y}$).

For these reasons, it is extremely important to know the goal of the analysis before beginning the modeling process. A good predictive model can have a looser fit to the data on which it is based, and a good explanatory model can have low prediction accuracy. In the remainder of this chapter, we focus on predictive models, because these are more popular in machine learning and because most statistics textbooks focus on explanatory modeling.

6.3 ESTIMATING THE REGRESSION EQUATION AND PREDICTION

Once we determine the input variables to include and their form, we estimate the coefficients of the regression formula from the data using a method called *ordinary least squares* (OLS). This method finds values $\hat{\beta}_0, \hat{\beta}_1, \hat{\beta}_2, \ldots, \hat{\beta}_p$ that minimize the sum of squared deviations between the actual values (Y) and their predicted values based on that model ($\hat{Y}$).

To predict the value of the outcome variable for a record with predictor values $x_1, x_2, \ldots, x_p$, we use the equation

$$\hat{Y} = \hat{\beta}_0 + \hat{\beta}_1 x_1 + \hat{\beta}_2 x_2 + \cdots + \hat{\beta}_p x_p. \tag{6.2}$$

[1] When we are comparing different model options (e.g., different predictors) or multiple models, the data should be partitioned into three sets: training, validation, and test. The validation set is used for selecting the model with the best performance, while the test set is used to assess the performance of the "best model" on new, unobserved data before model deployment.

Predictions based on this equation are the best predictions possible in the sense that they will be unbiased (equal to the true values on average) and will have the smallest average squared error compared to any unbiased estimates *if* we make the following assumptions:

1. The noise ϵ (or equivalently, Y) follows a normal distribution.
2. The choice of variables and their form is correct (*linearity*).
3. The cases are independent of each other.
4. The variability in the outcome (Y) values for a given set of predictors is the same regardless of the values of the predictors (*homoskedasticity*).

An important and interesting fact for the predictive goal is that *even if we drop the first assumption and allow the noise to follow an arbitrary distribution, these estimates are very good for prediction*, in the sense that among all linear models, as defined by equation (6.1), the model using the least squares estimates, $\hat{\beta}_0, \hat{\beta}_1, \hat{\beta}_2, \ldots, \hat{\beta}_p$, will have the smallest average squared errors. The assumption of a normal distribution is required in explanatory modeling, where it is used for constructing confidence intervals and statistical tests for the model parameters.

Even if the other assumptions are violated, it is still possible that the resulting predictions are sufficiently accurate and precise for predictive purposes. The key is to evaluate predictive performance of the model, which is the main priority. Satisfying assumptions is of secondary interest and residual analysis can give clues to potential improved models to examine.

Example: Predicting the Price of Used Toyota Corolla Automobiles

A large Toyota car dealership offers purchasers of new Toyota cars the option to buy their used car as part of a trade-in. In particular, a new promotion promises to pay high prices for used Toyota Corolla cars for purchasers of a new car. The dealer then sells the used cars for a small profit. To ensure a reasonable profit, the dealer needs to be able to predict the price that the dealership will get for the used cars. For that reason, data were collected on all previous sales of used Toyota Corollas at the dealership. The data include the sales price and other information on the car, such as its age, mileage, fuel type, and engine size. A description of each of the key variables in the dataset is given in Table 6.1. Several other

TABLE 6.1 Variables in the Toyota Corolla Example

Variable	Description
Price	Offer price in euros
Age	Age in months as of August 2004
Kilometers	Accumulated kilometers on odometer
Fuel type	Fuel type (*Petrol, Diesel, CNG*)
Horse Power	Horsepower
Metallic	Metallic color? (Yes = 1, No = 0)
Automatic	Automatic (Yes = 1, No = 0)
CC	Cylinder volume in cubic centimeters
Doors	Number of doors
Quart tax	Quarterly road tax in euros
Weight	Weight in kilograms

	Price	Age	Mileage	Fuel Type	Horse Power	Metalic	Automatic	CC	Doors	QuartTax	Weight	Validation
1	13500	23	46986	Diesel	90	1	0	2000	3	210	1165	Training
2	13750	23	72937	Diesel	90	1	0	2000	3	210	1165	Validation
3	13950	24	41711	Diesel	90	1	0	2000	3	210	1165	Validation
4	14950	26	48000	Diesel	90	0	0	2000	3	210	1165	Training
5	13750	30	38500	Diesel	90	0	0	2000	3	210	1170	Training
6	12950	32	61000	Diesel	90	0	0	2000	3	210	1170	Training
7	16900	27	94612	Diesel	90	1	0	2000	3	210	1245	Validation
8	18600	30	75889	Diesel	90	1	0	2000	3	210	1245	Validation
9	21500	27	19700	Petrol	192	0	0	1800	3	100	1185	Training
10	12950	23	71138	Diesel	69	0	0	1900	3	185	1105	Training
11	20950	25	31461	Petrol	192	0	0	1800	3	100	1185	Validation
12	19950	22	43610	Petrol	192	0	0	1800	3	100	1185	Training
13	19600	25	32189	Petrol	192	0	0	1800	3	100	1185	Validation
14	21500	31	23000	Petrol	192	1	0	1800	3	100	1185	Validation
15	22500	32	34131	Petrol	192	1	0	1800	3	100	1185	Validation

FIGURE 6.1   **Prices and attributes for sample of 15 cars**

variables, which won't be used in this example, have been grouped, hidden, and excluded in JMP. A sample of this dataset is shown in Figure 6.1.

The total number of records in the dataset is 1000 cars (we use the first 1000 cars from the dataset ToyotaCorolla.jmp). These data are found in the file ToyotaCorolla1000.jmp. We create a validation column using the *Make Validation Column* option (from *Analyze > Predictive Modeling > Make Validation Column*), assigning 60% to training set and 40% to the validation set. This validation column, which is used to assign records to the training, validation, or (if used) test partition, has been saved in the data table (it is the last column in Figure 6.1).

To fit a multiple linear regression model in JMP Pro, we use *Analyze > Fit Model*, with Price as the *Y, outcome* variable, Validation in the *Validation* field, and the other variables as model effects. This fits a model using only the training data.

Figure 6.2 shows the parameter estimates (i.e., the coefficients) from the *Fit Least Squares* analysis window. The regression coefficients are used to predict prices of individual

▼ **Parameter Estimates**

Term	Estimate	Std Error	t Ratio	Prob>|t|
Intercept	-1412.934	1571.885	-0.90	0.3691
Age	-134.1376	4.774744	-28.09	<.0001*
Mileage	-0.019905	0.002369	-8.40	<.0001*
Fuel Type[CNG]	-933.3714	325.1658	-2.87	0.0042*
Fuel Type[Diesel]	-804.1305	265.9956	-3.02	0.0026*
Horse Power	33.955119	5.375333	6.32	<.0001*
Metalic[0]	19.024549	60.16051	0.32	0.7519
Automatic	224.9384	269.0697	0.84	0.4035
CC	0.0209207	0.095982	0.22	0.8275
Doors	-3.003272	61.79519	-0.05	0.9613
QuartTax	22.903512	2.485834	9.21	<.0001*
Weight	12.938552	1.512499	8.55	<.0001*

FIGURE 6.2   **Estimated coefficients for regression model of price vs. car variables**

used Toyota Corolla cars based on their age, mileage, and so on. Notice that the Fuel Type predictor has three categories (Petrol, Diesel, and CNG) and we therefore have two indicator variables in the model: Fuel Type[CNG] and Fuel Type[Diesel]. Behind the scenes, JMP has created indicator variables for Fuel Type and has included only the first two levels (alphabetically) in the model. The indicator for Petrol is a perfect linear combination of the Diesel and CNG, and as a result Petrol is completely redundant to the other two and is not included in the model.[2]

CODING OF CATEGORICAL PREDICTORS IN JMP'S REGRESSION

In multiple regression, categorical predictors are added directly to the model and JMP applies indicator coding for these variables. The user need not dummy code categorical predictors prior to building the model. Note that JMP uses $-1/1$ effect coding in *Fit Model* rather than the typical 0/1 indicator or dummy coding.

With this $-1/1$ coding, we interpret each coefficient of a categorical variable as the difference between the average outcome for that category and the overall average (holding everything else constant). For example, the coefficient for Fuel Type[CNG] means that the average price of cars that use CNG is $933.37 lower than the average car price. To display coefficients using the traditional 0/1 indicator coding, select *Estimates > Indicator Parameterization Estimates* from the top red triangle.

If you'd prefer to use dummy-coded indicator variables instead, use the *Utilities > Make Indicator Columns* option under the *Cols* menu to create these variables. The modeling types for the indicator variables should be changed to continuous, and when fitting the model, you'll need to leave one indicator variable out of the model.

Since we used a validation column, validation statistics (RSquare and RASE) are also provided in the Fit Least Squares window under the *Crossvalidation* outline (see Figure 6.3). As expected, RASE is lower for the training set than the validation set, although RSquare for the training set is slightly lower in this example. See Chapter 5 for details on validation statistics.

▼ Crossvalidation			
Source	RSquare	RASE	Freq
Training Set	0.8613	1349.9	600
Validation Set	0.8694	1410.3	400

FIGURE 6.3   **Performance of regression model on training and validation sets**

To show predicted prices for cars in the dataset, we save the prediction formula to the data table (use *Save Columns > Prediction Formula* from the red triangle). We also save residuals to the data table (using *Save Columns > Residuals*). Figure 6.4 shows predicted prices (in the *Pred Formula Price* column) for the first 15 cars, using the estimated model. To view the prediction formula, right-click on the column header and select *Formula* (see Figure 6.5).

---

[2]Inclusion of this redundant variable would cause typical regression software to fail due to a *multicollinearity* error (see Section 4.8). In JMP, when there are such linear dependencies among model effects, a *Singularity Details Report* is provided. This report (when shown) provides a summary of the redundant terms in the model.

	Price	Validation	Pred Formula Price	Residual Price
1	13500	Training	16715.41	-3215.41
2	13750	Validation	16198.84	-2448.84
3	13950	Validation	16686.27	-2736.27
4	14950	Training	16330.86	-1380.86
5	13750	Training	16048.11	-2298.11
6	12950	Training	15331.96	-2381.96
7	16900	Validation	16265.92	634.08
8	18600	Validation	16236.20	2363.80
9	21500	Training	20500.30	999.70
10	12950	Training	14208.65	-1258.65
11	20950	Validation	20534.47	415.53
12	19950	Training	20695.05	-745.05
13	19600	Validation	20519.98	-919.98
14	21500	Validation	19860.02	1639.98
15	22500	Validation	19504.31	2995.69

FIGURE 6.4   **Predicted prices and residuals (errors) for the first 15 cars in the dataset**

$$-1412.934338$$
$$+ \; -134.1376156 \cdot Age$$
$$+ \; -0.019905497 \cdot Mileage$$
$$+ \; Match\left(Fuel\ Type \right) \begin{pmatrix} "CNG" & \Rightarrow -933.3714011 \\ "Diesel" & \Rightarrow -804.1304845 \\ "Petrol" & \Rightarrow 1737.5018856 \\ else & \Rightarrow \; . \end{pmatrix}$$
$$+ \; 33.955118877 \cdot Horse\ Power$$
$$+ \; Match\left(Metalic \right) \begin{pmatrix} 0 & \Rightarrow 19.024548709 \\ 1 & \Rightarrow -19.02454871 \\ else & \Rightarrow \; . \end{pmatrix}$$
$$+ \; 224.93840017 \cdot Automatic$$
$$+ \; 0.0209207013 \cdot CC$$
$$+ \; -3.003272273 \cdot Doors$$
$$+ \; 22.903511562 \cdot QuartTax$$
$$+ \; 12.938552059 \cdot Weight$$

FIGURE 6.5   **Prediction formula for price, saved to the data table**

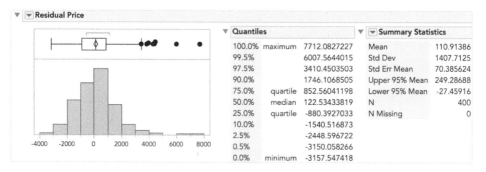

FIGURE 6.6   Histogram, boxplot, and summary statistics for model residuals (based on the validation set)

We create a histogram, boxplot, and summary statistics of the residuals for the validation set (see Figure 6.6) for additional measures of predictive accuracy using *Distribution*, with *Validation* in the *By* field. Note that the average error is $111. The boxplot and *Quantiles* boxplot of the residuals (Figure 6.6) shows that 50% of the errors are approximately between ±$850. This might be small relative to the car price but should be taken into account when considering the profit. Another observation of interest is the large positive residuals (underpredictions), which may or may not be a concern, depending on the application. Measures such as RASE (or RMSE) are used to assess the predictive performance of a model and to compare models. We discuss such measures in the next section.

This example also illustrates the point about the relaxation of the normality assumption. A histogram or probability plot (a *Normal Quantile Plot* in JMP) of prices shows a right-skewed distribution. In a descriptive/explanatory modeling case where the goal is to obtain a good fit to the data, the output variable would be transformed (e.g., by taking a logarithm) to achieve a more "normal" variable. Although the fit of such a model to the training data is expected to be better, it will not necessarily yield a significant predictive improvement.

ADDITIONAL OPTIONS FOR REGRESSION MODELS IN JMP

Many options for exploring results, diagnosing potential problems with the model, and saving results to the data table are available under the red triangles within the Fit Least Squares analysis window.

The regression model can be explored interactively using the *Prediction Profiler* (or, the *Profiler*). This graph is a red triangle option under *Factor Profiling*. To predict Price based on given values of the predictors, simply drag the vertical red lines for the predictors to the desired values. The predicted value for price, along with a confidence interval for the prediction, displays on the left of the profiler. The Profiler for the Price model is partially displayed in Figure 6.7.

The default output can be changed by using the JMP *Preferences*, found under the *File* menu. In Preferences, select *Platforms*, and then *Fit Least Squares* to view and change the default output for the platform.

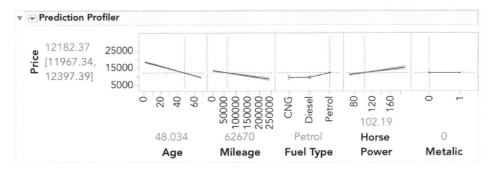

FIGURE 6.7   Prediction profiler for price model

6.4   VARIABLE SELECTION IN LINEAR REGRESSION

Reducing the Number of Predictors

A frequent problem in machine learning is that of using a regression equation to predict the value of a outcome variable when we have many variables available to choose as predictors in our model. Given the high speed of modern algorithms for multiple linear regression calculations, it is tempting in such a situation to take a kitchen-sink approach: why bother to select a subset? Just use all the variables in the model.

Another consideration favoring the inclusions of numerous variables is the hope that a previously hidden relationship will emerge. For example, a company found that customers who had purchased anti-scuff protectors for chair and table legs had lower credit risks. However, there are several reasons for exercising caution before throwing all possible predictors into a model.

- It may be expensive or not feasible to collect a full complement of predictors for future predictions.
- We may be able to measure fewer predictors more accurately (e.g., in surveys).
- The more predictors there are, the higher the chance of missing values in the data. If we delete or impute cases with missing values, multiple predictors will lead to a higher rate of case deletion or imputation.
- *Parsimony* is an important property of good models. We obtain more insight into the influence of predictors in models with few parameters.
- Estimates of regression coefficients are likely to be unstable, due to *multicollinearity* in models with many variables. (Multicollinearity is the presence of two or more predictors sharing the same linear relationship with the outcome variable.) Regression coefficients are more stable for parsimonious models. One very rough rule of thumb is to have a number of cases n larger than $5(p + 2)$, where p is the number of predictors.
- It can be shown that using predictors that are uncorrelated with the dependent variable increases the variance of predictions.
- It can be shown that dropping predictors that are actually correlated with the dependent variable can increase the average error (bias) of predictions.

The last two points mean that there is a trade-off between too few and too many predictors. In general, accepting some bias can reduce the variance in predictions. This *bias–variance trade-off* is particularly important for large numbers of predictors, since in that case it is very likely that there are variables in the model that have small coefficients relative to the standard deviation of the noise and also exhibit at least moderate correlation with other variables. Dropping such variables will improve the predictions, as it will reduce the prediction variance. This type of bias–variance trade-off is a basic aspect of most machine learning procedures for prediction and classification. In light of this, methods for reducing the number of predictors p to a smaller set are often used.

How to Reduce the Number of Predictors

The first step in trying to reduce the number of predictors should always be to use domain knowledge. It is important to understand what the various predictors are measuring and why they are relevant for predicting the outcome. With this knowledge, the set of predictors should be reduced to a sensible set that reflects the problem at hand. Some practical reasons for predictor elimination are expense of collecting this information in the future, inaccuracy, high correlation with another predictor, many missing values, or simply irrelevance. Also helpful in examining potential predictors are summary statistics and graphs, such as frequency and correlation tables, predictor-specific summary statistics and plots, and missing value counts.

One can either manually select the variables or make use of computational power and statistical performance metrics for reducing the predictors in a model.

Manual Variable Selection

We may want to manually explore the effect of removing or including a particular predictor or set of predictors. The *Effect Summary* table, which is provided at the top of the *Fit Least Squares* analysis window in JMP, provides an interactive summary of the terms in the model. The terms are sorted from the most significant to the least (as shown in Figure 6.8). This table allows you to remove terms from the model or add new terms, including interaction or quadratic effects. As you change terms in the model, all of the statistical output, including the parameter estimates and validation statistics, automatically updates. This is

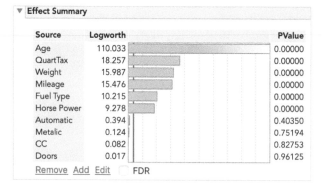

FIGURE 6.8   Effect summary table for price

useful to compare validation statistics for different models directly in the *Fit Least Squares* analysis window.

Automated Variable Selection

A more formal approach to reducing the number of predictors in the model is to use a built-in algorithm for model reduction. We'll discuss three general methods in this chapter. The first is an exhaustive search for the "best" subset of predictors by fitting regression models with all the possible combinations of predictors. In JMP, this is called *All Possible Models*. The second is to search through a partial set of models using algorithms that consider variables one at a time. These approaches are housed in the *Stepwise* platform, which is available from the *Analyze > Fit Model* dialog window (under *Personality*). The initial Stepwise Fit window for the Toyota Corolla data is shown in Figure 6.9. The third approach, called *regularization*, imposes a penalty on model complexity, resulting in shrinking coefficients for some predictors.

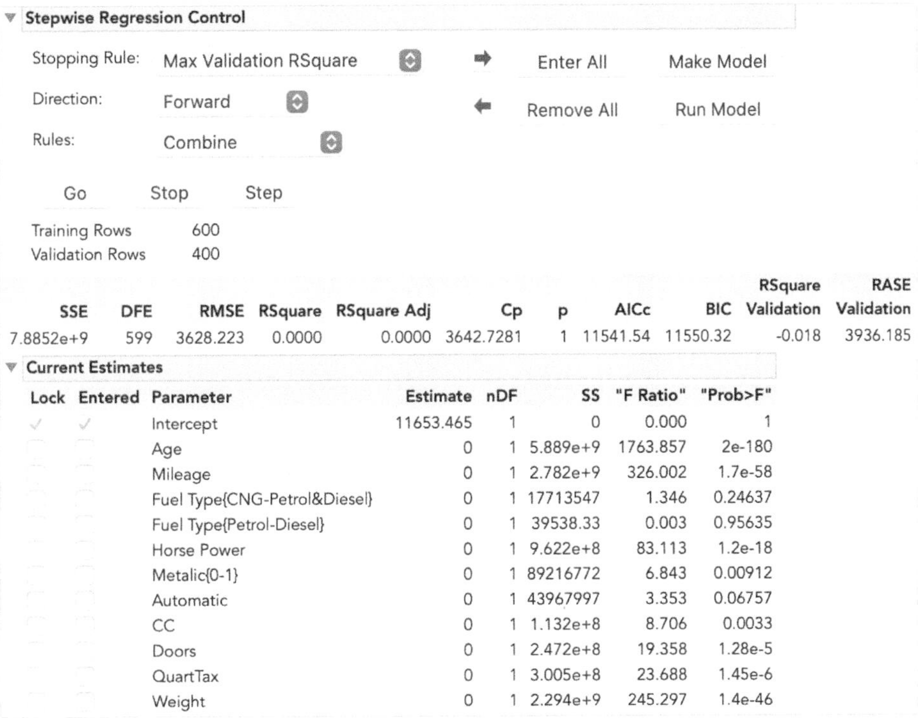

▼ Stepwise Regression Control

Stopping Rule:	Max Validation RSquare	🔼	➡	Enter All Make Model
Direction:	Forward 🔼		⬅	Remove All Run Model
Rules:	Combine 🔼			

Go	Stop	Step

Training Rows   600
Validation Rows   400

SSE	DFE	RMSE	RSquare	RSquare Adj	Cp	p	AICc	BIC	RSquare Validation	RASE Validation
7.8852e+9	599	3628.223	0.0000	0.0000	3642.7281	1	11541.54	11550.32	-0.018	3936.185

▼ Current Estimates

Lock	Entered	Parameter	Estimate	nDF	SS	"F Ratio"	"Prob>F"
✓	✓	Intercept	11653.465	1	0	0.000	1
		Age		0	1	5.889e+9	1763.857	2e-180
		Mileage		0	1	2.782e+9	326.002	1.7e-58
		Fuel Type{CNG-Petrol&Diesel}		0	1	17713547	1.346	0.24637
		Fuel Type{Petrol-Diesel}		0	1	39538.33	0.003	0.95635
		Horse Power		0	1	9.622e+8	83.113	1.2e-18
		Metalic{0-1}		0	1	89216772	6.843	0.00912
		Automatic		0	1	43967997	3.353	0.06757
		CC		0	1	1.132e+8	8.706	0.0033
		Doors		0	1	2.472e+8	19.358	1.28e-5
		QuartTax		0	1	3.005e+8	23.688	1.45e-6
		Weight		0	1	2.294e+9	245.297	1.4e-46

FIGURE 6.9   **Selection of the Stepwise personality from the Fit Model dialog window**

In any case, using computational variable selection methods involves comparing many models and choosing the best one. In such cases, it is advisable to have a *test set* in addition to the training and validation sets. The validation set is used to compare the models and select the best one. The test set is then used to evaluate the predictive performance of this selected model.

CODING OF CATEGORICAL VARIABLES IN STEPWISE REGRESSION

In the *Stepwise* platform, JMP applies a different type of coding for categorical variables. In this platform, $k - 1$ new indicator variables are created for each k-level categorical variable. When the outcome is continuous, these new variables represent hierarchical groups of levels, where the levels are grouped in way that maximizes the difference in the average outcome. This difference is measured in terms of the sum of squares between groups (SSB). In the Toyota Corolla example, Fuel Type has three levels. JMP has created two indicator variables (see Figure 6.9). The first variable, Fuel Type{CNG-Petrol&Diesel}, is an indicator variable for CNG vs. Petrol and Diesel. The second variable, Fuel Type{Petrol-Diesel}, is an indicator variable for Petrol vs. Diesel.

Exhaustive Search (All Possible Models)   The idea here is to evaluate all subsets. Since the number of subsets for even moderate values of p is very large, after the algorithm creates the subsets and runs all the models, we need some way to examine the most promising subsets and to select from them. The challenge is to select a model that is not too simplistic in terms of excluding important parameters (the model is *underfit*), nor overly complex thereby modeling random noise (the model is *overfit*).

Several criteria for evaluating and comparing models are based on metrics computed from the training data.

One popular criterion is the *adjusted* R^2, which is defined as

$$R_{adj}^2 = 1 - \frac{n-1}{n-p-1}(1 - R^2),$$

where R^2 is the proportion of explained variability in the model (in a model with a single predictor, this is the squared correlation). Like R^2, higher values of adjusted R^2 indicate better fit. Unlike R^2, which does not account for the number of predictors used, adjusted R^2 uses a penalty on the number of predictors. This avoids the artificial increase in R^2 that can result from simply increasing the number of predictors but not the amount of information. It can be shown that using R_{adj}^2 to choose a subset is equivalent to picking the subset that minimizes training MSE (or $\hat{\sigma}^2$).

A second popular set of criteria for balancing underfitting and overfitting are the *Akaike Information Criterion (AIC)* and the *Bayesian Information Criterion (BIC)*.[3] AIC and BIC measure the goodness of fit of a model but also include a penalty that is a function of the number of parameters in the model. As such, they can be used to compare various models for the same dataset. AIC and BIC are estimates of prediction error based on information theory. For linear regression, AIC and BIC can be computed from the formulas:

$$\text{AIC} = n \ln(\text{SSE}/n) + n(1 + \ln(2\pi)) + 2(p + 1), \tag{6.3}$$

$$\text{BIC} = n \ln(\text{SSE}/n) + n(1 + \ln(2\pi)) + \ln(n)(p + 1), \tag{6.4}$$

---

[3] AIC and BIC are especially effective with large datasets, when p-values are not as useful. In Stepwise, JMP reports AIC_c, the bias-corrected version of AIC, which is good for modeling with small datasets.

where SSE is the model's sum of squared errors. In general, models with smaller AIC and BIC values are considered better.

A third criterion often used for subset selection is *Mallow's C_p* (see formula below). This criterion assumes that the full model (with all predictors) is unbiased, although it may have predictors that if dropped would reduce prediction variability. With this assumption we can show that if a subset model is unbiased, the average C_p value equals the number of parameters $p + 1$ (= number of predictors + 1), the size of the subset. So a reasonable approach to identifying subset models with small bias is to examine those with values of C_p that are near $p + 1$. C_p is also an estimate of the error[4] for predictions at the x-values observed in the training set. Thus good models are those that have values of C_p near $p + 1$ and that have small p (i.e., are of small size). C_p is computed from the formula

$$C_p = \frac{SSE}{\hat{\sigma}^2_{\text{full}}} + 2(p + 1) - n, \tag{6.5}$$

where $\hat{\sigma}^2_{\text{full}}$ is the estimated value of σ^2 (MSE) in the full model that includes all predictors. It is important to remember that the usefulness of this approach depends heavily on the reliability of the estimate of σ^2 for the full model. This requires that the training set contain a large number of records relative to the number of predictors.

Note: It can be shown that for linear regression, in large samples Mallow's C_p is equivalent to AIC.

Finally, a useful point to note is that for a fixed size of subset, R^2, R^2_{adj}, C_p, and AIC_c all select the same subset. In fact, there is no difference between them in the order of merit they ascribe to subsets of a fixed size. This is good to know if comparing models with the same number of predictors, but often we want to compare models with different numbers of predictors.

The *All Possible Models* option is available under the red triangle in the Stepwise platform. The resulting window allows you to specify the maximum number of terms in the model and the best number of models of each size to display. For the Toyota Corolla price data (with the 11 predictors), we ask for the best two models for a single predictor, two predictors, and so on up to 11 predictors. The results of this search are shown in Figure 6.10. The C_p indicates a model with 7–11 predictors is good. It can be seen that both RMSE and AIC_c decrease until around seven predictors and then stabilize. Both RMSE and AIC_c point to a seven predictor model. The dominant predictor in all models is the age of the car, with horsepower and mileage (kilometers) playing important roles as well.

Note that selecting a model with the best validation performance runs the risk of overfitting, in the sense that we choose the best model that fits the validation set. Therefore, consider more than just the single top performing model, and among the good performers, favor models that have less predictors. Finally, remember to evaluate the performance of the selected model on the test set.

---

[4]In particular, it is the sum of the squared errors standardized by dividing by σ^2.

All Possible Models

Ordered up to best 2 models up to 11 terms per model.

Model	Number	RSquare	RMSE	AICc	BIC	Cp	Validation RSquare
Age	1	0.7468	1827.18	10719.4	10732.5	477.7124	0.7579
Mileage	1	0.3528	2921.26	11282.5	11295.6	2148.5336	0.3473
Age,Horse Power	2	0.7879	1673.79	10615.2	10632.7	305.5056	0.8057
Age,Weight	2	0.7742	1726.81	10652.6	10670.1	363.3937	0.8107
Age,Horse Power,Weight	3	0.8176	1553.43	10526.7	10548.6	181.4955	0.8573
Age,Mileage,Weight	3	0.8162	1559.52	10531.4	10553.3	187.5723	0.8538
Age,Mileage,Horse Power,Weight	4	0.8405	1454.09	10448.4	10474.7	86.5939	0.8774
Age,Mileage,Horse Power,QuartTax	4	0.8353	1477.39	10467.5	10493.7	108.4461	0.8434
Age,Mileage,Fuel Type{Petrol-Diesel},QuartTax,Weight	5	0.8509	1406.81	10409.8	10440.4	44.2411	0.8522
Age,Mileage,Horse Power,QuartTax,Weight	5	0.8492	1414.78	10416.6	10447.1	51.4216	0.8714
Age,Mileage,Fuel Type{Petrol-Diesel},Horse Power,QuartTax,Weight	6	0.8592	1368.31	10377.5	10412.5	11.1074	0.8665
Age,Mileage,Fuel Type{CNG-Petrol&Diesel},Fuel Type{Petrol-Diesel},QuartTax,Weight	6	0.8516	1404.97	10409.3	10444.2	43.5229	0.8532
Age,Mileage,Fuel Type{CNG-Petrol&Diesel},Fuel Type{Petrol-Diesel},Horse Power,QuartTax,Weight	7	0.8611	1360.07	10371.3	10410.6	4.9381	0.8693
Age,Mileage,Fuel Type{Petrol-Diesel},Horse Power,Automatic,QuartTax,Weight	7	0.8593	1368.89	10379.1	10418.4	12.6023	0.8667
Age,Mileage,Fuel Type{CNG-Petrol&Diesel},Fuel Type{Petrol-Diesel},Horse Power,Automatic,QuartTax,Weight	8	0.8613	1360.31	10372.6	10416.2	6.1495	0.8696
Age,Mileage,Fuel Type{CNG-Petrol&Diesel},Fuel Type{Petrol-Diesel},Horse Power,CC,QuartTax,Weight	8	0.8612	1361.07	10373.3	10416.9	6.8044	0.8692
Age,Mileage,Fuel Type{CNG-Petrol&Diesel},Fuel Type{Petrol-Diesel},Horse Power,Metalic{0-1},Automatic,QuartTax,Weight	9	0.8613	1361.34	10374.6	10422.5	8.0492	0.8694
Age,Mileage,Fuel Type{CNG-Petrol&Diesel},Fuel Type{Petrol-Diesel},Horse Power,Automatic,CC,QuartTax,Weight	9	0.8613	1361.41	10374.6	10422.5	8.1066	0.8695
Age,Mileage,Fuel Type{CNG-Petrol&Diesel},Fuel Type{Petrol-Diesel},Horse Power,Metalic{0-1},Automatic,CC,QuartTax,Weight	10	0.8613	1362.45	10376.6	10428.8	10.0024	0.8693
Age,Mileage,Fuel Type{CNG-Petrol&Diesel},Fuel Type{Petrol-Diesel},Horse Power,Metalic{0-1},Automatic,Doors,QuartTax,Weight	10	0.8613	1362.50	10376.6	10428.9	10.0475	0.8694
Age,Mileage,Fuel Type{CNG-Petrol&Diesel},Fuel Type{Petrol-Diesel},Horse Power,Metalic{0-1},Automatic,CC,Doors,QuartTax,Weight	11	0.8613	1363.60	10378.7	10435.2	12.0000	0.8694

FIGURE 6.10 Exhaustive search (all possible models) result for reducing predictors in Toyota Corolla price example

---

WORKING WITH THE ALL POSSIBLE MODELS OUTPUT

By default, the best model of each size will be selected (highlighted) and several statistics will be displayed for each model, as shown in Figure 6.10. To display Mallow's C_p, right-click on the table and select *Columns > Cp*. You can sort on any statistic in the output by right-clicking on the table and selecting *Sort by Column*. This is particularly useful when working with a large number of models. To select a model and view model fit statistics and estimates, use the radio button at the right of the table. R^2_{adj}, C_p, AIC_c, and a number of other fit statistics are reported for the selected model.

Popular Subset Selection Algorithms   The second method of finding the best model relies on an iterative search through the possible predictors. The end product is one best subset of predictors. This approach is computationally cheaper than the exhaustive search, but it has the potential of missing "good" combinations of predictors. In JMP Pro, different stopping rules are available to assist in finding the best model. These include *P-value Threshold*, *Max Validation RSquare*, *Minimum AIC$_c$*, and *Minimum BIC*. In this section, we limit our discussion to the first two stopping rules.[5]

Three popular iterative search algorithms are *Forward selection, Backward elimination*, and *Stepwise* (*mixed*) regression. In *Forward selection*, we start with no predictors and then add predictors one by one. Each predictor added is the one that is the most significant of the predictors not already in the model. The algorithm stops when the stopping rule is satisfied. When a validation column has been selected (in JMP Pro), the default stopping rule is *Max Validation RSquare*. When forward selection is used, and the model is built on the training data, the algorithm will select the model with the maximum RSquare for the validation set. The *Step History* report shows the order in which terms were added, the fit statistics, and the model that was selected as the best according to the stopping rule.

The main disadvantage of forward selection is that the algorithm will miss pairs or groups of predictors that perform very well together but perform poorly as single predictors. This is similar to interviewing job candidates for a team project one by one, thereby missing groups of candidates who may perform superiorly together ("colleagues"), but poorly on their own or with non-colleagues.

In *Backward elimination*, we start with all predictors in the model and then at each step eliminate the least useful predictor (according to statistical significance). The algorithm stops when the stopping rule has been satisfied.

Mixed stepwise regression is like *Forward selection* except that at each step we consider dropping predictors that are not statistically significant, as we do in *Backward elimination*. So, with *Mixed stepwise*, we both add and remove terms from the model. The stopping rule used with *Mixed stepwise* is *P-value Threshold*, and terms are added and then removed according to the p-values specified under *Prob to Enter* and *Prob to Leave*.

---

[5]For information on *Minimum AIC$_c$* and *Minimum BIC*, refer to the book *Fitting Linear Models* in the JMP Help (under JMP Documentation Library).

WHEN USING A STOPPING ALGORITHM IN JMP

- Select the *Stopping Rule*.
- Select the *Direction*. If using *Backward*, click the *Enter All* button to enter all terms into the model. Note that *Mixed* is only available with the *P-value Threshold* stopping rule.
- If your model includes categorical variables with more than two levels, JMP will recode the variables into groups that drive the greatest separation in the outcome.
- Use *Step* to step through one variable at a time, or *Go* to run the algorithm until the stopping rule is satisfied.
- Once you've selected a model, use the *Run Model* button to launch the *Fit Least Squares analysis* window.

Note: When using Stepwise with categorical predictors with more than two levels, change the *Rules* from *Combine* to *Whole Effects*. This adds the *whole effect* for the predictor to the model and makes model estimates easier to interpret.

For the Toyota Corolla price example, forward selection using the default *Max Validation RSquare* stopping rule yields a five parameter model (four terms plus the intercept). This model retains only Age, Mileage, Horsepower, and Weight (see Figure 6.11). In comparison, *Backward elimination* starts with the full model and then drops predictors one by one. Running this algorithm, again with the default *Max Validation RSquare* rule, yields the same model as the best eight predictor model from All Possible Models (as shown in Figure 6.12). For this model, Metallic, CC, and Doors are dropped from the model.

SSE	DFE	RMSE	RSquare	RSquare Adj	Cp	p	AICc	BIC	RSquare Validation	RASE Validation
1.2581e+9	595	1454.0948	0.8405	0.8394	86.59387	5	10448.41	10474.65	0.8774	1366.245

▼ **Current Estimates**

Lock	Entered	Parameter	Estimate	nDF	SS	"F Ratio"	"Prob>F"
✓	✓	Intercept	-682.78425	1	0	0.000	1
☐	☑	Age	-135.30498	1	1.54e+9	728.464	2e-105
☐	☑	Mileage	-0.0208229	1	1.802e+8	85.216	4.6e-19
☐	☐	Fuel Type{CNG-Petrol&Diesel}	0	1	2997672	1.419	0.23409
☐	☐	Fuel Type{Petrol-Diesel}	0	1	1524360	0.721	0.39629
☐	☑	Horse Power	41.1579977	1	1.915e+8	90.560	4.4e-20
☐	☐	Metalic{0-1}	0	1	689586.8	0.326	0.56838
☐	☐	Automatic	0	1	1238284	0.585	0.44457
☐	☐	CC	0	1	369366	0.174	0.67634
☐	☐	Doors	0	1	323881.4	0.153	0.69586
☐	☐	QuartTax	0	1	69118432	34.532	6.99e-9
☐	☑	Weight	14.9151065	1	3.448e+8	163.066	3.6e-33

FIGURE 6.11 Forward selection results in Toyota Corolla price example, using the *Max Validation RSquare* stopping rule

	SSE	DFE	RMSE	RSquare	RSquare Adj	Cp	p	AICc	BIC	RSquare Validation	RASE Validation
	1.0936e+9	591	1360.3081	0.8613	0.8594	6.149462	9	10372.59	10416.19	0.8696	1408.942

▼ Current Estimates

Lock	Entered	Parameter	Estimate	nDF	SS	"F Ratio"	"Prob>F"
✓	✓	Intercept	-1648.0331	1	0	0.000	1
	☑	Age	-134.0259	1	1.476e+9	797.639	1e-111
	☑	Mileage	-0.0198961	1	1.32e+8	71.333	2.3e-16
	☑	Fuel Type{CNG-Petrol&Diesel}	-706.77093	2	92715153	25.052	3.6e-11
	☑	Fuel Type{Petrol-Diesel}	1265.20038	1	74463635	40.241	4.5e-10
	☑	Horse Power	34.052001	1	75996454	41.069	3e-10
		Metalic{0-1}	0	1	186375.9	0.101	0.75126
	☑	Automatic	234.715517	1	1466437	0.792	0.37371
		CC	0	1	79726.28	0.043	0.83577
		Doors	0	1	10099.67	0.005	0.94118
	☑	QuartTax	22.933752	1	1.588e+8	85.823	3.6e-19
	☑	Weight	12.9338836	1	1.443e+8	77.960	1.2e-17

FIGURE 6.12   Backward elimination results using the *Max Validation RSquare* stopping rule

Changing the stopping rule can also produce different results. In Figure 6.13, we see the results of *Backward elimination* using the *P-value Threshold* stopping rule. The p-value to remove a term (*Prob to Leave*) is set at 0.15. With this stopping rule, terms are removed from the model one at a time until only terms with p-values less than the threshold (0.15 here) remain. *Forward selection* with the *P-value Threshold* works in much the same manner. Terms are added to the model one at a time based on the p-value. Only terms below the threshold *Prob to Enter* are entered.

	SSE	DFE	RMSE	RSquare	RSquare Adj	Cp	p	AICc	BIC	RSquare Validation	RASE Validation
	1.0951e+9	592	1360.0696	0.8611	0.8595	4.9381209	8	10371.33	10410.59	0.8693	1410.57

▼ Current Estimates

Lock	Entered	Parameter	Estimate	nDF	SS	"F Ratio"	"Prob>F"
✓	✓	Intercept	-1834.7531	1	0	0.000	1
	☑	Age	-133.7316	1	1.477e+9	798.292	8e-112
	☑	Mileage	-0.0199964	1	1.336e+8	72.245	1.5e-16
	☑	Fuel Type{CNG-Petrol&Diesel}	-693.5013	2	93869232	25.373	2.7e-11
	☑	Fuel Type{Petrol-Diesel}	1278.61028	1	76486799	41.349	2.6e-10
	☑	Horse Power	33.9189915	1	75463588	40.796	3.4e-10
		Metalic{0-1}	0	1	169580.7	0.092	0.76234
		Automatic	0	1	1466437	0.792	0.37371
		CC	0	1	248577.5	0.134	0.71426
		Doors	0	1	20437.61	0.011	0.91639
	☑	QuartTax	22.9071811	1	1.585e+8	85.667	3.9e-19
	☑	Weight	13.1274675	1	1.52e+8	82.149	1.8e-18

FIGURE 6.13   *Backward elimination* results using the *P-value Threshold* stopping rule

In this case, *Backward elimination*, *Forward selection*, and *Mixed stepwise* with the *P-value Threshold* (0.15 for both entering an leaving) all yield the same model, removing Metallic, CC, and Doors, and also removing Automatic. The R^2_{adj}, AIC_c, and other measures indicate that this is exactly the same subset as the best seven term model suggested by All Possible Models. In other words, these algorithms correctly identify Doors, CC, Metallic, and Automatic as the least useful predictors.

Comparing the reduced models with the full model, the model with the lowest RMSE on the training data is the full model, while the model with the highest R^2_{adj} is the eight parameter model. The five parameter model (from *Forward selection*, with *Max Validation RSquare*) has both the highest RSquare and the lowest RMSE of the models explored. However, this model has the largest C_p and AIC_c of the reduced models, indicating the the model is underfit. This example shows that the search algorithms yield fairly good solutions, but we need to carefully determine the number of predictors to retain. It also shows the merits of running a few searches and using the combined results to determine the subset to choose. There is a popular (but false) notion that *Mixed stepwise* regression is superior to *Backward elimination* and *Forward selection* because of its ability to add and to drop predictors. This example shows clearly that it is not always so.

Regularization (Shrinkage Models)

Selecting a subset of predictors is equivalent to setting some of the model coefficients to zero. This approach creates an interpretable results—we know which predictors were dropped and which were retained. A more flexible alternative, called *regularization* or *shrinkage*, "shrinks" the coefficients toward zero. Recall that adjusted-R^2 incorporates a penalty according to the number of predictors p. Shrinkage methods also impose a penalty on the model fit, except that the penalty is not based on the *number* of predictors but rather on some of the aggregation of the coefficient values (typically predictors are first standardized to have the same scale).

The reasoning behind constraining the magnitude of the ($\hat{\beta}$) coefficients is that highly correlated predictors will tend to exhibit coefficients with high standard errors, since small changes in the training data might radically shift which of the correlated predictors gets emphasized. This instability (high standard errors) leads to poor predictive power. By constraining the combined magnitude of the coefficients, this variance is reduced because the variance is a function of the coefficients' magnitudes.

The two most popular shrinkage methods are *ridge regression* and *lasso*. They differ in terms of the penalty used: in ridge regression, the penalty is based on the sum of squared coefficients $\sum_{j=1}^{p} \beta_j^2$ (called L2 penalty or L2 regularization), whereas lasso uses the sum of absolute values $\sum_{j=1}^{p} |\beta_j|$ (called L1 penalty or L1 regularization), for p predictors (excluding an intercept). It turns out that the lasso penalty effectively shrinks some of the coefficients to zero, thereby resulting in a subset of predictors.

Whereas in linear regression coefficients are estimated by minimizing the training data sum of squared errors (SSE), in ridge regression and lasso the coefficients are estimated by minimizing the training data SSE,

$$SSE = \sum_{i=1}^{N} \left(y_i - b_0 - \sum_{j=1}^{p} \beta_j x_{ij} \right),$$

subject to the penalty term.

Lasso: L1 regularization
$$SSE + \lambda \sum_{j=1}^{p} |\beta_j|,$$

Ridge regression: L2 regularization
$$SSE + \lambda \sum_{j=1}^{p} \beta_j^2,$$

The λ parameter controls the amount or strength of regularization applied to the model and can have values ≥ 0. When $\lambda = 0$, no regularization is applied, yielding ordinary linear regression.

The L1 and L2 regularization can be combined in an approach called *elastic net*.

Elastic net: L1 and L2 regularization
$$SSE + \lambda \left[\alpha \sum_{j=1}^{p} |\beta_j| + \frac{1}{2}(1 - \alpha) \sum_{j=1}^{p} \beta_j^2 \right].$$

The parameter α controls the penalty distribution between the L1 and L2 penalties and can have a value between 0 and 1. *Ridge regression* model is obtained with $\alpha = 0$ (only L2 penalty), while a *lasso* model is obtained with $\alpha = 1$ (only L1 penalty). Choosing $0 < \alpha < 1$ produces a model that is a combination of L1 and L2 penalties. As above, the λ parameter controls the amount or strength of regularization applied to the model.

The process of fitting a regularized (or *Generalized*) regression model in JMP Pro is similar to fitting a *Linear Regression* model, with the exception of selecting *Generalized Regression* from the *Personality* option from the Model Specification dialog (see Figure 6.14). The search for the optimum λ penalty parameter value is done automatically while building lasso, ridge, and elastic net regression models.

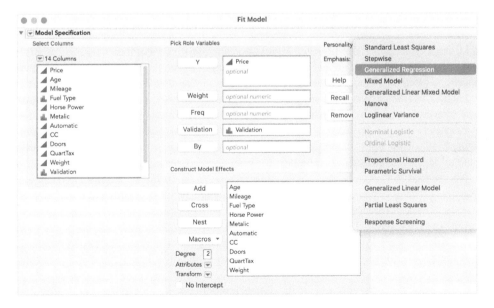

FIGURE 6.14 *Generalized Regression* option (*personality*) in the *Fit Model* specification dialog

---

FITTING GENERALIZED REGRESSION MODELS IN JMP Pro

The Generalized Regression platform in JMP Pro can be used to fit linear and regularized regression models for both continuous and categorical outcomes.

- Use *Analyze > Fit Model*, and specify your model.
- Under *Personality* in the Fit Model Specification window, select *Generalized Regression*.
- Select the outcome (response) distribution. The default is Normal, but you can fit Generalized Regression models for a number of common continuous and categorical distributions.
- Click *Run*. When the outcome distribution is Normal, JMP Pro will fit a Least Squares regression model.
- Select from the list of available *Estimation Methods*. The default is Lasso, but other options include Elastic Net, Ridge, and additional variable selection methods introduced earlier in the chapter. Click *Go* to fit the Lasso model and search for the optimum λ penalty value.

The Generalized Regression report (see Figure 6.15) includes the model summary (showing the optimum value of the λ penalty), parameter estimates for the model, and an interactive solution path for dynamically exploring the model for different values of λ. As you drag the slider in the solution path, JMP Pro updates the parameter estimates and model summary statistics, enabling you to see how the model coefficients and performance change as you change the value of the λ penalty.

An L1 penalty is applied to the coefficients when fitting a *lasso* regression model, and an L2 penalty is applied during *ridge* regression. The *elastic net* combines both L1 and L2 penalties. Figures 6.15, 6.16, and 6.17 show the output of the lasso, ridge, and elastic net models, respectively. Here, we have selected the option *Parameters Estimates for Centered and Scaled Predictors* from the red triangle menu to better compare the magnitudes of the parameter estimates, which are shown graphically in the *Solution Path*.

Because we have fit multiple models, a *Model Comparison* report is provided (see Figure 6.18). This report displays the model statistics for the lasso, ridge, elastic net, and standard least squares estimation models for easy comparison.

We see that in this case, the lasso performs best based on the *Validation Generalized RSquare* (higher is better) and *BIC* (lower is better), while the least squares and ridge models have the best performance based on *AICc* (lower is better). Looking at the coefficients (see Figure 6.15), we see that the lasso approach lead to a reduced model with six predictors (Age, Mileage, Fuel Type, Horse Power, Weight, and Quarterly Tax). The real strength of these methods becomes more evident when the dataset contains a large number of predictors with high correlation.

Finally, additional ways to reduce the dimension of the data, prior to building regression models, are by using principal components (Chapter 4) and regression trees (Chapter 9).

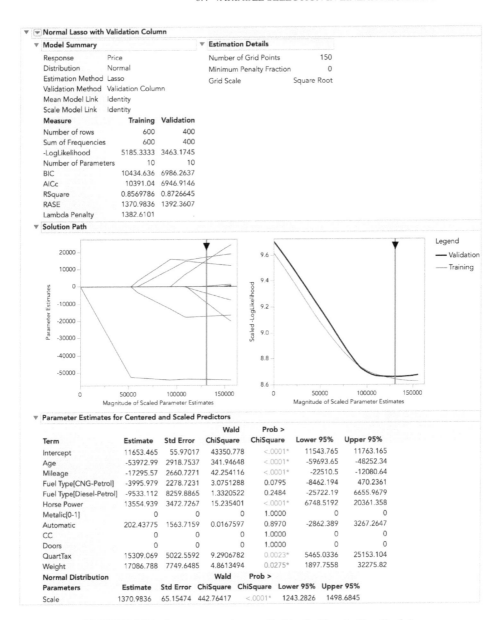

FIGURE 6.15   Lasso regression applied to the Toyota Corolla data

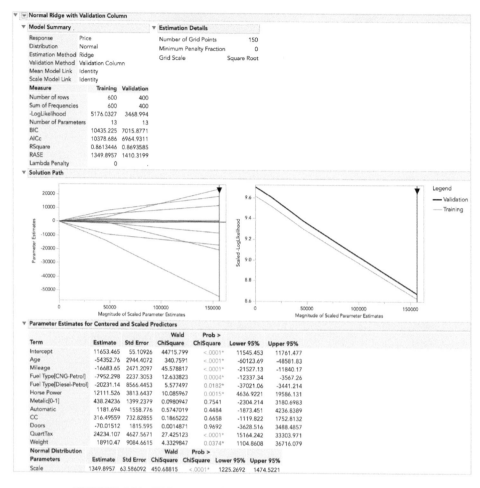

FIGURE 6.16   **Ridge regression applied to the Toyota Corolla data**

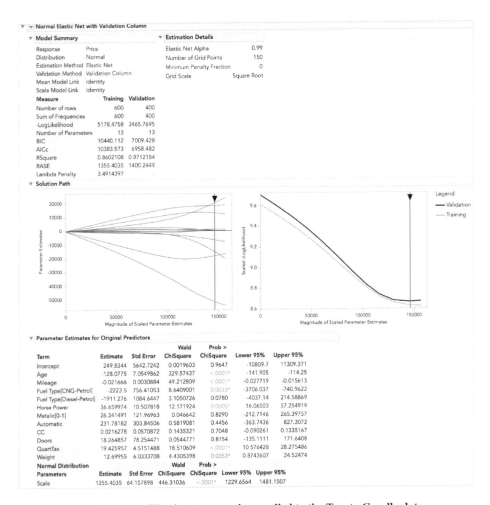

FIGURE 6.17   Elastic net regression applied to the Toyota Corolla data

▼ ⌄ Generalized Regression for Price

▼ Model Comparison

Show	Response Distribution	Estimation Method	Validation Method	Nonzero Parameters	AICc	BIC	Generalized RSquare	Validation Generalized RSquare
☑	Normal	Standard Least Squares	Validation Column	13	10378.686	10435.225	0.8613446	0.8693585
☑	Normal	Lasso	Validation Column	10	10391.04	10434.636	0.8569786	0.8726645
☑	Normal	Ridge	Validation Column	13	10378.686	10435.225	0.8613446	0.8693585
☑	Normal	Elastic Net	Validation Column	13	10383.573	10440.112	0.8602108	0.8712184

FIGURE 6.18   Comparison of lasso, ridge, elastic net, and least squares regression models

PROBLEMS

6.1   Predicting Boston Housing Prices. The file BostonHousing.jmp contains information collected by the US Bureau of the Census concerning housing in the area of Boston, Massachusetts. The dataset includes information on 506 census housing tracts in the Boston area. The goal is to predict the median house price in new tracts based on information such as crime rate, pollution, and number of rooms. The dataset contains 12 predictors, and the outcome variable is the median house price (MEDV). Table 6.2 describes each of the predictors and the outcome variable.

TABLE 6.2   Description of Variables for Boston Housing Example

CRIM	Per capita crime rate by town
ZN	Proportion of residential land zoned for lots over 25,000 ft^2
INDUS	Proportion of nonretail business acres per town
CHAS	Charles River dummy variable (= 1 if tract bounds river; = 0 otherwise)
NOX	Nitric oxide concentration (parts per 10 million)
RM	Average number of rooms per dwelling
AGE	Proportion of owner-occupied units built prior to 1940
DIS	Weighted distances to five Boston employment centers
RAD	Index of accessibility to radial highways
TAX	Full-value property-tax rate per $10,000
PTRATIO	Pupil/teacher ratio by town
LSTAT	% Lower status of the population
MEDV	Median value of owner-occupied homes in $1000s

a. Why should the data be partitioned into training and validation sets? What will the training set be used for? What will the validation set be used for?

b. Fit a multiple linear regression model to the median house price (MEDV) as a function of CRIM, CHAS, and RM. Write the equation for predicting the median house price from the predictors in the model.

c. Using the estimated regression model, what median house price is predicted for a tract in the Boston area that does not bound the Charles River, has a crime rate of 0.1, and where the average number of rooms per house is 6?

d. Reduce the 12 predictors:

i. Which predictors are likely to be measuring the same thing among the entire set of predictors? Discuss the relationships among INDUS, NOX, and TAX.

ii. Compute the correlation table for the 12 numerical predictors and search for highly correlated pairs. These have potential redundancy and can cause multicollinearity. Choose which ones to remove based on this table.

iii. Use three subset selection algorithms: *backward, forward,* and *stepwise* to reduce the remaining predictors. Compute the validation performance for each of the three selected models. Compare RMSE, C_p, AIC$_c$, and BIC. Finally, describe the best model.

6.2   Predicting Software Reselling Profits. Tayko Software is a software catalog firm that sells games and educational software. It started out as a software manufacturer and then added third-party titles to its offerings. It recently revised its collection of items in a new catalog, which it mailed out to its customers. This mailing yielded 1000 purchases. Based on these data, Tayko wants to devise a model for predicting the spending amount that a purchasing customer will yield. The file TaykoPurchasersOnly.jmp contains information on 1000 purchases. Table 6.3 describes the variables to be used in the problem (the file contains additional variables).

TABLE 6.3   Description of Variables for Tayko Software Example

FREQ	Number of transactions in the preceding year
LAST_UPDATE	Number of days since last update to customer record
WEB	Whether customer purchased by Web order at least once
GENDER	Female = 0, male = 1
ADDRESS_RES	Whether it is a residential address
ADDRESS_US	Whether it is a US address
SPENDING (outcome)	Amount spent by customer in test mailing (in dollars)

a. Explore the spending amount by creating a tabular summary of the categorical variables and computing the average and standard deviation of spending in each category.

b. Explore the relationship between spending and each of the two continuous predictors by creating two scatterplots (SPENDING vs. FREQ, and SPENDING vs. LAST_UPDATE). Does there seem to be a linear relationship?

c. To fit a predictive model for SPENDING:

i. Partition the 1000 records into training and validation sets. (Note that the dataset has a validation column—create your own validation column for this exercise.)

ii. Run a multiple linear regression model for SPENDING vs. all six predictors. Give the estimated predictive equation.

iii. Based on this model, what type of purchaser is most likely to spend a large amount of money?

iv. If we used backward elimination to reduce the number of predictors, which predictor would be dropped first from the model?

v. Show how the prediction and the prediction error are computed for the first purchase in the validation set.

vi. Evaluate the predictive accuracy of the model by examining its performance on the validation set.

vii. Create a histogram of the model residuals. Do they appear to follow a normal distribution? How does this affect the predictive performance of the model?

6.3 **Predicting Airfare on New Routes.** The following problem takes place in the United States in the late 1990s, when many major US cities were facing issues with airport congestion, partly as a result of the 1978 deregulation of airlines. Both fares and routes were freed from regulation, and low-fare carriers such as Southwest began competing on existing routes and starting nonstop service on routes that it previously lacked. Building completely new airports is generally not feasible, but sometimes decommissioned military bases or smaller municipal airports can be reconfigured as regional or larger commercial airports. There are numerous players and interests involved in the issue (airlines, city, state and federal authorities, civic groups, the military, airport operators), and an aviation consulting firm is seeking advisory contracts with these players. The firm needs predictive models to support its consulting service. One thing the firm might want to be able to predict is fares, in the event a new airport is brought into service. The firm starts with the file `Airfares.jmp`, which contains real data that were collected between Q3-1996 and Q2-1997. The variables in these data are listed in Table 6.4 and are believed to be important in predicting FARE. Some airport-to-airport data are available, but most data are at the city-to-city level. One question that will be of interest in the analysis is the effect that the presence or absence of Southwest (SW) has on FARE.

TABLE 6.4   Description of Variables for Airfare Example

S_CODE	Starting airport's code
S_CITY	Starting city
E_CODE	Ending airport's code
E_CITY	Ending city
COUPON	Average number of coupons (a one-coupon flight is a nonstop flight, a two-coupon flight is a one-stop flight, etc.) for that route
NEW	Number of new carriers entering that route between Q3-96 and Q2-97
VACATION	Whether (Yes) or not (No) a vacation route
SW	Whether (Yes) or not (No) Southwest Airlines serves that route
HI	Herfindahl index: measure of market concentration
S_INCOME	Starting city's average personal income
E_INCOME	Ending city's average personal income
S_POP	Starting city's population
E_POP	Ending city's population
SLOT	Whether or not either endpoint airport is slot controlled (this is a measure of airport congestion)
GATE	Whether or not either endpoint airport has gate constraints (this is another measure of airport congestion)
DISTANCE	Distance between two endpoint airports in miles
PAX	Number of passengers on that route during period of data collection
FARE	Average fare on that route

a. Explore the numerical predictors and outcome (FARE) by creating histograms, a correlation table and a scatterplot matrix. Examining potential relationships between FARE and those predictors. What seems to be the best single predictor of FARE?

b. Explore the categorical predictors (excluding the first four) by computing the percentage of flights in each category. Create a tabular summary with the average fare in each category. Which categorical predictor seems best for predicting FARE?

c. Find a model for predicting the average fare on a new route:

   i. Partition the data into training and validation sets. The model will be fit to the training data and evaluated on the validation set.

   ii. Use mixed stepwise regression to reduce the number of predictors. You can ignore the first four predictors (S_CODE, S_CITY, E_CODE, E_CITY). Report the estimated model selected.

   iii. Repeat (ii) using All Possible Models instead of *Mixed stepwise* regression. Compare the resulting best model to the one you obtained in (ii) in terms of the predictors that are in the model.

iv. Compare the predictive accuracy of both models (ii) and (iii) using measures such as RMSE, C_p, AIC_c, and validation RSquare.

v. Using model (iii), predict the average fare on a route with the following characteristics: COUPON = 1.202, NEW = 3, VACATION = No, SW = No, HI = 4442.141, S_INCOME = $28,760, E_INCOME = $27,664, S_POP = 4,557,004, E_POP = 3,195,503, SLOT = Free, GATE = Free, PAX = 12,782, DISTANCE = 1976 miles.

vi. Using model (iii), predict the reduction in average fare on the route in (v) if Southwest decides to cover this route.

vii. In reality, which of the factors will not be available for predicting the average fare from a new airport (i.e., before flights start operating on those routes)? Which ones can be estimated? How?

viii. Select a model that includes only factors that are available before flights begin to operate on the new route. Use an exhaustive search (All Possible Models) to find such a model.

ix. Use model (viii) to predict the average fare on a route with characteristics COUPON = 1.202, NEW = 3, VACATION = No, SW = No, HI = 4442.141, S_INCOME = $28,760, E_INCOME = $27,664, S_ POP = 4,557,004, E_POP = 3,195,503, SLOT = Free, GATE = Free, PAX = 12,782, DISTANCE = 1976 miles.

x. Compare the predictive accuracy of this model with model (iii). Is this model good enough, or is it worthwhile reevaluating the model once flights begin on the new route?

d. In competitive industries, a new entrant with a novel business plan can have a disruptive effect on existing firms. If a new entrant's business model is sustainable, other players are forced to respond by changing their business practices. If the goal of the analysis was to evaluate the effect of Southwest Airlines' presence on the airline industry rather than predicting fares on new routes, how would the analysis be different? Describe technical and conceptual aspects.

6.4   Predicting Prices of Used Cars. The file ToyotaCorolla.jmp contains data on used cars (Toyota Corolla) on sale during late summer of 2004 in the Netherlands. It has 1436 records containing details on 38 variables, including Price, Age, Kilometers, HP, and other specifications. The goal is to predict the price of a used Toyota Corolla based on its specifications. (The example in Section 6.3 is a subset of this dataset.)

Data preprocessing. Split the data into training (50%), validation (30%), and test (20%) datasets. Run a multiple linear regression with the output variable Price and input variables: Age_08_04, KM, Fuel_Type, HP, Automatic, Doors, Quarterly_Tax, Mfg_Guarantee, Guarantee_Period, Airco, Automatic_Airco, CD_Player, Powered_Windows, Sport_Model, and Tow_Bar.

a. What appear to be the three or four most important car specifications for predicting the car's price? Use the *Prediction Profiler* to explore the relationship between Price and these variables.

b. Use the *Stepwise* platform to develop a reduced predictive model for Price. What are the predictors in your reduced model?

c. Using metrics you consider useful, assess the performance of the reduced model in predicting prices.

d. Apply regularization techniques like ridge, lasso, and elastic net and compare the models with ordinary regression. What are the predictors in your high performing model?

7

k-NEAREST NEIGHBORS (k-NN)

In this chapter, we describe the k-nearest neighbors algorithm that can be used for classification (of a categorical outcome) or prediction (of a numerical outcome). To classify or predict a new record, the method relies on finding "similar" records in the training data. These "neighbors" are then used to derive a classification or prediction for the new record by voting (for classification) or averaging (for prediction). We explain how similarity is determined, how the number of neighbors is chosen, and how a classification or prediction is computed. k-NN is a highly automated data-driven method. We discuss the advantages and weaknesses of the k-NN method in terms of performance and practical considerations such as computational time.

k-NN in JMP: k-NN is only available in JMP Pro.

7.1 THE k-NN CLASSIFIER (CATEGORICAL OUTCOME)

The idea in k-nearest neighbors methods is to identify k records in the training dataset that are similar to a new record that we wish to classify. We then use these similar (neighboring) records to classify the new record into a class, assigning the new record to the predominant class among these neighbors. Denote by $(x_1, x_2, \ldots, x_p)$ the values of the predictors for this new record. We look for records in our training data that are similar or "near" the record to be classified in the predictor space (i.e., records that have values close to $x_1, x_2, \ldots, x_p$). Then, based on the classes to which those proximate records belong, we assign a class to the record that we want to classify.

Determining Neighbors

The k-nearest neighbors algorithm is a classification method that does not make assumptions about the form of the relationship between the class membership (Y) and the predictors

Machine Learning for Business Analytics: Concepts, Techniques, and Applications with JMP Pro®,
Second Edition. Galit Shmueli, Peter C. Bruce, Mia L. Stephens, Muralidhara Anandamurthy, and Nitin R. Patel.
© 2023 John Wiley & Sons, Inc. Published 2023 by John Wiley & Sons, Inc.

$X_1, X_2, \ldots, X_p$. This is a nonparametric method because it does not involve estimation of parameters in an assumed function form, such as the linear form assumed in linear regression (Chapter 6). Instead, this method draws information from similarities among the predictor values of the records in the dataset.

A central question is how to measure the distance between records based on their predictor values. The most popular measure of distance is the Euclidean distance. The Euclidean distance between two records $(x_1, x_2, \ldots, x_p)$ and $(u_1, u_2, \ldots, u_p)$ is

$$\sqrt{(x_1 - u_1)^2 + (x_2 - u_2)^2 + \cdots + (x_p - u_p)^2}. \tag{7.1}$$

You will find a host of other distance metrics in Chapters 12 and 16 for both numerical and categorical variables. However, the *k*-NN algorithm relies on many distance computations (between each record to be predicted and every record in the training set), and therefore the Euclidean distance, which is computationally cheap, is the most popular in *k*-NN.

To equalize the scales that the various predictors may have, note that in most cases predictors should first be standardized before computing a Euclidean distance.

Classification Rule

After computing the distances between the record to be classified and existing records, we need a rule to assign a class to the record to be classified, based on the classes of its neighbors. The simplest case is $k = 1$, where we look for the record that is closest (the nearest neighbor) and classify the new record as belonging to the same class as its closest neighbor. It is a remarkable fact that this simple, intuitive idea of using a single nearest neighbor to classify records can be very powerful when we have a large number of records in our training set. In turns out that the misclassification error of the 1-nearest neighbor scheme has a misclassification rate that is no more than twice the error when we know exactly the probability density functions for each class.

The idea of the *1-nearest neighbor* can be extended to $k > 1$ neighbors as follows:

1. Find the nearest k neighbors to the record to be classified.
2. Use a majority decision rule to classify the record, where the record is classified as a member of the majority class of the k neighbors.

Example: Riding Mowers

A riding mower manufacturer would like to find a way of classifying families in a city into those likely to purchase a riding mower and those not likely to buy one. A pilot random sample is undertaken of 12 owners and 12 non-owners in the city. The data are shown in Table 7.1. We first partition the data into training data (18 households) and holdout data (6 households). Obviously, this dataset is too small for partitioning, which can result in unstable results, but we continue with this for illustration purposes. A scatterplot for the training set is shown in Figure 7.1 (data are in `RidingMowerskNN.jmp`).

Now consider a new household with \$60,000 income and lot size 20,000 ft^2 (also shown in Figure 7.1 marked with an x). Among the households in the training set, the one closest to the new household (in Euclidean distance after normalizing income and lot size) is household 4, with \$61,500 income and lot size 20,800 ft^2. If we use a 1-NN classifier, we would classify the new household as an owner, like household 4. If we use $k = 3$, the three nearest

TABLE 7.1   Lot size, income, and ownership of a riding mower for 24 Households

Household number	Income ($000s)	Lot Size (000s ft^2)	Ownership of riding mower
1	60.0	18.4	Owner
2	85.5	16.8	Owner
3	4.8	21.6	Owner
4	61.5	20.8	Owner
5	87.0	23.6	Owner
6	110.1	19.2	Owner
7	108.0	17.6	Owner
8	82.8	22.4	Owner
9	69.0	20.0	Owner
10	93.0	20.8	Owner
11	51.0	22.0	Owner
12	81.0	20.0	Owner
13	75.0	19.6	Non-owner
14	52.8	20.8	Non-owner
15	64.8	17.2	Non-owner
16	43.2	20.4	Non-owner
17	84.0	17.6	Non-owner
18	49.2	17.6	Non-owner
19	59.4	16.0	Non-owner
20	66.0	18.4	Non-owner
21	47.4	16.4	Non-owner
22	33.0	18.8	Non-owner
23	51.0	14.0	Non-owner
24	63.0	14.8	Non-owner

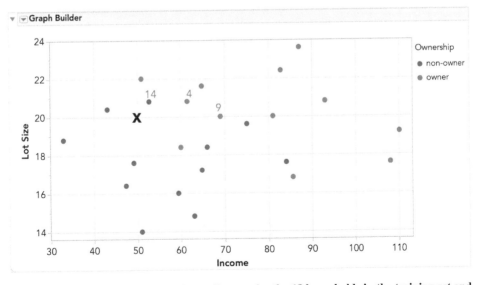

FIGURE 7.1   Scatterplot of Lot Size vs. Income for the 18 households in the training set and the new household to be classified

households are 4, 9, and 14. The first two are owners of riding mowers, and the last is a non-owner. The majority vote is therefore *owner*, and the new household would be classified as an owner.

Choosing Parameter *k*

The advantage of choosing $k > 1$ is that higher values of k provide smoothing that reduces the risk of overfitting due to noise in the training data. Generally speaking, if k is too low, we may be fitting to the noise in the data. However, if k is too high, we will miss out on the method's ability to capture the local structure in the data, one of its main advantages. In the extreme, $k = n =$ the number of records in the training dataset. In that case we simply assign all records to the majority class in the training data, regardless of the values of $(x_1, x_2, \ldots, x_p)$, which coincides with the naive rule! This is clearly a case of oversmoothing in the absence of useful information in the predictors about the class membership. In other words, we want to balance between overfitting to the predictor information and ignoring this information completely. A balanced choice depends greatly on the nature of the data. The more complex and irregular the structure of the data, the lower is the optimum value of k. Typically, values of k fall in the range 1–20. Often an odd number is chosen, to avoid ties.

K-NEAREST NEIGHBORS IN JMP Pro

k-Nearest Neighbors is an option under *Analyze > Predictive Modeling* in JMP Pro. It can be used for both classification of a categorical response and prediction of a continuous response. The dialog is specified as shown in Figure 7.2. Here are a few useful notes:

- The Validation Portion (the portion of the data to be used either as a validation set or as the holdout set) can be specified, or a prespecified column can be used for that purpose using the Validation field (as shown in Figure 7.2).
- Each continuous predictor is scaled by its standard deviation (so standardization of continuous predictors, in advance, is not required).
- Indicator coding is applied for each categorical predictor (so dummy coding of categorical predictors, in advance, is not required).
- Missing values for a continuous predictor are replaced by the mean of the predictor, while missing values for a categorical predictor are replaced by a value of zero.
- The default value of k is 10, but you can change this value.

After clicking OK, models are fit for one nearest neighbor up to the value that you specify for k.

So how is k chosen? Answer: We choose the k that has the best classification performance. We use the training data to classify the records in the validation data, then compute error rates for various choices of k. For our example, on the one hand, if we choose $k = 1$,

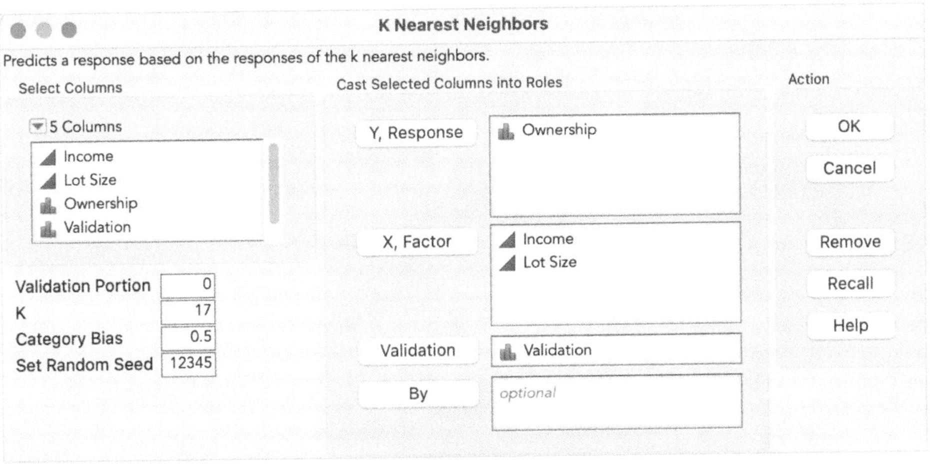

FIGURE 7.2 Selection of the *k-Nearest Neighbors* option from the *Analyze > Predictive Modeling* platform in JMP

we will classify in a way that is very sensitive to the local characteristics of the training data. On the other hand, if we choose a large value of k, such as $k = 18$, we would simply predict the most frequent class in the dataset in all cases. This is a very stable prediction, but it completely ignores the information in the predictors. To find a balance, we examine the misclassification rate (of the validation set) for different choices of k between 1 and 17. This is shown in Figure 7.3. JMP chooses $k = 8$, which minimizes the misclassification rate in the validation set.[1]

Since the validation set is used to determine the optimal value of k, ideally we would want a third holdout (test) set to evaluate the performance of the method on data that it did not see.

For categorical responses, the report includes a model selection graph, a summary table, confusion matrices, and mosaic plots of actual vs. predicted classifications. An asterisk marks the model for the value of k that has the smallest misclassification rate. By default, the slider on the model selection graph is placed on the value of k that corresponds to the best performing model. You can drag the slider to change the value of k in the report.

Once k is chosen, we can save the prediction formula to the data table to generate classifications for new records. An example is shown in Figure 7.4, where eight neighbors are used to classify a new household, which was added as a new row in the data table (row 25). This household was classified as an owner.

Setting the Threshold Value

k-NN uses a majority decision rule to classify a new record, where the record is classified as a member of the majority class of the k neighbors. The definition of "majority" is directly linked to the notion of a threshold value applied to the class membership probabilities. Let

---

[1] Partitioning such a small dataset into training and validation sets is unwise in practice, as results will heavily rely on the particular partition. For instance, if you use a different partitioning, you might obtain a different "optimal" k. We use this example for illustration only.

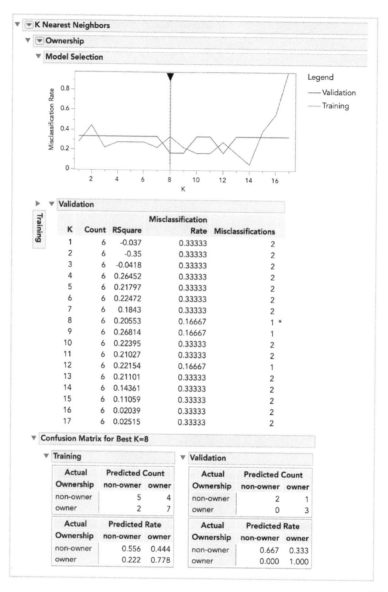

FIGURE 7.3 Model selection graph, holdout misclassification rates for various choices of *k*, and the confusion matrix for *k* = 8

us consider a binary outcome case. For a new record, the proportion of class 1 members among its neighbors is an estimate of its propensity (probability) of belonging to class 1. In the riding mowers example with $k = 3$, we found that the three nearest neighbors to the new household (with income = \$60,000 and lot size = 20,000 ft^2) are households 4, 9, and 14. Since 4 and 9 are owners and 14 is a non-owner, we can estimate for the new household a probability of 2/3 of being an owner (and 1/3 for being a non-owner). The predicted classification is shown in Figure 7.4. Using a simple majority rule is equivalent to setting the threshold value to 0.5.

	Income	Lot Size	Ownership	Validation	Predicted Formula Ownership 8
18	49.2	17.6	non-owner	Training	non-owner
19	59.4	16	non-owner	Training	non-owner
20	66	18.4	non-owner	Training	non-owner
21	47.4	16.4	non-owner	Training	non-owner
22	33	18.8	non-owner	Validation	non-owner
23	51	14	non-owner	Training	non-owner
24	63	14.8	non-owner	Validation	non-owner
25	60	40			• owner

FIGURE 7.4   Classifying a new household using the "best k"= 8

---

k-NN PREDICTIONS AND PREDICTION FORMULAS IN JMP Pro

Finding the nearest neighbor for every record in a dataset is time-consuming, particularly in large datasets. As a result, predictions aren't made until the prediction formula for a k-NN model is saved to the data table. The prediction formula contains all of the training data, so distances for each observation in the training set can be computed and predictions for new data can be made in more a timely manner. The fitted probabilities for categorical responses in k-NN models are not saved to the data table. As a result, the ability to change the threshold for classification in k-NN is not available.

Weighted k-NN

In using a majority vote approach, k-NN assigns equal weight to each of the k neighbors. For example, with $k = 3$, each neighbor's vote has a weight = 1/3. Weighted k-NN improves the k-NN algorithm by using an incredibly simple yet powerful idea that the closer the neighbor, the heavier its weight in the overall vote. Each neighbor's weight is determined by the inverse of its distance from the record to be predicted (normalized by the sum of inverse distances of all k neighbors). Thus, given $d_1, d_2, ..., d_k$ distances to each of the k neighbors, the weight for neighbor 1 is computed as

$$w_1 = \frac{1/d_1}{(1/d_1 + 1/d_2 + ... + 1/d_k)}. \tag{7.2}$$

Then, to compute the confidence (probability) of belonging to a certain class, we sum up the weights of the neighbors with that class.

For example, returning to our example with $k = 3$, suppose the distances from the new record to its three nearest neighbors (and their labels) were $d_1 = 0.5$ (Non-owner), $d_2 = 0.2$ (Non-owner), and $d_3 = 0.1$ (Owner). Using equation (7.2), we get weights $w_1 = 0.12$, $w_2 = 0.29$, and $w_3 = 0.59$. The confidence (probability) of belonging to the Owner class is then 0.59 (and for belonging to Non-owner class is $0.12 + 0.29 = 0.41$). Assuming a 0.5 threshold, the predicted class for the record would be Owner even though the majority vote is Non-owner.

Note: JMP does not support weighted k-NN.

k-NN with More Than Two Classes

The *k*-NN classifier can easily be applied to an outcome with m classes, where $m > 2$. The "majority rule" means that a new record is classified as a member of the majority class of its k neighbors. An alternative, when there is a specific class that we are interested in identifying (and are willing to "overidentify" records as belonging to this class), is to calculate the proportion of the k neighbors that belong to this class of interest, use that as an estimate of the probability (propensity) that the new record belongs to that class, and then refer to a user-specified threshold value to decide whether to assign the new record to that class. For more on the use of threshold value in classification where there is a single class of interest, see Chapter 5.

Working with Categorical Predictors

It usually does not make sense to calculate Euclidean distance between two non-numeric categories (e.g., cookbooks and maps, in a bookstore). Therefore, before *k*-NN can be applied, categorical variables must be converted to binary dummies. In contrast to the situation with statistical models such as regression, all m binaries should be created and used with *k*-NN. While mathematically this is redundant, since $m - 1$ dummies contain the same information as m dummies, this redundant information does not create the multicollinearity problems that it does for linear models. Moreover, in *k*-NN, the use of $m - 1$ dummies can yield different classifications than the use of m dummies and lead to an imbalance in the contribution of the different categories to the model.

JMP does not require converting categorical variables to binary dummy variables. This is done behind the scenes. In JMP, both numeric and categorical predictors can be used in *k*-NN.

PANDORA

Pandora is an Internet music radio service that allows users to build customized "stations" that play music similar to a song or artist that they have specified. Pandora uses a *k*-NN style clustering/classification process called the Music Genome Project to locate new songs or artists that are close to the user-specified song or artist.

Pandora was the brainchild of Tim Westergren, who worked as a musician and a nanny when he graduated from Stanford in the 1980s. Together with Nolan Gasser, who was studying medieval music, he developed a "matching engine" by entering data about a song's characteristics into a spreadsheet. The first result was surprising—a Beatles song matched to a Bee Gees song, but they built a company around the concept. The early days were hard—Westergren racked up over $300,000 in personal debt, maxed out 11 credit cards, and ended up in the hospital once due to stress-induced heart palpitations. A venture capitalist finally invested funds in 2004 to rescue the firm, and as of 2013 it is listed on the NY Stock Exchange.

In simplified terms, the process works roughly as follows for songs:

1. Pandora has established hundreds of variables on which a song can be measured on a scale from 0 to 5. Four such variables from the beginning of the list are:
   - Acid Rock Qualities
   - Accordion Playing

- Acousti-Lectric Sonority
- Acousti-Synthetic Sonority.

2. Pandora pays musicians to analyze tens of thousands of songs, and rate each song on each of these variables. Each song will then be represented by a row vector of values between 0 and 5. For example, for Led Zeppelin's *Kashmir*:

   Kashmir 4 0 3 3 ... (high on acid rock variables, no accordion, etc.)

   This step represents a costly investment, and lies at the heart of Pandora's value. That is to say, these variables have been tested and selected because they accurately reflect the essence of a song, and provide a basis for defining highly individualized preferences.

3. The online user specifies a song that s/he likes (the song must be in Pandora's database).

4. Pandora then calculates the statistical distance[2] between the user's song and the songs in its database. It selects a song that is close to the user-specified song and plays it.

5. The user then has the option of saying "I like this song," "I don't like this song," or saying nothing.

6. If "like" is chosen, the original song plus the new song are merged into a 2-song cluster[3] that is represented by a single vector comprised of means of the variables in the original two song vectors.

7. If "dislike" is chosen, the vector of the song that is not liked is stored for future reference. (If the user does not express an opinion about the song, in our simplified example here, the new song is not used for further comparisons.)

8. Pandora looks in its database for a new song, one whose statistical distance is close to the "like" song cluster,[4] and not too close to the "dislike" song. Depending on the user's reaction, this new song might be added to the "like" cluster or "dislike" cluster.

Over time, Pandora develops the ability to deliver songs that match a particular taste of a particular user. A single user might build up multiple stations around different song clusters. Clearly, this is a less limiting approach than selecting music in terms of which "genre" it belongs to.

While the process described above is a bit more complex than the basic "classification of new data" process described in this chapter, the fundamental process—classifying a record according to its proximity to other records—is the same at its core. Note the role of domain knowledge in this machine learning process—the variables have been tested and selected by the project leaders, and the measurements have been made by human experts.

Further reading: See www.pandora.com, Wikipedia's article on the Music Genome Project, and Joyce John's article "Pandora and the Music Genome Project," in the September 2006 *Scientific Computing* [23 (10): 14, 40–41].

---

[2] See Chapter 12 for an explanation of statistical distance.

[3] See Chapter 16 for more on clusters.

[4] See Case 22.6 "Segmenting Consumers of Bath Soap" for an exercise involving the identification of clusters, which are then used for classification purposes.

7.2   *K*-NN FOR A NUMERICAL RESPONSE

The idea of *k*-NN can readily be extended to predicting a continuous value (as is our aim with multiple linear regression models). The first step of determining neighbors by computing distances remains unchanged. The second step, where a majority vote of the neighbors is used to determine class, is modified such that we take the average response value of the *k*-nearest neighbors to determine the prediction. Often this average is a weighted average, with the weight decreasing with increasing distance from the point at which the prediction is required.

Another modification is in the error metric used for determining the "best *k*." Rather than the overall error rate used in classification, RMSE and SSE are used to determine the best *k* in when the response is continuous (see Chapter 5).

7.3   ADVANTAGES AND SHORTCOMINGS OF *K*-NN ALGORITHMS

The main advantage of *k*-NN methods is their simplicity and lack of parametric assumptions. In the presence of a large enough training set, these methods perform surprisingly well, especially when each class is characterized by multiple combinations of predictor values. For instance, in real estate databases there are likely to be multiple combinations of {home type, number of rooms, neighborhood, asking price, etc.} that characterize homes that sell quickly as opposed to those that remain for a long period on the market.

There are three difficulties with the practical exploitation of the power of the *k*-NN approach. First, although no time is required to estimate parameters from the training data, as we discussed earlier, the time to find the nearest neighbors in a large training set can be prohibitive. A number of ideas have been implemented to overcome this difficulty. The main ideas are as follows:

- Reduce the time taken to compute distances by working in a reduced dimension using dimension reduction techniques such as principal components analysis (Chapter 4).
- Use sophisticated data structures such as search trees to speed up identification of the nearest neighbor. This approach often settles for an "almost nearest" neighbor to improve speed. An example is using bucketing, where the records are grouped into buckets so that records within each bucket are close to each other. For a to-be-predicted record, buckets are ordered by their distance to the record. Starting from the nearest bucket, the distance to each of the records within the bucket is measured. The algorithm stops when the distance to a bucket is larger than the distance to the closest record thus far.

Second, the number of records required in the training set to qualify as large increases exponentially with the number of predictors *p*. This is because the expected distance to the nearest neighbor goes up dramatically with *p* unless the size of the training set increases exponentially with *p*. This phenomenon is known as the *curse of dimensionality*, a fundamental issue pertinent to all classification, prediction, and clustering techniques. This is why we often seek to reduce the number of predictors through methods such as selecting subsets of the predictors for our model or by combining them using methods such as princi-pal component analysis, singular value decomposition, and factor analysis (see Chapter 4).

Third, *k*-NN is a "lazy learner": the time-consuming computation is deferred to the time of prediction. For every record to be predicted, we compute its distances from the entire set of training records only at the time of prediction. This behavior prohibits using this algorithm for real-time prediction of a large number of records simultaneously.

PROBLEMS

7.1   **Calculating Distance with Categorical Predictors.** This exercise with a tiny dataset illustrates the calculation of Euclidean distance and the creation of binary dummies. (Note that creating dummies is not required in JMP, but it is useful to understand how they are derived and used.)

The online education company Statistics.com segments its customers and prospects into three main categories: IT professionals (IT), statisticians (Stat), and other (Other). It also tracks, for each customer, the number of years since first contact (years). Consider the following customers; information about whether they have taken a course or not (the outcome to be predicted) is included:

Customer 1: Stat, 1 year, did not take course
Customer 2: Other, 1.1 year, took course

a. Consider now the following new prospect:
Prospect 1: IT, 1 year
Using the information given above on the two customers and one prospect, create one dataset for all three with the categorical predictor variable transformed into two dummy variables, and a similar dataset with the categorical predictor variable transformed into three dummy variables.

b. For each derived dataset, calculate the Euclidean distance between the prospect and each of the other two customers.

c. Using *k*-NN with $k = 1$, classify the prospect as taking or not taking a course using each of the two derived datasets. Does it make a difference whether you use two or three dummies?

7.2   **Personal Loan Acceptance.** Universal Bank is a relatively young bank growing rapidly in terms of overall customer acquisition. The majority of these customers are liability customers (depositors) with varying sizes of relationship with the bank. The customer base of asset customers (borrowers) is quite small, and the bank is interested in expanding this base rapidly to bring in more loan business. In particular, it wants to explore ways of converting its liability customers to personal loan customers (while retaining them as depositors).

A campaign that the bank ran last year for liability customers showed a healthy conversion rate of over 9% success. This has encouraged the retail marketing department to devise smarter campaigns with better target marketing. The goal is to use *k*-NN to predict whether a new customer will accept a loan offer. This will serve as the basis for the design of a new campaign.

The file `UniversalBank.jmp` contains data on 5000 customers. The data include customer demographic information (age, income, etc.), the customer's relationship with the bank (mortgage, securities account, etc.), and the customer response to the last personal loan campaign (*Personal Loan*). Among these 5000 customers, only 480 ($= 9.6\%$) accepted the personal loan that was offered to them in the earlier campaign.

Partition the data into training (60%) and validation (40%) sets.

a. Consider the following customer:
Age = 40, Experience = 10, Income = 84, Family = 2, CCAvg = 2, Education = 2, Mortgage = 0, Securities Account = 0, CD Account = 0, Online = 1, and Credit Card = 1. Perform a *k*-NN classification with all predictors except ID and ZIP code using $k = 10$. How would this customer be classified?

b. What is a choice of k that balances between overfitting and ignoring the predictor information?

c. Show the classification matrix for the validation data that results from using the best k.

d. Consider the following customer: Age = 40, Experience = 10, Income = 84, Family = 2, CCAvg = 2, Education = 2, Mortgage = 0, Securities Account = 0, CD Account = 0, Online = 1, and Credit Card = 1. Classify the customer using the best k.

e. Repartition the data, this time into training, validation, and test sets (50% : 30% : 20%). Apply the k-NN method with the k chosen above. Compare the classification matrix of the test set with that of the training and validation sets. Comment on the differences and their reason.

7.3 **Predicting Housing Median Prices.** The file `BostonHousing.jmp` contains information on over 500 census tracts in Boston, where for each tract multiple variables are recorded. The last column (CAT.MEDV) was derived from the median value (MEDV), such that it obtains the value 1 if MEDV>30 and 0 otherwise. Consider the goal of predicting the MEDV of a tract, given the information in the first 12 columns. Partition the data into training (60%) and validation (40%) sets.

a. Perform a k-NN prediction with all 12 predictors (ignore the CAT.MEDV column), trying values of k from 1 to 5. What is the best k chosen? What does it mean?

b. Predict the MEDV for a tract with the following information, using the best k:

CRIM	ZN	INDUS	CHAS	NOX	RM	AGE	DIS	RAD	TAX	PTRATIO	LSTAT
0.2	0	7	0	0.538	6	62	4.7	4	307	21	10

(Save the prediction formula to the data table, add a new row, and enter these values.)

c. Why is the validation data error overly optimistic compared to the error rate when applying this k-NN predictor to new data?

d. If the purpose is to predict MEDV for several thousands of new tracts, what would be the disadvantage of using k-NN prediction? List the operations that the algorithm goes through in order to produce each prediction.

8

THE NAIVE BAYES CLASSIFIER

In this chapter, we introduce the naive Bayes classifier, which can be applied to data with categorical predictors. We review the concept of conditional probabilities, then present the complete, or exact, Bayesian classifier. We next see how it is impractical in most cases, and learn how to modify it and instead use the "naive Bayes" classifier, which is more generally applicable.

Naive Bayes in JMP: Naive Bayes is only available in JMP Pro.

8.1 INTRODUCTION

The naive Bayes method (and, indeed, an entire branch of statistics) is named after the Reverend Thomas Bayes (1702–1761). To understand the naive Bayes classifier, we first look at the complete, or exact, Bayesian classifier. The basic principle is simple.

For each new record to be classified:

1. Find all of the training records with the same predictor profile (i.e., records having the same predictor values).
2. Determine what classes the records belong to and which class is most prevalent.
3. Assign that class to the new record.

Alternatively (or in addition), it may be desirable to adjust the method so that it answers the question: "What is the propensity of belonging to the class of interest?" instead of "Which class is the most probable?" Obtaining class probabilities allows using a sliding threshold to classify a record as belonging to class i, even if i is not the most probable class for that record. This approach is useful when there is a specific class of interest that we want to identify and we are willing to "overidentify" records as belonging to this class.

Machine Learning for Business Analytics: Concepts, Techniques, and Applications with JMP Pro®,
Second Edition. Galit Shmueli, Peter C. Bruce, Mia L. Stephens, Muralidhara Anandamurthy, and Nitin R. Patel.
© 2023 John Wiley & Sons, Inc. Published 2023 by John Wiley & Sons, Inc.

(See Chapter 5 for more details on the use of thresholds for classification and on asymmetric misclassification costs.)

Threshold Probability Method

For each new record to be classified:

1. Establish a threshold probability for the class of interest (C_1) above which we consider that a record belongs to that class.
2. Find all the records with the same predictor profile as the new record (i.e., records having the same predictor values).
3. Determine the probability that these records belong to the class of interest.
4. If this probability is above the threshold probability, assign the new record to the class of interest.

Conditional Probability

Both procedures incorporate the concept of *conditional probability*, or the probability of event A given that event B has occurred, which is denoted $P(A|B)$. We will be looking at the probability of the record belonging to class i given that its predictor values are $x_1, x_2, \ldots, x_p$. In general, for a response with m classes $C_1, C_2, \ldots, C_m$, and the predictor values $x_1, x_2, \ldots, x_p$, we want to compute

$$P(C_i|x_1, \ldots, x_p). \tag{8.1}$$

To classify a record, we compute its probability of belonging to each of the classes in this way, then either classify the record to the class that has the highest probability or use the threshold probability to decide whether it should be assigned to the class of interest.

From this definition, we see that the Bayesian classifier works only with categorical predictors. If we use a set of numerical predictors, then it is highly unlikely that multiple records will have identical values on these numerical predictors.

In that case, one option is to convert numerical predictors to categorical predictors by binning or to use a related method called *Gaussian naive Bayes classifier* that can incorporate numerical predictors. Another option is to estimate *conditional marginal density* of continuous predictors and is discussed later in this chapter. For now, we will focus on categorical predictors only.

Example 1: Predicting Fraudulent Financial Reporting

An accounting firm has many large companies as customers. Each customer submits an annual financial report to the firm, which is then audited by the accounting firm. For simplicity, we will designate the outcome of the audit as "fraudulent" or "truthful," referring to the accounting firm's assessment of the customer's financial report. The accounting firm has a strong incentive to be accurate in identifying fraudulent reports—passing a fraudulent report as truthful could have legal consequences.

The accounting firm notes that in addition to all the financial records, it has information on whether the customer has had prior legal trouble (criminal or civil charges of any nature filed against it). This information has not been used in previous audits, but the accounting firm is wondering whether it could be used in the future to identify reports that merit more intensive review. Specifically, it wants to know whether having had prior legal trouble is predictive of fraudulent reporting.

Each customer is a record, and the response of interest has two classes into which a company can be classified: C_1 = fraudulent and C_2 = truthful. The predictor variable—"prior legal trouble"—has two values: 0 (no prior legal trouble) and 1 (prior legal trouble).

The accounting firm has data on 1000 companies that it has investigated in the past. For each company it has information on whether the financial report was judged fraudulent or truthful and whether the company had prior legal trouble. The counts in the training set are shown in Table 8.1. The raw data are available in `FinancialReportingRaw.jmp`.

TABLE 8.1 Tabular Summary of Financial Reporting Data

	Prior legal ($X = 1$)	No prior legal ($X = 0$)	Total
Fraudulent (C_1)	50	50	100
Truthful (C_2)	180	720	900
Total	230	770	1000

8.2 APPLYING THE FULL (EXACT) BAYESIAN CLASSIFIER

Now consider the financial report from a new company, which we wish to classify as fraudulent or truthful by using these data. To do this, we compute the probabilities, as above, of belonging to each of the two classes.

If the new company had prior legal trouble, the probability of belonging to the fraudulent class would be P(fraudulent | prior legal) = 50/230 (there were 230 companies with prior legal trouble in the training set, and 50 of them had fraudulent financial reports). The probability of belonging to the other class, "truthful," is, of course, the remainder = 180/230. (These numbers can be verified using the raw data and the *Distribution* platform, with Outcome as *Y*, *Columns* and filtering on Prior Legal Trouble using the *Local Data Filter*.)

Using the "Assign to the Most Probable Class" Method

If the new company had prior legal trouble, we assign it to the "truthful" class. Similar calculations for the case of no prior legal trouble are left as an exercise to the reader. In this example, using the rule "assign to the most probable class," all records are assigned to the "truthful" class. This is the same result as the naive rule of "assign all records to the majority class."

Using the Threshold Probability Method

In this example, we are more interested in identifying the fraudulent reports—these are the ones that can land the auditor in jail. We recognize that in order to identify the fraudulent reports, some truthful reports will be misidentified as fraudulent, and the overall classification accuracy may decline. Our approach is therefore to establish a threshold value for the probability of being fraudulent, and classify all records above that value as fraudulent.

The Bayesian formula for the calculation of this probability that a record belongs to class C_i is as follows:

$$P(C_i|x_1,\ldots,x_p) = \frac{P(x_1,\ldots,x_p|C_i)P(C_i)}{P(x_1,\ldots,x_p|C_1)P(C_1) + \cdots + P(x_1,\ldots,x_p|C_m)P(C_m)}.$$

(8.2)

In this example (where frauds are more rare), if the threshold were established at 0.20, we would classify a new company with prior legal trouble as fraudulent because P(fraudulent | prior legal) $= 50/230 = 0.22$. The user can treat this threshold as a "slider" to be adjusted to optimize performance, like other parameters in any classification model.

Practical Difficulty with the Complete (Exact) Bayes Procedure

The approach outlined above amounts to finding all the records in the sample that have the same predictor values as the new record. This was easy in the small example presented above, where there was just one predictor.

As the number of predictors grows (even to a modest number like 20), many of the records to be classified will be without exact matches. This can be understood in the context of a model to predict voting on the basis of demographic variables. For example, even a sizable sample may not contain even a single match for a new record that is a male Hispanic with medium income from the US midwest who voted in the last election, did not vote in the prior election, has three daughters and one son, and is divorced. And this is just eight variables, a small number for most data mining exercises. The addition of just a single new variable with five equally frequent categories reduces the probability of a match by a factor of 5.

8.3 SOLUTION: NAIVE BAYES

In the naive Bayes solution, we no longer restrict the probability calculation to those records that exactly match the record to be classified. Instead, we use the entire dataset.

Returning to our original basic classification procedure outlined at the beginning of the chapter, we recall the procedure for classifying a new record:

1. Find all the training records with the same predictor profile (i.e., having the same predictor values).
2. Determine what classes the records belong to and which class is most prevalent.
3. Assign that class to the new record.

The naive Bayes modification (to the basic classification procedure) is as follows:

1. For class C_1, estimate the individual conditional probabilities for each predictor $P(x_j|C_1)$—these are the probabilities that the predictor value in the record to be classified occurs in class C_1. For example, for X_1 this probability is estimated by the proportion of x_1 values among the records belonging to class C_1 in the training set.

2. Multiply the proportion of records in class C_1 by the product of these conditional probabilities (see the numerator in the equation below).

3. Repeat steps 1 and 2 for each class.

4. For each class, estimate a probability for class i by taking the value calculated in step 2 for class i and dividing it by the sum of such values for all classes (see the denominator in the equation below).

5. Assign the record to the class with the highest estimated probability for this set of predictor values.

The above steps lead to the naive Bayes formula for calculating the probability that a record with a given set of predictor values $x_1, \ldots, x_p$ belongs to class C_1 among m classes. The formula can be written as follows:

$$P_{nb}(C_1 \mid x_1, \ldots x_p) =$$

$$\frac{P(C_1)[P(x_1 \mid C_1)P(x_2 \mid C_1) \cdots P(x_p \mid C_1)]}{P(C_1)[P(x_1 \mid C_1)P(x_2 \mid C_1) \cdots P(x_p \mid C_1)] + \cdots + P(C_m)[P(x_1 \mid C_m)P(x_2 \mid C_m) \cdots P(x_p \mid C_m)]}.$$

$$(8.3)$$

This is a somewhat formidable formula; see Example 2 for a simpler numerical version. Note that all the necessary quantities can be obtained from tabular summaries of Y vs. each of the categorical predictors (using the *Tabulate* platform).

The Naive Bayes Assumption of Conditional Independence

In probability terms, we have made the simplifying assumption that the exact *conditional probability* of seeing a record with predictor profile $x_1, x_2, \ldots, x_p$ within a certain class, $P(x_1, x_2, \ldots, x_p | C_i)$, is well approximated by the product of the individual conditional probabilities $P(x_1|C_i) \times P(x_2|C_i) \cdots \times P(x_p|C_i)$. The two quantities are identical when the predictors are independent within each class.

For example, suppose that "lost money last year?" is an additional variable in the accounting fraud example. The simplifying assumption we make with naive Bayes is that, within a given class, we no longer need to look for the records characterized both by "prior legal trouble" and "lost money last year." Rather, assuming that the two are independent, we can simply multiply the probability of "prior legal trouble" by the probability of "lost money last year." Of course, complete independence is unlikely in practice, where some correlation between predictors is expected.

In practice, despite the assumption violation, the procedure works quite well—primarily because what is usually needed is not a propensity for each record that is accurate in absolute terms but just a reasonably accurate *rank ordering* of propensities. Even when the independence assumption is violated, the rank ordering of the records' propensities is typically preserved.

Note that if all we are interested in is a rank ordering, and the denominator remains the same for all classes, it is sufficient to concentrate only on the numerator. The disadvantage of this approach is that the probability values (the propensities) obtained, while ordered correctly, are not on the same scale as the exact values that the user would anticipate.

Using the Threshold Probability Method

The above procedure is for the basic case where we seek maximum classification accuracy for all classes. In the case of the *relatively rare class of special interest*, the procedure is as follows:

1. Establish a threshold probability for the class of interest (C_1) above which we consider a record to belong to this class.
2. For the class of interest, compute the probability that each individual predictor value in the record to be classified occurs in the training data.
3. Multiply the proportion of records in class C_1 by the product of the probabilities obtained in step 2.
4. Estimate the probability for the class of interest by taking the value calculated in step 3 for the class of interest and dividing it by the sum of the similar values for all classes.
5. If this value falls above the threshold, assign the new record to the class of interest; otherwise not.
6. Adjust the threshold value as needed.

Example 2: Predicting Fraudulent Financial Reports, Two Predictors

Let us expand the financial reports example to two predictors and, using a small subset of data, compare the complete (exact) Bayes calculations to the naive Bayes calculations.

Consider the 10 customers of the accounting firm listed in Table 8.2 (data are available in `FinancialReporting10Companies.jmp`). For each customer, we have information on whether it had prior legal trouble, whether it is a small or large company, and whether the financial report was found to be fraudulent or truthful. Using this information, we will calculate the conditional probability of fraud, given each of the four possible combinations {*yes, small*}, {*yes, large*}, {*no, small*}, {*no, large*}.

TABLE 8.2   Information on 10 Companies

Prior legal trouble	Company size	Status
Yes	Small	Truthful
No	Small	Truthful
No	Large	Truthful
No	Large	Truthful
No	Small	Truthful
No	Small	Truthful
Yes	Small	Fraudulent
Yes	Large	Fraudulent
No	Large	Fraudulent
Yes	Large	Fraudulent

Complete (Exact) Bayes Calculations   The probabilities are computed as

$$P(\text{fraudulent} \mid \text{PriorLegal} = yes, \quad \text{Size} = small) = 1/2 = 0.5$$

$$P(\text{fraudulent} \mid \text{PriorLegal} = yes, \quad \text{Size} = large) = 2/2 = 1$$

$$P(\text{fraudulent} \mid \text{PriorLegal} = no, \quad \text{Size} = small) = 0/3 = 0$$

$$P(\text{fraudulent} \mid \text{PriorLegal} = no, \quad \text{Size} = large) = 1/3 = 0.33$$

Naive Bayes Calculations   Now we can compute the naive Bayes probabilities. For the conditional probability of fraudulent behaviors given {Prior Legal = *yes*, Size = *small*}, the numerator is the product of the proportion of {Prior Legal= *yes*} instances among the fraudulent companies and the proportion of {Size = *small*} instances among the fraudulent companies, times the proportion of fraudulent companies: $(3/4)(1/4)(4/10) = 0.075$. To get the actual probabilities, we must also compute the numerator for the conditional probability of *truthful* behaviors given {Prior Legal = *yes*, Size = *small*}: $(1/6)(4/6)(6/10) = 0.067$. The denominator is then the sum of these two conditional probabilities ($0.075 + 0.067 = 0.14$). The conditional probability of fraudulent behaviors given {Prior Legal = *yes*, Size = *small*} is therefore $0.075/0.14 = 0.53$.

In a similar way, we compute all four conditional probabilities:

$$P_{\text{nb}}(\text{fraudulent} \mid \text{PriorLegal} = yes, \quad \text{Size} = small) = \frac{(3/4)(1/4)(4/10)}{(3/4)(1/4)(4/10) + (1/6)(4/6)(6/10)}$$

$$= 0.53$$

$$P_{\text{nb}}(\text{fraudulent} \mid \text{PriorLegal} = yes, \quad \text{Size} = large) = 0.87$$

$$P_{\text{nb}}(\text{fraudulent} \mid \text{PriorLegal} = no, \quad \text{Size} = small) = 0.07$$

$$P_{\text{nb}}(\text{fraudulent} \mid \text{PriorLegal} = no, \quad \text{Size} = large) = 0.31$$

Note how close these naive Bayes probabilities are to the exact Bayes probabilities. Although they are not equal, both would lead to exactly the same classification for a threshold of 0.5 (and many other values). It is often the case that the rank ordering of probabilities is even closer to the exact Bayes method than are the probabilities themselves, and for classification purposes it is the rank orderings that matter.

We now consider a larger example, where flight data is used to predict delays (data are in the file `FlightDelaysNB.jmp`).

Example 3: Predicting Delayed Flights

Predicting flight delays can be useful to a variety of organizations: airport authorities, airlines, and aviation authorities. At times, joint task forces have been formed to address the problem. If such an organization were tasked with providing ongoing, real-time assistance with flight delays, it would benefit from being able to predict which flights are likely to be delayed.

In this simplified illustration, we look at five predictors (see Table 8.3). The outcome of interest is whether the flight is delayed (*delayed* here means arrived more than 15 minutes late). Our data consist of all flights from the Washington, DC area into the New York City area during January 2004. A record is a particular flight. The percentage of delayed flights among these 2201 flights is 19.5%. The data were obtained from the Bureau of Transportation Statistics (available on the web at `www.transtats.bts.gov`). The goal is to accurately predict whether or not a new flight (not in this dataset) will be delayed. The response "Delayed" is whether the flight was delayed or not (*1 = Delayed* and *0 = On Time*). The variable Departure Time Block was created by binning the departure times into hourly intervals between 6:00 AM and 10:00 PM.

TABLE 8.3 Description of Variables for Flight Delays Example

Delayed	Coded as: 1 = Delayed and 0 = On Time
Day of Week	Coded as: 1 = Monday, 2 = Tuesday, ..., 7 = Sunday
Departure Time Block	Broken down into 16 intervals between 6:00 AM and 10:00 PM
Origin	Three airport codes: DCA (Reagan National), IAD (Dulles), BWI (Baltimore–Washington Int'l)
Destination	Three airport codes: JFK (Kennedy), LGA (LaGuardia), EWR (Newark)
Carrier	Eight airline codes: CO (Continental), DH (Atlantic Coast), DL (Delta), MQ (American Eagle), OH (Comair), RU (Continental Express), UA (United), and US (USAirways)

The data were partitioned into training (60%) and validation (40%) by using the *Make Validation Column* utility from the *Analyze > Predictive Modeling* menu. Instead of splitting the data randomly, Delayed was used as a stratification column. This method partitions the data into sets balanced on the levels of the specified stratification column.

USING NAIVE BAYES IN JMP Pro

Naive Bayes is available under *Analyze > Predictive Modeling* in JMP Pro. The dialog is specified as shown in Figure 8.1. You can use a validation column or specify the portion of the data to be used as holdout validation.

Before we look at the JMP output, let us see how the algorithm works. Figure 8.2 shows conditional probabilities for each class, as a function of the predictor values. These conditional probabilities were computed using the *Tabulate* platform in JMP, where *Delayed* is used as the column variable, each of the predictors is used as a row variable, and *Column %* is the summary statistic. The *Local Data Filter* was used to compute conditional probabilities for only the training data.

Using these conditional probabilities, we can classify a new flight by comparing the probability that a flight will be delayed with the probability that it will be on-time. Recall that since both probabilities will have the same denominator, we can just compare the numerators. Each numerator is computed by forming the product of the conditional probabilities of the relevant predictor values, and multiplying this result by the proportion of this

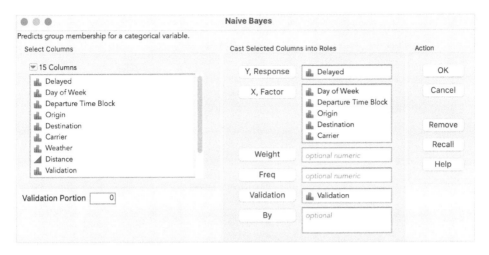

FIGURE 8.1   The *Naive Bayes* launch dialog

class in the dataset (in this case $\hat{P}$(delayed) $= 0.195$). For example, to classify a Delta flight from DCA to LGA between 10 AM and 11 AM on a Sunday, we compute the numerators using the values from Figure 8.2:

$$\hat{P}(\text{delayed} \mid \text{Carrier} = \text{DL}, \text{DayofWeek} = 7, \text{DepartureTime} = 1000 - 1059,$$
$$\text{Destination} = \text{LGA}, \text{Origin} = \text{DCA}$$
$$\propto (0.109)(0.16)(0.027)(0.409)(0.483)(0.195) = 0.0000181,$$

$$\hat{P}(\text{ontime} \mid \text{Carrier} = \text{DL}, \text{DayofWeek} = 7, \text{DepartureTime} = 1000 - 1059,$$
$$\text{Destination} = \text{LGA}, \text{Origin} = \text{DCA}$$
$$\propto (0.201)(0.105)(0.052)(0.563)(0.671)(0.805) = 0.000333,$$

The symbol $\propto$ means "is proportional to," reflecting the fact that this calculation deals only with the numerator in the naive Bayes formula (equation (8.3)).

Comparing the numerators, it is therefore more likely that the flight will be on-time. Note that a record with such a combination of predictor values does not exist in the training set, and therefore we use naive Bayes rather than exact Bayes. To compute the actual probabilities, we divide each of the numerators by their sum:

$$\hat{P}(\text{delayed}) \mid \text{Carrier} = \text{DL}, \text{DayofWeek} = 7, \text{DepartureTime} = 1000 - 1059,$$
$$\text{Destination} = \text{LGA}, \text{Origin} = \text{DCA}$$
$$= \frac{0.0000181}{0.0000181 + 0.000333} = 0.052,$$
$$\hat{P}(\text{ontime} \mid \text{Carrier} = \text{DL}, \text{DayofWeek} = 7, \text{DepartureTime} = 1000 - 1059,$$
$$\text{Destination} = \text{LGA}, \text{Origin} = \text{DCA}$$
$$= \frac{0.000333}{0.0000181 + 0.000333} = 0.948.$$

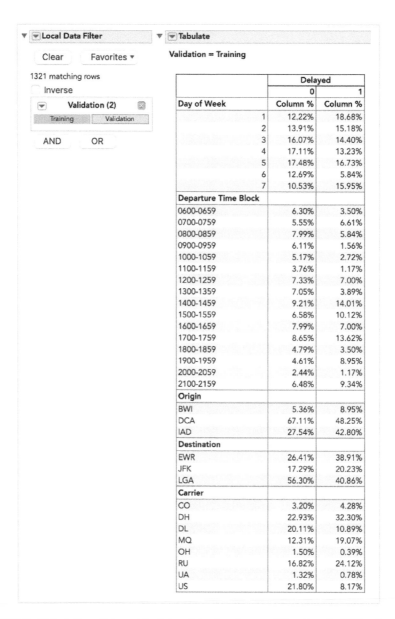

FIGURE 8.2 Tabulation of airport training data showing conditional probabilities for delayed and on-time flights for each class

Of course, we rely on the software to compute these probabilities for any records of interest (in the training set, validation set, or for scoring new data).

Evaluating the Performance of Naive Bayes output from JMP

By default, the naive Bayes report includes confusion matrices, ROC curves, and lift curves. For this example, the results are shown in Figure 8.3. We see that the misclassification rate for the validation set is 23% (the accuracy is 77%). In comparison, a naive rule that would

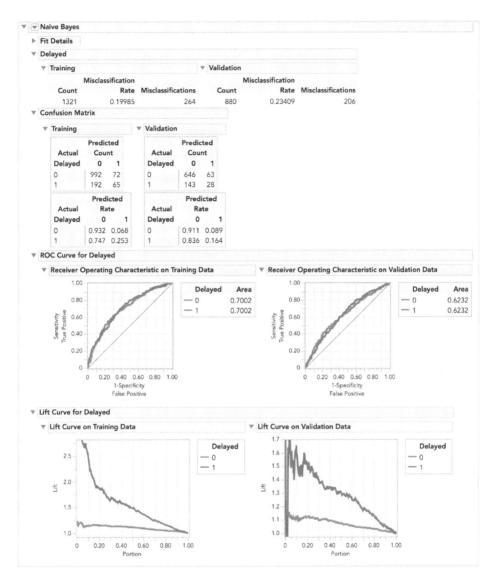

FIGURE 8.3   **Confusion matrix, ROC curves, and lift curves**

classify all 880 flights in the validation set as "ontime" would have missed the 171 delayed flights, resulting in an 80.6% accuracy. Thus, by a simple accuracy or misclassification measure, the naive Bayes model does only slightly worse than the naive rule. However, the lift chart (Figure 8.3), shows the strength of the naive Bayes in capturing the delayed flights effectively, when the goal is ranking.

Working with Continuous Predictors

Naive Bayes can be extended to numeric (continuous) predictor variables, most commonly by assuming a Gaussian (normal) distribution for these variables. This is useful when numeric predictors are present in the data and transforming them to categorical data types can lead to significant information loss. In applying the *Gaussian naive Bayes* method, each numeric variable is summarized in the form of its mean and standard deviation, for each outcome class. If we assume a normal (Gaussian) distribution for the numerical variables, we can use the normal curve to "discretize" the numerical values. This is done by replacing μ and σ with the predictor mean and standard deviation in the Gaussian probability density function $f(x)$:

$$f(x) = \frac{1}{\sigma\sqrt{2\pi}} \exp\left(-\frac{1}{2}\left(\frac{x-\mu}{\sigma}\right)^2\right). \tag{8.4}$$

In JMP Pro, you can fit a naive Bayes classifier with both categorical and continuous predictors. Using equation (8.4), probability density values are computed for each numeric variable, for each outcome class. The idea is that the probability of any exact value is zero for a continuous variable, but by using the probability density function, we are essentially considering a small range around the required exact value and since this range is the same while computing the density values for each class label, it is possible to use the probability density values in a manner similar to the conditional probability values for categorical predictors in the naive Bayes probability computation.

To illustrate how the naive Bayes computations work for continuous predictors, suppose Distance were to be considered as the predictor for whether the flight was delayed or not. The data from the summary statistics shown in Figure 8.4 can be plugged into equation (8.4) to compute probability density values for any specific distance value.

▼ ▼ Tabulate

		Delayed	
		0	1
Distance	Mean	211.99	211.36
	Std Dev	12.79	15.315

FIGURE 8.4 Descriptive statistics for Distance as a continuous predictor for flight delays

For example, for Distance = 213, we get:

$$\hat{P}(\text{Distance} = 213 \mid \text{delayed}) = 0.025$$
$$\hat{P}(\text{Distance} = 213 \mid \text{ontime}) = 0.031.$$

As before, we compute only the numerator in the naive Bayes formula (equation (8.3)), indicated with the $\propto$ symbol, as follows:

$$\hat{P}(\text{delayed} \mid \text{Distance} = 213) \propto (0.025)(0.195) = 0.005$$
$$\hat{P}(\text{ontime} \mid \text{Distance} = 213) \propto (0.031)(0.805) = 0.025.$$

By dividing each of the numerators by their sum, we obtain the actual probability values:

$$\hat{P}(\text{delayed} \mid \text{Distance} = 213) = \frac{0.005}{0.005 + 0.025} = 0.167,$$
$$\hat{P}(\text{ontime} \mid \text{Distance} = 213) = \frac{0.025}{0.005 + 0.025} = 0.833.$$

8.4 ADVANTAGES AND SHORTCOMINGS OF THE NAIVE BAYES CLASSIFIER

The naive Bayes classifier's beauty is in its simplicity, computational efficiency, reasonably good classification performance, and ability to handle categorical predictor variables directly. In fact, it often outperforms more sophisticated classifiers even when the underlying assumption of independent predictors is far from true. This advantage is especially pronounced when the number of predictors is very large.

There are three main issues that should be kept in mind, however. First, the naive Bayes classifier requires a very large number of records to obtain good results.

Second, where a predictor category is not present in the training data, naive Bayes assumes that a new record with that category of the predictor has zero probability. This can be a problem if this rare (or unobserved) predictor value is important. One example is the binary predictor *Weather* in the flights delay dataset, which we did not use for analysis, and which denotes bad weather. When the weather was bad, all flights were delayed. Consider another example, where the outcome variable is *bought high-value life insurance* and a predictor category is *owns yacht*. If the training data have no records with *owns yacht* = 1, for any new records where *owns yacht* = 1, naive Bayes will assign a probability of 0 to the outcome variable *bought high-value life insurance*. With no training records with *owns yacht* = 1, of course, no machine learning technique will be able to incorporate this potentially important variable into the classification model—it will be ignored. With naive Bayes, however, the absence of this predictor actively "outvotes" any other information in the record to assign a 0 to the outcome value (when, in this case, it has a relatively good chance of being a 1). The presence of a large training set (and judicious binning of continuous variables, if required) helps mitigate this effect.

Finally, good performance is obtained when the goal is *classification* or *ranking* of records according to their probability of belonging to a certain class. However, when the goal is actually to *estimate* the probability of class membership (propensity), this method provides very biased results. For this reason, the naive Bayes method is rarely used in credit scoring (Larsen, 2005).

SPAM FILTERING

Filtering spam in email has long been a widely familiar application of machine learning. Spam filtering, which is based in large part on natural language vocabulary, is a natural fit for a naive Bayes classifier, which uses exclusively categorical variables. Most spam filters are based on this method, which works as follows:

1. Humans review a large number of emails, classify them as "spam" or "not spam," and from these select an equal (also large) number of spam emails and nonspam emails. This is the training data.

2. These emails will contain thousands of words; for each word compute the frequency with which the word occurs in the spam dataset and the frequency with which it occurs in the nonspam dataset. Convert these frequencies into estimated probabilities (i.e., if the word "free" occurs in 500 out of 1000 spam emails, and only 100 out of 1000 nonspam emails, the probability that a spam

email will contain the word "free" is 0.5, and the probability that a nonspam email will contain the word "free" is 0.1).

3. If the only word in a new message that needs to be classified as spam or not-spam is "free," we would classify the message as spam, since the Bayesian posterior probability is 0.5/(0.5+0.1) or 5/6 that, given the appearance of "free," the message is spam.

4. Of course, we will have many more words to consider. For each such word, the probabilities described in step 2 are calculated, and multiplied together, and formula (8.3) is applied to determine the naive Bayes probability of belonging to the classes. In the simple version, class membership (spam or not spam) is determined by the higher probability.

5. In a more flexible interpretation, the ratio between the "spam" and "not-spam" probabilities is treated as a score, for which the operator can establish (and change) a cutoff threshold—anything above that level is classified as spam.

6. Users have the option of building a personalized training database by classifying incoming messages as spam or not spam, and adding them to the training database. One person's spam may be another person's substance.

It is clear that, even with the "naive" simplification, this is an enormous computational burden. Spam filters now typically operate at two levels—at servers (intercepting some spam that never makes it to your computer) and on individual computers (where you have the option of reviewing it). Spammers have also found ways to "poison" the vocabulary-based Bayesian approach, by including sequences of randomly selected irrelevant words. Since these words are randomly selected, they are unlikely to be systematically more prevalent in spam than in nonspam, and they dilute the effect of key spam terms such as "Viagra" and "free." For this reason, sophisticated spam classifiers also include variables based on elements other than vocabulary, such as the number of links in the message, the vocabulary in the subject line, determination of whether the "From:" email address is the real originator (anti-spoofing), use of HTML and images, and origination at a dynamic or static IP address (the latter are more expensive and cannot be set up quickly).

PROBLEMS

8.1 **Personal Loan Acceptance.** The file `UniversalBank.jmp` contains data on 5000 customers of Universal Bank. The data include customer demographic information (age, income, etc.), the customer's relationship with the bank (mortgage, securities account, etc.), and the customer response to the last personal loan campaign (Personal Loan). Among these 5000 customers, only 480 (= 9.6%) accepted the personal loan that was offered to them in the earlier campaign. In this exercise we focus on two predictors: Online (whether or not the customer is an active user of online banking services) and Credit Card (abbreviated CC below) (does the customer hold a credit card issued by the bank), and the outcome Personal Loan (abbreviated Loan below).

  a. Partition the data into training (60%) and validation (40%) sets. (*Hint*: Prior to partitioning the data, review the summary statistics using distribution platform. Use personal loan variable as stratification column. This will partition the data into balanced sets based on the levels of the personal loan variable.) Check to make sure that the variables are coded as Nominal and that none of the variables has the *Value Labels* column property (remove this column property if needed).

  b. Create a summary of the training data using *Tabulate*, with Online as a column variable, CC as a row variable, and Loan as a secondary row variable. The values inside the cells should convey the count (how many records are in that cell).

  c. Consider the task of classifying a customer that owns a bank credit card and is actively using online banking services. Looking at the tabulation, what is the probability that this customer will accept the loan offer? (This is the probability of loan acceptance (Loan = 1) conditional on having a bank credit card (CC = 1) and being an active user of online banking services (Online = 1)).

  d. Create two tabular summaries of the data. One will have Loan (rows) as a function of Online (columns) and the other will have Loan (rows) as a function of CC.

  e. Compute the following quantities [$P(A|B)$ means "the probability of A given B"]:
   i. $P(CC = 1 \mid Loan = 1)$ (the proportion of credit card holders among the loan acceptors)
   ii. $P(Online = 1 \mid Loan = 1)$
   iii. $P(Loan = 1)$ (the proportion of loan acceptors)
   iv. $P(CC = 1 \mid Loan = 0)$
   v. $P(Online = 1 \mid Loan = 0)$
   vi. $P(Loan = 0)$

  f. Use the quantities computed above to compute the naive Bayes probability $P(Loan = 1 \mid CC = 1, Online = 1)$.

  g. Compare this value with the one obtained from the tabulation in (b). Which is a more accurate estimate of $P(Loan = 1 \mid CC = 1, Online = 1)$?

  h. In JMP, run naive Bayes predictive modeling on the data. Examine the model output on the training data, save the prediction formula to the data table, and find the records that correspond(s) to $P(Loan = 1 \mid CC = 1, Online = 1)$. Compare this to the number you obtained in (f).

8.2 **Automobile Accidents.** The file `Accidents.jmp` contains information on 42,183 actual automobile accidents in 2001 in the United States that involved one of three levels of injury: NO INJURY, INJURY, or FATALITY. For each accident, additional information is recorded, such as day of week, weather conditions, and road type. A firm might be interested in developing a system for quickly classifying the severity of an accident based on initial reports and associated data in the system (some of which rely on GPS-assisted reporting).

Our goal here is to predict whether an accident just reported will involve an injury (MAX_SEV_IR = 1 or 2) or will not (MAX_SEV_IR = 0). For this purpose, use *Recode* to create a new variable called INJURY that takes the value "yes" if MAX_SEV_IR = 1 or 2, and otherwise "no." (See Chapter 3 for information on using *Recode*.)

a. Using the information in this dataset, if an accident has just been reported and no further information is available, what should the prediction be? (INJURY = Yes or No?) Why?

b. Create a subset of the first 12 records in the dataset and look only at the response (INJURY) and the two predictors WEATHER_R and TRAF_CON_R.

    i. Create a tabular summary that examines INJURY as a function of the two predictors for these 12 records.

    ii. Compute the exact Bayes conditional probabilities of an injury (INJURY = Yes) given the six possible combinations of the predictors.

    iii. Classify the 12 accidents using these probabilities and a threshold of 0.5.

    iv. Compute manually the naive Bayes conditional probability of an injury given WEATHER_R = 1 and TRAF_CON_R = 1.

    v. Run a naive Bayes classifier on the 12 records and two predictors using the *Naive Bayes* platform. Save the prediction formula to the data table to classify the 12 records. Compare this to the exact Bayes classification. Are the resulting classifications equivalent?

c. Now return to the entire dataset. Partition the data into training (60%) and validation (40%)

    i. Assuming that no information or initial reports about the accident itself are available at the time of prediction (only location characteristics, weather conditions, etc.), which predictors can we include in the analysis? (Hold your mouse over the column names in *Columns Panel* in the data table for descriptions of the predictors.)

    ii. Run a naive Bayes classifier on the data set with the relevant predictors (and INJURY as the response). Show the confusion matrix.

    iii. What is the overall error for the validation set?

    iv. What is the percent improvement relative to the naive rule (using the validation set)?

9

CLASSIFICATION AND REGRESSION TREES

This chapter describes a flexible data-driven method that can be used for both classification (called *classification tree*) and prediction (called *regression tree*). Among the data-driven methods, trees (also known as *decision trees*) are the most transparent and easy to interpret. Trees are based on separating records into subgroups by creating splits on predictors. These splits create logical rules that are transparent and easily understandable, such as "IF Age < 55 AND Education > 12, THEN class = 1." The resulting subgroups should be more homogeneous in terms of the outcome variable, thereby creating useful prediction or classification rules. We discuss the two key ideas underlying trees: *recursive partitioning* (for constructing the tree) and *pruning* (for cutting the tree back). In the context of tree construction, we also describe a few metrics of homogeneity that are popular in tree algorithms for determining the homogeneity of the resulting subgroups of records. We explain that pruning is a useful strategy for avoiding overfitting and describe alternative strategies for avoiding overfitting. As with other data-driven methods, trees require large amounts of data. However, once constructed, they are computationally cheap to deploy even on large samples. They also have other advantages such as being highly automated, robust to outliers, and able to handle missing values. In addition to prediction and classification, we describe how trees can be used for dimension reduction. Finally, we introduce *random forests* and *boosted trees*, which combine results from multiple trees to improve predictive power.

Classification and Regression Trees in JMP: Classification and regression trees are available in the standard version of JMP. JMP Pro is required to fit decision trees with a validation column or to fit random forests or boosted trees.

Machine Learning for Business Analytics: Concepts, Techniques, and Applications with JMP Pro®,
Second Edition. Galit Shmueli, Peter C. Bruce, Mia L. Stephens, Muralidhara Anandamurthy, and Nitin R. Patel.
© 2023 John Wiley & Sons, Inc. Published 2023 by John Wiley & Sons, Inc.

9.1   INTRODUCTION

If one had to choose a classification technique that performs well across a wide range of situations without requiring much effort from the analyst while being readily understandable by the consumer of the analysis, a strong contender would be a tree. Trees, also called *decision trees*, can be used for both classification and prediction. A procedure developed by Breiman et al. (1984) is called CART (classification and regression trees). A related procedure is called C4.5. JMP uses a tree approach called *recursive partitioning*.

What is a classification tree? Figure 9.1 shows a tree for classifying bank customers who receive a loan offer. Customers either accept (1) or don't accept (0) the offer. The tree shows the probability of accepting or not accepting the loan, as a function of information such as income, education level, and average credit card expenditure.

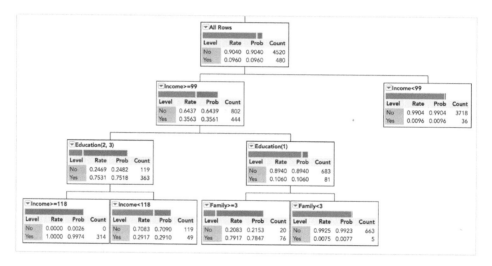

FIGURE 9.1   Example of a tree for classifying bank customers as loan acceptors or nonacceptors

Tree Structure

We have two types of nodes in a tree: decision (= splitting) nodes and leaf nodes. Nodes that have successors are called *decision nodes* because if we were to use a tree to classify a new record for which we knew only the values of the predictor variables, we would "drop" the record down the tree so that at each decision node, the appropriate branch is taken until we get to a node that has no successors. Such nodes are called the *leaf nodes* (or *terminal nodes* or *leaves* of the tree) and represent the partitioning of the data by predictors.

Decision trees in JMP, for both classification and prediction, are fit using the *Partition* platform under *Analyze > Predictive Modeling*. For decision nodes, the name of the predictor variable chosen for splitting appears at the top. Of the two child nodes connected below a decision node, the right box is for records that meet the splitting condition ("< splitting value"), while the left box is for records that meet the complementary condition (">= splitting value").

Decision Rules

One of the reasons that classification trees are very popular is that they provide easily understandable decision rules (at least if the trees are not too large). Consider the tree in Figure 9.1. The splitting node lists the predictor name. The colored bar within each node gives a visual indication of the proportion and counts of the two classes in that node. This tree can easily be translated into a set of rules for classifying a bank customer. For example, the bottom node under the "Family < 3" in this tree gives us the following rule:

IF(*Income* ≥ 99) AND (*Education* = 1) AND (*Family* < 3),
    THEN *Prob(Class = 1)* = 0.0077 (classify as 0, don't accept).

Classifying a New Record

To classify a new record, it is "dropped" down the tree. When it has dropped all the way down to a leaf node, we can assign its class simply by taking a "vote" (or average, if the outcome is numerical) of all the training data that belonged to the leaf node when the tree was grown. The class with the highest vote is assigned to the new record. For instance, a new record reaching the leftmost leaf node in Figure 9.1, which has a majority of records that belong to the acceptor class (1), would be classified as "acceptor." Alternatively, we can look at the proportion of acceptors in the node (probability) and compare it to a user-specified threshold value. For example, applying a threshold of 0.5 to the probability of acceptance in the leftmost node (0.7518) would lead to a classification of "acceptor."

See Chapter 5 for further discussion of the use of a threshold value in classification, for cases where a single class is of interest.

In the following sections, we show how trees are constructed and evaluated.

9.2   CLASSIFICATION TREES

The key idea underlying tree construction is *recursive partitioning* of the space of the predictor variables. (This is the root of the JMP platform name, *Partition*.) A second important issue is avoiding overfitting. We start by explaining recursive partitioning and then describe approaches for evaluating and fine-tuning trees while avoiding overfitting.

Recursive Partitioning

Let us denote the outcome (target, response) variable by Y and the predictor (input) variables by $X_1, X_2, X_3, \ldots, X_p$. In classification, the outcome will be a categorical variable, sometimes called a class label. Specifically, in binary classification, the outcome will be binomial (it will have a *nominal* modeling type in JMP). Recursive partitioning divides up the p-dimensional space of the X variables into nonoverlapping rectangles. The predictor variables here can be continuous (numeric) or categorical (*nominal* or *ordinal* modeling type). This division is accomplished recursively (i.e., operating on the results of prior divisions).

First, one of the variables is selected, say X_i, and a value of X_i, say s_i, is chosen to split the p-dimensional space into two parts: one part that contains all the points with $X_i < s_i$ and the other with all the points with $X_i \geq s_i$. Then one of these two parts is divided in a similar manner by choosing a variable again (it could be X_i or another variable) and a split value for the variable. This results in three rectangular regions. This process is continued so that we get smaller and smaller rectangular regions. The idea is to divide the entire X-space up into rectangles such that each rectangle is as homogeneous or "pure" as possible. By *pure*, we mean containing records that belong to just one class. (Of course, this is not always possible, as there may be records that belong to different classes but have exactly the same values for every one of the predictor variables.)

Let us illustrate recursive partitioning with an example.

Example 1: Riding Mowers

We again use the riding mower example presented in Chapter 7. A riding mower manufacturer would like to find a way of classifying families in a city into those likely to purchase a riding mower and those not likely to buy one. A pilot random sample of 12 owners and 12 nonowners in the city is collected. The data are shown and plotted in Table 9.1 and in Figure 9.2, respectively.

TABLE 9.1   Lot Size, Income, and Ownership of a Riding Mower for 24 Households

Household number	Income ($000s)	Lot size (000s ft^2)	Ownership of riding mower
1	60.0	18.4	Owner
2	85.5	16.8	Owner
3	64.8	21.6	Owner
4	61.5	20.8	Owner
5	87.0	23.6	Owner
6	110.1	19.2	Owner
7	108.0	17.6	Owner
8	82.8	22.4	Owner
9	69.0	20.0	Owner
10	93.0	20.8	Owner
11	51.0	22.0	Owner
12	81.0	20.0	Owner
13	75.0	19.6	Non-owner
14	52.8	20.8	Non-owner
15	64.8	17.2	Non-owner
16	43.2	20.4	Non-owner
17	84.0	17.6	Non-owner
18	49.2	17.6	Non-owner
19	59.4	16.0	Non-owner
20	66.0	18.4	Non-owner
21	47.4	16.4	Non-owner
22	33.0	18.8	Non-owner
23	51.0	14.0	Non-owner
24	63.0	14.8	Non-owner

If we apply the classification tree procedure to these data, the procedure will choose *Income* for the first split with a splitting value of 85.5. The Income, Lot_Size space is now

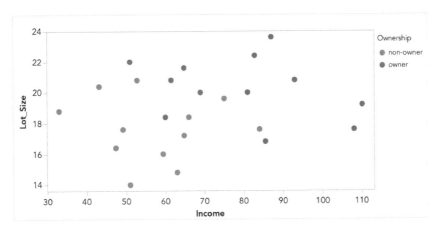

FIGURE 9.2 **Scatterplot of** *Lot Size* **vs.** *Income* **for 24 owners and non-owners of riding mowers**

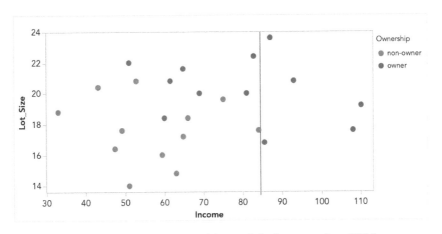

FIGURE 9.3   **Splitting the 24 records by income value of 85.5**

divided into two rectangles, one with Income $<$ 85.5 and the other with Income $\geq$ 85.5. This is illustrated in Figure 9.3.

Notice how the split has created two rectangles, each of which is much more homogeneous than the rectangle before the split. The rectangle on the right contains points that are all owners (five owners) and the rectangle on the left contains mostly non-owners (twelve non-owners and seven owners).

How was this particular split selected? The algorithm examined each predictor variable (in this case, Income and Lot Size) and all possible split values for each variable to find the best split. What are the possible split values for a variable? They are simply each value for the variable. There are 22 unique values for Income and 18 unique values for Lot Size. All of these split points, or *candidates*, are tested to see which split separates the probability of ownership the most. The split that drives the greatest dissimilarity in the probabilities (or proportions) for the two rectangles is selected as the split point.

Categorical Predictors

The riding mower example used numerical predictors. However, categorical predictors can also be used in the recursive partitioning context. To handle categorical predictors, the split choices for a categorical predictor are all ways in which the set of categories can be divided into two subsets. For example, a categorical variable with four categories, say $\{a, b, c, d\}$, can be split in seven ways into two subsets: $\{a\}$ and $\{b, c, d\}$, $\{b\}$ and $\{a, c, d\}$, $\{c\}$ and $\{a, b, d\}$, $\{d\}$ and $\{a, b, c\}$, $\{a, b\}$ and $\{c, d\}$, $\{a, c\}$ and $\{b, d\}$, and $\{a, d\}$ and $\{b, c\}$. When the number of categories is large, the number of splits becomes very large.[1] For ordinal predictors, JMP provides an option to restrict the ordering of the splits. So, for an ordinal variable with four categories, $\{1, 2, 3, 4\}$, the only possible splits are $\{1\}$ and $\{2, 3, 4\}$, $\{1, 2\}$ and $\{3, 4\}$, and $\{1, 2, 3\}$ and $\{4\}$.

Standardization

Whether predictors are numerical or categorical, it does not make any difference if they are standardized (normalized) or not.

9.3   GROWING A TREE FOR RIDING MOWERS EXAMPLE

The initial recursive partition output for the mower data is shown in Figure 9.4. The graphic displays the overall proportion of non-owners and owners in the dataset, where the y-axis

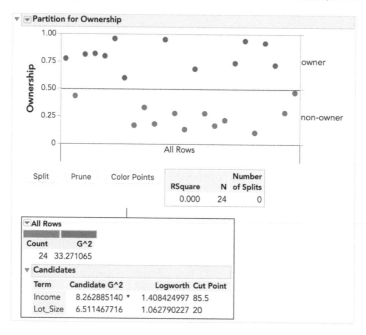

FIGURE 9.4   The initial recursive partition output in JMP

---

[1]Recursive partitioning in JMP (and CART) differs from that in C4.5: JMP performs only binary splits, leading to binary trees, whereas C4.5 performs splits as large as the number of categories, leading to "bushlike" structures.

value for the horizontal line is the proportion of non-owners (0.50). This shows the situation prior to any splitting. For each predictor, JMP tests all possible splits and measures the difference in the proportion of owners and non-owners between each pair of split groups. For each predictor, the optimal split, or *Cut Point*, is listed under *Candidates*.

The G^2 statistic,[2] reported under Candidates, provides a measure of the dissimilarity in the proportions at the cut point for each variable. The *LogWorth*, which is based on the p-value of this statistic, is used to determine the optimal split location.[3] The larger the *LogWorth* value, the lower the p-value and the better the split (Sall, 2002). In this example, the predictor with the highest *LogWorth* is Income, and the cut point for Income is 85.5.

Choice of First Split

The tree after one split is shown in Figure 9.5. It has two leaf nodes. The graph at the top is updated to show the two split groups (Income $\geq$ 85.5 and Income < 85.5). The horizontal line is the proportion of non-owners in each split group. As we saw in Figure 9.3, when Income $\geq$ 85.5, all of the records (5) are owners, and when Income < 85.5, 12 out of the 17 are non-owners.

Alternatively, we can look at the results of splitting in terms of the proportions of the two classes in each node. The red triangle option *Display Options > Show Split Prob* displays both *Rate*, which is the proportion of records in a node, and *Prob*, the predicted probabilities for the node. These numbers may differ (as they do in this example; see Figure 9.5).[4] JMP uses the predicted probability *Prob* values for classification.

FITTING CLASSIFICATION TREES IN JMP

Trees are fit using *Analyze > Predictive Modeling > Partition*. Use the *Split* button to grow the tree and the *Prune* button to prune the tree back. A few options of note:

1. To show split statistics within each node, select *Display Options > Show Split Prob* or *Show Split Count* from the top red triangle.
2. To show a high-level summary of the tree, select *Show Leaf Report* or *Small Tree View* from the top red triangle.
3. The minimum split size can be changed using a red triangle option.
4. To save the formula for the tree to the data table, select *Save Columns > Save Prediction Formula*.
5. To change the threshold value for classification, use the *Decision Threshold* from the top red triangle (see Chapter 5 for details).

---

[2] G^2 is the likelihood-ratio chi-squared test statistic.
[3] Because several different splits are tested, the p-value is adjusted. The *LogWorth* is equal to $-\log_{10}$ (p-value).
[4] Predicted probabilities are calculated using a weighting factor to ensure probabilities are always nonzero. This allows logs of probabilities to be calculated on validation sets (this is needed to compute the validation RSquare).

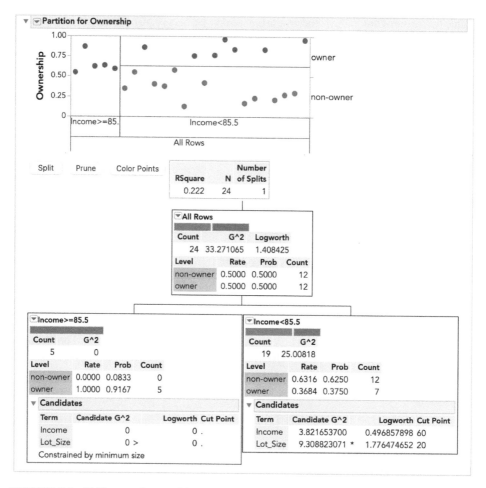

FIGURE 9.5   JMP recursive partition output after 1 split. *Display Options > Show Split Prob* **was selected to show** *Rates* **and** *Probs*

Choice of Second Split

If Income ≥ 85.5, all the records are owners (no further splitting in that branch is possible), so the next candidate split is under Income < 85.5 (see the bottom of Figure 9.5). The split variable is Lot Size, and the cut point is 20.

Final Tree

We change the minimum split size to 1 record (this is a top red triangle option—the default is 5) and continue to split until no further splits are permitted. Figure 9.6 shows the final tree. Each of the leaf nodes contains either all owners or all non-owners, so no additional splits are possible.

To help interpret the tree, we use the *Leaf Report* (see Figure 9.7). This provides a summary of our tree and displays the rules (the split *Prob* values) for classifying the outcome. The leaf node on the right in the tree corresponds to the last row on the leaf report: if Income < 85.5 and Lot Size < 17.6, then the probability P(Ownership = owner) is 0.0752.

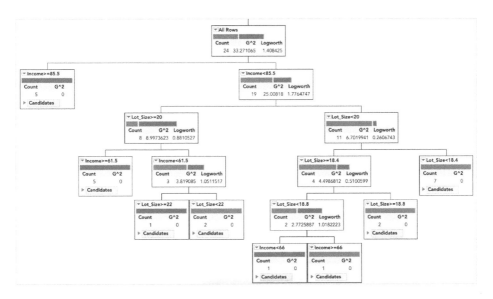

FIGURE 9.6   JMP recursive partition output for full tree

▼ **Leaf Report**

Response Prob

Leaf Label	non-owner	.2 .4 .6 .8	owner
Income>=85.5	0.0833		0.9167
Income<85.5&Lot_Size>=20&Income>=61.5	0.0815		0.9185
Income<85.5&Lot_Size>=20&Income<61.5&Lot_Size>=22	0.2512		0.7488
Income<85.5&Lot_Size>=20&Income<61.5&Lot_Size<22	0.8342		0.1658
Income<85.5&Lot_Size<20&Lot_Size>=18.4&Lot_Size<18.8&Income<66	0.2803		0.7197
Income<85.5&Lot_Size<20&Lot_Size>=18.4&Lot_Size<18.8&Income>=66	0.7803		0.2197
Income<85.5&Lot_Size<20&Lot_Size>=18.4&Lot_Size>=18.8	0.8550		0.1450
Income<85.5&Lot_Size<20&Lot_Size<18.4	0.9436		0.0564

Response Counts

Leaf Label	non-owner		owner
Income>=85.5	0		5
Income<85.5&Lot_Size>=20&Income>=61.5	0		5
Income<85.5&Lot_Size>=20&Income<61.5&Lot_Size>=22	0		1
Income<85.5&Lot_Size>=20&Income<61.5&Lot_Size<22	2		0
Income<85.5&Lot_Size<20&Lot_Size>=18.4&Lot_Size<18.8&Income<66	0		1
Income<85.5&Lot_Size<20&Lot_Size>=18.4&Lot_Size<18.8&Income>=66	1		0
Income<85.5&Lot_Size<20&Lot_Size>=18.4&Lot_Size>=18.8	2		0
Income<85.5&Lot_Size<20&Lot_Size<18.4	7		0

FIGURE 9.7   JMP leaf report for full tree

Using a Tree to Classify New Records

As explained earlier, we can use the tree to classify an existing or new record by dropping the record down the tree until it reaches a leaf node. The leaf node's probability is then used to classify the record. Figure 9.8 shows the probabilities and classifications for the 24 (training) records, resulting in perfect classification. Such perfect performance is indicative of overfitting. We discuss this next.

CLASSIFYING A NEW RECORD IN JMP

In building a classification tree, we build a model. We can save the prediction formula for this model to the data table (see Figure 9.8). This creates columns with the estimated probabilities for each class, along with the most likely class (the classification). When a new record is added to the data table, the estimated probability for each class is calculated and the most likely class is provided (using a threshold of 0.5). Note that after the prediction formula is saved in JMP data table, *Graph > Profiler* can be used to explore predicted probabilities for given values of the predictors. (*Profiler* is also a red triangle option in the Partition output.)

	Income	Lot_Size	Ownership	Prob(Ownership ==non-owner)	Prob(Ownership ==owner)	Most Likely Ownership
1	60	18.4	owner	0.2803	0.7197	owner
2	85.5	16.8	owner	0.0833	0.9167	owner
3	64.8	21.6	owner	0.0815	0.9185	owner
4	61.5	20.8	owner	0.0815	0.9185	owner
5	87	23.6	owner	0.0833	0.9167	owner
6	110.1	19.2	owner	0.0833	0.9167	owner
7	108	17.6	owner	0.0833	0.9167	owner
8	82.8	22.4	owner	0.0815	0.9185	owner
9	69	20	owner	0.0815	0.9185	owner
10	93	20.8	owner	0.0833	0.9167	owner
11	51	22	owner	0.2512	0.7488	owner
12	81	20	owner	0.0815	0.9185	owner
13	75	19.6	non-owner	0.8550	0.1450	non-owner
14	52.8	20.8	non-owner	0.8342	0.1658	non-owner
15	64.8	17.2	non-owner	0.9436	0.0564	non-owner
16	43.2	20.4	non-owner	0.8342	0.1658	non-owner
17	84	17.6	non-owner	0.9436	0.0564	non-owner
18	49.2	17.6	non-owner	0.9436	0.0564	non-owner
19	59.4	16	non-owner	0.9436	0.0564	non-owner
20	66	18.4	non-owner	0.7803	0.2197	non-owner
21	47.4	16.4	non-owner	0.9436	0.0564	non-owner
22	33	18.8	non-owner	0.8550	0.1450	non-owner
23	51	14	non-owner	0.9436	0.0564	non-owner
24	63	14.8	non-owner	0.9436	0.0564	non-owner

FIGURE 9.8   Riding mower data table with saved prediction formula and classifications from the full tree model

9.4   EVALUATING THE PERFORMANCE OF A CLASSIFICATION TREE

We have seen with previous methods that the modeling job is not completed by fitting a model to training data; we need out-of-sample data to assess and tune the model. This is particularly true with classification and regression trees, for two reasons:

1. Tree structure can be quite unstable, shifting substantially depending on the sample chosen.
2. A fully-fit tree will invariably lead to overfitting.

To visualize the first challenge, potential instability, imagine that we partition the data randomly into two samples, A and B, and we build a tree with each. If there are several predictors of roughly equal predictive power, you can see that it would be easy for samples A and B to select different predictors for the top level split, just based on which records ended up in which sample. And a different split at the top level would likely cascade down and yield completely different sets of rules. So we should view the results of a single tree with some caution. To illustrate the second challenge, overfitting, let's examine another example.

Example 2: Acceptance of Personal Loan

Universal Bank is a relatively young bank that is growing rapidly in terms of overall customer acquisition. The majority of these customers are liability customers with varying sizes of relationship with the bank. The customer base of asset customers is quite small, and the bank is interested in growing this base rapidly to bring in more loan business. In particular, it wants to explore ways of converting its liability (deposit) customers to personal loan customers.

A campaign the bank ran for liability customers showed a healthy conversion rate of over 9% successes. This has encouraged the retail marketing department to devise smarter campaigns with better target marketing. The goal of our analysis is to model the previous campaign's customer behavior in order to analyze what combination of factors make a customer more likely to accept a personal loan. This will serve as the basis for the design of a new campaign.

The bank's dataset includes data on 5000 customers. The data include customer demographic information (age, income, etc.), customer response to the last personal loan campaign (Personal Loan), and the customer's relationship with the bank (mortgage, securities account, etc.). Among these 5000 customers, only 480 (= 9.6%) accepted the personal loan that was offered to them in the earlier campaign. Figure 9.9 contains a sample of the bank's customer database for 20 customers (the data are in `UniversalBankCT.jmp`).

ID	Person al Loan	Age	Experi ence	Inco me	ZIP Code	Family	CCAvg	Educat ion	Mortga ge	Securities Account	CD Account	Online	Credit Card
1	No	25	1	49	91107	4	1.6	1	0	1	0	0	0
2	No	45	19	34	90089	3	1.5	1	0	1	0	0	0
3	No	39	15	11	94720	1	1	1	0	0	0	0	0
4	No	35	9	100	94112	1	2.7	2	0	0	0	0	0
5	No	35	8	45	91330	4	1	2	0	0	0	0	1
6	No	37	13	29	92121	4	0.4	2	155	0	0	1	0
7	No	53	27	72	91711	2	1.5	2	0	0	0	1	0
8	No	50	24	22	93943	1	0.3	3	0	0	0	0	1
9	No	35	10	81	90089	3	0.6	2	104	0	0	1	0
10	Yes	34	9	180	93023	1	8.9	3	0	0	0	0	0
11	No	65	39	105	94710	4	2.4	3	0	0	0	0	0
12	No	29	5	45	90277	3	0.1	2	0	0	0	1	0
13	No	48	23	114	93106	2	3.8	3	0	1	0	0	0
14	No	59	32	40	94920	4	2.5	2	0	0	0	1	0
15	No	67	41	112	91741	1	2	1	0	1	0	0	0
16	No	60	30	22	95054	1	1.5	3	0	0	0	1	1
17	Yes	38	14	130	95010	4	4.7	3	134	0	0	0	0
18	No	42	18	81	94305	4	2.4	1	0	0	0	0	0
19	Yes	46	21	193	91604	2	8.1	3	0	0	0	0	0
20	No	55	28	21	94720	1	0.5	2	0	1	0	0	1

FIGURE 9.9   Sample of 20 customers of universal bank

First, we create a validation column (*Analyze > Predictive Modeling > Make Validation Column*) and divide the data into three random subsets using a 0.50/0.25/0.25 split: training (2500 records), validation (1500 records), and test (1000 records). This validation column has been saved to the data table. We set the minimum split size to 1 and then construct a full-grown tree (clicking *Split* repeatedly until the tree stops splitting). A full-grown tree is shown in Figure 9.10. The tree after 10 splits is shown in Figure 9.11.

A look at the top tree node reveals that the first predictor that is chosen to split the data is Income. The split value, shown in the next two nodes (the second row in the tree), is 99 ($000s).

Since the full-grown tree leads to completely pure leaf nodes, it is 100% accurate in classifying the training data. This can be seen in Figure 9.12. In contrast, the confusion matrix for the validation and test data (which were not used to construct the full-grown tree) show lower classification accuracy. The main reason is that the full-grown tree overfits the training data (to complete accuracy!). This motivates the next section, where we describe ways to avoid overfitting either by stopping the growth of the tree before it is fully grown or by pruning the full-grown tree.

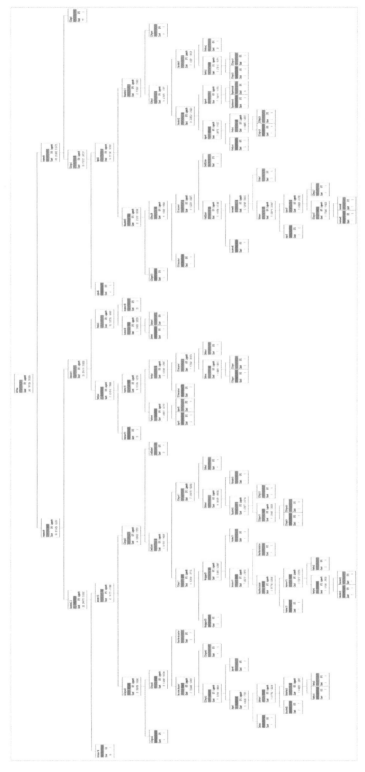

FIGURE 9.10   A full tree for the universal bank data using the training set (2500 records)

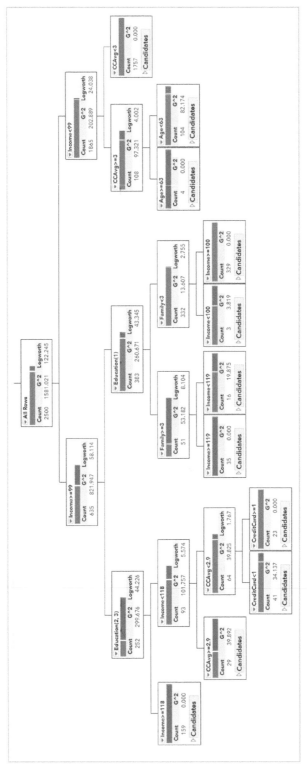

FIGURE 9.11   Tree view of first 10 splits for the universal bank data using the training set (2500 records)

▼ **Fit Details**

Measure	Training	Validation	Test	Definition
Entropy RSquare	0.9767	0.8424	0.7224	1-Loglike(model)/Loglike(0)
Generalized RSquare	0.9832	0.8812	0.7825	$(1-(L(0)/L(model))^{\wedge}(2/n))/(1-L(0)^{\wedge}(2/n))$
Mean -Log p	0.0074	0.0498	0.0878	$\sum -Log(\rho[j])/n$
RASE	0.0369	0.1076	0.1424	$\sqrt{\sum(y[j]-\rho[j])^2/n}$
Mean Abs Dev	0.0065	0.0165	0.0246	$\sum |y[j]-\rho[j]|/n$
Misclassification Rate	0.0000	0.0147	0.0240	$\sum (\rho[j] \neq \rho Max)/n$
N	2500	1500	1000	n

▼ **Confusion Matrix**

	Training			Validation			Test	
Actual Personal Loan	**Predicted Count**		**Actual Personal Loan**	**Predicted Count**		**Actual Personal Loan**	**Predicted Count**	
	No	**Yes**		**No**	**Yes**		**No**	**Yes**
No	2260	0	No	1347	9	No	897	7
Yes	0	240	Yes	13	131	Yes	17	79

Actual Personal Loan	**Predicted Rate**		**Actual Personal Loan**	**Predicted Rate**		**Actual Personal Loan**	**Predicted Rate**	
	No	**Yes**		**No**	**Yes**		**No**	**Yes**
No	1.000	0.000	No	0.993	0.007	No	0.992	0.008
Yes	0.000	1.000	Yes	0.090	0.910	Yes	0.177	0.823

FIGURE 9.12   Universal bank data: classification matrix and error rates for the training, validation, and test data using the full tree

9.5 AVOIDING OVERFITTING

As the last example illustrated, using a full-grown tree (based on the training data) leads to overfitting of the data. As discussed in Chapter 5, overfitting will lead to poor performance on new data. If we look at the overall error at the various levels of the tree, it is expected to decrease as the number of levels grows until the point of overfitting. Of course, for the training data the overall error decreases more and more until it is zero (or close to zero[5]) at the maximum depth of the tree. However, for new data, the overall error is expected to decrease until the point where the tree models the relationship between class and the predictors. After that, the tree starts to model the noise in the training set, and we expect the overall error for the validation set to start increasing. This is depicted in Figure 9.13. One intuitive reason for the overfitting at the high levels of the tree is that these splits are based on very small numbers of records. In such cases, class difference is likely to be attributed to noise rather than predictor information.

Two ways to try and avoid exceeding this level, thereby limiting overfitting, are by setting rules to stop tree growth called *prepruning* or, alternatively, by *pruning* the full-grown tree back to a level where it does not overfit. These solutions are discussed below.

---

[5]Nonzero leaf nodes can result if the training data contains records that have identical values for all predictors, but have different outcome classes.

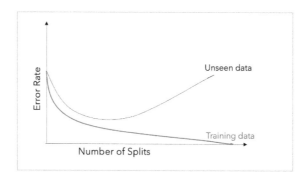

FIGURE 9.13   **Error rate as a function of the number of splits for training vs. holdout data**

Stopping Tree Growth: CHAID

One can think of different criteria for stopping the tree growth before it starts overfitting the data. Examples are tree depth (i.e., number of splits), minimum number of records in a leaf node, and minimum reduction in impurity. The problem is that it is not simple to determine what is a good stopping point using such rules.

A number of methods have been developed based on the idea of using rules to prevent the tree from growing excessively and overfitting the training data. One popular method called *CHAID* (chi-squared automatic interaction detection) is a recursive partitioning method that predates classification and regression tree (CART) procedures and is widely used in database marketing applications. It uses a well-known statistical test (the chi-square test for independence) to assess whether splitting a node improves the purity by a statistically significant amount. In particular, at each node we split on the predictor that has the strongest association with the response variable. The strength of association is measured by the p-value of a chi-squared test of independence. If, for the best predictor, the test does not show a significant improvement, the split is not carried out and the tree is terminated. This method is more suitable for categorical predictors, but it can be adapted to continuous predictors by binning the continuous values into categorical bins.

Growing a Full Tree and Pruning It Back

An alternative solution to stopping tree growth is pruning the full-grown tree. This is the basis of methods such as CART (developed by Breiman et al., implemented in multiple machine learning software packages such as SAS Enterprise Miner, R's *rpart* package, CART, and MARS) and C4.5 (developed by Quinlan and implemented in, e.g., IBM SPSS Modeler, Weka, and RapidMiner). In C4.5, the training data are used both for growing and pruning the tree. In CART, the innovation is to use the validation data to prune back the tree that is grown from training data. CART and CART-like procedures use validation data to prune back the tree that has deliberately been overgrown using the training data.

The idea behind pruning is to recognize that a very large tree is likely to be overfitting the training data, and that the weakest branches, which hardly reduce the error rate, should be removed. In the riding mower example, the last few splits resulted in nodes with very few points. We can see intuitively that these last splits are likely simply to be capturing noise in the training set rather than reflecting patterns that would occur in future data, such as the validation data. Pruning consists of successively lopping off the branches, thereby reducing the size of the tree. The pruning process trades off misclassification error in the validation dataset against the number of decision nodes in the pruned tree to arrive at a tree that captures the patterns, but not the noise, in the training data. We would like to find the point where the curve in Figure 9.13, for the unseen data, begins to increase.

Returning to the Universal Bank example, we expect that the classification accuracy of the validation set using the pruned tree would be higher than using the full-grown tree (although that is not the case here). However, the performance of the pruned tree on the validation data is not fully reflective of the performance on completely new data because the validation data were used for the pruning. This is a situation where it is particularly useful to evaluate the performance of the chosen model, whatever it may be, on a third set of data, the test set, which has not been used. In our example, the pruned tree applied to the test data yields an overall error rate of 1.9% (compared to 1.27% for the validation data). This is a highly accurate model, yielding only 19 errors in the test set. This example also shows the typical tendency for the test data to show higher error rates than the validation data, since the latter are, in effect, part of the model building process.

How JMP Pro Limits Tree Size

To prevent overfitting, JMP Pro uses a combination of limiting the initial growth of the tree through setting the minimum split size and pruning a tree after it has grown.

When a validation column is used, JMP Pro provides automated tree building and pruning via the *Go* button. The tree is built, one split at a time, and the RSquare for the training and validation sets is tracked. If the tree reaches a point where the validation RSquare stops increasing, the software continues building the tree for 10 additional splits. If no further increase in validation RSquare is attained after these additional splits, the tree is pruned back to the point of the highest validation RSquare value.[6]

The *Split History* report in JMP Pro tracks the path of the RSquare measures for the training, validation, and test sets. The report for the Universal Banking example is shown in Figure 9.14. While the training RSquare continues to climb after eight splits, the validation RSquare (the middle line) actually starts to decrease. At 18 splits, we continue to see no improvement in the validation RSquare, so the tree is pruned back to eight splits. The fit statistics for the final eight-split tree are shown at the bottom of Figure 9.14. (*Split History* and *Show Fit Details* are top red triangle options.)

While the misclassification rate for the training set is higher than what we observed for the full tree, the rate for the validation set is improved [0.0193 for the full model (see Figure 9.12) vs. 0.0127 for the pruned model].

---

[6]RSquare is typically not used in machine learning software for detecting overfitting. The RSquare computed in JMP for classification trees is the *Entropy RSquare*, defined as $1 - \frac{\log(\text{Likelihood}_{\text{Full}})}{\log(\text{Likelihood}_{\text{Reduced}})}$.

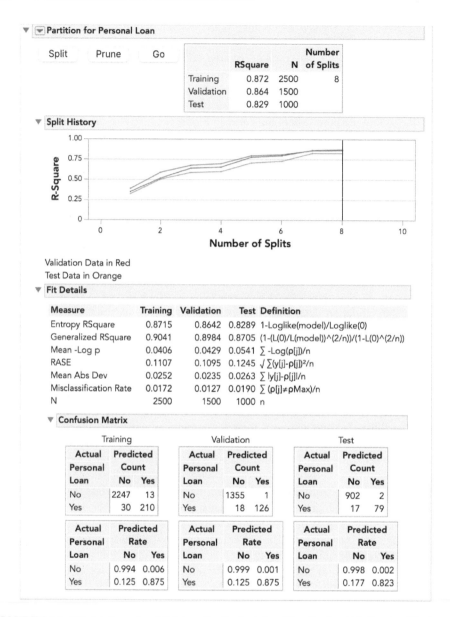

FIGURE 9.14   **Fit details and split history for the training (blue line), validation (red line), and test (orange line) sets for the universal bank data**

9.6   CLASSIFICATION RULES FROM TREES

As described in Section 9.1, classification trees provide easily understandable *classification rules* (if the trees are not too large). Each leaf node is equivalent to a classification rule. Returning to the example above, our pruned tree has eight splits. The leaf report provides rules for classification corresponding to the leaf nodes in our tree. The bottom row in the leaf report (in Figure 9.15), which represents the right-most leaf node, gives us the rule

Leaf Report

Response Prob

Leaf Label	No		Yes	
Income>=99&Education(2, 3)&Income>=118	0.0051		0.9949	
Income>=99&Education(2, 3)&Income<118&CCAvg>=2.9	0.4605		0.5395	
Income>=99&Education(2, 3)&Income<118&CCAvg<2.9	0.9048		0.0952	
Income>=99&Education(1)&Family>=3&Income>=119	0.0226		0.9774	
Income>=99&Education(1)&Family>=3&Income<119	0.6950		0.3050	
Income>=99&Education(1)&Family<3	0.9966		0.0034	
Income<99&CCAvg>=3&Income>=82	0.7205		0.2795	
Income<99&CCAvg>=3&Income<82	0.9447		0.0553	
Income<99&CCAvg<3	1.0000		0.0000	

Response Counts

Leaf Label	No		Yes	
Income>=99&Education(2, 3)&Income>=118	0		159	
Income>=99&Education(2, 3)&Income<118&CCAvg>=2.9	13		16	
Income>=99&Education(2, 3)&Income<118&CCAvg<2.9	58		6	
Income>=99&Education(1)&Family>=3&Income>=119	0		35	
Income>=99&Education(1)&Family>=3&Income<119	11		5	
Income>=99&Education(1)&Family<3	331		1	
Income<99&CCAvg>=3&Income>=82	38		15	
Income<99&CCAvg>=3&Income<82	52		3	
Income<99&CCAvg<3	1757		0	

FIGURE 9.15   Universal bank data: leaf report for pruned tree

IF (*Income* < 99) AND (*CCAvg* < 3), THEN the probability that *Class* = 1 is 0.

However, in many cases, the number of rules can be reduced by removing redundancies. For example, the top rule in the leaf report, corresponding to the left-most leaf node, is

IF (*Income* ≥ 99) AND (*Education* is 2 or 3) AND (*Income* ≥ 118),
THEN the probability that *Class* = 1 is 0.9949.

This rule can be simplified to

IF (*Income* ≥ 118) AND (*Education* is 2 or 3),
THEN the probability that *Class* = 1 is 0.9949.

This transparency in the process and understandability of the algorithm that leads to classifying a record as belonging to a certain class is advantageous in settings where the final classification is not solely of interest. Berry and Linoff (2000) give the example of health insurance underwriting, where the insurer is required to show that coverage denial is not based on discrimination. By showing rules that led to denial (e.g., income < $20K AND low credit history), the company can avoid law suits. Compared to the output of other classifiers, such as discriminant functions, tree-based classification rules are easily explained to managers and operating staff. As we will see, the decision tree structure and resulting model are certainly far more transparent and are easier to interpret (and explain!) than neural networks!

9.7   CLASSIFICATION TREES FOR MORE THAN TWO CLASSES

Classification trees can be used with an outcome that has more than two classes. In terms of determining optimal split or cut points, the split statistics were defined for m classes and hence can be used for any number of classes. The tree itself will have the same structure, except that its nodes include statistics for each of the m-class labels, and classifications are into one of the m-classes based on the highest estimated probability.

9.8   REGRESSION TREES

The tree method can also be used for a numerical outcome variable. Regression trees for prediction operate in much the same fashion as classification trees. The outcome variable, Y, is a numerical variable in this case, but both the principle and the procedure are the same. Many splits are attempted, and for each split, we compute a test statistic and use the *LogWorth* to determine the best cut point. For classification trees, the test statistic is G^2, the chi-square statistic. In regression trees, the test statistic is the *Sum of Squares*, which measures the difference in the means of the groups at the best cut point.

To illustrate a regression tree, consider the example of predicting prices of Toyota Corolla automobiles (from Chapter 6). The dataset includes information on 1000 sold Toyota Corolla cars (we use the dataset ToyotaCorolla1000.jmp). The goal is to find a predictive model of price as a function of 10 predictors (including mileage, horsepower, number of doors, etc.). A regression tree will be built using a training set of 600 records. The first split, shown in Figure 9.16, is on *Age*, and the cut point is 32. The difference in the means at this cut point, $7893.24, is reported in the first node. The horizontal line in the graph represents the overall mean price for each of the split groups, and the position of each point relative to the y-axis is the price for the particular record.

The small tree and split history for the final tree are shown in Figure 9.17. The tree was pruned back to 14 splits, using the validation set. We see that from the 10 predictor variables only six are useful for predicting price. This information is displayed in the *Column Contribution* report in Figure 9.18 (this is a top red triangle option). This report displays the number of splits and the total sum of squares attributed to each of the variables.

Prediction

Predicting the value of the response Y for a record is performed in a fashion similar to the classification case: each leaf node gives a predicted price. The *Leaf Report* (Figure 9.18, bottom) provides a summary of the regression tree, with rules for predicting the outcome. For instance, to predict the price of a Toyota Corolla with $Age = 30$ and $Weight = 1300$, we use the last row in the leaf report (or the right-most leaf node in the tree) and see the predicted mean price is $26,140. In regression trees, the leaf node is determined by the average outcome value of the training records in that leaf node. In the example above, the value $26,140 is the average of the five cars in the training set that fall in the category of Age < 32 and Weight ≥ 1265.

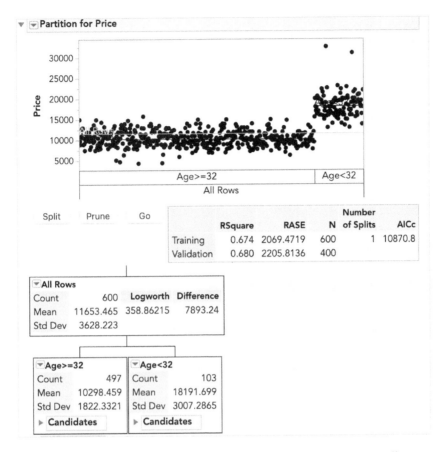

FIGURE 9.16   **Regression tree for Toyota Corolla prices after one split**

Classification trees produce models for making classifications, while regression trees produce models for making predictions. Formulas for these models can be saved to the data table. In this example, the saved formula can be used to predict prices for new records. The *Profiler* from the *Graph* menu can be used to explore this formula and make predictions. The model can also be explored using the *Profiler* from the top red triangle in the Partition output.

Evaluating Performance

As stated above, predictions are obtained by averaging the values of the outcome variable in the nodes. We therefore have the usual definition of predictions and errors. The predictive performance of regression trees can be measured in the same way that other predictive methods are evaluated, using summary measures such as RMSE. Additional statistics are available in the *Model Screening* and *Model Comparison* platforms.

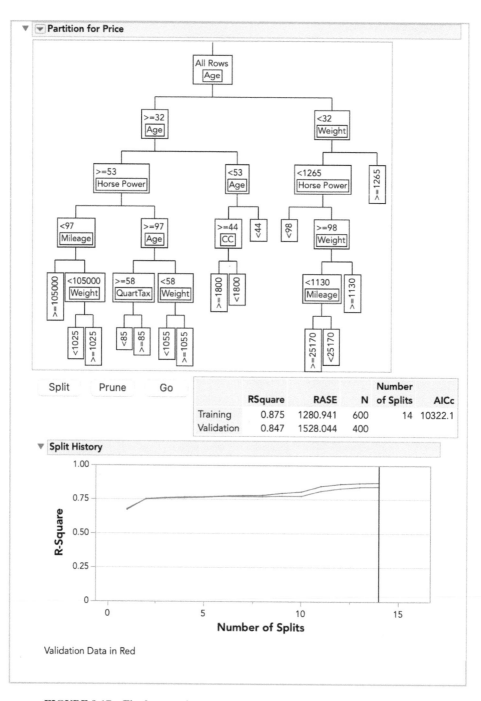

FIGURE 9.17   Final regression tree for Toyota Corolla prices, small tree view

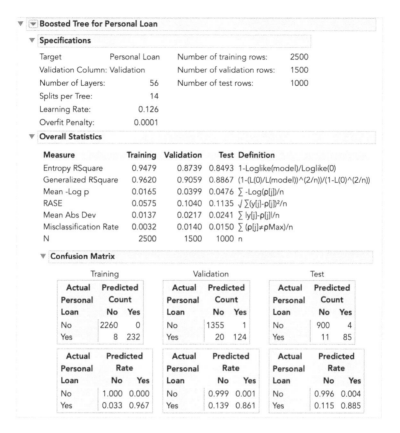

FIGURE 9.20   Universal bank data: confusion matrix and error rates for the training, validation, and test data based on boosted tree

---

FITTING BOOTSTRAP (RANDOM) FORESTS AND BOOSTED TREES IN
JMP Pro

In JMP Pro, we choose *Bootstrap Forest* or *Boosted Tree* from the *Analyze > Predictive Modeling* menu. A variety of options for controlling tree growth and the size of the trees can be selected for each ensemble method (see Figure 9.21). These methods also provide options for randomly sampling both records and predictors. So the random seed should be set in order to produce repeatable results.

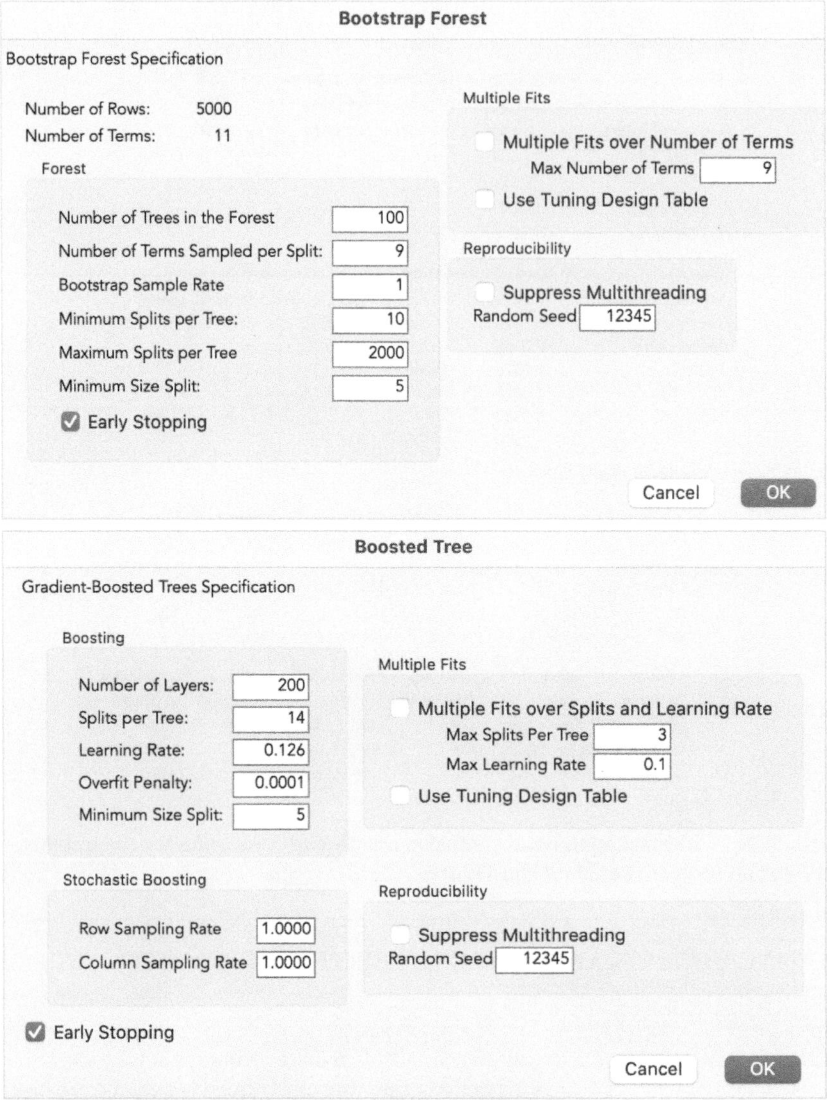

Bootstrap Forest

Bootstrap Forest Specification

Number of Rows: 5000
Number of Terms: 11

Forest

Number of Trees in the Forest   100
Number of Terms Sampled per Split: 9
Bootstrap Sample Rate   1
Minimum Splits per Tree: 10
Maximum Splits per Tree   2000
Minimum Size Split: 5

☑ Early Stopping

Multiple Fits

○ Multiple Fits over Number of Terms
    Max Number of Terms   9
○ Use Tuning Design Table

Reproducibility

○ Suppress Multithreading
Random Seed   12345

Cancel   OK

Boosted Tree

Gradient-Boosted Trees Specification

Boosting

Number of Layers: 200
Splits per Tree: 14
Learning Rate: 0.126
Overfit Penalty: 0.0001
Minimum Size Split: 5

Stochastic Boosting

Row Sampling Rate   1.0000
Column Sampling Rate   1.0000

☑ Early Stopping

Multiple Fits

○ Multiple Fits over Splits and Learning Rate
    Max Splits Per Tree   3
    Max Learning Rate   0.1
○ Use Tuning Design Table

Reproducibility

○ Suppress Multithreading
Random Seed   12345

Cancel   OK

FIGURE 9.21   Options for *Bootstrap Forest* and *Boosted Tree* in JMP Pro

PROBLEMS

9.1 **Competitive Auctions on eBay.com.** The file `eBayAuctions.jmp` contains information on 1972 auctions that transacted on eBay.com during May–June 2004. The goal is to use these data to build a model that will classify auctions as competitive or noncompetitive. A *competitive auction* is defined as an auction with at least two bids placed on the item auctioned. The data include variables that describe the item (auction category), the seller (his/her eBay rating), and the auction terms that the seller selected (auction duration, opening price, currency, day-of-week of auction close). In addition, we have the price at which the auction closed. The task is to predict whether or not the auction will be competitive.

Data preprocessing. Split the data into training and validation datasets using a 60% : 40% ratio.

    a. Fit a classification tree using all predictors and validation, using the *Go* button. Display the leaf report, and write down the first four branches in the leaf report in terms of rules.

    b. Is this model practical for predicting the outcome of a new auction?

    c. Describe the interesting and uninteresting information that these rules provide.

    d. Fit another classification tree, this time only with predictors that can be used for predicting the outcome of a new auction. Describe the resulting tree in terms of rules.

    e. Examine the lift curve and the classification matrix for the tree. What can you say about the predictive performance of this model?

    f. Plot the resulting tree on a scatterplot: use the two axes for the two best (quantitative) predictors. Each auction will appear as a point, with coordinates corresponding to its values on those two predictors. Use different colors or symbols to separate competitive and noncompetitive auctions. Draw lines (you can use the line tool in JMP or an axis reference line) at the values that create splits. Does this splitting seem reasonable with respect to the meaning of the two predictors? Does it seem to do a good job of separating the two classes?

    g. Based on this last tree (d), what can you conclude from these data about the chances of an auction obtaining at least two bids and its relationship to the auction settings set by the seller (duration, opening price, ending day, currency)? What would you recommend for a seller as the strategy that will most likely lead to a competitive auction?

9.2 **Predicting Delayed Flights.** The file `FlightDelays.jmp` contains information on all commercial flights departing the Washington, DC area and arriving at New York during January 2004. For each flight there is information on the departure and arrival airports, the distance of the route, the scheduled time and date of the flight, and so on. The variable that we are trying to predict is whether or not a flight is delayed. A delay is defined as an arrival that is at least 15 minutes later than scheduled.

Data preprocessing. Bin the scheduled departure time (CRS_DEP_TIME) into eight bins. This will avoid treating the departure time as a continuous variable, because it is reasonable that delays are related to rush-hour times. (Note that these data are not stored in JMP with a time format, so you'll need to explore the best way to bin this data—two options are (1) via the formula editor and (2) using the *Make Binning Formula* column utility.)

Partition the data into training (60%) and validation (40%) sets. Save this file under a new name (`FlightDelaysPrepared.jmp`).

a. Fit a classification tree to the flight delay variable using all the relevant predictors (use the binned version of the departure time) and the validation column. Do not include DEP_TIME (actual departure time) in the model because it is unknown at the time of prediction (unless we are doing our predicting of delays after the plane takes off, which is unlikely).

   i. How many splits are in the final model?

   ii. How many variables are involved in the splits?

   iii. Which variables contribute the most to the model?

   iv. Which variables were not involved in any of the splits?

   v. Express the resulting tree as a set of rules.

   vi. If you needed to fly between DCA and EWR on a Monday at 7 AM, would you be able to use this tree to predict whether the flight will be delayed? What other information would you need? Is this information available in practice? What information is redundant?

b. Fit another tree, this time using the original scheduled departure time rather than the binned version. Save the formula for this model to the data table (we'll return to this in a future exercise).

   i. Compare this tree to the original, in terms of the number of splits and the number of variables involved. What are the key differences?

   ii. Does it make more sense to use the original scheduled departure time variable or the binned version? Why?

9.3 **Predicting Prices of Used Cars (Regression Trees).** The file `ToyotaCorolla.jmp` contains the data on used cars (Toyota Corolla) on sale during late summer of 2004 in The Netherlands. It has 1436 records containing details on 38 variables, including *Price, Age, Kilometers, HP,* and other specifications. The goal is to predict the price of a used Toyota Corolla based on its specifications. (The example in Section 9.8 is a subset of this dataset.)

Data preprocessing. Split the data into training (50%), validation (30%), and test (20%) datasets.

a. Run a regression tree with the outcome variable Price and input variables Age_08_04, KM, Fuel_Type, HP, Automatic, Doors, Quarterly_Tax, Mfg_Guarantee, Guarantee_Period, Airco, Automatic_Airco, CD_Player, Powered_Windows, Sport_Model, and Tow_Bar. Set the minimum split size to 1, and use the split button repeatedly to create a full tree (*Hint:* Use the red triangle options to hide the tree and the graph). As you split, keep an eye on RASE and RSquare for the training, validation, and test sets.

   i. Describe what happens to the RSquare and RASE for the training, validation, and test sets as you continue to split the tree.

   ii. How does the performance of the test set compare to the training and validation sets on these measures? Why does this occur?

iii. Based on this tree, which are the most important car specifications for predicting the car's price?

iv. Refit this model, and use the *Go* button to automatically split and prune the tree based on the validation RSquare. Save the prediction formula for this model to the data table.

v. How many splits are in the final tree?

vi. Compare RSquare and RMSE for the training, validation, and test sets for the reduced model to the full model.

vii. Which model is better for making predictions? Why?

10

LOGISTIC REGRESSION

In this chapter, we describe the highly popular and powerful classification method called logistic regression. Like linear regression, it relies on a specific model relating the predictors with the outcome. The user must specify the predictors to include and their form (e.g., including any interaction terms). This means that even small datasets can be used for building logistic regression classifiers and that, once the model is estimated, it is computationally fast and cheap to classify even large samples of new records. We describe the logistic regression model formulation and its estimation from data. We also explain the concepts of "logit," "odds," and "probability" of an event that arise in the logistic model context and the relations among the three. We discuss variable importance and coefficient interpretation, variable selection for dimension reduction, and extensions to multi-class classification.

Logistic Regression in JMP: Logistic regression models can be fit using the standard version of JMP. However, to compute validation statistics using a validation column or to fit regularized regression models, JMP Pro is required.

10.1 INTRODUCTION

Logistic regression extends the ideas of linear regression to the situation where the outcome variable Y, is categorical. We can think of a categorical variable as dividing the records into classes. For example, if Y denotes a recommendation on holding/selling/buying a stock, we have a categorical variable with three categories. We can think of each of the stocks in the dataset (the records) as belonging to one of three classes: the *hold* class, the *sell* class, and the *buy* class. Logistic regression can be used for classifying a new record, where the class is unknown, into one of the classes, based on the values of its predictor variables (called *classification*). It can also be used in data where the class is known to find factors

Machine Learning for Business Analytics: Concepts, Techniques, and Applications with JMP Pro®,
Second Edition. Galit Shmueli, Peter C. Bruce, Mia L. Stephens, Muralidhara Anandamurthy, and Nitin R. Patel.
© 2023 John Wiley & Sons, Inc. Published 2023 by John Wiley & Sons, Inc.

distinguishing between records in different classes in terms of their predictor variables, or "predictor profile" (called *profiling*). Logistic regression is used in applications such as:

1. Classifying customers as returning or nonreturning (classification)
2. Finding factors that differentiate between male and female top executives (profiling)
3. Predicting the approval or disapproval of a loan based on information such as credit scores (classification).

The logistic regression model is used in a variety of fields: whenever a structured model is needed to explain or predict categorical (in particular, binary) outcomes. One such application is in describing consumer choice behavior in econometrics.

In this chapter, we focus on the use of logistic regression for classification. We deal mostly with a binary outcome variable having two possible classes. In Section 10.5, we show how the results can be extended to the case where Y assumes more than two possible classes. Popular examples of binary outcomes are success/failure, yes/no, buy/don't buy, default/don't default, and survive/die. For convenience we often code the values of a binary outcome variable Y as 0 and 1.

Note that in some cases we may choose to convert a numeric (continuous) outcome variable or a categorical outcome with multiple classes into a binary outcome variable for the purpose of simplification, reflecting the fact that decision making may be binary (approve /don't approve the loan, make/don't make an offer). As with multiple linear regression, the predictors $X_1, X_2, \ldots, X_k$ may be categorical or continuous variables or a mixture of these two types. While in multiple linear regression the aim is to predict the value of the continuous Y for a new record, in logistic regression the goal is to predict which class a new observation will belong to, or simply to *classify* the record into one of the classes. In the stock example, we would want to classify a new stock into one of the three recommendation classes: sell, hold, or buy. Or, we might want to compute for a new record its *propensity* (= its probability) to belong to each class and then possibly rank a set of new records from highest to lowest propensity in order to act on those with the highest propensity.[1]

In logistic regression we take two steps: the first step yields estimates of the *propensities* or *probabilities* of belonging to each class. In the binary case we get an estimate of $p = P(Y = 1)$, the probability of belonging to class 1 (which also tells us the probability of belonging to class 0). In the next step, we use a threshold value (also called *decision threshold* in JMP) on these probabilities in order to classify each case into one of the classes. For example, in a binary case, a threshold of 0.5 means that cases with an estimated probability of $P(Y = 1) \geq 0.5$ are classified as belonging to class 1, whereas cases with $P(Y = 1) < 0.5$ are classified as belonging to class 0. This threshold does not need be set at 0.5. When the event in question is a low-probability but notable event or important event (say, 1 = fraudulent transaction), a lower threshold may be used to classify more cases as belonging to class 1.

---

[1] Potentially, one could use linear regression for classification, by training a linear regression on a 0/1 outcome variable (called a *Linear Probability Model*). The model is then used to generate numerical predictions which are converted into binary classifications using a threshold. However, linear probability models, despite their name, do not produce proper predicted probabilities. The numerical predictions they produce are useful for comparison to the classification threshold but are otherwise meaningless.

10.2   THE LOGISTIC REGRESSION MODEL

The idea behind logistic regression is straightforward: instead of using Y as the outcome variable we use a function of it, which is called the *logit*. The logit, it turns out, can be modeled as a linear function of the predictors. Once the logit has been predicted, it can be mapped back to a probability.

To understand the logit, we take several intermediate steps. First, we look at $p = P(Y = 1)$, the probability of belonging to class 1 (as opposed to class 0). In contrast to Y, the class label, which only takes the values 0 and 1, p can take any value in the interval $[0, 1]$. However, if we express p as a linear function of the q predictors[2] in the form

$$p = \beta_0 + \beta_1 x_1 + \beta_2 x_2 + \cdots + \beta_q x_q, \tag{10.1}$$

it is not guaranteed that the right-hand side will lead to values within the interval $[0, 1]$. The solution is to use a nonlinear function of the predictors in the form

$$p = \frac{1}{1 + e^{-(\beta_0 + \beta_1 x_1 + \beta_2 x_2 + \cdots + \beta_q x_q)}}. \tag{10.2}$$

This is called the *logistic response function*. For any values of $x_1, \ldots, x_q$, the right-hand side will always lead to values in the interval $[0, 1]$. Next, we look at a different measure of belonging to a certain class, known as *odds*. The odds of belonging to class 1 ($Y = 1$) is defined as *the ratio of the probability of belonging to class 1 to the probability of belonging to class 0*:

$$\text{Odds}(Y = 1) = \frac{p}{1 - p}. \tag{10.3}$$

This metric is very popular in horse races, sports, gambling in general, epidemiology, and other areas. Instead of talking about the *probability* of winning or contacting a disease, people talk about the *odds* of winning or contacting a disease. How are these two different? If, for example, the probability of winning is 0.5, the odds of winning are $0.5/0.5 = 1$. We can also perform the reverse calculation. That is, given the odds of an event, we can compute its probability by manipulating equation (10.3):

$$p = \frac{\text{Odds}}{1 + \text{Odds}}. \tag{10.4}$$

Substituting (10.2) into (10.4), we can write the relationship between the odds and the predictors as

$$\text{Odds}(Y = 1) = e^{\beta_0 + \beta_1 x_1 + \beta_2 x_2 + \cdots + \beta_q x_q}. \tag{10.5}$$

This last equation describes a multiplicative (proportional) relationship between the predictors and the odds. Such a relationship is interpretable in terms of percentages, for example, a unit increase in predictor x_j is associated with an average increase of $e^{\beta_j} \times 100\%$ in the odds (holding all other predictors constant).

---

[2]Unlike elsewhere in the book, where p denotes the number of predictors, in this chapter, we indicate predictors by q to avoid confusion with the probability p.

Now, if we take a natural logarithm[3] on both sides, we get the standard formulation of a logistic model:

$$\log(\text{odds}) = \beta_0 + \beta_1 x_1 + \beta_2 x_2 + \cdots + \beta_q x_q. \tag{10.6}$$

The log(odds) is called the *logit*, and it takes values from $-\infty$ (very low odds) to ∞ (very high odds).[4] A logit of 0 corresponds to even odds of 1 (probability = 0.50). Thus our final formulation of the relationship between the outcome and the predictors uses the logit as the outcome variable and models it as a *linear function* of the q predictors.

To see the relationship between the probability, odds, and logit of belonging to class 1, look at Figure 10.1, which shows the odds (top) and logit (bottom) as a function of p. Notice that the odds can take any no-negative value and that the logit can take any real value.

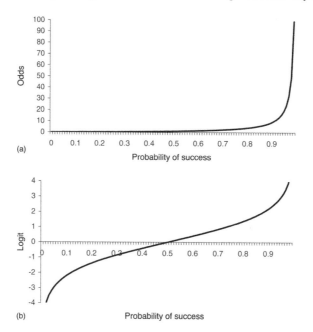

FIGURE 10.1   Odds (top) and logit (bottom) as a function of p

10.3   EXAMPLE: ACCEPTANCE OF PERSONAL LOAN

Recall the example described in Chapter 9, of acceptance of a personal loan by Universal Bank. The bank's dataset includes data on 5000 customers (data are in `UniversalBank.jmp`). The data include customer demographic information (*Age, Income,* etc.), customer response to the last personal loan campaign (*Personal Loan*), and the customer's relationship with the bank (mortgage, securities account, etc.). Table 10.1 provides descriptions of the different variables. Among these 5000 customers, only 480 (= 9.6%) accepted the personal loan that was offered to them in a previous campaign. The goal is to

---

[3]The natural logarithm function is typically denoted *ln()* or *log()*. In this book, we use *log()*.
[4]We use the terms *odds* and *odds* ($Y = 1$) interchangeably.

build a model that identifies customers who are most likely to accept the loan offer in future mailings.

TABLE 10.1   Description of Predictors for Acceptance of Personal Loan Example

Age	Customer's age in completed years
Experience	Number of years of professional experience
Income	Annual income of the customer ($000s)
Family	Family size of the customer
CCAvg	Average spending on credit cards per month ($000s)
Education	Education Level (Undergrad, Graduate, Professional)
Mortgage	Value of house mortgage if any ($000s)
Securities Account	Coded as 1 if customer has a securities account in bank, or 0 otherwise
CD Account	Coded as 1 if customer has a CD account in bank, or 0 otherwise
Online	Coded as 1 if customer uses online banking, or 0 otherwise
CreditCard	Coded as 1 if customer holds Universal Bank credit card, or 0 otherwise

Data Preprocessing   We start by partitioning the data randomly using a standard 60% : 40% ratio, into training and validation sets. We use the training set to fit a model and the validation set to assess the model's performance.

In this dataset, dummy (or indicator) variables have been created for the four two-level categorical predictors (shown on the next page). These indicator variables are coded as *Continuous* variables in JMP.

In JMP, both the outcome and the predictors can remain in text form (Yes, No, etc.), and it is not necessary to code categorical predictor variables into continuous indicator (dummy) variables. So the variable Education, which has three categories, can be used in its original form without dummy coding.

INDICATOR (DUMMY) VARIABLES IN JMP

By default, JMP creates indicator variables behind the scenes for all categorical predictors. The dropped category (the reference category) for each categorical variable is the last category in alphabetical order or the highest level alphanumerically. Another point to note is that instead of applying the typical 0/1 coding, JMP applies $-1/+1$ coding. This results in different model coefficients compared to 0/1 coding but produces *identical* model predictions and classifications.

If desired, dummy variables can be used instead of the original categorical variables. To manually create indicator variables for categorical predictors, use the *Make Indicator Columns* feature under *Cols > Utilities*.

Note: Coefficients using the 0/1 indicator coding can be requested using a red triangle option in the logistic analysis window (select *Indicator Parameterization Estimates*).

Model with a Single Predictor

Consider a simple logistic regression model with just one predictor. This is conceptually analogous to the simple linear regression model in which we fit a straight line to relate the outcome Y to a single predictor X.

Let us construct a simple logistic regression model for classification of customers using the single predictor Income. The equation relating the outcome variable to the predictor in terms of probabilities is

$$P(\text{Personal Loan} = Yes \mid \text{Income} = x) = \frac{1}{1 + e^{-(\beta_0 + \beta_1 x)}},$$

or equivalently, in terms of odds,

$$\text{Odds}(\text{Personal Loan} = Yes) = e^{\beta_0 + \beta_1 x}. \tag{10.7}$$

Suppose the estimated coefficients for the model are $b_0 = -6.3525$ and $b_1 = 0.0392$. So the fitted model is

$$P(\text{Personal Loan} = Yes \mid \text{Income} = x) = \frac{1}{1 + e^{6.3525 - 0.0392x}}. \tag{10.8}$$

For the single predictor case, we can plot the data and graphically display the logistic model. Examine Figure 10.2. The curve represents the probability that someone will accept a loan as Income increases. The y-axis on the left represents the probability of acceptance, and the y-axis on the right acts as a stacked bar chart, providing a breakdown of the overall probability of Yes vs. No (there were fewer accepted loans than nonaccepted). The points are plotted at the income level (on the x-axis) and appear either above or below the curve based on actual acceptance. Accepted loans (Yes) are plotted below the curve and are randomly scattered relative to the y-axis (this is done to provide an indication of scatter relative to Income in larger datasets).

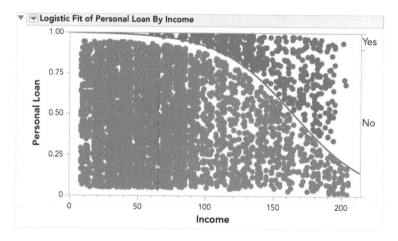

FIGURE 10.2   The logistic curve for personal loan as a function of income. A row legend was added to color and mark records in each personal loan group. To add a row legend, right-click over the plot and select *row legend*

Although logistic regression can be used for prediction in the sense that we predict the *probability* of a categorical outcome, it is most often used for classification. To see the difference between the two, consider predicting the probability of a customer accepting the loan offer as opposed to classifying the customer as an acceptor/nonacceptor. From Figure 10.2 it can be seen that the loan acceptance probabilities produced by the logistic regression model (the s-shaped curve in the Figure 10.2) can yield values between 0 and 1. To end up with classifications into either Yes or No (e.g., a customer either accepts the loan offer or not), we need a threshold, or cutoff value (see section on "Propensities and threshold for Classification" in Chapter 5). This is true in the case of multiple predictors as well.

In the Universal Bank example, in order to classify a new customer as an acceptor/ nonacceptor of the loan offer, we use the information on his/her income by plugging it into the fitted equation in (10.8). This yields an estimated probability of accepting the loan offer. We then compare it to the threshold value. The customer is classified as an acceptor if the probability of his/her accepting the offer is above the threshold.[5]

FITTING ONE PREDICTOR LOGISTIC MODELS IN JMP

Logistic regression models in JMP with one predictor variable are fit using *Analyze > Fit Y by X* or *Analyze > Fit Model*. This displays the logistic curve shown in Figure 10.2. In the Universal Bank example, we model the probability of Personal Loan = Yes (or 1, accepted). By default, JMP will model the probability that Personal Loan = Yes. The outcome category can be selected directly in both *Fit Y by X* and *Fit Model* dialogs before running the model.

Estimating the Logistic Model from Data: Multiple Predictors

In logistic regression, the relationship between Y and the β parameters is nonlinear. For this reason the β parameters are not estimated using the method of least squares (as in multiple linear regression). Instead, a method called *maximum likelihood* is used. The idea, in brief, is to find the estimates that maximize the chance of obtaining the data that we have.[6]

Algorithms to compute the coefficient estimates in logistic regression are less robust than algorithms for linear regression. Computed estimates are generally reliable for well-behaved datasets where the number of records with outcome variable values of both 0 and 1 is large, their ratio is "not too close" to either 0 or 1, and when the number of coefficients in the logistic regression model is small relative to the sample size (e.g., no more than 10%). As with linear regression, collinearity (strong correlation among the predictors) can lead to computational difficulties. Computationally intensive algorithms have been developed recently that circumvent some of these difficulties. For technical details on the maximum likelihood estimation in logistic regression, see Hosmer and Lemeshow (2000).

---

[5]Here, we compared the probability to a threshold c (the threshold is 0.5 by default). If we prefer to look at *odds* of accepting rather than the probability, an equivalent method is to use the equation (10.7) and compare the odds to $c/(1-c)$. If the odds are higher than this number, the customer is classified as an acceptor. If the odds are lower, we classify the customer as a nonacceptor.

[6]The method of maximum likelihood ensures good asymptotic (large sample) properties for the estimates. Under very general conditions, maximum likelihood estimators are (1) *consistent* (the probability of the estimator differing from the true value approaches zero with increasing sample size), (2) *asymptotically efficient* (the variance is the smallest possible among consistent estimators), and (3) *asymptotically normally distributed* (the confidence intervals and perform statistical tests are to be computed in a manner analogous to the analysis of linear multiple regression models, provided that the sample size is *large*).

To illustrate a typical output from such a procedure, look at the output in Figure 10.3 for the logistic model fitted to the training set of 3000 Universal Bank customers. The outcome variable is Personal Loan, with Yes defined as the *success* (this is equivalent to setting the outcome variable to 1 for an acceptor and 0 for a nonacceptor). Here we use all 12 predictors, omitting ZIP Code (note that for each categorical predictor with m categories JMP creates $m - 1$ indicator variables).

▼ **Parameter Estimates**

Term	Estimate	Std Error	ChiSquare	Prob>ChiSq
Intercept	-10.164069	2.4497848	17.21	<.0001*
Age	-0.044547	0.090961	0.24	0.6243
Experience	0.05658147	0.0900536	0.39	0.5298
Income	0.06576067	0.0042213	242.68	<.0001*
Family	0.57155568	0.1011896	31.90	<.0001*
CCAvg	0.18723439	0.0615372	9.26	0.0023*
Education[Undergrad]	-3.0372506	0.2432931	155.85	<.0001*
Education[Graduate]	1.55179759	0.1752704	78.39	<.0001*
Mortgage	0.00175308	0.0008038	4.76	0.0292*
Securities Account	-0.8548708	0.4186376	4.17	0.0411*
CD Account	3.46902866	0.4489309	59.71	<.0001*
Online	-0.843563	0.2283237	13.65	0.0002*
CreditCard	-0.9640741	0.2825423	11.64	0.0006*

For log odds of Yes/No

FIGURE 10.3   Logistic regression parameter estimates (coefficients) table for personal loan acceptance as a function of 12 predictors based on training data

Ignoring p-values for the coefficients, a model based on all 12 predictors would have the estimated logistic equation shown in Figure 10.4. This equation, the logit, is stored in JMP as *Lin[Yes]* when the probability formula is saved to the data table. Note that the coefficient for the third level of the categorical variable Education is provided in this formula.

The positive coefficients for CD Account and Education[Graduate] mean that holding a CD account and having a graduate education are associated with higher probabilities of accepting the loan offer. On the other hand, having a securities account, holding an undergraduate degree, using online banking, and owning a Universal Bank credit card are associated with lower acceptance rates. For the continuous predictors, positive coefficients indicate that a higher value on that predictor is associated with a higher probability of accepting the loan offer (e.g., income: higher income customers tend more to accept the offer). Similarly, negative coefficients indicate that a higher value on that predictor is associated with a lower probability of accepting the loan offer (e.g., Age: older customers are less likely to accept the offer).

The probability of accepting the offer (the *propensity*) for a customer with given values of the 12 predictors can be calculated using equation (10.2).[7] In JMP, when the probability formula is saved to the data table, the formula for calculating propensities is saved as *Prob[Yes]*, as shown in Figure 10.5.

---

[7] If all q predictors are categorical, each having m_q categories, we need not compute probabilities/odds for each of the n records. The number of different probabilities/odds is exactly $m_1 \times m_2 \times \cdots \times m_q$.

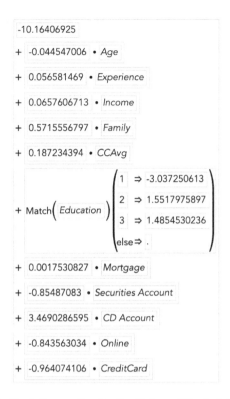

FIGURE 10.4   **Estimated logistic equation for the Universal Bank data based on training data**

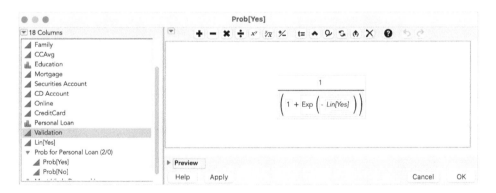

FIGURE 10.5   **The equation for calculating the probability of acceptance, Prob[Yes], saved to the data table as a formula column**

FITTING LOGISTIC MODELS IN JMP WITH MULTIPLE PREDICTORS

To run a logistic regression model with more than one predictor or with a validation column, use *Analyze > Fit Model*. The resulting model and the relationships between the predictors and the outcome variable can be graphically explored using the
Profiler.

To look at the equation for the model, use *Save Probability Formula* from the top
red triangle. This saves the logit (Under *Lin[1]* or *Lin[Yes]*), along with predicted
probabilities for each class of the response and the predicted outcome, using a 0.50
threshold or cutoff value.

Interpreting Results in Terms of Odds (for a Profiling Goal)

Logistic models, when they are appropriate for the data, can give useful information about
the roles played by different predictor variables. For example, suppose we want to know
how increasing family income by one unit will affect the probability of loan acceptance.
This can be found straightforwardly if we consider not probabilities, but odds.

Recall that the odds are given by

$$\text{Odds} = e^{\beta_0 + \beta_1 x_1 + \beta_2 x_2 + \cdots + \beta_q x_q}.$$

At first, let us return to the single predictor example, where we model a customer's acceptance of a personal loan offer as a function of his/her income:

$$\text{Odds}(\text{Personal Loan} = Yes \mid \text{Income}) = e^{\beta_0 + \beta_1 \text{Income}}.$$

We can think of the model as a multiplicative model of odds. The odds that a customer with
income zero will accept the loan is estimated by $e^{-6.535 + (0.039)(0)} = 0.0017$. These are the
base case odds. In this example, it is obviously economically meaningless to talk about a
zero income; the value zero and the corresponding base case odds could be meaningful,
however, in the context of other predictors. The odds of accepting the loan with an income
of \$100K will increase by a multiplicative factor of $e^{(0.039)(100)} = 50.5$ over the base case,
so the odds that such a customer will accept the offer are $e^{-6.535 + (0.039)(100)} = 0.088$.

Suppose that the value of Income, or in general X_1, is increased by one unit from x_1 to
$x_1 + 1$, while the other predictors are held at their current value $(x_2, \ldots, x_{12})$. We get the
odds ratio

$$\frac{\text{Odds}(x_1 + 1, x_2, \ldots, x_{12})}{\text{Odds}(x_1, \ldots, x_{12})} = \frac{e^{\beta_0 + \beta_1(x_1+1) + \beta_2 x_2 + \cdots + \beta_{12} x_{12}}}{e^{\beta_0 + \beta_1 x_1 + \beta_2 x_2 + \cdots + \beta_{12} x_{12}}} = e^{\beta_1}.$$

This tells us that a single unit increase in X_1, holding $X_2, \ldots, X_{12}$ constant, is associated
with an increase in the odds that a customer accepts the offer by a factor of e^{β_1}. In other
words, e^{β_1} is the multiplicative factor by which the odds (of belonging to class 1) increase
when the value of X_1 is increased by 1 unit, *holding all other predictors constant*. If $\beta_1 < 0$,
an increase in X_1 is associated with a decrease in the odds of belonging to class 1, whereas
a positive value of β_1 is associated with an increase in the odds.

When a predictor is a dummy variable, the interpretation is technically the same but has a
different practical meaning. For instance, the coefficient for *CD Account* in the 12-predictor

model was estimated from the data to be 3.469. Recall that the reference group is customers not holding a CD account. We interpret this coefficient as follows: $e^{3.469} = 32.106$ are the odds that a customer who has a CD account will accept the offer relative to a customer who does not have a CD account, holding all other variables constant. This means that customers who hold CD accounts at Universal Bank are more likely to accept the offer than customers without a CD account (holding all other variables constant).

The advantage of reporting results in odds as opposed to probabilities is that statements such as those above are true for any value of X_1. Unless X_1 is a dummy variable, we cannot apply such statements about the effect of increasing X_1 by a single unit to probabilities. This is because the result depends on the actual value of X_1. So if we increase X_1 from, say, 3 to 4, the effect on p, the probability of belonging to class 1, will be different than if we increase X_1 from 30 to 31. In short, the change in the probability, p, for a unit increase in a particular predictor variable, while holding all other predictors constant, is not a constant—it depends on the specific values of the predictor variables. We therefore talk about probabilities only in the context of specific records.

10.4 EVALUATING CLASSIFICATION PERFORMANCE

The general measures of performance that were described in Chapter 5 are used to assess how well the logistic model does. Recall that there are several performance measures, the most popular being those based on the classification matrix (accuracy alone or combined with costs) and the lift curve. As in other classification methods, the goal is to find a model that accurately classifies records to their class, using only the predictor information. A variant of this goal is *ranking*, or finding a model that does a superior job of identifying the members of a particular class of interest for a new set of records (which might come at some cost to overall accuracy). Since the training data are used for selecting the model, we expect the model to perform quite well for those data, and therefore we prefer to test its performance on the holdout set. Recall that the data in the holdout set were not involved in the model building, and thus we can use them to test the model's ability to classify data that it has not "seen" before.

Note that in JMP, when you fit a logistic model with data partitioned into training and validation sets, the training data are used to fit the model, and the validation data are used to assess model performance. As a result, the validation data serve as holdout data.

To obtain the classification matrix from a logistic regression analysis, we use the estimated equation to predict the probability of class membership (the *propensities*) for each record in the validation set and use the threshold value to decide on the class assignment of these records. We then compare these classifications to the actual class memberships of these records. In the Universal Bank case, we use the estimated model to predict the probability of offer acceptance in a validation set that contains 2000 customers (these data were not used in the modeling step). Technically, this is done (by the software) by predicting the logit using the estimated model in Figure 10.4 and then obtaining the probabilities p through the equation shown in Figure 10.5. We then compare these probabilities to our chosen threshold value in order to classify each of the 2000 validation records as acceptors or nonacceptors.

JMP Pro provides the misclassification rates for the training and validation sets automatically, and the classification (or confusion) matrix is a red triangle option. Detailed probabilities and the classification for each record are obtained by saving the probability formula to the data table. Figure 10.6 shows a partial output of scoring the validation set (these are the shaded rows). We see that the first four customers in the validation set have a probability of accepting the offer that is lower than the threshold of 0.5, and therefore they are classified as nonacceptors (0). The fifth customer in the validation set has a probability of acceptance estimated by the model to exceed 0.5, and he or she is therefore classified as an acceptor (Yes), which in fact is a misclassification. Figure 10.7 displays the confusion matrix and decision threshold report for the validation data, using the 0.5 threshold.

	Personal Loan	Validation	Lin[Yes]	Prob[Yes]	Prob[No]	Most Likely Personal Loan
1	No	Training	-9.3052	0.0001	0.9999	No
2	No	Validation	-10.7544	0.0000	1.0000	No
3	No	Validation	-12.6078	0.0000	1.0000	No
4	No	Training	-2.0090	0.1183	0.8817	No
5	No	Training	-5.2502	0.0052	0.9948	No
6	No	Training	-5.8286	0.0029	0.9971	No
7	No	Validation	-4.1304	0.0158	0.9842	No
8	No	Validation	-8.4376	0.0002	0.9998	No
9	No	Training	-3.1132	0.0426	0.9574	No
10	Yes	Training	4.3909	0.9878	0.0122	Yes
11	No	Validation	0.2730	0.5678	0.4322	Yes
12	No	Training	-5.7722	0.0031	0.9969	No
13	No	Validation	-1.0191	0.2652	0.7348	No
14	No	Validation	-4.8888	0.0075	0.9925	No
15	No	Validation	-6.4098	0.0016	0.9984	No

FIGURE 10.6   Scoring the data: Output for the first 15 Universal Bank customers. Rows in the validation set are shaded

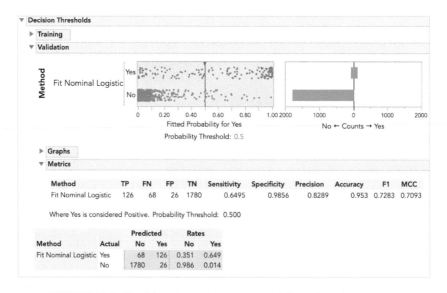

▼ Decision Thresholds
  ▶ Training
  ▼ Validation

Method   Fit Nominal Logistic

Yes / No   Fitted Probability for Yes   0   0.20   0.40   0.60   0.80   1.00 2000   1000   0   1000   2000
No ← Counts → Yes
Probability Threshold: 0.5

  ▶ Graphs
  ▼ Metrics

Method	TP	FN	FP	TN	Sensitivity	Specificity	Precision	Accuracy	F1	MCC
Fit Nominal Logistic	126	68	26	1780	0.6495	0.9856	0.8289	0.953	0.7283	0.7093

Where Yes is considered Positive. Probability Threshold: 0.500

Method	Actual	Predicted No	Predicted Yes	Rates No	Rates Yes
Fit Nominal Logistic	Yes	68	126	0.351	0.649
	No	1780	26	0.986	0.014

FIGURE 10.7   Decision threshold report for Universal Bank loan offer

Another useful tool for assessing model classification performance is the lift curve (see Chapter 5). Figure 10.8 illustrates the lift curve obtained for the Personal Loan offer model using the validation set. The lift indicates the additional responders that you can identify, for a given portion of cases (read on the x-axis), by using the model. Taking the 10% (portion = 0.1) of the records that are ranked by the model as most probable responders yields roughly 7.7 times as many responders as would simply selecting 10% of the records at random.

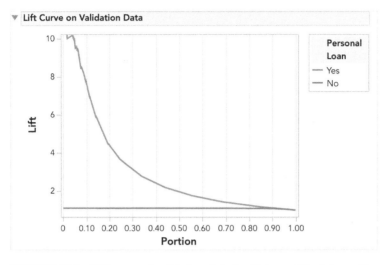

FIGURE 10.8   Lift curve of validation data for Universal Bank loan offer

10.5   VARIABLE SELECTION

The next step includes searching for alternative models. One option is to look for simpler models by trying to reduce the number of predictors used. We can also build more complex models that reflect interactions among predictors by including new variables that are derived from the predictors. For example, if we hypothesize that there is an interactive effect between income and family size, we should add an interaction term of the form *Income*Family*. The choice among the set of alternative models is guided primarily by subject matter knowledge and performance on the validation data. For models that perform roughly equally well, simpler models are generally preferred over more complex models.

Note also that performance on the validation data may be overly optimistic when it comes to predicting performance on data that have not been exposed to the model at all. This is because, when the validation data are used to select a final model, we are selecting based on how well the model performs with those data and therefore may be incorporating some of the random idiosyncrasies of those data into the judgment about the best model. The model still may be the best among those considered, but it will probably not do as well with the unseen data. Therefore, it is useful to evaluate the chosen model on a new test set to get a sense of how well it will perform on new data. In addition, one must consider practical issues such as costs of collecting variables, error-proneness, and model complexity in the selection of the final model.

As in linear regression, in logistic regression we can use automated variable selection heuristics such as mixed stepwise selection, forward selection, and backward elimination. In JMP (and JMP Pro), an exhaustive search using *All Possible Models* is not available for logistic regression models. (See Section 6.4 in Chapter 6 for details on stepwise variable selection.)

10.6   LOGISTIC REGRESSION FOR MULTI-CLASS CLASSIFICATION

The logistic model for a binary outcome variable can be extended for more than two classes. Suppose that there are m classes. Using a logistic regression model, for each record we would have m probabilities of belonging to each of the m classes. Since the m probabilities must add up to 1, we need to estimate only $m - 1$ probabilities.

When classes are distinct from one another and are not related (e.g., by order), we have nominal classes. When the classes have a meaningful order such as *low, medium, high*, then we have ordinal classes. JMP supports logistic regression for both nominal and ordinal classes.

Logistic Regression for Nominal Classes

An example for nominal is the choice between several brands of cereal. A simple way to verify that the classes are nominal is when it makes sense to tag them as $A, B, C, \ldots$, and the assignment of letters to classes does not matter. For simplicity, let us assume that there are $m = 3$ brands of cereal that consumers can choose from (assuming that each consumer chooses one). Then, we estimate the probabilities $P(Y = A)$, $P(Y = B)$, and $P(Y = C)$. If we know two of the probabilities, the third probability is determined. We therefore use one of the classes as the reference class. Let us use C as the reference class.

The goal is to model the class membership as a function of predictors. So in the cereals example, we might want to predict which cereal will be chosen if we know the cereal's price x.

Next, we form $m - 1$ pseudologit equations that are linear in the predictors. In our example, we would have

$$\text{logit}(A) = \log \frac{P(Y = A)}{P(Y = C)} = \alpha_0 + \alpha_1 x,$$

$$\text{logit}(B) = \log \frac{P(Y = B)}{P(Y = C)} = \beta_0 + \beta_1 x.$$

Once the four coefficients are estimated from the training set, we can estimate the class membership probabilities:

$$P(Y = A) = \frac{e^{a_0 + a_1 x}}{1 + e^{a_0 + a_1 x} + e^{b_0 + b_1 x}},$$

$$P(Y = B) = \frac{e^{b_0 + b_1 x}}{1 + e^{a_0 + a_1 x} + e^{b_0 + b_1 x}},$$

$$P(Y = C) = 1 - P(Y = A) - P(Y = B),$$

where a_0, a_1, b_0, and b_1 are the coefficient estimates obtained from the training set.[8] Finally, a record is assigned to the class that has the highest probability.

Logistic Regression for Ordinal Classes

Ordinal classes are classes that have a meaningful order. For example, in stock recommendations, the three classes *buy*, *hold*, and *sell* can be treated as ordered. As a simple rule, if classes can be numbered in a meaningful way, we consider them ordinal. When the number of classes is large (typically, more than 5), we can treat the outcome variable as continuous and perform multiple linear regression. When $m = 2$, the nominal logistic model described above is used. We therefore need an extension of the logistic regression for a small number of ordinal classes ($3 \leq m \leq 5$). There are several ways to extend the binary class case. Here, we describe the *proportional odds* or *cumulative logit method*. For other methods, see Hosmer and Lemeshow (2000).

For simplicity of interpretation and computation, we look at *cumulative* probabilities of class membership. For example, in the stock recommendations we have $m = 3$ classes. Let us denote them by $1 = buy$, $2 = hold$, and $3 = sell$. The probabilities that are estimated by the model are $P(Y \leq 1)$, which is the probability of a *buy* recommendation, and $P(Y \leq 2)$, which is the probability of a *buy* or *hold* recommendation. The three noncumulative probabilities of class membership can easily be recovered from the two cumulative probabilities:

$$P(Y = 1) = P(Y \leq 1),$$
$$P(Y = 2) = P(Y \leq 2) - P(Y \leq 1),$$
$$P(Y = 3) = 1 - P(Y \leq 2).$$

Next we want to model each logit as a function of the predictors. Corresponding to each of the $m - 1$ cumulative probabilities is a logit. In our example we would have

$$\text{logit(buy)} = \log \frac{P(Y \leq 1)}{1 - P(Y \leq 1)},$$
$$\text{logit(buy or hold)} = \log \frac{P(Y \leq 2)}{1 - P(Y \leq 2)}.$$

Each of the logits is then modeled as a linear function of the predictors (as in the two-class case). If in the stock recommendations we have a single predictor x, we have two equations:

$$\text{logit(buy)} = \alpha_0 + \beta_1 x,$$
$$\text{logit(buy or hold)} = \beta_0 + \beta_1 x.$$

---

[8] From the two logit equations, we see that $P(Y = A) = P(Y = C) \cdot e^{\alpha_0 + \alpha_1 x}$ and $P(Y = B) = P(Y = C) \cdot e^{\beta_0 + \beta_1 x}$. Since $P(Y = A) + P(Y = B) + P(Y = C) = 1$, we get

$$P(Y = C) = 1 - P(Y = C) \cdot e^{\alpha_0 + \alpha_1 x} - P(Y = C) \cdot e^{\beta_0 + \beta_1 x} = \frac{1}{e^{\alpha_0 + \alpha_1 x} + e^{\beta_0 + \beta_1 x}}.$$

By plugging this form into the two equations above it, we also obtain the membership probabilities in classes A and B.

This means that both lines have the same slope (β_1) but different intercepts. Once the coefficients $\alpha_0, \beta_0, \beta_1$ are estimated, we can compute the class membership probabilities by rewriting the logit equations in terms of probabilities. For the three-class case, for example, we would have

$$P(Y = 1) = P(Y \leq 1) = \frac{1}{1 + e^{-(a_0 + b_1 x)}},$$

$$P(Y = 2) = P(Y \leq 2) - P(Y \leq 1) = \frac{1}{1 + e^{-(b_0 + b_1 x)}} - \frac{1}{1 + e^{-(a_0 + b_1 x)}},$$

$$P(Y = 3) = 1 - P(Y \leq 2) = 1 - \frac{1}{1 + e^{-(b_0 + b_1 x)}}.$$

where a_0, b_0, and b_1 are the estimates obtained from the training set.

For each record we now have the estimated probabilities that it belongs to each of the classes. In our example, each stock would have three probabilities: for a *buy* recommendation, a *hold* recommendation, and a *sell* recommendation. The last step is to classify the record into one of the classes. This is done by assigning it to the class with the highest membership probability. So, if a stock had estimated probabilities $P(Y = 1) = 0.2$, $P(Y = 2) = 0.3$, and $P(Y = 3) = 0.5$, we would classify it as getting a *sell* recommendation.

Example: Accident Data

Let's take, for example, data on accidents from the US Bureau of Transportation Statistics. These data can be used to predict whether an accident will result in injuries or fatalities based on predictors such as alcohol involvement, time of day, and road condition. The severity of the accident has three classes: 0 = No-Injury, 1 = Non-fatal Injuries, 2 = Fatalities. The estimated logistic models will differ, depending on whether the outcome variable is coded as ordinal or nominal.

Let us first consider severity as an ordinal variable, and fit a simple model with two nominal predictors, alcohol and weather. Each of these predictors is coded as No (not involved in the accident) or Yes (involved in the accident). Next, we change the modeling type for severity to nominal, and fit a model with the same predictors. The reference severity level, in both models, is severity = 2.

The parameter estimates for the resulting models are shown in Figures 10.9 and 10.10. The ordinal logistic model has estimates for two intercepts (the first for severity = 0 and the second for severity = 1), but one estimate for the slopes (the coefficients for each predictor). In contrast, the nominal logistic model has estimates for the two intercepts and also has separate estimates for the slopes for each level of the outcome.

▼ Parameter Estimates

Term	Estimate	Std Error	ChiSquare	Prob>ChiSq
Intercept[0]	-0.1163012	0.0203592	32.63	<.0001*
Intercept[1]	4.42547899	0.0496809	7934.9	<.0001*
WEATHER_R[1]	-0.1293157	0.0139465	85.97	<.0001*
ALCHL_I[1]	-0.2090611	0.0174487	143.56	<.0001*

FIGURE 10.9   Parameter estimates for *ordinal* logistic regression model for accident severity with two predictors

▼ Parameter Estimates

Term	Estimate	Std Error	ChiSquare	Prob>ChiSq
Intercept	3.65328871	0.0953827	1467.0	<.0001*
WEATHER_R[1]	-0.3868429	0.0873699	19.60	<.0001*
ALCHL_I[1]	-0.591057	0.0591182	99.96	<.0001*
Intercept	3.73181765	0.0952068	1536.4	<.0001*
WEATHER_R[1]	-0.264972	0.0874458	9.18	0.0024*
ALCHL_I[1]	-0.4075349	0.0586724	48.25	<.0001*

For log odds of 0/2, 1/2

FIGURE 10.10 Parameter estimates for *nominal* logistic regression model for accident severity with two predictors

---

Logistic regression for *ordinal* and *multinomial* (more than two unordered classes) data is performed in *Fit Y by X* (for one predictor) and *Fit Model* (for more than one predictor). The type of model generated is determined by the modeling type of the outcome variable in JMP.

10.7 EXAMPLE OF COMPLETE ANALYSIS: PREDICTING DELAYED FLIGHTS

Predicting flight delays would be useful to a variety of organizations: airport authorities, airlines, and aviation authorities. At times, joint task forces have been formed to address the problem. Such an organization, if it were to provide ongoing real-time assistance with flight delays, would benefit from some advance notice about flights that are likely to be delayed. Data are in `FlightDelaysLR.jmp`.

In this simplified illustration, we look at six predictors (see Table 10.2). The outcome of interest is whether the flight is delayed or not (*delayed* means more than 15 minutes late). Our data consist of flights from the Washington, DC, area into the New York City area during January 2004. The percent of delayed flights among these 2201 flights is 19.45%. The data were obtained from the Bureau of Transportation Statistics website (www.transtats.bts.gov).

TABLE 10.2 Description of Predictors for Flight Delays Example

Delayed	Coded as: 1 = Yes, 0 = No
Carrier	Eight airline codes: CO (Continental), DH (Atlantic Coast), DL (Delta), MQ (American Eagle), OH (Comair), RU (Continental Express), UA (United), and US (USAirways)
Day of Week	Coded as: 1 = Monday, 2 = Tuesday,..., 7 = Sunday
Departure Time	Broken down into 16 intervals
Destination	Three airport codes: JFK (Kennedy), LGA (LaGuardia), EWR (Newark)
Origin	Three airport codes: DCA (Reagan National), IAD (Dulles), BWI (Baltimore—Washington Int'l)
Weather	Coded as 1 if there was a weather-related delay

The goal is to predict accurately whether a new flight, not in this dataset, will be delayed. Our dependent variable is a binary variable called *Delayed*, coded as 1 for a delayed flight and 0 otherwise.

Other information that is available on the website, such as distance and arrival time, is irrelevant because we are looking at a certain route (distance, flight time, etc., should be approximately equal). A sample of the data for 20 flights is shown in Figure 10.11. Figures 10.12 and 10.13 show visualizations of the relationships between flight delays and different predictors or combinations of predictors. From Figure 10.12 we see that Sundays (7) and Mondays (1) have the largest proportion of delays. Delay rates also seem to differ by carrier, time of day, origin and destination airports. For weather, we see a strong distinction between delays when Weather = 1 (in that case, there is always a delay) and Weather = 0.

	Delayed	Carrier	Day of Week	Departure Time	Destination	Origin	Weather
110	0	RU	5	1300-1359	EWR	DCA	0
111	1	RU	5	1500-1559	EWR	IAD	0
112	0	RU	5	1500-1559	EWR	DCA	0
113	0	DH	6	1200-1259	LGA	IAD	0
114	0	DH	6	1700-1759	LGA	IAD	0
115	0	DH	6	0659 or earlier	LGA	IAD	0
116	0	DH	6	1000-1059	LGA	IAD	0
117	0	DH	6	0800-0859	JFK	IAD	0
118	0	DH	6	1200-1259	JFK	IAD	0
119	0	DH	6	1400-1459	JFK	IAD	0
120	0	DH	6	1600-1659	JFK	IAD	0
121	0	DH	6	1700-1759	JFK	IAD	0
122	1	DH	6	2100 or later	JFK	IAD	0
123	0	DH	6	1600-1659	JFK	IAD	0
124	0	DH	6	2100 or later	LGA	IAD	0
125	1	DL	6	1500-1559	JFK	DCA	0
126	0	DL	6	0800-0859	LGA	DCA	0
127	0	DL	6	1000-1059	LGA	DCA	0
128	0	DL	6	1200-1259	LGA	DCA	0
129	0	DL	6	1400-1459	LGA	DCA	0
130	0	DL	6	1600-1659	LGA	DCA	0

FIGURE 10.11   Sample of 20 flights

The heatmap in Figure 10.13 reveals some specific combinations with high delay rates, such as Sunday flights by Continental Express (Carrier RU) departing from BWI or Sunday flights by American Eagle (MQ) departing from DCA. We can also see combinations with very low delay rates.[9]

Our main goal is to find a model that can obtain accurate classifications of new flights based on their predictor information. An alternative goal is finding a certain percentage of flights that are most/least likely to get delayed (*ranking*). And a third, different, goal is profiling flights: finding out which factors distinguish between ontime and delayed flights (not only in this sample but also in the entire population of flights on this route), and for those factors we would like to quantify these effects. A logistic regression model can be used for all these goals, albeit in different ways.

---

[9]Before creating the heatmap in Figure 10.13, data were first summarized using *Tables > Summary*. To create this summary, select Delayed, and under *Statistics* select *Mean*, then select Carrier, Day of Week, and Origin, and then select *Group*.

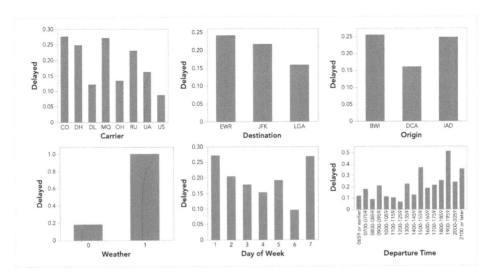

FIGURE 10.12   **Proportion of delayed flights by each of the predictors. The modeling type for** *Delayed* **was first changed to continuous**

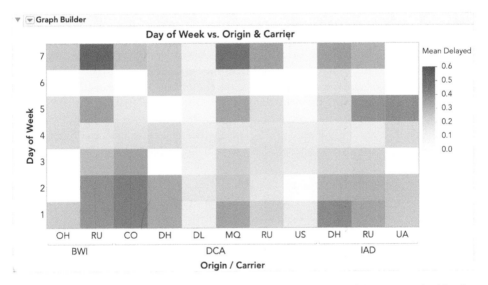

FIGURE 10.13   **Heatmap of proportion of delayed flights (darker = higher proportion) by day of week, origin, and carrier using** *Graph Builder*

Data Preprocessing

To prepare our data for modeling, we use graphical and numeric summaries, along with subject matter knowledge, to identify variables in need of recoding or reformatting. Here are some of the steps we've taken in this example:

1. To produce the bar charts and heat map in Figures 10.12 and 10.13, we first changed the modeling type for the outcome to continuous. We then changed the modeling type back to nominal before fitting logistic models.

2. Since we are interested in predicting delayed flights, we need to tell JMP to model the probability of Delayed = 1. To do this, we used the *Value Ordering* column property.

3. Departure Time is derived from another column, Time of Day (not shown). This is a numeric column, but the time data aren't coded in JMP using a time format (JMP provides many date and time formats). We have binned the time data into one hour increments (using the *Formula Editor*).

4. We have grouped variables that won't be used in the analysis and have hidden and excluded these variables (using options under the *Cols* menu).

We then partitioned the data using a 60% : 40% ratio into training and validation sets. We use the training set to fit a model and the validation set to assess the model's performance.

Model Fitting, Estimation, and Interpretation—A Simple Model

For this example, we have several categorical predictors. Instead of using indicator (dummy) variables, we rely on JMP to create indicator variables behind the scenes for each of these categorical predictors. Recall that JMP uses $-1/1$ indicator coding rather than the traditional 0/1 coding. The resulting coefficients for these indicator variables *sum to zero*, so the coefficient for the last level is not displayed. (Note that logistic models using either $-1/1$ or 0/1 indicator coding produce the same estimated probabilities.)

To demonstrate how JMP codes categorical predictors, we fit a model for Delayed with one predictor, Destination. In the top of Figure 10.14, we see the parameter estimates for this model. The coefficient for arrival airport JFK is estimated from the data to be 0.0822, and the coefficient for arrival at EWR is 0.2207. Since the coefficients for the categorical indicator variables in JMP sum to zero, the coefficient for arrival at LGA is $-[(0.0822) + (0.2207)] = -0.3029$. This can be seen by saving the probability formula to the data table. The logit, saved as Lin[1], is shown in Figure 10.15. The coefficients tell us that, ignoring all other factors, flights to LGA are slightly less likely to be delayed than those to JFK or EWR.

▼ Parameter Estimates				
Term	Estimate	Std Error	ChiSquare	Prob>ChiSq
Intercept	-1.3618315	0.0576483	558.05	<.0001*
Destination[EWR]	0.22065958	0.0778145	8.04	0.0046*
Destination[JFK]	0.08222126	0.091626	0.81	0.3695
For log odds of 1/0				

FIGURE 10.14   Estimated logistic regression model for delayed flights with one categorical predictor (destination) based on the training set

$$-1.361831482 + \text{Match}\left(\textit{Destination}\right)\begin{pmatrix} \text{"EWR"} & \Rightarrow & 0.2206595786 \\ \text{"JFK"} & \Rightarrow & 0.0822212631 \\ \text{"LGA"} & \Rightarrow & -0.302880842 \\ \text{else} & \Rightarrow & . \end{pmatrix}$$

FIGURE 10.15   Logit for delayed flights with one predictor, *Destination*

The estimates for *Weather* and *Carrier* are marked as *unstable*. This means that the predicted probability (of delays, in this case) is 1 or near 1 for values of these factors or that there is not enough information to estimate the coefficients. All of the occurrences of *Weather* = 1 were delayed. As a result, JMP is providing a warning that estimate of the coefficient for weather is unstable (note that Weather is not significant even though the coefficient is large). For *Carrier*, the warning is related to the fact that there are relatively few flights for some of the carriers. For example, OH and UA had a combined 61 flights, but only 36 of these are in the training set. In this case, the *unstable* warning is a hint that data may need to be coded or grouped differently. We revisit these issues later in this example.

Model Fitting, Estimation, and Interpretation—The Full Model

We now fit a full model with all six predictors. The estimated model with 33 parameters (plus the intercept) is given in Figure 10.16. For each categorical variable, JMP has created indicator variables, and as we have seen, the highest level alphanumerically (or last alphabetically) is kept out of the model.

With other predictors in the model, the coefficients for the arrival airports have changed slightly. The coefficient for Destination[JFK] is now −0.06799, and the coefficient for Destination[EWR] is −0.0113. So the coefficient for arrival at LGA can be computed as −[(−0.06799) + (−0.0113)] = 0.0793. If we take into account the statistical significance of the coefficients, we see that in general the destination airport is not associated with the chance of delays. However, the departure airport is statistically significant.

The day of the week is also significant, with flights leaving on Monday (Day of Week[1]) and Friday (Day of Week[5]) having positive coefficients (higher probability of delays) and flights on Thursdays (Day of Week[4]) and Saturdays (Day of Week[6]) having negative coefficients (lower probability of delays). Also, the coefficients appear to change over the course of the day, with lower probability of delays mostly occurring earlier in the day.

The *Profiler* (Figure 10.17) provides a visual representation of the model coefficients and the relationship between the different predictors and the probability of flight delays. We can see, for example, that the probability of a delayed flight generally increases throughout the day, peaking in the evening. We can also see that the delay rate is roughly the same for the different destinations, while BWI appears to have a higher delay rate than DCA and IAD.

Model Performance

How should we measure the performance of logistic models? One possible measure is "percent of flights correctly classified." This measure, along with several other fit statistics, are provided for the training and validation sets under *Whole Model Test*. The misclassification rate for the validation set is 0.1557 (see Figure 10.18). This is slightly better than the misclassification rate for the training set (0.1612).

The confusion matrix gives a sense of what type of misclassification is more frequent. From the confusion matrix at the bottom of Figure 10.19, it can be seen that the model does better in classifying nondelayed flights correctly and is less accurate in classifying flights that were delayed. (*Note*: The same pattern appears in the classification matrix for the training data, so it is not surprising to see it emerge for new data.) If there is an asymmetric cost

▼ **Parameter Estimates**

Term		Estimate	Std Error	ChiSquare	Prob>ChiSq
Intercept	Unstable	11.8788584	553.54246	0.00	0.9829
Carrier[CO]	Unstable	2.59702148	101.19343	0.00	0.9795
Carrier[DH]	Unstable	2.24635404	101.1929	0.00	0.9823
Carrier[DL]	Unstable	1.27877004	101.1929	0.00	0.9899
Carrier[MQ]	Unstable	2.4932625	101.19282	0.00	0.9803
Carrier[OH]	Unstable	0.68811571	101.19539	0.00	0.9946
Carrier[RU]	Unstable	2.12976488	101.19293	0.00	0.9832
Carrier[UA]	Unstable	-12.010284	708.34745	0.00	0.9865
Day of Week[1]		0.30157894	0.1811195	2.77	0.0959
Day of Week[2]		0.07546288	0.196388	0.15	0.7008
Day of Week[3]		-0.1589367	0.1944871	0.67	0.4138
Day of Week[4]		-0.3842139	0.1886445	4.15	0.0417*
Day of Week[5]		0.22013162	0.170299	1.67	0.1961
Day of Week[6]		-0.4786919	0.2401693	3.97	0.0462*
Departure Time[0659 or earlier]		-0.5474193	0.2819076	3.77	0.0522
Departure Time[0700-0759]		-0.6301708	0.4818879	1.71	0.1910
Departure Time[0800-0859]		-1.0669539	0.3485066	9.37	0.0022*
Departure Time[0900-0959]		0.39007625	0.4551093	0.73	0.3914
Departure Time[1000-1059]		-1.038575	0.4390541	5.60	0.0180*
Departure Time[1100-1159]		0.19465575	0.5541449	0.12	0.7254
Departure Time[1200-1259]		-1.5173006	0.3847497	15.55	<.0001*
Departure Time[1300-1359]		0.41106926	0.3017861	1.86	0.1732
Departure Time[1400-1459]		-0.3596341	0.272698	1.74	0.1872
Departure Time[1500-1559]		0.80398207	0.2379079	11.42	0.0007*
Departure Time[1600-1659]		-0.2156358	0.261717	0.68	0.4100
Departure Time[1700-1759]		0.11350714	0.2485986	0.21	0.6480
Departure Time[1800-1859]		0.66798459	0.2691259	6.16	0.0131*
Departure Time[1900-1959]		1.79857406	0.354208	25.78	<.0001*
Departure Time[2000-2059]		0.65792876	0.3463971	3.61	0.0575
Destination[EWR]		-0.0112986	0.2004744	0.00	0.9551
Destination[JFK]		-0.0679927	0.1568678	0.19	0.6647
Origin[BWI]		0.64053176	0.2429902	6.95	0.0084*
Origin[DCA]		-0.5265972	0.2174133	5.87	0.0154*
Weather[0]	Unstable	-14.965294	614.88872	0.00	0.9806

For log odds of 1/0

FIGURE 10.16   Estimated logistic regression model for delayed flights with all predictors (based on the training set)

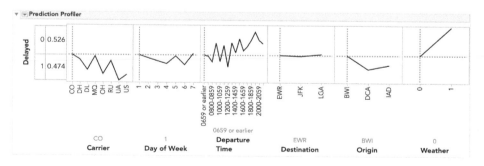

FIGURE 10.17   Prediction profiler for flight delays. The contours show how the predicted probability of delays changes for different values of the predictors

Whole Model Test

Model	-LogLikelihood	DF	ChiSquare	Prob>ChiSq
Difference	131.62702	33	263.254	<.0001*
Full	533.26770			
Reduced	664.89473			

RSquare (U)	0.1980
AICc	1136.39
BIC	1310.86
Observations (or Sum Wgts)	1321

Fit Details

Measure	Training	Validation	Definition
Entropy RSquare	0.1980	0.0112	1-Loglike(model)/Loglike(0)
Generalized RSquare	0.2847	0.0173	$(1-(L(0)/L(model))^{(2/n)})/(1-L(0)^{(2/n)})$
Mean -Log p	0.4037	0.4705	$\sum$ -Log(ρ[j])/n
RASE	0.3531	0.3604	$\sqrt{\sum(y[j]-\rho[j])^2/n}$
Mean Abs Dev	0.2525	0.2593	$\sum |y[j]-\rho[j]|/n$
Misclassification Rate	0.1612	0.1557	$\sum (\rho[j]\neq\rho Max)/n$
N	1321	880	n

FIGURE 10.18   Fit statistics, including misclassification rates, for the flight delay data

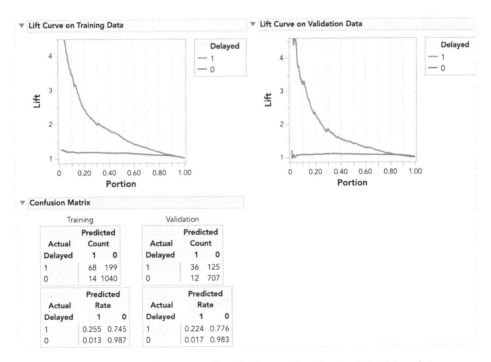

Lift Curve on Training Data

Lift Curve on Validation Data

Confusion Matrix

Training

	Predicted Count	
Actual Delayed	1	0
1	68	199
0	14	1040

Validation

	Predicted Count	
Actual Delayed	1	0
1	36	125
0	12	707

	Predicted Rate	
Actual Delayed	1	0
1	0.255	0.745
0	0.013	0.987

	Predicted Rate	
Actual Delayed	1	0
1	0.224	0.776
0	0.017	0.983

FIGURE 10.19   Lift curves and confusion matrices for the flight delays data

structure such that one type of misclassification is more costly than the other, the threshold value can be selected to minimize the cost (via the *Profit Matrix*). Of course, this tweaking should be carried out on the training data and assessed only using the validation data.

In most conceivable situations, it is likely that the purpose of the model will be to identify those flights most likely to be delayed so that resources can be directed toward either reducing the delay or mitigating its effects. Air traffic controllers might work to open up additional air routes, or to allocate more controllers to a specific area for a short time. Airlines might bring on personnel to rebook passengers and to activate standby flight crews and aircraft. Hotels might allocate space for stranded travelers. In all cases the resources available are going to be limited, and resources might vary over time and from organization to organization. In this situation the most useful model would provide an ordering of flights by their probability of delay, letting the model users decide how far down that list to go in taking action. Therefore model lift is a useful measure of performance—as you move down that list of flights, ordered by their delay probability, how much better does the model do in predicting delay than would a naive model, which is simply the average delay rate for all flights? From the lift curves for the validation data (Figure 10.19), we see that our model is superior to the baseline (simple random selection of flights).

Variable Selection

From the initial exploration of predictors shown in Figures 10.12 and 10.13, the parameter estimates table in Figure 10.16, and the *Prediction Profiler* shown in Figure 10.17, it appears that several of the variables might be dropped or coded differently. Additionally, we look at the number of flights in different categories to identify categories with very few or no flights—such categories are candidates for removal or merger.

First, we find that most carriers depart from a single airport (DCA). For those that depart from all three airports, the delay rates are similar regardless of airport. We use the *Effect Summary* table (at the top of the Fit Nominal Logistic window) to remove the variable Origin from the model and find that the model performance on the validation set is not harmed. We find that the destination airport is not significant (see Figure 10.20) and decide to drop it for a practical reason: not all carriers fly to all airports (this can be confirmed with *Tabulate*). Our model would then be invalid for prediction in nonexistent combinations of carrier and destination airport.

Source	Logworth		PValue
Weather	13.592		0.00000
Departure Time	12.595		0.00000
Carrier	8.041		0.00000
Day of Week	1.949		0.01124
Destination	0.072		0.84804

Remove Add Edit Undo   FDR

FIGURE 10.20   **Effect Summary table, after the removal of *Origin***

Earlier we discussed the issue with instability of the parameter estimates for *Carrier* (see Figure 10.16). To resolve this, we try grouping the carriers into fewer categories. We also try to group *Departure Time* and *Day of Week* into fewer categories that are more distinguishable with respect to delays. From the *Prediction Profiler* in Figure 10.17, we see that groups of carriers have similar delay rates (e.g., DL, OH, UA, and US) and that Sunday and Monday seem to have similar delay rates. The grouping of Sunday and Monday, using the JMP *Recode* utility, is shown in Figure 10.21.

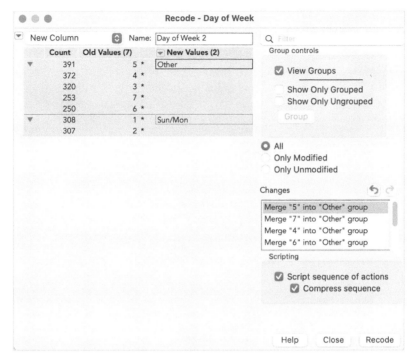

FIGURE 10.21   Using *Recode*, from the *Cols > Utilities* menu to group day of the week

---

REGROUPING AND RECODING VARIABLES IN JMP

Variable recoding in JMP can be done manually using the *Formula Editor* or using *Recode* from the column utilities menu. New recoded variables can also be created directly from Stepwise. Recall that the stepwise platform creates coded variables that maximize the difference in the outcome—these coded variables can be saved to the data table and used to build new models.

Another popular approach to identify potential groupings, which works particularly well when there is a large number of categorical predictors with many levels, is to use the *Partition* platform to create a classification tree (see Chapter 9). Tree-based methods can also be used to identify important variables (using *Column Contributions*), to identify threshold's for continuous variables (for binning), and to identify potential interactions among predictors.

One potential reduced model, shown in Figure 10.22, includes only seven parameter estimates and has the advantage of being more parsimonious. Figure 10.22 displays the training and validation fit statistics (top) and the confusion matrices and the lift curves (bottom). It can be seen that the misclassification rate for this model competes well with the larger model in terms of both accurate classification and lift. (Note that there are many other possible models that we might consider.)

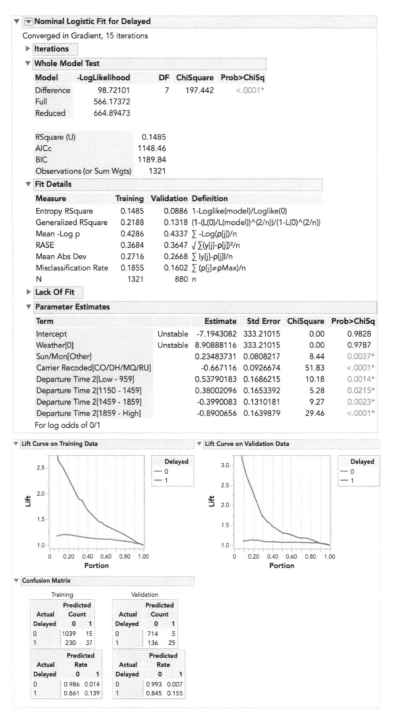

FIGURE 10.22 Output for logistic regression model with only seven predictors. Estimated model (top) and performance (bottom)

We therefore conclude with a seven-predictor model that requires only knowledge of the carrier, the day of week, the hour of the day, and whether it is likely that there will be a delay due to weather. However, this weather variable refers to actual weather at flight time, not a forecast, and is not known in advance! If the aim is to predict in advance whether a particular flight will be delayed, a model without weather must be used. In contrast, if the goal is profiling delayed vs. nondelayed flights, we can include weather in the model to allow evaluating the impact of the other factors only when there *isn't* a weather-related delay. Since Weather = 1 corresponds to a delayed flight, we cannot explore the impact of the other factors when there *is* a weather-related delay (the probability of a delayed flight will be 1.0 under all conditions). If profiling flight delays in the event of different weather conditions is of primary concern, compiling additional information related to weather is recommended.

To conclude, based on the data from January 2004 and our final model, the highest chance of a nondelayed flight is Tuesday through Saturday (recoded as *Other*), and flying in the morning, on Delta, Comair, United, or US Airways (see the bottom *Prediction Profiler* in Figure 10.23). And clearly, good weather is advantageous!

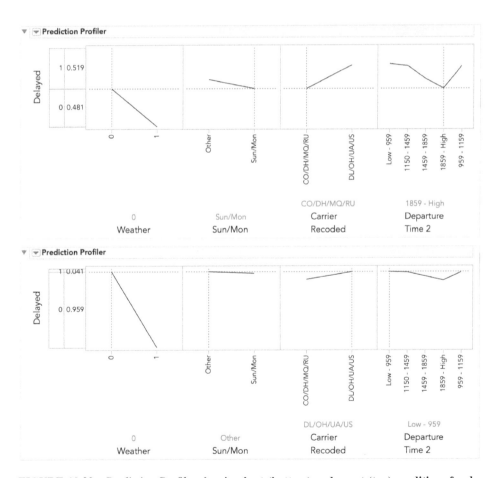

FIGURE 10.23 *Prediction Profiler* showing best (bottom) and worst (top) conditions for delayed flights, assuming there are no delays due to weather

PROBLEMS

10.1 Financial Condition of Banks. The file Banks.jmp includes data on a sample of 20 banks. The "Financial Condition" column records the judgment of an expert on the financial condition of each bank. This dependent variable takes one of two possible values—*weak* or *strong*—according to the financial condition of the bank. The predictors are two ratios used in the financial analysis of banks: TotLns&Lses/Assets is the ratio of total loans and leases to total assets, and TotExp/Assets is the ratio of total expenses to total assets. The goal is to use the two ratios for classifying the financial condition of a new bank.

Run a logistic regression model (on the entire dataset) that models the status of a bank as a function of the two financial measures provided. Use the *Value Ordering* column property to specify the *success* class as *weak* (i.e., move it to the top of the value list).

a. Write the estimated equation that associates the financial condition of a bank with its two predictors in two forms:

  i. The logit as a function of the predictors.

  ii. The probability as a function of the predictors.

b. Consider a new bank whose total loans and leases/assets ratio = 0.6 and total expenses/assets ratio = 0.11. From your logistic regression model, estimate the following quantities for this bank (save the probability formula to the data table, and enter these values in a new row): the logit, the probability of being financially weak, and the classification of the bank. Confirm the probability using the *Profiler*.

c. The threshold probability value of 0.5 is used to classify banks as being financially weak or strong. What is the misclassification rate for weak banks that are incorrectly classified as strong?

d. When a bank that is in poor financial condition is misclassified as financially strong, the misclassification cost is much higher than when a financially strong bank is misclassified as weak. To minimize the expected cost of misclassification, should the threshold value for classification (which is currently at 0.5) be increased or decreased?

e. Interpret the estimated coefficient for the total loans and leases-to-total assets ratio (TotLns&Lses/Assets) in terms of the odds of being financially weak.

10.2 Identifying Good System Administrators. A management consultant is studying the roles played by experience and training in a system administrator's ability to complete a set of tasks in a specified amount of time. In particular, she is interested in discriminating between administrators who are able to complete given tasks within a specified time and those who are not. Data are collected on the performance of 75 randomly selected administrators. They are stored in the file SystemAdministrators.jmp.

The variable Experience measures months of full-time system administrator experience, while Training measures the number of relevant training credits. The dependent variable Completed task is either Yes or No, according to whether or not the administrator completed the tasks.

a. Create a scatterplot of Experience vs. Training using color or symbol to differentiate programmers who complete the task from those who did not complete it. Which predictor(s) appear(s) potentially useful for classifying task completion?

b. Run a logistic regression model with both predictors using the entire dataset as training data. Model the probability of Completed Task = Yes. Among those who complete the task, what is the percentage of programmers who are incorrectly classified as failing to complete the task?

c. To decrease the percentage in part (b), should the threshold probability be increased or decreased?

d. How much experience must be accumulated by a programmer with four years of training before his or her estimated probability of completing the task exceeds 50%?

10.3 **Sales of Riding Mowers.** A company that manufactures riding mowers wants to identify the best sales prospects for an intensive sales campaign. In particular, the manufacturer is interested in classifying households as prospective owners or non-owners on the basis of Income (in $1000s) and Lot Size (in 1000 ft^2). The marketing expert looked at a random sample of 24 households, given in the file `RidingMowers.jmp`. Use all the data to fit a logistic regression of ownership on the two predictors.

a. What percentage of households in the study were owners of a riding mower?

b. Create a scatterplot of Income vs. Lot Size using color or symbol to differentiate owners from non-owners. From the scatterplot, which class seems to have the higher average income, owners or non-owners?

c. Among non-owners, what is the percentage of households classified correctly?

d. To increase the percentage of correctly classified non-owners, should the threshold probability be increased or decreased?

e. What are the odds that a household with a $60K income and a lot size of 20,000 ft^2 is an owner?

f. What is the classification of a household with a $60K income and a lot size of 20,000 ft^2?

g. What is the minimum income that a household with 16,000 ft^2 lot size should have before it is classified as an owner?

10.4 **Competitive Auctions on eBay.com.** The file `eBayAuctions.jmp` contains information on 1972 auctions transacted on eBay.com during May–June 2004. The goal is to use these data to build a model that will distinguish competitive auctions from noncompetitive ones. A competitive auction is defined as an auction with at least two bids placed on the item being auctioned. The data include variables that describe the item (auction category), the seller (his or her eBay rating), and the auction terms that the seller selected (auction duration, opening price, currency, day of week of auction close). In addition we have the price at which the auction closed. The goal is to predict whether or not the auction will be competitive.

Data Preprocessing. Split the data into training and validation datasets using a 60 : 40% ratio.

a. Create a tabular summary of the average of the binary dependent variable (Competitive?) as a function of the various categorical variables. Use the information in the tables to reduce the number of categories for the categorical variables.

For example, categories that appear most similar with respect to the distribution of competitive auctions could be combined using *Recode*.

b. Run a logistic model with all predictors (or recoded predictors). Model the probability that Competitive? = 1.

c. If we want to predict at the start of an auction whether it will be competitive, we cannot use the information on the closing price. Run a logistic model with all predictors as above, excluding closing price. How does this model compare to the full model with respect to accurate prediction?

d. Interpret the meaning of the coefficient for closing price. Does closing price have a practical significance? Is it statistically significant for predicting competitiveness of auctions? (Use a 10% significance level.)

e. Use mixed stepwise selection to find the model with the best fit to the training data. Which predictors are used?

f. Use stepwise selection to find the model with the lowest predictive error rate (use the validation data). Which predictors are used?

g. What is the danger in the best predictive model that you found?

h. Explain why the best-fitting model and the best predictive models are the same or different.

i. If the major objective is accurate classification, what threshold value should be used? (*Hint*: Use the *Decision Threshold*.)

j. Based on these data, what auction settings by the seller (duration, opening price, ending day, currency) would you recommend as being most likely to lead to a competitive auction?

11

NEURAL NETS

In this chapter, we describe neural networks, a flexible data-driven method that can be used for classification or prediction. Although considered a "blackbox" in terms of interpretability, neural nets have been highly successful in terms of predictive accuracy. We discuss the concepts of "nodes" and "layers" (input layers, output layers, and hidden layers) and how they connect to form the structure of a network. We then explain how a neural network is fitted to data using a numerical example. Because overfitting is a major danger with neural nets, we present a strategy for avoiding it. We describe the different parameters that a user must specify and explain the effect of each on the process. Finally, we move from a detailed description of a basic neural net to a more general discussion of the deeper and more complex neural nets that power deep learning.

Neural Nets in JMP: Basic neural network models can be fit using the standard version of JMP. However, JMP Pro is required for advanced functionality. Deep learning is not currently available in JMP or JMP Pro.

11.1 INTRODUCTION

Neural networks, also called *artificial neural networks*, are models for classification and prediction. The neural network is based on a model of biological activity in the brain, where neurons are interconnected and learn from experience. Neural networks mimic the way that human experts learn. The learning and memory properties of neural networks resemble the properties of human learning and memory, and neural nets also have a capacity to generalize from particulars.

Neural nets began to be applied successfully in the late 1990s in areas such as bankruptcy trading and detecting fraud in credit card and monetary transactions. They really took off in the early 2000s with the development of more complex networks and the growing power of computers. Often called "deep learning," these complex networks have been primarily

Machine Learning for Business Analytics: Concepts, Techniques, and Applications with JMP Pro®,
Second Edition. Galit Shmueli, Peter C. Bruce, Mia L. Stephens, Muralidhara Anandamurthy, and Nitin R. Patel.
© 2023 John Wiley & Sons, Inc. Published 2023 by John Wiley & Sons, Inc.

responsible for a powerful revolution in machine learning. Indeed, much of what the public thinks of as "artificial intelligence" has at its heart deep (multilayer) neural networks. Text classification and image recognition are perhaps the most ubiquitous examples, but what has captured the public imagination the most is the promise of self-driving cars. It might surprise many machine learning practitioners today to learn that an application of neural nets from the 1980s was ALVINN, an autonomous vehicle driving application for normal speeds on highways. Using as input a 30×32 grid of pixel intensities from a fixed camera on the vehicle, the classifier provides the direction of steering. The outcome variable is a categorical one with 30 classes, such as *sharp left*, *straight ahead*, and *bear right*.

The main strength of neural networks is their high predictive performance. Their structure supports capturing very complex relationships between predictors and an outcome variable, which is often not possible with other predictive models.

11.2   CONCEPT AND STRUCTURE OF A NEURAL NETWORK

The idea behind neural networks is to combine the input information in a very flexible way that captures complicated relationships among these variables and between them and the outcome variable. For instance, recall that in linear regression models the form of the relationship between the outcome and the predictors is specified directly by the user. In many cases, the exact form of the relationship is very complicated or is generally unknown. In linear regression modeling we might try different transformations of the predictors, interactions between predictors, and so on, but the specified form of the relationship remains linear. In comparison, in neural networks, the user is not required to specify the correct form. Instead, the network tries to learn about such relationships from the data. In fact, linear regression and logistic regression can be thought of as special cases of very simple neural networks that have only input and output layers and no hidden layers.

Although researchers have studied numerous different neural network architectures, the most successful applications of neural networks in machine learning have been *multilayer feedforward networks*. These are networks in which there is an *input layer* consisting of nodes (sometimes called *neurons*) that simply accept the input values and successive layers of nodes that receive input from the previous layers. The outputs of nodes in each layer are inputs to nodes in the next layer. The last layer is called the *output layer*. Layers between the input and output layers are known as *hidden layers*. A feedforward network is a fully connected network with a one-way flow and no cycles. Figure 11.1 shows a diagram for

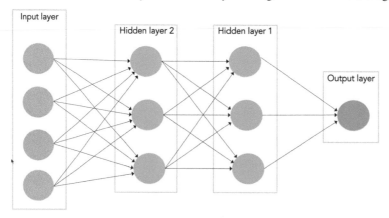

FIGURE 11.1   Multilayer feedforward neural network

this architecture, with two hidden layers and one node in the output layer representing the outcome variable value to be predicted. In a classification problem with m classes, there would be m output nodes.

11.3   FITTING A NETWORK TO DATA

To illustrate how a neural network is fitted to data, we start with a very small illustrative example. Although the method is by no means useful in such a small example, it is useful for explaining the main steps and operations, for showing how computations are done, and for integrating all the different aspects of neural network data fitting. We will later discuss a more realistic setting.

Example 1: Tiny Dataset

Consider the following very small dataset. Table 11.1 includes information on a tasting score for a certain processed cheese. The two predictors are scores for fat and salt, indicating the relative presence of fat and salt in the particular cheese sample (where 0 is the minimum amount possible in the manufacturing process, and 1 the maximum). The outcome variable is the cheese sample's consumer taste preference, where *like* or *dislike* indicates whether the consumer likes the cheese or not. Data are in `TinyDataset.jmp`.

TABLE 11.1   Tiny Example on Tasting Scores for Six Consumers and Two Predictors

Obs.	Fat score	Salt score	Acceptance
1	0.2	0.9	like
2	0.1	0.1	dislike
3	0.2	0.4	dislike
4	0.2	0.5	dislike
5	0.4	0.5	like
6	0.3	0.8	like

Figure 11.2 describes a typical neural net that could be used for predicting cheese preference (like/dislike) by new consumers, based on these data. We numbered the nodes in the example from N1 to N7. Nodes N1 and N2 belong to the input layer, nodes N3–N5 belong to the hidden layer, and nodes N6 and N7 belong to the output layer. The values on the connecting arrows are called *weights*, and the weight on the arrow from node i to node j is denoted by $w_{i,j}$. The *bias* values, denoted by θ_j, serve as an intercept for the output from node j (JMP reports this as the *intercept*). These are all explained in further detail below.

Computing Output of Nodes

We discuss the input and output of the nodes separately for each of the three types of layers (input, hidden, and output). The main difference is the function used to map from the input to the output of the node.

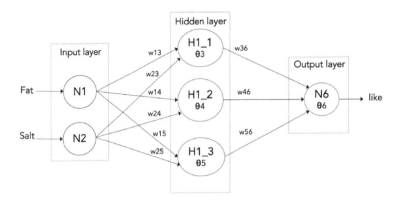

FIGURE 11.2   Neural network for the tiny example. Circles represent nodes ("neurons"), $w_{i,j}$ on arrows are weights, and θ_j inside nodes are bias values

Input nodes take as input the values of the predictors. Their output is the same as the input. If we have p predictors, the input layer will include p nodes. In our example there are two predictors, and therefore the input layer (shown in Figure 11.2) has two nodes, each feeding into each node of the hidden layer. Consider the first observation: the input into the input layer is Fat = 0.2 and Salt = 0.9, and the output of this layer is also $x_1 = 0.2$ and $x_2 = 0.9$.

Hidden layer nodes take as input the output values from the input layer. The hidden layer in this example consists of three nodes, each receiving input from all the input nodes. To compute the output of a hidden layer node, we compute a weighted sum of the inputs and apply a certain function to it. More formally, for a set of input values $x_1, x_2, \ldots, x_p$, we compute the output of node j by taking the weighted sum[1] $\theta_j + \sum_{i=1}^{p} w_{ij} x_i$, where $\theta_j, w_{1,j}, \ldots, w_{p,j}$ are weights that are initially set randomly, then adjusted as the network "learns." Note that θ_j, the *bias* or *intercept* of node j, is a constant that controls the level of contribution of node j.

In the next step, we take a function g of this sum. The function g, also called a *transfer function* or *activation function*, is some monotone function. Examples include the linear function $[g(s) = bs]$, an exponential function $[g(s) = \exp(bs)]$, a logistic function $\left[g(s) = \frac{1}{1+e^{-s}}\right]$, and a hyperbolic tangent (TanH) function $\left[g(s) = -1 + \frac{2}{1+e^{-s}}\right]$. The last two functions are by far the most popular in neural networks. Both are sigmoidal functions, namely they have an S shape. The logistic function maps values to the interval [0, 1]. The practical value of the logistic function arises from the fact that it has a squashing effect on very small or very large values but is almost linear in the range where the value of the function is between 0.1 and 0.9. The TanH maps values to the interval [−1, 1] and is a centered and scaled version of the logistic function. Deep learning uses mostly the ReLU (rectified linear unit) activation function or variants of it. This function is identical to the linear function, but set to zero for $s < 0$.

---

[1] Other options exist for combining inputs, such as taking the maximum or minimum of the weighted inputs rather than their sum, but they are much less popular.

If we use a logistic function, we can write the output of node j in the hidden layer as

$$\text{Output}_j = g\left(\theta_j + \sum_{i=1}^{p} w_{ij}x_i\right) = \frac{1}{1 + e^{-(\theta_j + \Sigma_{i=1}^{p} w_{ij}x_i)}}. \tag{11.1}$$

Initializing the Weights   The values of θ_j and w_{ij} are initialized to small, usually random, numbers (typically, but not always, in the range 0.00 ± 0.05). In JMP, normally distributed random values are used for initial weights. Such values represent a state of no knowledge by the network, similar to a model with no predictors. The initial weights are used in the first round of training.

Returning to our example, suppose that the initial weights for node H1_1 are $\theta_3 = -0.3$, $w_{1,3} = 0.05$, and $w_{2,3} = 0.01$ (as shown in Figure 11.3). Using the logistic function, we can compute the output of node H1_1 in the hidden layer (using the first observation) as

$$\text{Output}_{H1\_1} = \frac{1}{1 + e^{-[-0.3+(0.05)(0.2)+(0.01)(0.9)]}} = 0.43.$$

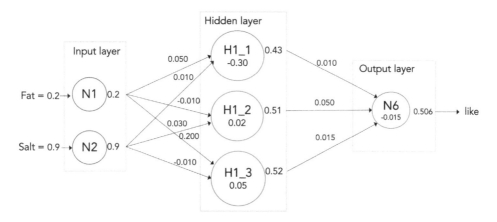

FIGURE 11.3   Computing node outputs (values are to the right of each node), using the first observation in the tiny example and a logistic function

Figure 11.3 shows the initial weights, inputs, and outputs for first record in our tiny example. If there is more than one hidden layer, the same calculation applies, except that the input values for the second, third, and so on hidden layers would be the output of the preceding hidden layer. This means that the number of input values into a certain node is equal to the number of nodes in the preceding layer. (If there were an additional hidden layer in our example, its nodes would receive input from the three nodes in the first hidden layer.) Finally, the output layer obtains input values from the (last) hidden layer. It applies the same function as above to create the output. In other words, it takes a weighted average of its input values and then applies the function g. In our example, output node N6 receives input from the three hidden layer nodes. We can compute the output of this node by

$$\text{Output}_{N6} = \frac{1}{1 + e^{-[-0.015+(0.010)(0.430)+(0.050)(0.511)+(0.015)(0.520)]}} = 0.506.$$

This is the propensity P(*Y=like*) for this record. For classification, we use a threshold value (for a binary outcome) on the propensity. Using a threshold of 0.5, we would classify this record as *like*. For applications with more than two classes, we choose the output node with the largest value.

Relation to Linear and Logistic Regression   Consider a neural network with a single output node and no hidden layers. For a dataset with p predictors, the output node receives $x_1, x_2, \ldots, x_p$, takes a weighted sum of these and applies the g function. The output of the neural network is therefore $g\left(\theta + \sum_{i=1}^{p} w_i x_i\right)$.

First, consider a numerical outcome variable y. If g is the identity function [$g(s) = s$], the output is simply

$$\hat{y} = \theta + \sum_{i=1}^{p} w_i x_i.$$

This is exactly equivalent to the formulation of a multiple linear regression! This means that a neural network with no hidden layers, a single output node, and an identity function g searches only for linear relationships between the outcome variable and the predictors.

Now consider a binary outcome variable Y. If g is the logistic function, the output is simply

$$\hat{P}(Y = 1) = \frac{1}{1 + e^{-\left(\theta + \sum_{i=1}^{p} w_i x_i\right)}},$$

which is equivalent to the logistic regression formulation!

In both cases, although the formulation is equivalent to the linear and logistic regression models, the resulting estimates for the weights (*coefficients* in linear and logistic regression) will differ. The neural net estimation method is different from the method used to calculate coefficients in linear regression (least squares), or the method used in logistic regression (maximum likelihood). We explain below the method by which the neural network learns.

Preprocessing the Data

When using sigmoidal activation functions such as logistic and hyperbolic tangent function (often called the *TanH* function), neural networks perform best when the predictors are on the same scale, like [0, 1] or [−1, 1]. For this reason, all variables should be scaled to a [0, 1] or a [−1, 1] interval before training the network. For a numerical variable X that takes values in the range $[a, b]$ where $a < b$, we rescale the measurements to a [0, 1] scale by subtracting a and dividing by $b - a$. The normalized [0, 1] measurement is then

$$X_{\text{norm}} = \frac{X - a}{b - a}.$$

Another operation that can improve the performance of the neural network is transforming highly skewed predictors. In business applications there tend to be many highly right-skewed variables (e.g., income). Taking a log transform of a right-skewed variable (before rescaling) will usually spread out the values more symmetrically.

ACTIVATION FUNCTIONS AND DATA PROCESSING FEATURES IN JMP Pro

The three activation functions used in the hidden layer in JMP Pro are *TanH* (the default), *Linear*, and *Gaussian*. In the Model Launch dialog, selecting the *Transform Covariates* checkbox (under *Fitting Options*) transforms all continuous variables to near normality, so transforming the measurements in advance is not required. The *Robust Fit* option can also be used to minimize the impact of response outliers (available for continuous outcome variables only). For categorical variables, the neural platform automatically creates indicators or dummy variables (again, this is done behind the scenes).

Training the Model

Training the model involves estimating the weights θ_j and w_{ij} that lead to the best predictive results. The process that we described earlier (Section 11.3) for computing the neural network output for an observation is repeated for all the observations in the training set. For each record, the model produces a prediction which is then compared with the actual outcome value. Their difference is the error for the output node.

Popular approaches to estimating weights in neural networks involve using the errors iteratively to update the estimated weights. In particular, the error for the output node is distributed across all the hidden nodes that led to it so that each node is assigned "responsibility" for part of the error. Each of these node-specific errors is then used for updating the weights. The most popular method for using model errors to update weights is *Back Propagation of Errors*.

In the next section, we describe *Back Propagation of Errors*. Then we discuss neural model fitting in JMP Pro and illustrate with the tiny data example. Rather than using back propagation of error, JMP Pro uses a global function of the errors (e.g., sum of squared errors) for estimating the weights.

Back Propagation of Error   The most popular method for using model errors to update weights ("learning") is an algorithm called *back propagation*. As the name implies, errors are computed from the last layer (the output layer) back to the hidden layers.

Let us denote by $\hat{y}_k$ the output from output node k. y_k is 1 or 0 depending on whether the actual value of the observation coincides with node k's value or not. For example, in Figure 11.3, for a person with output value *like* we have $y_6 = 1$.

The error associated with output node k is computed by

$$\text{err}_k = \hat{y}_k(1 - \hat{y}_k)(y_k - \hat{y}_k).$$

Notice that this is similar to the ordinary definition of an error $(y_k - \hat{y}_k)$ multiplied by a correction factor. The weights are then updated as follows:

$$\theta_j^{\text{new}} = \theta_j^{\text{old}} + l \times \text{err}_j, \tag{11.2}$$
$$w_{i,j}^{\text{new}} = w_{i,j}^{\text{old}} + l \times \text{err}_j,$$

where l is a *learning rate* or *weight decay* parameter, a constant ranging typically between 0 and 1, which controls the amount of change in weights from one iteration to the other.

In our example, the error associated with output node N6 for the first record is $(0.506)(1 - 0.506)(1 - 0.506) = 0.123$. This error is then used to compute the errors associated with the hidden layer nodes, and those weights are updated accordingly using a formula similar to equation (11.2).

Two methods for updating the bias and weights are case updating and batch updating. In *case updating*, the weights are updated after each observation is run through the network (called a *trial*). For example, if we used case updating in the tiny example, the weights would first be updated after running observation 1 as follows: using a learning rate of 0.5, the weights θ_6, $w_{3,6}$, $w_{4,6}$, and $w_{5,6}$ are updated to

$$\theta_6 = -0.015 + (0.5)(0.123) = 0.047,$$
$$w_{3,6} = 0.01 + (0.5)(0.123) = 0.072,$$
$$w_{4,6} = 0.05 + (0.5)(0.123) = 0.112,$$
$$w_{5,6} = 0.015 + (0.5)(0.123) = 0.077.$$

These new weights are next updated after the second observation is run through the network, the third, and so on, until all observations are used. This is called one *epoch*, *sweep*, or *iteration,* through the data. Typically, there are many iterations.

In *batch updating*, the entire training set is run through the network before each updating of weights takes place. As a result, the errors err_k in the updating equation is the sum of the errors from all observations. In practice, case updating tends to yield more accurate results than batch updating, but it requires a longer runtime. This is a serious consideration, since even in batch updating, hundreds or even thousands of sweeps through the training data are executed.

When does the updating stop? The most common conditions are one of the following:

1. When the new values of the bias and weights are only incrementally different from those of the preceding iteration,
2. When the misclassification rate reaches a required threshold or if there is no improvement in the misclassification rate over the prior iteration (also called early stopping strategy),
3. The limit on the number of iterations is reached.

FITTING A NEURAL NETWORK MODEL IN JMP Pro

Choose the *Neural* option from the *Analyze > Predictive Modeling* menu to fit a neural network. There are several options that the user can choose, which are described in more detail in the JMP documentation:

- Some form of model validation is required in the neural platform. A validation column can be specified in the launch dialog. If a validation column is not used, the default is a one-third random holdout set. (Refer to Chapter 5 for evaluating performance using a validation column.)

- The default neural network has one hidden layer with three nodes and uses the TanH activation function.
- Options to transform covariates, to minimize the influence of response outliers (robust fit, with continuous responses), to apply different penalty methods, and to start the fitting algorithm multiple times (each start is a tour) are found under *Fitting Options*.
- Options for *Boosting*, which fits an additive sequence of models, are also provided (see the boosting discussion in Section 11.4).
- The red triangle for the model provides additional options, such as *Diagram* (to display the neural model), *Estimates* (to show the estimated weights), and *Save Formula* (to save the prediction formula, probabilities, and classifications or predicted values to the data table).

Neural Network Model Fitting in JMP   The neural platform in JMP uses a different approach to optimizing the weights. JMP treats this as a nonlinear optimization problem and finds optimal values for the weights and bias values that minimize a function of the combined errors.[2] This approach is used, in part, because it produces similar results to back propagation of error but is generally much faster. It can also be used for both continuous and categorical responses.

Neural networks are highly flexible and generally have excellent predictive capabilities. However, a drawback is that neural networks have a tendency to overfit the data. JMP minimizes overfitting in two ways: utilizing a penalty parameter and requiring model validation. Both the penalty parameter and validation are integrated into the fitting process. The neural fitting algorithm searches iteratively for the optimal value of the penalty parameter AND the optimal weights that minimize error.[3] The algorithm stops searching when the function of the errors on the validation data is no longer improving.

While we omit the technical details, the iterative process works as follows:

1. Set the penalty parameter to 0.
2. Set the starting values for the weights (typically random normal values).
3. Vary the penalty parameter within an increasingly narrowing range of values.
4. For each value of the penalty parameter, the algorithm searches for weights that minimize the model errors.
5. The model error for the cross-validation data is monitored, and the algorithm stops when the cross-validation error is no longer improving. The model with the lowest error on the cross-validation data is selected.

---

[2]JMP Pro uses the *maximum likelihood function* to help find the optimal weight and bias values. In general, the likelihood function represents the agreement between the parameter estimates and the observed data. Maximizing the likelihood function provides parameter estimates that are most consistent with the data (Sall et al., 2017). The algorithm used in the neural platform searches for weights that minimize the *negative loglikelihood* of the data.

[3]We use the term error in this section to refer to the negative loglikelihood, which is the function of the errors that is minimized.

Additional information is available in the JMP documentation under the *Help* menu (search for *neural networks* under *Predictive and Specialized Modeling*).

Fitting a Neural Network for the Tiny Data Example in JMP   Let us examine the output from running a neural network in JMP on the tiny data. Following Figures 11.2 and 11.3, we used a single hidden layer with three nodes and the default TanH activation function. Since validation is required in the JMP Pro neural platform, we fit this model with a validation column, with four observations in the training set and two observations in the validation set.

Confusion matrices for the training and validation sets are shown in Figure 11.4. We can see that the network correctly classifies all the observations in the validation set. In Figure 11.5, we see the saved formulas and probabilities and the resulting model classifications (under *Most Likely Acceptance*). The number of observations is far too small to estimate the 13 weights. However, for purposes of illustration, we discuss the remainder of the output.

The final network weights are shown on the bottom left in Figure 11.4, under *Estimates*. The first group of estimates, starting with H1, show the weights and bias values that connect the input layer and the hidden layer. The first three estimates, starting with H1_1, are the weights and bias value for the first node, the second three values (starting with H1_2) are for the second node, and the last three values (starting with H1_3) are the estimates for the third node. The *bias values* (intercepts) are the weights θ_3, θ_4, and θ_5 in Figure 11.2. These estimated weights are used to compute the output of the hidden layer nodes. The last four values, starting with *Acceptance*, are the weights and intercept for the output node.

The neural diagram provided by JMP is shown on the right in Figure 11.4. The symbols in the nodes in the hidden layer indicate the activation functions used for each node. In this case, we used the default TanH function for all three nodes.[4] JMP doesn't display the final model weights on the diagram. For illustration, we have added the estimated weights to the diagram (this was done manually, using a graphics editor). JMP uses these weights to compute the hidden layer's output for each observation. When we save the formula for the model, these values are saved in columns H1_1, H1_2, and H1_3 (see Figure 11.5). The formula for computing the output from the first hidden layer node is shown in Figure 11.6.

To compute the output for the Acceptance output node for each observation, we use the outputs from the hidden layer nodes. The formula for *P(Acceptance = like)* provided by JMP is shown in Figure 11.7. This formula is mathematically equivalent to the formula shown below. For observation 1, this probability is

$$\text{Output} = \frac{1}{1 + e^{-[0.437+(27.40)(0.182)+(8.92)(0.248)+(17.65)(0.196)]}} = 0.999.$$

The probabilities for the other five observations are calculated in the same way, computing the hidden layer outputs and then plugging these outputs into the computation for the output layer.

---

[4]When the response is categorical, the logistic function is used to calculate the estimated probability for each class from the output of the hidden layer(s). When the response is continuous, a linear function of the output from the hidden layer(s) is used to calculate the predicted response.

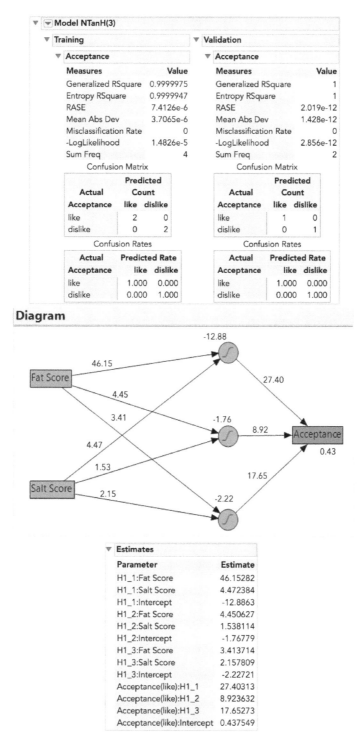

FIGURE 11.4 Output for neural network applied to the tiny data example. Estimates and diagram obtained using red triangle options. The final weights in the neural network diagram were added manually

Obs	Fat Score	Salt Score	Acceptance	Validation	H1_1	H1_2	H1_3	Probability(Acceptance =like)	Probability(Acceptance =dislike)	Most Likely Acceptance
1	1	0.2	0.9	like	Training	0.1826	0.2480	0.1962	1.0000	0.0000	like
2	2	0.1	0.1	dislike	Training	-0.9992	-0.5259	-0.6832	0.0000	1.0000	dislike
3	3	0.2	0.4	dislike	Validation	-0.7322	-0.1305	-0.3281	0.0000	1.0000	dislike
4	4	0.2	0.5	dislike	Training	-0.6105	-0.0543	-0.2287	0.0000	1.0000	dislike
5	5	0.4	0.5	like	Training	0.9992	0.3720	0.1082	1.0000	0.0000	like
6	6	0.3	0.8	like	Validation	0.9788	0.3790	0.2558	1.0000	0.0000	like

FIGURE 11.5   Saved formulas and classifications for the tiny data example

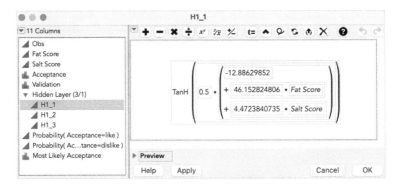

FIGURE 11.6   Formula to compute output from the first hidden layer node

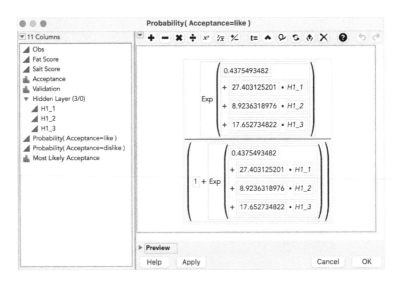

FIGURE 11.7   Formula to compute Prob(Acceptance=*like*) in JMP

Using the Output for Prediction and Classification

In the case of a binary response ($m = 2$), JMP uses just one output node with a threshold value to map a numerical output value to one of the two classes. Although we typically use a threshold of 0.5 with other classifiers, in neural networks there is a tendency for values to cluster around 0.5 (from above and below). A more suitable alternative is therefore to use the validation set to determine a threshold that produces reasonable predictive performance.

When the neural net is used for multi-class classification with m classes, the diagram displays only one node, but JMP estimates the weights for $m - 1$ classes and provides confusion matrices, lift curves, and ROC outputs for all m classes. How do we translate these m outputs into a classification rule? Usually the class with the largest probability value determines the neural network's classification.

When the neural network is used for predicting a numerical rather than a categorical response, the same fitting process is used and the model launch options are identical. But, rather than calculating the probability of an outcome the model calculates a predicted value.

Example 2: Classifying Accident Severity

Let us apply the network training process to some real data: US automobile accidents that have been classified by their level of severity as *no-injury*, *injury*, or *fatality*. A firm might be interested in developing a system for quickly classifying the severity of an accident, based on initial reports and associated data in the system (some of which rely on GPS-assisted reporting). Such a system could be used to assign emergency response team priorities. Figure 11.8 shows a small extract (15 records, four predictor variables) from a US government database. The data are in `Accidents NN.jmp`.

	Row Id.	ALCHL_I	PROFIL_I_R	SUR_COND	VEH_INVL	MAX_SEV_IR	Validation
1	1	2	0	1	1	0	Training
2	2	2	1	1	1	2	Validation
3	3	1	0	1	1	0	Training
4	4	2	0	2	2	1	Validation
5	5	2	1	1	2	1	Validation
6	6	2	0	1	1	0	Training
7	7	2	0	2	1	2	Validation
8	8	2	1	2	1	1	Training
9	9	2	1	1	1	1	Training
10	10	2	0	1	1	0	Validation
11	11	2	1	1	1	1	Training
12	12	1	1	1	2	1	Validation
13	13	1	1	1	2	2	Training
14	14	2	0	1	2	2	Validation
15	15	2	0	1	1	0	Training

FIGURE 11.8 Subset from the accident data, for a high-fatality region

The explanation of the four predictor variables and outcome variable is given in Table 11.2.

TABLE 11.2   Description of Variables for Automobile Accident Example

ALCHL_I	Presence (1) or absence (2) of alcohol
PROFIL_I_R	Profile of the roadway: level (1), other (0)
SUR_COND	Surface condition of the road: dry (1), wet (2), snow/slush (3), ice (4), unknown (9)
VEH_INVL	Number of vehicles involved
MAX_SEV_IR	Presence of injuries/fatalities: no-injuries (0), injury (1), fatality (2)

With the exception of alcohol involvement and a few other variables in the larger database, most of the variables are ones that we might reasonably expect to be available at the time of the initial accident report, before accident details, and severity have been determined by first responders. A machine learning model that could predict accident severity on the basis of these initial reports would have value in allocating first responder resources.

We partition the data into training and validation sets (60%–40%). (To replicate the results, use random seed 12345). We fit the neural network, using a training set with 599 observations and a validation set with 400 observations. The neural net architecture for this classification problem has four nodes in the input layer, one for each of the four predictors (note that JMP will create indicator variables for categorical predictors behind the scenes). Many different neural models are possible. We use a single hidden layer and experiment with the number of nodes using the default TanH activation function. We run several models, starting with three nodes and slowly increasing the number of nodes in the model. Our largest models has eight nodes. We examine the resulting confusion matrices and find that five nodes gives a good balance between improving the predictive performance on the training set without deteriorating the performance on the validation set. In other words, networks with more than five nodes in the hidden layer performed as well as the five-node network, but add undesirable complexity.

The *Model Launch* dialog in the neural platform in JMP Pro allows you to easily compare models of different complexity. For example, we can run models with different numbers of hidden layers, different numbers of nodes in the hidden layers, and different activation functions and compare the misclassification rates or other validation statistics. For illustration, in Figure 11.9, we show the results of the single layer five-node model (top) described above and the results for a much more complex model (bottom). The complex model has two hidden layers, nine nodes in each layer, and uses the three activation functions available in JMP Pro (TanH, Linear, and Gaussian). The validation misclassification rate for the second model is almost the same as the first model, but with a great deal of added complexity.

To put model complexity into some perspective, let us consider the model weights that must be estimated for each of these models (in Figure 11.9). For our five-node model, there are five connections from each of the four nodes in the input layer to each node in the hidden layer, for a total of $4 \times 5 = 20$ connections between the input layer and the hidden layer. There are also five connections between the hidden layer and the output layer. However, two of our predictors are binary categorical variables, and the other two predictors and the response have more than two levels. This results in a total of 67 parameters that must be estimated for this model. (To display these parameter estimates, select *Show Estimates*

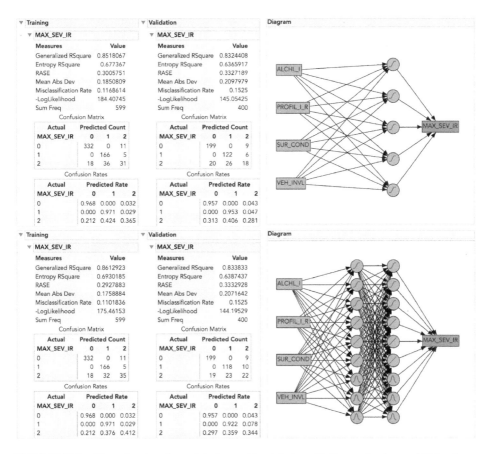

▼ Training

▼ MAX_SEV_IR

Measures	Value
Generalized RSquare	0.8518067
Entropy RSquare	0.677367
RASE	0.3005751
Mean Abs Dev	0.1850809
Misclassification Rate	0.1168614
-LogLikelihood	184.40745
Sum Freq	599

Confusion Matrix

Actual	Predicted Count		
MAX_SEV_IR	0	1	2
0	332	0	11
1	0	166	5
2	18	36	31

Confusion Rates

Actual	Predicted Rate		
MAX_SEV_IR	0	1	2
0	0.968	0.000	0.032
1	0.000	0.971	0.029
2	0.212	0.424	0.365

▼ Validation

▼ MAX_SEV_IR

Measures	Value
Generalized RSquare	0.8324408
Entropy RSquare	0.6365917
RASE	0.3327189
Mean Abs Dev	0.2097979
Misclassification Rate	0.1525
-LogLikelihood	145.05425
Sum Freq	400

Confusion Matrix

Actual	Predicted Count		
MAX_SEV_IR	0	1	2
0	199	0	9
1	0	122	6
2	20	26	18

Confusion Rates

Actual	Predicted Rate		
MAX_SEV_IR	0	1	2
0	0.957	0.000	0.043
1	0.000	0.953	0.047
2	0.313	0.406	0.281

Diagram

▼ Training

▼ MAX_SEV_IR

Measures	Value
Generalized RSquare	0.8612923
Entropy RSquare	0.6930185
RASE	0.2927883
Mean Abs Dev	0.1758884
Misclassification Rate	0.1101836
-LogLikelihood	175.46153
Sum Freq	599

Confusion Matrix

Actual	Predicted Count		
MAX_SEV_IR	0	1	2
0	332	0	11
1	0	166	5
2	18	32	35

Confusion Rates

Actual	Predicted Rate		
MAX_SEV_IR	0	1	2
0	0.968	0.000	0.032
1	0.000	0.971	0.029
2	0.212	0.376	0.412

▼ Validation

▼ MAX_SEV_IR

Measures	Value
Generalized RSquare	0.833833
Entropy RSquare	0.6387437
RASE	0.3332928
Mean Abs Dev	0.2071642
Misclassification Rate	0.1525
-LogLikelihood	144.19529
Sum Freq	400

Confusion Matrix

Actual	Predicted Count		
MAX_SEV_IR	0	1	2
0	199	0	9
1	0	118	10
2	19	23	22

Confusion Rates

Actual	Predicted Rate		
MAX_SEV_IR	0	1	2
0	0.957	0.000	0.043
1	0.000	0.922	0.078
2	0.297	0.359	0.344

Diagram

FIGURE 11.9 **Neural network models for accident data, with five nodes in one hidden layer (top) and two hidden layers with several nodes and different activation functions (bottom)**

from the red triangle for the model.) To estimate all of the weights in the more complicated model (bottom, in Figure 11.9), we must estimate 209 parameters! (This can be seen by right-clicking on the estimates table and selecting *Make Into Data Table.*)

Note: If you are following along in JMP Pro, your model results might be different. You can obtain reproducible results by setting the random seed of 12345 prior to creating neural models. In JMP, an option to set the random seed is available in the Neural Model Launch dialog.

For all neural models, our results depend largely on how we build the model, and there are a few pitfalls to avoid. We discuss these next.

Avoiding Overfitting

As previously discussed, a weakness of the neural network is that it can easily overfit the data, causing the error rate on validation data (and most important, on new data) to be too large. JMP guards against overfitting by using a penalty parameter in the optimization process and by requiring model cross-validation. Recall that the neural fitting algorithm searches for both optimal weights and the optimal value of the penalty parameter. At each iteration, the algorithm keeps track of the errors on the cross-validation data. JMP uses an early stopping rule to terminate the fitting algorithm when the cross-validation error is no longer improving.

To illustrate the effect that the penalty parameter has on preventing overfitting, compare the confusion matrices of the five-node network (top, in Figure 11.9) with those from another five-node neural network fit with no penalty parameter (Figure 11.10). The initial model performs slightly better on the training data (a misclassification rate of 0.117 vs. 0.124) but has the same misclassification rate on the validation set.

▼ ☑ Model NTanH(5)						

▼ Training				▼ Validation		

▼ MAX_SEV_IR				▼ MAX_SEV_IR		

Measures	Value			Measures	Value	
Generalized RSquare	0.8493759			Generalized RSquare	0.8309402	
Entropy RSquare	0.6734303			Entropy RSquare	0.6342825	
RASE	0.30425			RASE	0.3310608	
Mean Abs Dev	0.1817176			Mean Abs Dev	0.203247	
Misclassification Rate	0.1235392			Misclassification Rate	0.1525	
-LogLikelihood	186.65752			-LogLikelihood	145.97598	
Sum Freq	599			Sum Freq	400	

Confusion Matrix

Actual	Predicted Count			Actual	Predicted Count		
MAX_SEV_IR	0	1	2	MAX_SEV_IR	0	1	2
0	332	0	11	0	199	0	9
1	0	165	6	1	0	124	4
2	18	39	28	2	19	29	16

Confusion Rates

Actual	Predicted Rate			Actual	Predicted Rate		
MAX_SEV_IR	0	1	2	MAX_SEV_IR	0	1	2
0	0.968	0.000	0.032	0	0.957	0.000	0.043
1	0.000	0.965	0.035	1	0.000	0.969	0.031
2	0.212	0.459	0.329	2	0.297	0.453	0.250

FIGURE 11.10   JMP Pro neural network for accident data with five nodes and no penalty

11.4   USER INPUT IN JMP Pro

One of the time-consuming and complex aspects of training a model is that we first need to decide on a network architecture. This means specifying the number of hidden layers, the number of nodes in each layer, and the activation function. The usual procedure is to make intelligent guesses using past experience and to conduct several trial-and-error runs on different architectures. Algorithms exist that grow the number of nodes selectively during training or trim them in a manner analogous to what is done in classification and regression trees (see Chapter 9). Research continues on such methods. As of now, no automatic method seems clearly superior to the trial-and-error approach. The Neural Model Launch dialog, with available options, is shown in Figure 11.11. A few general guidelines for choosing an architecture follow.

Number of Hidden Layers: The most popular choice for the number of hidden layers is one. A single hidden layer is usually sufficient to capture even very complex relationships between the predictors. (JMP Pro supports up to two hidden layers.)

Number of Nodes: The number of nodes in the hidden layers also determines the level of complexity of the relationship between the predictors that the network captures. The trade-off is between under- and overfitting. On the one hand, using too few nodes might not be sufficient to capture complex relationships (e.g., recall the special cases

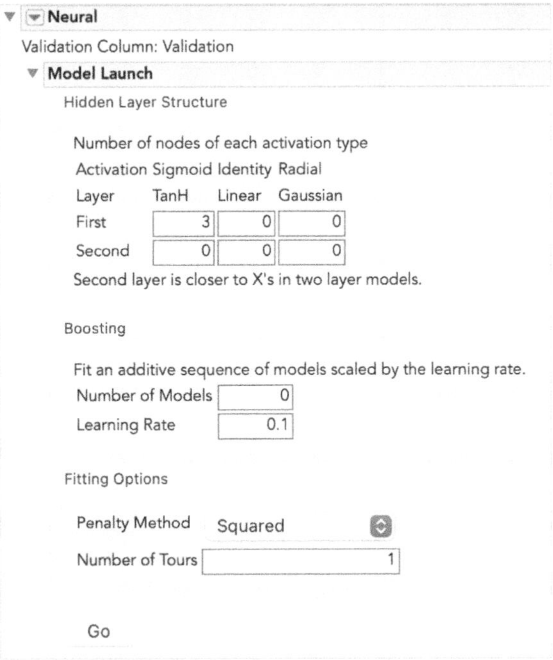

FIGURE 11.11   JMP Pro *Neural Model Launch* dialog

of a linear relationship in linear and logistic regression, in the extreme case of zero nodes or no hidden layer). On the other hand, too many nodes might lead to overfitting. A rule of thumb is to start with p (number of predictors) nodes and gradually decrease or increase a bit while checking for overfitting. Another approach is to start with the default neural model, with one layer and three nodes and then run a much more complex model with two layers and several nodes and different activation functions. If the fit statistics don't improve substantially with the more complex model, then a simpler model may suffice.

Number of Tours: This field, in the *Fitting Options* section of the Model Launch dialog, tells JMP how many times to restart the model-fitting algorithm. Since the initial weights are based on random starting values, the algorithm can lead to different final models. When a value (other than 1) is entered into this field, JMP keeps track of the best model after each *tour*, and the model with the lowest −loglikelihood on the validation data is chosen.

Informative Missing: This option, which is a check box in the initial Neural dialog window, tells JMP how to handle missing values. If this option is selected, missing values for categorical predictor variables are treated as a separate category. For continuous variables, the mean is imputed for missing values, and a new column, indicating whether or not the value is missing, is included in the model as another predictor.

In addition to the choice of architecture, the user should pay attention to the *choice of predictors*. Since neural networks are highly dependent on the quality of their input, the choice of predictors should be done carefully using domain knowledge, variable selection and dimension reduction techniques before using the network. We return to this point in the discussion of advantages and weaknesses.

Another option that can be selected in JMP Pro's neural platform is *Boosting*, which is also available for classification and regression trees (see Chapter 9) and is discussed further in Chapter 13. Boosting is an automated way to quickly construct a large additive neural network model by fitting a sequence of smaller models. First, a very simple model is fit, with one layer and one or two nodes. Then, the simple model is fit on the scaled residuals from this model, and so on. All of the models are combined to form the larger final model. The number of models can be specified, and the validation data is used to stop the fitting process when an optimal model is found (or the specified number of models is reached). A *learning rate* for boosted models can also be specified. This parameter takes a value in the range $[0, 1]$ and is used to avoid overfitting. Learning rates close to 1 not only converge more quickly on a final model but also have a higher tendency to overfit data.

11.5   EXPLORING THE RELATIONSHIP BETWEEN PREDICTORS AND OUTCOME

Neural networks are known to be "blackboxes" in the sense that their output does not shed light on the patterns in the data that it models (like our brains). In fact, that is one of the biggest criticism of the method. However, in some cases it is possible to learn more about the relationships that the network captures, by conducting a sensitivity analysis on a validation set. This is done by setting all predictor values to their mean (using the *Categorical Profiler*, available from the red triangle for the model) and obtaining the network's prediction. Then the process is repeated by setting each predictor sequentially to its minimum (and then maximum) value. By comparing the predictions from different levels of the predictors, we can get a sense of which predictors affect predictions more and in what way.

UNDERSTANDING NEURAL MODELS IN JMP Pro

As we have seen with other types of models, the JMP *Profiler*, or *Categorical Profiler* for categorical outcome variables, provides a graphical way to explore the relationships between the predictors and the outcome variable(s). This option is available under the red triangle for the model, and from within the *Model Comparison* platform if prediction formulas for the model have been saved to the data table.

The *Assess Variable Importance* option in the *Profiler* for neural (or other) models can be used to explore the variables in the model, and helps to understand which predictor variables are the most important. This feature, available for models with continuous or binary outcomes, uses simulation to estimate the variability in the outcome variable caused by the variables in the model.

When formulas are saved for different neural (or other) models, the *Model Comparison* platform (from *Analyze > Predictive Modeling*) provides an efficient way to compare validation or holdout statistics for the different models.

11.6   DEEP LEARNING[5]

Neural nets had their greatest impact in the period from 2006 onward, with a paper published by AI researcher Geoffrey Hinton, and the rapidly growing popularity of what came to be called *deep learning*. Deep learning involves complex networks with many layers, incorporating processes for dimension reduction and feature discovery.

The data we have dealt with so far in this book are structured data, typically from organizational databases. The predictor variables, or features, that we think might be meaningful in predicting an outcome of interest already exist in our data. In predicting possible bank failure, for example, we would guess that certain financial ratios (return on assets, return on equity, etc.) might have predictive value. In predicting insurance fraud, we might guess that policy age would be predictive. We are not limited, of course, to variables that we know are predictive; sometimes a useful predictor emerges in an unexpected way.

With tasks like voice and image recognition, structured high-level predictor information like this is not available. All we have are individual "low-level" sound wave frequency and amplitude or pixel values indicating intensity and color. You'd like to be able to tell the computer "just look for two eyes" and then provide further detail on how an eye appears—a small solid circle (pupil), surrounded by a ring (iris), surrounded by a white area. But, again, all the computer has are columns of (low-level) pixel values—you'd need to do a lot of extra work to define all the different (higher-level) pixel patterns that correspond to eyes. That's where deep learning comes in—it can "learn" how to identify these higher level features by itself.

Deep learning is a rapidly growing and evolving field, with many aspects that distinguish it from simpler predictive models. A full examination of the field of deep learning is beyond the scope of this book, but let us consider one key innovation in the area of algorithmic design that is associated with deep learning and extends the ability of simple neural networks to perform the tasks of supervised and unsupervised learning that we have been discussing. This innovation is "convolution." Convolutional neural networks are used in many different applications and fields and have enabled major advances in voice recognition and the extraction of meaning from text via natural language processing. The basic idea of convolution is illustrated well in the context of image recognition.

Convolutional Neural Networks (CNNs)

In a standard neural network, each predictor gets its own weight at each layer of the network. A convolution, by contrast, selects a subset of predictors (pixels) and applies the same operation to the entire subset. It is this grouping that fosters the automated discovery of features. Recall that the data in image recognition tasks consist of a large number of pixel values, which, in a black and white image, range from 0 (black) to 255 (white). Since we are interested in detecting the black lines and shadings, we will reverse this to 255 for black and 0 for white.

Consider the line drawing in Figure 11.12, from a 1893 Funk and Wagnalls publication. Before the computer can begin to identify complex features like eyes, ears, noses, heads, it needs to master very simple features like lines and borders. For example, the line of the man's chin (Figure 11.13).

---

[5]This section copyright ©2019 Datastats, LLC, Galit Shmueli, and Peter Gedeck. Used by permission.

FIGURE 11.12   Line drawing, from a 1893 Funk and Wagnalls publication

FIGURE 11.13   Focusing on the line of the man's chin

In a typical convolution, the algorithm considers a small area at a time, say, three pixels × 3 pixels. Figure 11.14 illustrates such a case. The line at the chin might look like the Figure 11.14(a); the matrix in Figure 11.14(b) might be our pixel values. In its first convolution operation, the network could apply a filter operation by multiplying the pixel values by a 3 × 3 matrix of values that happens to be good at identifying vertical lines, such as the matrix in Figure 11.14(c).

(a)

(b)

25	200	25
25	225	25
25	225	25

(c)

0	1	0
0	1	0
0	1	0

FIGURE 11.14   3 × 3 pixel representation of line on man's chin using shading (a) and values (b). The filter (c) identifies vertical lines

The sum of the individual cell multiplications is $[0 + 0 + 0 + 200 + 225 + 225 + 0 + 0 + 0] = 650$. This is a relatively high value, compared to what another arrangement of the filter matrix might produce, because both the image section and the filter have high values in the center column and low values elsewhere. (A little experimentation with other 3 × 3 arrangements of three 1's and six 0's will reveal why this particular matrix is good at identifying vertical lines.) So, for this initial filter action, we can say that the filter has detected a vertical line, and thus we can consolidate the initial nine values of the image section into a single value (say, a value between 0 and 1 to indicate the absence or presence of a vertical line).

Local Feature Map

The "vertical line detector" filter moves across and down the original image matrix, recalculating and producing a single output each time. We end up with a smaller matrix; how much smaller depends on whether the filter moves one pixel at a time, two, or more (called the *stride*). While the original image values were simply individual pixel values, the new, smaller matrix is a map of features, answering the question, "is there a vertical line in this section?"

The fact that the frame for the convolution is relatively small means that the overall operation can identify features that are local in character. We could imagine other local filters to discover horizontal lines, diagonal lines, curves, boundaries, etc. Further layers of different convolutional operations, taking these local feature maps as inputs, can then successively build up higher level features (corners, rectangles, circles, etc.).

A Hierarchy of Features

The first feature map is of vertical lines; we could repeat the process to identify horizontal lines and diagonal lines. We could also imagine filters to identify boundaries between light and dark areas. Then, having produced a set of initial low-level feature maps, the process could repeat, except this time working with these feature maps instead of the original pixel values. This iterative process continues, building up multidimensional matrix maps, or tensors, of higher and higher level features. As the process proceeds, the matrix representation of higher level features becomes somewhat abstract, so it is not necessarily possible to peer into a deep network and identify, for example, an eye.

In this process, the information is progressively compressed (simplified) as the higher level features emerge, as illustrated in Figure 11.15.

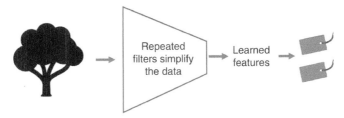

FIGURE 11.15   Convolution network process, supervised learning: the repeated filtering in the network convolution process produces high-level features that facilitate the application of class labels, which are compared to actual classes to train the network

The Learning Process

How does the net know which convolutional operations to do? Put simply, it retains the ones that lead to successful classifications. In a basic neural net, the individual weights are what get adjusted in the iterative learning process. In a convolutional network, the net also learns which convolutions to do.

In a supervised learning setting, the network keeps building up features up to the highest level, which might be the goal of the learning task. Consider the task of deciding whether an image contains a face. You have a training set of labeled images with faces and images without faces. The training process yields convolutions that identify hierarchies of features (e.g., edges > circles > eyes) that lead to success in the classification process. Other hierarchies that the net might conceivably encounter (e.g., edges > rectangles > houses) get dropped because they do not contribute to success in the identification of faces. Sometimes it is the case that the output of a single node in the network is an effective classifier, an indication that this node codes for the feature you are focusing on.

Unsupervised Learning

The most magical-seeming accomplishment of deep learning is its ability to identify features and, hence, objects in an unsupervised setting. Famous examples include identifying images with faces and identifying dogs and cats in images (see, e.g., Le et al., 2012). How is this done?

One method is to use a so-called *autoencoder* network. These networks are trained to reproduce the input that is fed into them, by first creating a lower-dimensional representation of the data and then using the created representation to reproduce the original data. The network is trained to retain the features that facilitate accurate reproduction of the input.

Viewed in the context of our image example, autoencoders have the high-level architecture shown in Figure 11.16.

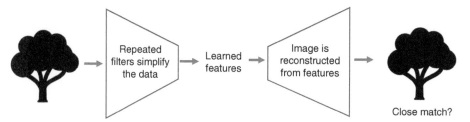

FIGURE 11.16   Autoencoder network process: the repeated filtering in the network convolution process produces high-level features that are then used to reconstruct the image. The reconstructed image is compared to the original; network architectures that produce close matches are saved as the network is "trained" without benefit of labeled data

Up to the learned features point (the bottleneck in the image), the network is similar to the supervised network. Once it has developed the learned features, which are a low-dimensional representation of the data, it expands those features into an image by a reverse process. The output image is compared to the input image, and if they are not similar, the network keeps working (using the same backpropagation method we discussed earlier). Once the network reliably produces output images that are similar to the inputs, the process stops.

This internal representation at the bottleneck now has useful information about the general domain (here, images) on which the network was trained. It turns out that the learned features (outputs of nodes at the bottleneck) that emerge in this process are often useful. Since the features can be considered a low-dimensional representation of the original data (similar in spirit to principal components in PCA—see Chapter 4), they can be used, for example, for building a supervised predictive model or for unsupervised clustering.

Conclusion

The key to convolutional networks' success is their ability to build, through iteration, multi-dimensional feature maps of great complexity (requiring substantial computing power and capacity), leading to the development of learned features that form a lower-dimensional representation of the data. You can see that different convolutional architectures (e.g., types of filtering operations) would be suitable for different tasks. The AI community shares pre-trained networks that allow analysts to short-circuit the lengthy and complex training process that is required, as well as datasets that allow training and benchmarking to certain tasks (e.g., datasets of generic images in many different classes, images specifically of faces, satellite imagery, text data, voice data, etc.).

Finally, you may have heard of other types of deep learning. At the time of writing, Recurrent Neural Networks (RNN) are very popular. They are usually implemented with either Long Short-Term Memory (LSTM) or Gated Recurrent Units (GRU) building blocks. These networks are especially useful for data that are sequential in nature or have temporal behavior, such as time series, music, text, and speech (where the order of words in a sentence matters), and user or robot behavior (e.g., clickstream data, where the order of a user's actions can be important). These networks' memory helps them capture such sequences.

In deciding whether to use such methods in a given machine learning application, a key consideration is their performance relative to other methods and their required resources (in terms of data and computing).

Note: JMP Pro supports up to two layers of hidden nodes, with TanH, Linear, and Gaussian activation functions. JMP Pro does not support deep learning as of Version 17.

11.7   ADVANTAGES AND WEAKNESSES OF NEURAL NETWORKS

The most prominent advantage of neural networks is their good predictive performance. They are known to have high tolerance to noisy data and the ability to capture highly complicated relationships between the predictors and a response. Their weakest point is in providing insight into the structure of the relationship, hence their blackbox reputation.

Several considerations and dangers should be kept in mind when using neural networks. First, although they are capable of generalizing from a set of examples, extrapolation is still a serious danger. If the network sees only cases in a certain range, its predictions outside this range can be completely invalid.

Second, the extreme flexibility of the neural network relies heavily on having sufficient data for training purposes. As our tiny example shows, a neural network performs poorly when the training set size is insufficient, even when the relationship between the response and predictors is very simple. A related issue is that in classification problems, the network requires sufficient records of the minority class. This is achieved by oversampling, as explained in Chapter 5.

Finally, a practical consideration that can determine the usefulness of a neural network is the timeliness of computation. Neural networks are relatively heavy on computation time, requiring a longer runtime than other classifiers. This runtime grows greatly when the number of predictors is increased (as there will be many more weights to compute). The fitting algorithm in JMP Pro (a quasi Newton nonlinear optimization algorithm) was implemented to reduce overall computation time. In applications where real-time or near-real-time prediction is required, runtime should be measured to make sure that it does not cause unacceptable delay in the decision-making.

PROBLEMS

11.1 Credit Card Use. Consider the following hypothetical bank data on consumers' use of credit card credit facilities in Table 11.3. Create JMP data, and create a neural network like that used for the tiny data example. Use the default validation method (Holdback Portion).

 a. How does the model perform on the validation set?

 b. How many parameters are estimated to build this model?

 c. Use the *Categorical Profiler* to explore how the predicted response changes as you change values of the predictors. Describe what you observe.

 d. Fit another neural network to this same data.

   i. Describe the differences between this model and the first model you generated.

   ii. Why do the results differ?

TABLE 11.3 Data for Credit Card Example and Variable Descriptions

Years	Salary	Used credit
4	43	0
18	65	1
1	53	0
3	95	0
15	88	1
6	112	1

Note: Years = Number of years that a customer has been with the bank; Salary = customer's salary (in thousands of dollars); Used credit 1 = customer has left an unpaid credit card balance at the end of at least one month in the prior year; and 0 = balance was paid off at the end of each month.

11.2 Neural Net Evolution. A neural net typically starts out with random weights; hence it produces essentially random predictions in the first iteration. Describe how the neural net evolves (in JMP) to produce a more accurate prediction.

11.3 Car Sales. Consider again the data on used cars (`ToyotaCorolla.jmp`) with 1436 records and details on 38 variables, including Price, Age, KM, HP, and other specifications. The goal is to predict the price of a used Toyota Corolla based on its specifications.

 a. Determine which variables to include, and use the neural platform in JMP Pro to fit a model. Use the validation column for validation, and use the default values in the Neural Model Launch dialog. Record the RMSE for the training data and the validation data, and save the formula for the model to the data table (use the *Save Fast Formulas* option, which will save the formula as one column in the data table). Repeat the process, changing the number of nodes (and only this) to 5, 10, and 25.

   i. Using your recorded values, what happens to the RMSE for the training data as the number of nodes increases?

   ii. What happens to the RMSE for the validation data?

   iii. Comment on the appropriate number of nodes for the model.

     iv. Use the *Model Comparison* platform to compare these four models (use the Validation column as either a *By* variable or as a *Group* variable, and focus only on the validation data). Here, *RASE* is reported rather than RMSE. Compare RASE and AAE (average absolute error) values for these four models. Which model has the lowest "error"?

  b. Conduct a similar experiment to assess the effect of changing the number of layers in the network as well as the activation functions.

11.4 **Direct Mailing to Airline Customers.** East–West Airlines has entered into a partnership with the wireless phone company Telcon to sell the latter's service via direct mail. The file `EastWestAirlinesNN.jmp` contains a subset of a data sample of who has already received a test offer. About 13% accepted.

You are asked to develop a model to classify East–West customers as to whether they purchased a wireless phone service contract (outcome variable Phone_Sale), a model that can be used to predict classifications for additional customers.

  a. Create a validation column (stratified on Phone_Sale). Then run a neural net model on these data. Request lift curves, and interpret the meaning (in business terms) of the lift curve for the validation set.

  b. Comment on the difference between the training and validation lift curves.

  c. Run a second neural net model on the data, this time setting the number of tours to 20. Comment now on the difference between this model and the model you ran earlier in terms of lift.

  d. Run a third neural net model on the data, with one tour. This time add a second hidden layer, different activation functions, and several nodes. Comment on the difference between this model and the first model you ran, and how overfitting might have affected results.

  e. What sort of information, if any, is provided about the effects and importance of the various variables?

  f. For this assignment, we did not ask you to set the random seed when creating the validation column or when building the models.

    i. Comment on why we might set the random seed before creating the validation column.

    ii. Comment on why we might set the random seed before building each neural network model.

12

DISCRIMINANT ANALYSIS

In this chapter, we describe the method of discriminant analysis, which is a model-based approach to classification. We discuss the main principle where classification is based on the distance of an observation from each of the class averages. We explain the underlying measure of "statistical distance," which takes into account the correlation between predictors. The output of a discriminant analysis procedure generates estimated "discriminant functions," which are then used to produce discriminant scores that can be translated into classifications or propensities (probabilities of class membership). Finally, we discuss the underlying model assumptions, the practical robustness to some, and the advantages of discriminant analysis when the assumptions are reasonably met (e.g., the sufficiency of a small training sample).

Discriminant analysis in JMP: The methods discussed in this chapter are available in the standard version of JMP. However, to compute validation statistics JMP Pro is required.

12.1 INTRODUCTION

Discriminant analysis is a classification method. Like logistic regression, it is a classical statistical technique that can be used for classification and profiling. It uses sets of measurements on different classes of records to classify new records into one of those classes (*classification*). Common uses of the method have been in classifying organisms into species and subspecies; classifying applications for loans, credit cards, and insurance into low- and high-risk categories; classifying customers of new products into early adopters, early majority, late majority, and laggards; classifying bonds into bond rating categories; classifying skulls of human fossils; as well as in research studies involving disputed authorship, decision on college admission, medical studies involving alcoholics and nonalcoholics, and methods to identify human fingerprints. Discriminant analysis can also be used to highlight aspects that distinguish the classes (*profiling*).

Machine Learning for Business Analytics: Concepts, Techniques, and Applications with JMP Pro®,
Second Edition. Galit Shmueli, Peter C. Bruce, Mia L. Stephens, Muralidhara Anandamurthy, and Nitin R. Patel.
© 2023 John Wiley & Sons, Inc. Published 2023 by John Wiley & Sons, Inc.

We return to two examples that were described in earlier chapters, riding mowers and personal loan acceptance (Universal Bank). In each of these, the outcome variable has two classes. We close with a third example involving more than two classes.

Example 1: Riding Mowers

We return to the example from Chapter 7, where a riding mower manufacturer would like to find a way to classify families in a city into those likely to purchase a riding mower and those not likely to buy one. A pilot random sample of 12 owners and 12 non-owners in the city is undertaken. The data are given in Chapter 7 (Table 7.1), and a scatterplot is shown in Figure 12.1.

We can think of a linear classification rule as a line that separates the two-dimensional region into two parts, with most of the owners in one half-plane and most non-owners in the complementary half-plane. A good classification rule would separate the data so that the fewest points are misclassified: the ad hoc line shown in Figure 12.1 seems to do a good job in discriminating between the two classes as it makes four misclassifications out of 24 points. Can we do better?

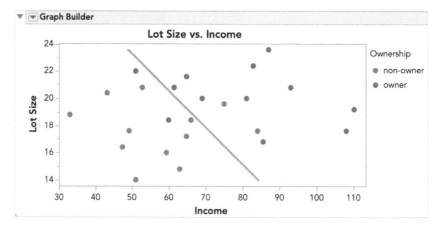

FIGURE 12.1 **Scatterplot of** *Lot Size* **vs.** *Income* **for 24 owners and non-owners of riding mowers. The (ad hoc) line tries to separate owners from non-owners**

Example 2: Personal Loan Acceptance (Universal Bank)

Riding mowers is a classic example and is useful in describing the concept and goal of discriminant analysis. However, in today's business applications, the number of observations is generally much larger, and the separation into classes is much less distinct. To illustrate this, we return to the Universal Bank example described in Chapter 9, where the bank's goal is to identify new customers most likely to accept a personal loan. For simplicity, we consider only two predictor variables: the customer's annual income (Income, in $000s), and the average monthly credit card spending (CCAvg, in $000s). The top plot in Figure 12.2 shows the acceptance of a personal loan by a random subset of 200 customers from the bank's database as a function of Income and CCAvg (the random sample was selected in *Graph Builder* using the *Sampling* option from the red triangle). We use a logarithmic scale on both axes to enhance visibility because there are many points condensed in the

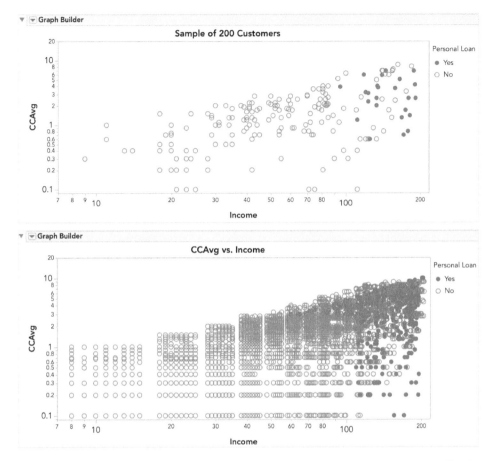

FIGURE 12.2   **Personal loan acceptance as a function of income and credit card spending for 200 (top) and 5000 (bottom) customers of the Universal Bank. Axes are in logarithmic scale**

low-income, low-CCAvg spending area (right-click on an axis and use *Axis Settings* to change the scale). Even for this small subset, the separation is not clear. The bottom plot shows all 5000 customers and the added complexity of dealing with large numbers of records.

12.2   DISTANCE OF AN OBSERVATION FROM A CLASS

Finding the best separation between observations involves measuring their distance from their class. The general idea is to classify an observation to the class to which it is closest. Suppose that we are required to classify a new customer of Universal Bank as being an acceptor or a nonacceptor of their personal loan offer, based on an income of x. From the bank's database, we find that the average income for loan acceptors was $144.75K and for nonacceptors $66.24K. We can use Income as a predictor of loan acceptance via a simple *Euclidean distance rule*: if x is closer to the average income of the acceptor class than to the average income of the nonacceptor class, classify the customer as an acceptor; otherwise, classify the customer as a nonacceptor. In other words, if $|x - 144.75| < |x - 66.24|$, classification = acceptor; otherwise, nonacceptor. Moving from a single predictor variable (Income) to two or more predictor variables, the equivalent of the mean of a class is the *centroid* of a class. This is simply the vector of means $\bar{\mathbf{x}} = [\bar{x}_1, \ldots, \bar{x}_p]$. The Euclidean distance between an observation with p measurements $\mathbf{x} = [x_1, \ldots, x_p]$ and the centroid $\bar{\mathbf{x}}$ is defined as the root of the sum of the squared differences between the individual values and the means:

$$D_{\text{Euclidean}}(\mathbf{x}, \bar{\mathbf{x}}) = \sqrt{(x_1 - \bar{x}_1)^2 + \cdots + (x_p - \bar{x}_p)^2}. \tag{12.1}$$

Using the Euclidean distance has three drawbacks. First, the distance depends on the units we choose to measure the predictor variables. We will get different answers if we decide to measure income in dollars, for instance, rather than in thousands of dollars.

Second, Euclidean distance does not take into account the variability of the predictor variables. For example, when we compare the variability in income in the two classes, we find that for acceptors the standard deviation is lower than for nonacceptors ($31.6K vs. $40.6K). Therefore, the income of a new customer might be closer to the acceptors' average income in dollars, but because of the large variability in income for nonacceptors, this customer is just as likely to be a nonacceptor. For this reason, we want the distance measure to take into account the variance of the different variables and measure a distance in standard deviations rather than in the original units. This is equivalent to z-scores.

Third, Euclidean distance ignores the correlation between the predictor variables. This is often a very important consideration, especially when we are using many predictor variables to separate classes. In this case, there will often be variables which by themselves are useful discriminators between classes, but in the presence of other predictor variables are practically redundant, as they capture the same effects as the other variables.

A solution to these drawbacks is to use a measure called *statistical distance* (or *Mahalanobis distance*). Let us denote by S the covariance matrix between the p-variables. The definition of a statistical distance is

$$\begin{aligned} D_{\text{Statistical}}(\mathbf{x}, \bar{\mathbf{x}}) &= \sqrt{[\mathbf{x} - \bar{\mathbf{x}}]' S^{-1} [\mathbf{x} - \bar{\mathbf{x}}]} \\ &= \sqrt{[(x_1 - \bar{x}_1), (x_2 - \bar{x}_2), \ldots, (x_p - \bar{x}_p)] S^{-1} \begin{bmatrix} x_1 - \bar{x}_1 \\ x_2 - \bar{x}_2 \\ \vdots \\ x_p - \bar{x}_p \end{bmatrix}}. \end{aligned} \tag{12.2}$$

(The notation $\prime$, which represents the *transpose operation*, simply turns the column vector into a row vector. S^{-1} is the inverse matrix of S, which is the p-dimension extension to division.)

When there is a single predictor ($p = 1$), this reduces to a z-score, since we subtract the mean and divide by the standard deviation. The statistical distance takes into account not only the predictor averages but also the spread of the predictor values and the correlations between the different predictors. To compute a statistical distance between an observation and a class, we must compute the predictor averages (the centroid) and the covariances between each pair of predictors. These are used to construct the distances. The method of discriminant analysis uses *squared* statistical distance as the basis for finding a separating line (or, if there are more than two variables, a separating hyperplane) that is equally distant from the different class means.[1] It is based on measuring the squared statistical distances of an observation to each of the classes and allocating it to the closest class.

12.3 FROM DISTANCES TO PROPENSITIES AND CLASSIFICATIONS

Linear classification functions were suggested in 1936 by the noted statistician R. A. Fisher as the basis for improved separation of observations into classes. The idea is to find linear functions of the measurements that maximize the ratio of between-class variability to within-class variability. In other words, we would obtain classes that are very homogeneous and differ the most from each other. For each observation, these functions are used to compute scores that measure the proximity of that observation to each of the classes. An observation is classified as belonging to the class for which it has the highest classification score (equivalent to the smallest statistical distance).

In JMP, discriminant functions are estimated to compute the squared distance of each record from the centroids of each class. Returning to the riding mowers example, in Figure 12.3, we see the formula for the squared distances for owner (*SqDist[owner]*). To view these functions in JMP, the formulas must be saved to the data table (from the red triangle select *Score Options > Save Formulas*). Note that the values in column *SqDist[0]* are the squared distances of each observation to the overall centroid of the data (this formula is stored in a format that allows JMP to calculate the squared distances more quickly).

FIGURE 12.3 **Discriminant analysis output for riding mower data, displaying the formula for the squared distance for owner (saved as a formula to the data table)**

A household is classified into the class of *owners* if the owners squared distance (*or distance score*) is lower than the non-owner squared distance, and into *non-owners* if the reverse is the case. These discriminant functions are specified in a way that can be generalized easily to more than two classes. The values given for the functions are simply the weights to be associated with each variable in the linear function in a manner analogous to multiple linear regression. For instance, the first household (highlighted in Figure 12.3)

---

[1] An alternative approach finds a separating line or hyperplane that is "best" at separating the different clouds of points. In the case of two classes, the two methods coincide.

has an income of \$60K and a lot size of 18.4K ft^2. The *owner* distance score is therefore $110.54 + (-0.86)(60) + (-10.93)(18.4) + 144.93 = 2.75$. The non-owner score for this record is 0.19. Since the distance score for non-owner is lower than the distance score for owner, the household is (mis)classified by the model as a non-owner. The squared distances for all 24 households are given in Figure 12.4.

▼ Squared Distances to Each Group

Row	Actual	non-owner	owner
1	owner	0.1948191	2.7499086
2	owner	2.8564614	2.8123966
3	owner	4.4321438	0.9998403
4	owner	2.6833323	1.1688444
5	owner	14.276475	3.2532075
6	owner	12.1519	3.4076181
7	owher	9.657628	3.846904
8	owner	9.5429788	1.2460889
9	owner	2.2294449	0.4677979
10	owner	8.6820571	0.85306
11	owher	4.4655294	3.1714345
12	owner	4.1881874	0.0210131
13	noñ-owner	2.558766	0.2225031
14	non-owner	2.3486534	2.5781727
15	non-owner	0.2120977	3.6894197
16	non-owner	2.1349846	4.9170735
17	non-owner	2.6637394	1.6640699
18	non-owner	0.2579409	6.2343006
19	non-owner	0.6264304	7.0706829
20	non-owner	0.50748	1.8598059
21	non-owner	0.9075089	9.129141
22	non-owner	2.2209497	9.5603279
23	non-owner	3.6821947	14.951055
24	non-owner	1.8811406	9.4881782

FIGURE 12.4 Squared distances for riding mower data. To view this table in JMP, select *Score Options > Show Distances to Each Group* **from the red triangle**

An alternative way for classifying an observation into one of the classes is to compute the probability of belonging to each of the classes and assigning the observation to the most likely class. If we have two classes, we need only compute a single probability for each observation (e.g., of belonging to *owners*). Using a threshold of 0.5 is equivalent to assigning the observation to the class with the lowest distance score. The advantage of this approach is that we obtain propensities, which can be used for goals such as ranking. For example, we can generate ROC curves, and can save the formulas and use the *Model Comparison* platform to compare lift curves and other statistics (the options for ROC curves and saving formulas are available from *Score Options*, under the top red triangle).

Let us assume that there are m classes. To compute the probability of belonging to a certain class k for an observation, we need to compute the squared distances for the different classes, $d_1^2, d_2^2, \ldots, d_m^2$ for that observation and combine them using the following formula:

$$P(\text{Observation belongs to class } k) = \frac{1}{1 + \sum_{k \neq m} e^{(-[d_k^2 - d_m^2]/2)}}. \quad (12.3)$$

In JMP, these probabilities are computed automatically, as can be seen in the *Discriminant Scores* table in Figure 12.5. Probabilities for all groups can be displayed by selecting *Show Probabilities to Each Group* from *Score Options* under the red triangle, or by saving the formulas to the data table.

▼ **Discriminant Scores**

Row	Actual	SqDist(Actual)	Prob(Actual)	-Log(Prob)		Predicted	Prob(Pred)	Others
1	owner	2.749909	0.2180	1.523	*	non-owner	0.7820	
2	owner	2.812397	0.5055	0.682		owner	0.5055	non-owner 0.49
3	owner	0.999840	0.8476	0.165		owner	0.8476	non-owner 0.15
4	owner	1.168844	0.6808	0.385		owner	0.6808	non-owner 0.32
5	owner	3.253207	0.9960	0.004		owner	0.9960	
6	owner	3.407618	0.9875	0.013		owner	0.9875	
7	owner	3.846904	0.9481	0.053		owner	0.9481	
8	owner	1.246089	0.9845	0.016		owner	0.9845	
9	owner	0.467798	0.7070	0.347		owner	0.7070	non-owner 0.29
10	owner	0.853060	0.9804	0.020		owner	0.9804	
11	owner	3.171435	0.6563	0.421		owner	0.6563	non-owner 0.34
12	owner	0.021013	0.8893	0.117		owner	0.8893	non-owner 0.11
13	non-owner	2.558766	0.2372	1.439	*	owner	0.7628	
14	non-owner	2.348653	0.5287	0.637		non-owner	0.5287	owner 0.47
15	non-owner	0.212098	0.8505	0.162		non-owner	0.8505	owner 0.15
16	non-owner	2.134985	0.8008	0.222		non-owner	0.8008	owner 0.20
17	non-owner	2.663739	0.3776	0.974	*	owner	0.6224	
18	non-owner	0.257941	0.9520	0.049		non-owner	0.9520	
19	non-owner	0.626430	0.9617	0.039		non-owner	0.9617	
20	non-owner	0.507480	0.6629	0.411		non-owner	0.6629	owner 0.34
21	non-owner	0.907509	0.9839	0.016		non-owner	0.9839	
22	non-owner	2.220950	0.9751	0.025		non-owner	0.9751	
23	non-owner	3.682195	0.9964	0.004		non-owner	0.9964	
24	non-owner	1.881141	0.9782	0.022		non-owner	0.9782	

'*' indicates misclassified

FIGURE 12.5   Output for riding mower data, displaying the propensities (estimated probability) of ownership (under *Prob(Actual)*)

We now have three misclassifications, compared to four in our original (ad hoc) classification (see Figure 12.1). This can be seen in the scatterplot matrix in Figure 12.6, which was

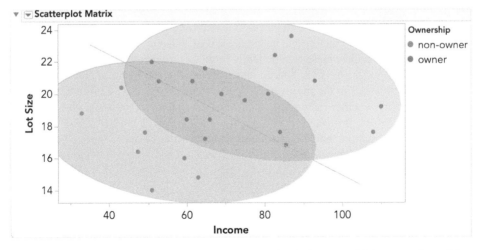

FIGURE 12.6   Class separation obtained from the discriminant model (compared to ad hoc line)

produced using the *Scatterplot Matrix* option from the top red triangle in the Discriminant analysis window.[2]

LINEAR DISCRIMINANT ANALYSIS IN JMP

The *Discriminant* platform in JMP is available from the *Analyze > Multivariate Methods* platform. The predictor variables and the validation column must be coded as numeric. The default output includes:

- The *Canonical Plot*, which shows the squared distances from each observation to the centroids for each group.
- The table of *Discriminant Scores*, which includes squared distances, probabilities, and predicted classes.
- The *Score Summaries* table, which provides overall statistics and confusion matrices.

Additional options, such as distances to each group and probabilities for each group, are available under the red triangle. To save the probability formulas and the predicted classifications to the data table, use *Score Options > Save Formulas* from the top red triangle.

12.4 CLASSIFICATION PERFORMANCE OF DISCRIMINANT ANALYSIS

The discriminant analysis method relies on two main assumptions: first, it assumes that the predictor variables in all classes come from a multivariate normal distribution. When this assumption is reasonably met, discriminant analysis is a more powerful tool than other classification methods, such as logistic regression. In fact, Efron (1975) showed that discriminant analysis is 30% more efficient than logistic regression if the data are multivariate normal, in the sense that we require 30% fewer observations to arrive at the same results. In practice, it has been shown that this method is relatively robust to departures from normality in the sense that predictors can be nonnormal or coded as dummy variables. This is true as long as the smallest class is sufficiently large (approximately more than 20 observations). This method is also known to be sensitive to outliers in both the univariate space of single predictors and in the multivariate space. Exploratory analysis should therefore be used to identify whether there are any extreme cases that need to be addressed (i.e., corrected, re-coded, or eliminated).

The second assumption behind discriminant analysis is that the correlation structure between the different predictors within a class is the same across classes. This can be roughly checked by computing the correlation matrix between the predictors for each class and

---

[2] The density ellipses graphically display the correlation between the two variables for each group. The covariance of the data in each ellipse is a weighted average of the covariances for the two groups. The discriminant function is the line that best separates the two density ellipses. The scatterplot matrix in Figure 12.6 was modified for illustration. It was rescaled, a legend was added to better visualize the points (right-click in the middle of the graph and select *Row Legend*), marker colors were changed, and the line was drawn (using the *Line* tool from the toolbar) to better visualize the class separation.

comparing matrices. If the correlations differ substantially across classes, the classifier will tend to classify cases into the class with the largest variability. When the correlation structure differs significantly and the dataset is very large, an alternative is to use quadratic discriminant analysis (QDA) (under *Discriminant Method* select *Quadratic, Different Covariances*).[3]

Notwithstanding the caveats embodied in these statistical assumptions, recall that in a predictive modeling environment, the ultimate test is whether the model works effectively. A reasonable approach is to conduct some exploratory analysis with respect to normality and correlation, train and then evaluate a model, and then, depending on classification accuracy and what you learned from the initial exploration, circle back and explore further to learn whether outliers should be examined or choice of predictor variables revisited.

With respect to the evaluation of classification accuracy, we once again use the general measures of performance that were described in Chapter 5 (judging the performance of a classifier), with the principal measure based on the misclassification rate and the confusion matrix. The same argument for using the validation set for evaluating performance still holds. For example, in the riding mower example, families 1, 13, and 17 are misclassified, as can be seen in Figure 12.5 (misclassified observations are marked with an asterisk). This means that the model yields an error rate of 12.5% for these data. However, this rate is a biased estimate—it is overly optimistic because we have used the same data for fitting the classification functions and for estimating the error. Therefore, as with all other models, we test performance on a validation set that includes data that were not involved in estimating the classification functions.

The confusion matrix for the mower data, along with other summary statistics, is shown in Figure 12.7. Since our dataset is small, we have not used a validation set in this example.

▼ **Score Summaries**

Source	Count	Number Misclassified	Percent Misclassified	Entropy RSquare	-2LogLikelihood
Training	24	3	12.5000	0.53433	15.4934

Training

Actual Ownership	Predicted Count non-owner	owner
non-owner	10	2
owner	1	11

Actual Ownership	Predicted Rate non-owner	owner
non-owner	0.833	0.167
owner	0.083	0.917

FIGURE 12.7   **Score summaries and confusion matrix for mower data**

12.5 PRIOR PROBABILITIES

So far we have assumed that our objective is to minimize the classification error. The method presented above assumes that the chances of encountering an observation from either class

---

[3] In practice, quadratic discriminant analysis has not been found useful except when the difference in the correlation matrices is large and the number of observations available for training and testing is large. The reason is that the quadratic model requires estimating many more parameters that are all subject to error [for c classes and p variables, the total number of parameters to be estimated for all the different correlation matrices is $cp(p + 1)/2$].

are the same. If the probability of encountering an observation for classification in the future is not equal for the different classes, we should modify our functions to reduce our expected (long-run average) error rate. The modification is done as follows: let us denote by p_k the prior or future probability of membership in class k (in the two-class case we have p_1 and $p_2 = 1 - p_1$). We modify[4] the formula for the estimated probabilities for each class by adding $-2\log(p_k)$. The formula for computing probabilities, where p_k is the prior probability of membership for group k, becomes

$$P(\text{Observation belongs to class } k) = \frac{1}{1 + \sum_{k \neq m} e^{(-[(d_k^2 - 2log(p_k)) - (d_m^2 - 2log(p_m))]/2)}}.$$

$$(12.4)$$

To illustrate this, suppose that the percentage of riding mower owners in the population is 15%, compared to 50% in the sample. This means that the model should classify fewer households as owners. To account for this distortion, we adjust the constants in the discriminant functions (see Figure 12.3) by adding $-2\log(0.15) = 3.794$ to the squared distance formula for owner and $-2\log(0.85) = 0.325$ to the formula for non-owner. For example, for family 1, the adjusted owner distance score is now $110.54 + (-0.86)(60) + (-10.93)(18.4) + 144.93 + 3.794 = 6.54$ (see Figure 12.8), and the adjusted non-owner score is now 0.519. It is therefore classified as non-owner.

▼ Discriminant Scores

Row	Actual	SqDist(Actual)	Prob(Actual)	-Log(Prob)		Predicted	Prob(Pred)	Others
1	owner	6.544149	0.0469	3.060		* non-owner	0.9531	
2	owner	6.606637	0.1528	1.878		* non-owner	0.8472	
3	owner	4.794080	0.4954	0.702		* non-owner	0.5046	
4	owner	4.963084	0.2734	1.297		* non-owner	0.7266	
5	owner	7.047447	0.9776	0.023		owner	0.9776	
6	owner	7.201858	0.9332	0.069		owner	0.9332	
7	owner	7.641144	0.7633	0.270		owner	0.7633	non-owner 0.24
8	owner	5.040329	0.9179	0.086		owner	0.9179	
9	owner	4.262038	0.2986	1.209		* non-owner	0.7014	
10	owner	4.647300	0.8984	0.107		owner	0.8984	non-owner 0.10
11	owner	6.965675	0.2521	1.378		* non-owner	0.7479	
12	owner	3.815253	0.5864	0.534		owner	0.5864	non-owner 0.41
13	non-owner	2.883804	0.6379	0.449		non-owner	0.6379	owner 0.36
14	non-owner	2.673691	0.8641	0.146		non-owner	0.8641	owner 0.14
15	non-owner	0.537136	0.9699	0.031		non-owner	0.9699	
16	non-owner	2.460022	0.9579	0.043		non-owner	0.9579	
17	non-owner	2.988777	0.7747	0.255		non-owner	0.7747	owner 0.23
18	non-owner	0.582979	0.9912	0.009		non-owner	0.9912	
19	non-owner	0.951468	0.9930	0.007		non-owner	0.9930	
20	non-owner	0.832518	0.9176	0.086		non-owner	0.9176	
21	non-owner	1.232547	0.9971	0.003		non-owner	0.9971	
22	non-owner	2.545988	0.9955	0.004		non-owner	0.9955	
23	non-owner	4.007233	0.9994	0.001		non-owner	0.9994	
24	non-owner	2.206178	0.9961	0.004		non-owner	0.9961	

'*' indicates misclassified

FIGURE 12.8   **Discriminant scores for mower data, with the prior probability for owners set at 0.15**

[4]JMP sets the prior class probabilities as the proportions that are encountered in the dataset. This is based on the assumption that a random sample will yield a reasonable estimate of membership probabilities. However, the priors can be specified within the discriminant analysis platform using *Specify Priors*.

To see how the adjustment can affect classifications, consider family 13, which was misclassified as an owner in the case involving equal probability of class membership. When we account for the lower probability of owning a mower in the population, family 13 is classified properly as a non-owner (its owner distance score is above the non-owner score). Figure 12.8 shows the distance scores for all households, adjusting for priors. Note that 6 of the 24 observations are now misclassified.

12.6   CLASSIFYING MORE THAN TWO CLASSES

Example 3: Medical Dispatch to Accident Scenes

Ideally, every automobile accident call to 911 results in the immediate dispatch of an ambulance to the accident scene. However, in some instances, the dispatch might be delayed (e.g., at peak accident hours or in some resource-strapped towns or shifts). In such an event, the 911 dispatchers must make decisions about which units to send based on sketchy information. It is useful to augment the limited information provided in the initial call with additional information in order to classify the accident as minor injury, serious injury, or death. For this purpose, we can use data that were collected on automobile accidents in the United States in 2001 that involved some type of injury. For each accident, additional information is recorded, such as day of week, weather conditions, and road type. Figure 12.9 shows a small sample of observations with 10 measurements of interest. These measurements were extracted from a larger set of possible measurements, some of which were dummy coded. Data are in `Accidents1000DA.jmp`.

USING CATEGORICAL PREDICTORS IN DISCRIMINANT ANALYSIS IN JMP

All predictors in discriminant analysis must be continuous (numeric)—categorical predictors must first be converted to continuous indicator (dummy) variables. To create indicator columns in JMP, use the *Make Indicator Columns* utility under *Cols > Utilities*, and make sure the modeling type is continuous.

The goal is to see how well the predictors can be used to classify injury type correctly. To evaluate this, a sample of 1000 records was drawn and partitioned into training and validation sets, and a discriminant analysis was performed. Discriminant scores for the first 15 observations are shown in Figure 12.10.

The output structure is very similar to that for the two-class case. The only difference is that each observation now has three discriminant functions (one for each injury type), and the confusion matrices for the training and validation sets are 3×3 to account for all the combinations of correct and incorrect classifications (see *Score Summaries* at the top of Figure 12.11).

The rule for classification is still to classify an observation to the class that has the lowest corresponding distance score (or highest probability). The distance scores are computed, as before, using the discriminant function coefficients. The saved formula for the squared distances for *fatal* accidents is shown in Figure 12.12. For each accident, the propensity

	RushH our	WRK_ZO NE	WKDY	INT_ HWY	LGTCON _day	LEVEL	SPD_ LIM	SUR_COND _dry	TRAF_WAY _two_way	WEATHER _adverse	MAX_SEV
1	1	0	1	1	0	1	70	0	0	1	no-injury
2	1	0	1	0	0	0	70	0	0	1	no-injury
3	1	0	1	0	0	0	65	0	0	1	non-fatal
4	1	0	1	0	0	0	55	0	1	0	non-fatal
5	1	0	0	0	0	0	35	0	0	1	no-injury
6	1	0	1	0	0	1	35	0	0	1	no-injury
7	0	0	1	1	0	1	70	0	0	1	non-fatal
8	0	0	1	0	0	1	35	0	1	1	no-injury
9	1	0	1	0	0	0	25	0	0	1	non-fatal
10	1	0	1	0	0	0	35	0	0	1	non-fatal
11	1	0	1	0	0	0	30	0	0	1	non-fatal
12	1	0	1	0	0	0	60	0	0	0	no-injury
13	1	0	1	0	0	0	40	0	1	0	no-injury
14	0	0	1	0	0	1	45	1	0	0	non-fatal
15	1	0	1	0	0	1	35	1	1	0	non-fatal

FIGURE 12.9   Sample of 15 automobile accidents from the 2001 Department of Transportation database. Each accident is classified as one of three injury types (no-injury, non-fatal, or fatal)

▼ Discriminant Scores

Row	Actual	SqDist(Actual)	Prob(Actual)	-Log(Prob)			Predicted	Prob(Pred)	Others
1	no-injury	15.3283	0.5454	0.606			no-injury	0.5454	fatal 0.13 non-fatal 0.32
2	no-injury	14.8251	0.6751	0.393		v	no-injury	0.6751	non-fatal 0.30
3	non-fatal	14.5141	0.3115	1.166		v *	no-injury	0.6687	
4	non-fatal	21.5454	0.0159	4.142		*	fatal	0.9548	
5	no-injury	15.6533	0.5332	0.629			no-injury	0.5332	non-fatal 0.44
6	no-injury	13.5957	0.5754	0.553			no-injury	0.5754	non-fatal 0.41
7	non-fatal	15.4285	0.2691	1.313		v *	no-injury	0.4343	fatal 0.30
8	no-injury	11.6947	0.5265	0.642		v	no-injury	0.5265	non-fatal 0.40
9	non-fatal	15.9223	0.4029	0.909		*	no-injury	0.5940	
10	non-fatal	12.7587	0.3799	0.968		*	no-injury	0.6152	
11	non-fatal	14.1062	0.3914	0.938		v *	no-injury	0.6047	
12	no-injury	21.0122	0.0491	3.015		*	fatal	0.9258	
13	no-injury	18.5104	0.0538	2.922		v *	fatal	0.9125	
14	non-fatal	10.0670	0.4702	0.755		v	non-fatal	0.4702	no-injury 0.43
15	non-fatal	9.8636	0.5094	0.674		v	non-fatal	0.5094	no-injury 0.44

FIGURE 12.10   Discriminant scores for the first 15 observations of the accidents data (validation data are marked with v)

(probability) of each class can also be calculated. The saved formula for the probability for *fatal* accidents is shown in Figure 12.13. The probabilities of an accident involving no injuries or non-fatal injuries are computed in a similar manner.

Squared distances and membership probabilities can be obtained directly in the discriminant platform in JMP for all observations and classes (see Figure 12.11). For the first accident in the training set, the highest probability (and the lowest squared distance) is for no injuries, and therefore it is classified as a *no-injury* accident. The actual and predicted classifications are shown in Figure 12.10 (and are provided when the formulas are saved to the data table). The first two observations are correctly classified as no-injury, while the third accident, which was a non-fatal accident, is incorrectly classified as no-injury.

Score Summaries

Source	Count	Number Misclassified	Percent Misclassified	Entropy RSquare	-2LogLikelihood
Training	600	303	50.5000	-0.3226	1167.13
Validation	400	228	57.0000	-0.4788	

Training

Actual	Predicted Count		
MAX_SEV	fatal	no-injury	non-fatal
fatal	3	0	2
no-injury	30	117	145
non-fatal	27	99	177

Validation

Actual	Predicted Count		
MAX_SEV	fatal	no-injury	non-fatal
fatal	1	0	2
no-injury	23	68	109
non-fatal	24	70	103

Actual	Predicted Rate		
MAX_SEV	fatal	no-injury	non-fatal
fatal	0.600	0.000	0.400
no-injury	0.103	0.401	0.497
non-fatal	0.089	0.327	0.584

Actual	Predicted Rate		
MAX_SEV	fatal	no-injury	non-fatal
fatal	0.333	0.000	0.667
no-injury	0.115	0.340	0.545
non-fatal	0.122	0.355	0.523

Probabilities to Each Group

Row	Actual	fatal	no-injury	non-fatal
1	no-injury	0.133	0.54538	0.32162
2	no-injury	0.025	0.67508	0.30006
3	non-fatal	0.020	0.66869	0.31152
4	non-fatal	0.955	0.02926	0.01589
5	no-injury	0.022	0.53316	0.44493
6	no-injury	0.013	0.57541	0.41133
7	non-fatal	0.297	0.43433	0.26905
8	no-injury	0.073	0.52649	0.40042
9	non-fatal	0.003	0.59399	0.40294
10	non-fatal	0.005	0.61519	0.37990
11	non-fatal	0.004	0.60472	0.39140
12	no-injury	0.926	0.04906	0.02510
13	no-injury	0.913	0.05383	0.03366
14	non-fatal	0.100	0.43019	0.47020
15	non-fatal	0.051	0.44001	0.50942

Squared Distances to Each Group

Row	Actual	fatal	no-injury	non-fatal
1	no-injury	18.151	15.328269	16.384485
2	no-injury	21.428	14.825068	16.446744
3	non-fatal	20.026	12.98638	14.514108
4	non-fatal	13.354	20.324376	21.54541
5	no-injury	22.037	15.653277	16.015088
6	no-injury	21.136	13.595659	14.267062
7	non-fatal	15.233	14.470694	15.428501
8	no-injury	15.644	11.694723	12.242125
9	non-fatal	25.678	15.146165	15.9223
10	non-fatal	21.454	11.794671	12.758704
11	non-fatal	23.332	13.236122	14.106207
12	no-injury	15.137	21.012248	22.352824
13	no-injury	12.850	18.510388	19.449576
14	non-fatal	13.171	10.244805	10.066957
15	non-fatal	14.483	10.156557	9.8636269

FIGURE 12.11 Score Summaries (with confusion matrices), probabilities, and squared distances for the first 15 records

SqDist0

+ -1.845124702 • *RushHour*

+ -1.035721888 • *WRK_ZONE*

+ -9.560298894 • *WKDY*

+ 3.6837565313 • *INT_HWY*

+ -7.414024843 • *LGTCON_day*

+ -5.253787546 • *LEVEL*

+ -1.010263444 • *SPD_LIM*

+ -19.99772015 • *SUR_COND_dry*

+ -14.21595323 • *TRAF_WAY_two_way*

+ -19.3760422 • *WEATHER_adverse*

+ 41.616635636

FIGURE 12.12 Formula for the squared distance for fatal accidents, saved to the data table

$$\frac{1}{\left(\begin{array}{l}1\\ +\operatorname{Exp}\left(-0.5\cdot\left(\mathit{SqDist[no\text{-}injury]}-\mathit{SqDist[fatal]}\right)\right)\\ +\operatorname{Exp}\left(-0.5\cdot\left(\mathit{SqDist[non\text{-}fatal]}-\mathit{SqDist[fatal]}\right)\right)\end{array}\right)}$$

FIGURE 12.13 Formula for the probability of fatal accidents, saved to the data table

12.7 ADVANTAGES AND WEAKNESSES

Discriminant analysis is considered more of a statistical classification method than a machine learning method. This is reflected in its absence or short mention in many machine learning resources. However, it is very popular in social sciences and has shown good performance. The use and performance of discriminant analysis are similar to those of multiple linear regression. The two methods therefore share several advantages and weaknesses.

Like linear regression, discriminant analysis searches for the optimal weighting of predictors. In linear regression the weighting is with relation to the outcome variable, whereas in discriminant analysis it is with relation to separating the classes. Both use the same estimation method of least squares, and the resulting estimates are robust to local optima.

In both methods, an underlying assumption is normality. In discriminant analysis, we assume that the predictors are approximately from a multivariate normal distribution. Although this assumption is violated in many practical situations (e.g., with commonly used binary predictors), the method is surprisingly robust. According to Hastie et al. (2001), the reason might be that data can usually support only simple separation boundaries, such as linear boundaries. However, for continuous variables that are found to be very skewed (as can be seen through a histogram), transformations such as the log transform can improve performance. In addition, the method's sensitivity to outliers commands exploring the data for extreme values and removing those observations from the analysis.

An advantage of discriminant analysis as a classifier (it is like logistic regression in this respect) is that it provides estimates of single-predictor contributions.[5] This is useful for obtaining a ranking of the importance of predictors and for variable selection.

Finally, the method is computationally simple, parsimonious, and especially useful for small datasets. With its parametric form, discriminant analysis makes the most out of the data and is therefore especially useful where the data are few (as explained in Section 12.4).

[5]Comparing predictor contribution requires normalizing all the predictors before running discriminant analysis. Then compare each coefficient across the discriminant functions: coefficients with large differences indicate a predictor with high separation power.

PROBLEMS

12.1 Personal Loan Acceptance. Universal Bank is a relatively young bank growing rapidly in terms of overall customer acquisition. The majority of these customers are liability customers with varying sizes of relationship with the bank. The customer base of asset customers is quite small, and the bank is interested in expanding this base rapidly to bring in more loan business. In particular, it wants to explore ways of converting its liability customers to personal loan customers.

A campaign the bank ran for liability customers last year showed a healthy conversion rate of over 9% successes. This has encouraged the retail marketing department to devise smarter campaigns with better target marketing. The goal of our analysis is to model the previous campaign's customer behavior to analyze what combination of factors make a customer more likely to accept a personal loan. This will serve as the basis for the design of a new campaign.

The file `UniversalBank.jmp` contains data on 5000 customers. The data include customer demographic information (e.g., age, income), the customer's relationship with the bank (e.g., mortgage, securities account), and the customer response to the last personal loan campaign (Personal Loan). Among these 5000 customers, only 480 (= 9.6%) accepted the personal loan that was offered to them in the previous campaign.

Data preparation: Partition the data (60% training and 40% validation), and turn categorical predictors into continuous dummy variables.

a. Compute summary statistics for the predictors separately for loan acceptors and nonacceptors. For continuous predictors, compute the mean and standard deviation. For the original categorical predictors, compute the percentages or proportions. Are there predictors where the two classes differ substantially?

b. Perform a discriminant analysis that models Personal Loan as a function of the remaining predictors (excluding ZIP Code). The *success* class as 1 (loan acceptance). Examine the model performance on the validation set.

    i. What is the misclassification rate on the validation set?

    ii. Is one type of misclassification more likely than the other?

    iii. Select three customers who were misclassified as acceptors and three who were misclassified as nonacceptors. The goal is to determine why they are misclassified. First, examine their probability of being classified as acceptors: is it close to the threshold of 0.5? If not, compare their predictor values to the summary statistics of the two classes to determine why they were misclassified.

c. As in many marketing campaigns, it is more important to identify customers who will accept the offer rather than customers who will not accept it. Therefore, a good model should be especially accurate at detecting acceptors. Examine the ROC curve and lift curve for the validation set and interpret them in light of this ranking goal. (*Hint*: Save the formulas to the data table, and use *Analyze > Modeling > Model Comparison*, with the validation column in the *By* field.)

d. Compare the results from the discriminant analysis with those from a logistic regression. Examine the confusion matrices, the ROC and lift curves (again, using

Model Comparison). Which method performs better on your validation set in detecting the acceptors?

e. The bank is planning to continue its campaign by sending its offer to 1000 additional customers. Suppose that the cost of sending the offer is $1 and the profit from an accepted offer is $50. What is the expected profitability of this campaign? (Compute this manually—refer to the discussion in Chapter 5.)

f. The cost of misclassifying a loan acceptor customer as a nonacceptor is much higher than the opposite misclassification cost. To minimize the expected cost of misclassification, should the threshold value for classification (which is currently at 0.5) be increased or decreased?

12.2 **Identifying Good System Administrators.** A management consultant is studying the roles played by experience and training in a system administrator's ability to complete a set of tasks in a specified amount of time. In particular, she is interested in discriminating between administrators who are able to complete given tasks within a specified time and those who are not. Data are collected on the performance of 75 randomly selected administrators. They are stored in the file SystemAdministrators.jmp.

Using these data, the consultant performs a discriminant analysis. The variable Experience measures months of full time system administrator experience, while Training measures number of relevant training credits. The dependent variable Completed is either *Yes* or *No*, according to whether or not the administrator completed the tasks.

a. Create a scatterplot of Experience vs. Training using color or symbol to differentiate administrators who completed the tasks from those who did not complete them. See if you can identify a line that separates the two classes with minimum misclassification.

b. Run a discriminant analysis with both predictors using the entire dataset as training data. Among those who completed the tasks, what is the percentage of administrators who are classified incorrectly as failing to complete the tasks?

c. How would you classify an administrator with four months of experience and six credits of training?

d. How much experience must be accumulated by a administrator with four training credits before his or her estimated probability of completing the tasks exceeds 50%?

e. Compare the classification accuracy of this model to that resulting from a logistic regression model.

12.3 **Detecting Spam Email (from the UCI Machine Learning Repository).** A team at Hewlett–Packard collected data on a large number of email messages from their postmaster and personal email for the purpose of finding a classifier that can separate email messages that are spam vs. nonspam (aka "ham"). The spam concept is diverse: it includes advertisements for products or websites, "make money fast" schemes, chain letters, pornography, and so on. The definition used here is "unsolicited commercial email."

The file Spambase.jmp contains information on 4601 email messages, among which 1813 are tagged "spam." The predictors include 57 variables, most of them are the average number of times a certain word (e.g., mail, George) or symbol (e.g., #, !) appears in the email. A few predictors are related to the number and length of capitalized words.

a. To reduce the number of predictors to a manageable size, examine how each predictor differs between the spam and nonspam emails by comparing the spam-class average and nonspam-class average. (*Hint*: Use *Tabulate* and the *Graph Builder* with the *Column Switcher*.) Which are the 11 predictors that appear to vary the most between spam and nonspam emails? From these 11, which words or signs occur more often in spam?

b. Partition the data into training and validation sets, then perform a discriminant analysis on the training data using only the 11 predictors.

c. If we are interested mainly in detecting spam messages, is this model useful? Use the confusion matrix, ROC curve and lift curve for the validation set for the evaluation.

d. In the sample, almost 40% of the email messages were tagged as spam. However, suppose that the actual proportion of spam messages in these email accounts is 10%. How does this information change the distance scores, the predicted probabilities, and the misclassifications?

e. A spam filter that is based on your model is used, so that only messages that are classified as nonspam are delivered while messages that are classified as spam are quarantined. Consequently, misclassifying a nonspam email (as spam) has much heftier results. Suppose that the cost of quarantining a nonspam email is 20 times that of not detecting a spam message. Show how the distance formula can be adjusted to account for these costs (assume that the proportion of spam is reflected correctly by the sample proportion).

13

GENERATING, COMPARING, AND COMBINING MULTIPLE MODELS

The previous chapters in this part of the book introduced different supervised methods for prediction and classification. Earlier, in Chapter 5, we learned about evaluating predictive performance and introduced metrics for comparing competing models and selecting the best one. This is facilitated by the *Model Comparison* platform in JMP Pro.

In this chapter, we look at collections of supervised models. First, we look at an approach for handling multiple models called *ensembles*, which combines multiple supervised models into a "super-model." Instead of choosing a single predictive model, we can combine several models to achieve improved predictive accuracy. We explain the underlying logic of why ensembles can improve predictive accuracy and introduce popular approaches for combining models, including simple averaging, bagging, boosting, and stacking.

Secondly, we introduce the idea of *automated machine learning*, or *AutoML*, which allows us to automatically train many supervised models and see their resulting performance. While in previous chapters we have shown how to manually tune parameters for each supervised method, AutoML automates this process by automatically fitting multiple models and generating a rank-ordered list of candidate models. We illustrate this process and its results using the *Model Screening* platform in JMP Pro.

Ensembles and AutoML in JMP: The methods introduced in this chapter are only available in JMP Pro.

13.1 ENSEMBLES[1]

Ensembles, in which predictions from multiple models are joined together, is one of the ways to harness the power of multiple models. Ensembles played a major role in the

[1] This and subsequent sections in this chapter copyright © 2019 Datastats, LLC, and Galit Shmueli. Used by permission.

Machine Learning for Business Analytics: Concepts, Techniques, and Applications with JMP Pro®,
Second Edition. Galit Shmueli, Peter C. Bruce, Mia L. Stephens, Muralidhara Anandamurthy, and Nitin R. Patel.
© 2023 John Wiley & Sons, Inc. Published 2023 by John Wiley & Sons, Inc.

million-dollar Netflix Prize contest that started in 2006. At the time, Netflix, the largest DVD rental service in the United States, wanted to improve their movie recommendation system (from www.netflixprize.com):

> Netflix is all about connecting people to the movies they love. To help customers find those movies, we've developed our world-class movie recommendation system: Cinematch^{SM}...And while Cinematch is doing pretty well, it can always be made better.

In a bold move, the company decided to share a large amount of data on movie ratings by their users, and set up a contest, open to the public, aimed at improving their recommendation system:

> We provide you with a lot of anonymous rating data, and a prediction accuracy bar that is 10% better than what Cinematch can do on the same training data set.

During the contest, an active leader-board showed the results of the competing teams. An interesting behavior started appearing: different teams joined forces to create combined, or *ensemble* predictions, which proved more accurate than the individual predictions. The winning team, called "BellKor/s Pragmatic Chaos," combined results from the "BellKor" and "Big Chaos" teams alongside additional members. In a 2010 article in *Chance* magazine, the Netflix Prize winners described the power of their ensemble approach:

> An early lesson of the competition was the value of combining sets of predictions from multiple models or algorithms. If two prediction sets achieved similar RMSEs, it was quicker and more effective to simply average the two sets than to try to develop a new model that incorporated the best of each method. Even if the RMSE for one set was much worse than the other, there was almost certainly a linear combination that improved on the better set.

Why Ensembles Can Improve Predictive Power

The principle of combining methods is popular for reducing risk. For example, in finance, portfolios are created for reducing investment risk. The return from a portfolio is typically less risky because the variation is smaller than each of the individual components.

In predictive modeling, "risk" is equivalent to variation in prediction error. The more variable our prediction errors, the more volatile our predictive model. Consider predictions from two different models for a set of n observations. $e_{1,i}$ is the prediction error for the ith observation by method 1, and $e_{2,i}$ is the prediction for the same observation by method 2.

Suppose that each model produces prediction errors and that are, on average, zero (for some observations the model overpredicts and for some it underpredicts, but on average, the error is zero):

$$E(e_{1,i}) = E(e_{2,i}) = 0.$$

If, for each observation, we take an average of the two predictions, $\overline{y}_i = \hat{y}_{1,i} + \hat{y}_{2,i}/2$, then the expected average error will also be zero:

$$E\left(y_i - \overline{y}_i\right) = E\left(y_i - \frac{\hat{y}_{1,i} + \hat{y}_{2,i}}{2}\right) \tag{13.1}$$

$$= E\left(\frac{y_i - \hat{y}_{1,i}}{2} + \frac{y_i - \hat{y}_{2,i}}{2}\right) = E\left(\frac{e_{1,i} + e_{2,i}}{2}\right) = 0.$$

This means that the ensemble has the same average error as the individual models. Now let us examine the variance of the ensemble's prediction errors:

$$\mathrm{Var}\left(\frac{e_{1,i} + e_{2,i}}{2}\right) = \frac{1}{4}\left(\mathrm{Var}(e_{1,i}) + \mathrm{Var}(e_{2,i})\right) + 2 \times \frac{1}{2}\mathrm{Cov}\left(e_{1,i}, e_{2,i}\right). \tag{13.2}$$

This variance can be lower than each of the individual variances $\mathrm{Var}(e_{1,i})$ and $\mathrm{Var}(e_{2,i})$ under some circumstances. A key component is the covariance (or equivalently, correlation) between the two prediction errors. The case of no correlation leaves us with a quantity that can be smaller than each of the individual variances. The variance of the average prediction error will be even smaller when the two prediction errors are negatively correlated.

In summary, using an average of two predictions can potentially lead to smaller error variance, and therefore better predictive power. These results generalize to more than two methods; you can combine results from multiple prediction methods or classifiers.

THE WISDOM OF CROWDS

In his book *The Wisdom of Crowds*, James Surowiecki recounts how Francis Galton, a prominent statistician from the 19th century, watched a contest at a county fair in England. The contest's objective was to guess the weight of an ox. Individual contest entries were highly variable, but the mean of all the estimates was surprisingly accurate—within 1% of the true weight of the ox. On balance, the errors from multiple guesses tended to cancel one another out. You can think of the output of a predictive model as a more informed version of these guesses. Averaging together multiple guesses will yield a more accurate answer than the vast majority of the individual guesses. Note that in Galton's story there were a few (lucky) individuals who scored better than the average. An ensemble estimate will not always be more accurate than all the individual estimates in all cases, but it will be more accurate most of the time.

Simple Averaging or Voting

The simplest approach for creating an ensemble is to combine the predictions, classifications, or propensities from multiple models. For example, we might have a linear regression model, a regression tree, and a k-NN algorithm. We use all of three methods to score, say, a test set. We then combine the three sets of results.

The three models can also be variations that use the same algorithm. For example, we might have three linear regression models, each using a different set of predictors.

Combining Predictions   In prediction tasks where the outcome variable is numerical, we can combine the predictions from the different methods by simply taking an average. In the example above, for each observation in the test set we have three predictions (one from each model). The ensemble prediction is then the average of the three values.

One alternative to a simple average is taking the median prediction, which would be less affected by extreme predictions. Another possibility is computing a weighted average, where weights are proportional to a quantity of interest. For instance, weights can be proportional to the accuracy of the model, or if different data sources are used, the weights can be proportional to the quality of the data.

Ensembles for prediction are useful not only in cross-sectional prediction, but also in time series forecasting (see Chapters 15–17). In forecasting, the same approach of combining future forecasts from multiple methods can lead to more precise predictions. One example is the weather forecasting mobile app Dark Sky (formerly Forecast.io, `www.darksky.net/app`), which described their algorithm as follows:

> Forecast.io is backed by a wide range of data sources, which are aggregated together statistically to provide the most accurate forecast possible for a given location.

Combining Classifications   Combining the results from multiple classifiers can be done using "voting." If for each record we have multiple classifications, a simple rule would be to choose the most popular class among these classifications. For example, we could use a classification tree, a Naive Bayes classifier, and discriminant analysis for classifying a binary outcome variable. For each observation, we then generate three predicted classes. Simple voting would choose the most common class among the three.

As in prediction, we can assign heavier weights to scores from some models, based on considerations such as model accuracy or data quality. This would be done by setting a "majority rule" that is different from 50%.

Combining Propensities   Similar to predictions, propensities can be combined by taking a simple (or weighted) average. Recall that some algorithms, such as naive Bayes, produce biased propensities and should therefore not be simply averaged with propensities from other methods.

Bagging

Another form of ensembles is based on averaging across multiple random data samples. Bagging, short for "Bootstrap aggregating," comprises two steps:

1. Generating multiple random samples (by sampling with replacement from the original data). This method is called "bootstrap sampling."
2. Running an algorithm on each sample and producing scores.

Bagging improves the performance stability of a model and helps avoid overfitting by separately modeling different data samples and then combining the results. It is therefore especially useful for algorithms such as trees and neural networks. In Chapter 9, we described random forests, an ensemble based on bagged trees.

Boosting

Boosting is a slightly different approach to creating ensembles that sequentially builds better models by leveraging the prediction errors from the previous model. Here, the goal is to directly improve areas in the data where our model makes errors by forcing the model to pay more attention to those records. The steps in Boosting are:

1. Fit a model to the data.
2. Draw a sample from the data so that misclassified records (or records with large prediction errors) have higher probabilities of selection.
3. Fit the model to the new sample.
4. Repeat steps 2–3 multiple times.

CREATING ENSEMBLE MODELS IN JMP Pro

There are a number of ways of creating ensemble models in JMP Pro:

- To combine predictions, classifications, or propensities, formulas for models can be saved to the data table. These saved models (or their results) can then be averaged or combined using a formula column in the data table.
- Simple model averaging is available in the *Model Comparison* platform under *Analyze > Predictive Modeling*. This can be used when formulas for models have been saved to the data table. (Model averaging via *Model Comparison* is also available for models published to the *Formula Depot*.)
- JMP provides a variant of model averaging in the *Fit Model > Stepwise* platform. Here several models are fit, and the coefficients in the final model are weighted averages of the coefficients from each of the fitted models. The weights are used to minimize overfitting.
- Bagging, or bootstrap aggregation, and boosting are available for trees in JMP Pro under *Analyze > Predictive Modeling*. (For a discussion of *Bootstrap Forest* and *Boosted Tree*, see Chapter 9.)
- Boosting is also an option in the *Neural* platform in JMP Pro.

Although bagging and boosting can be applied to any supervised machine learning method, most applications are for trees, where they have proved extremely effective. In Chapter 9, we described random (*Bootstrap*) forests—an ensemble based on bagged trees, and gradient boosted trees—an ensemble based on boosted trees. We illustrated a random forest implementation for the personal loan example. Here, we demonstrate ensembles with decision tree classifiers, again using the personal loan example.

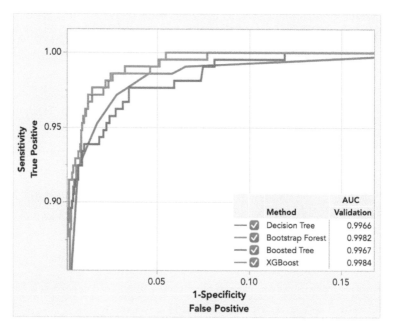

FIGURE 13.1   Validation ROC curves (zooming in to top left part) and AUC values for a single tree, a bagged tree, and two boosted classification trees (personal loan data)

Figure 13.1 compares the ROC curves and the AUC for a single classification tree, a bootstrap forest, a boosted tree, and an extreme gradient boosted tree using the validation data. These models were fit using the *Model Screening* platform (under *Analyze > Predictive Modeling*). The results are from the *ROC curve* (a red triangle option in the *Model Screening* platform for classifier models).

Looking at the ROC curves and AUC values, all models appear to perform well and similarly for this dataset. From the zoomed-in ROC chart shown in Figure 13.1, it appears the bagged tree (bootstrap forest) and XGBoost (extreme gradient boosting) show slightly better performance over the single tree and boosted tree.

Stacking

Yet another form of ensembles is based on the idea of taking heterogeneous "base learners," i.e., different model algorithms, and then using their predictions as predictors for training a meta-learner model. This is called *stacking*, short for "stacked generalization." The steps in stacking are:

1. Fit different models to the data; these are called "base learners."
2. Create a new dataset containing just the predictions from each model and the target attribute (i.e., omitting the original predictors).
3. Fit a new meta-model called a "stacking model learner," to the new dataset.

Note: Stacking is not directly available in JMP Pro.

Advantages and Weaknesses of Ensembles

Combining scores from multiple models is aimed at generating more precise predictions (lowering the prediction error variance). The approach is most useful when the combined models generate prediction errors that are negatively associated, but it can also be useful when the correlation is low. Ensembles can use simple averaging, weighted averaging, voting, medians, and so forth. Models can be based on the same algorithm or on different algorithms, using the same sample or different samples. Ensembles have become a major strategy for participants in machine learning contests (e.g., kaggle), where the goal is to optimize some predictive measure. In that sense, ensembles also provide an operational way to obtain solutions with high predictive power in a fast way, by engaging multiple teams of "data crunchers" working in parallel and combining their results.

Ensembles that are based on different data samples help avoid overfitting. However, remember that you can also overfit the data with an ensemble if you tweak it (e.g., choosing the "best" weights when using a weighted average).

The major disadvantage of an ensemble is the resources that it requires: computationally, as well as in terms of software availability and the analyst's skill and time investment. Implementing ensembles that combine results from different algorithms requires developing each of the models and evaluating them. Boosting-type ensembles and bagging-type ensembles do not require such effort, but they do have a computational cost (although boosting has been parallelized in JMP Pro and computation time is generally not an issue). Ensembles that rely on multiple data sources require collecting and maintaining multiple data sources. And finally, ensembles are "black-box" methods, in that the relationship between the predictors and the outcome variable usually becomes nontransparent.

13.2 AUTOMATED MACHINE LEARNING (AUTOML)

Machine learning platforms have now started incorporating features for automating the machine learning workflow from data loading to generating a rank-ordered list of candidate models with automatic parameter tuning. This is called *automated machine learning* or *AutoML*. As we have seen in previous chapters, JMP Pro provides capabilities to fit a large variety of machine learning models individually. With the *Model Screening* platform, it is straightforward to train a variety of models and rank them based on their performance. The steps in this example follow the general machine learning steps shown earlier in Figure 2.1 and are discussed next in the context of the personal loan example.

AutoML: Explore and Clean Data

Once the data relevant to the business purpose is loaded into JMP Pro, some tasks related to data transformation, preparation, and feature selection are usually required. Since these steps usually require domain knowledge, they are handled separately or in a semi-automatic manner.

The *Workflow Builder* (under *File > New > Workflow*) can be used to capture data exploration and cleaning tasks, such as exploring missing values, identifying outliers, transforming variables, creating visualizations, adding formulas, and formatting columns. This is particularly useful for recording repetitive and sequential tasks so that you can replay them and share with others. The Workflow Builder can also capture all of the steps in the machine learning workflow, providing a log of steps taken and enabling reproducibility of results. Figure 13.2 shows a part of workflow created for exploring and cleaning the personal loan data.

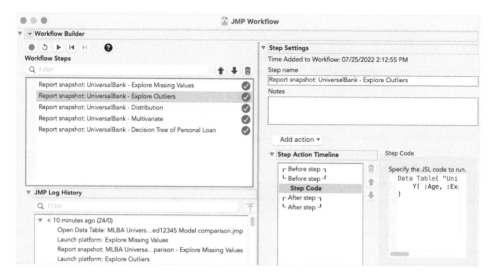

FIGURE 13.2   *Workflow Builder* **example, with recorded data exploration and preparation steps**

AutoML: Determine Machine Learning Task

Model Screening supports classifications (including binary and multi-class classification) and prediction of a numerical variable. The specific task is automatically selected based on the type of the outcome variable.

In classification, if one class has many more records than the other, we call the dataset *imbalanced*. For imbalanced datasets, one often wants to oversample the minor class (as discussed in Chapter 5). In our case, 90% of the records belong to the *No* class, and only 10% belong to the *Yes* class. We could potentially build models with and without balancing during training. Here we will build the models only without balancing.

AutoML: Choose Features and Machine Learning Methods

At this step, we select the features to be used as predictors in the supervised model. As is the case when fitting individual models, the decision should be guided by considerations of data quality and regulatory/ethical requirements (in some applications some variables are forbidden, such as using gender in loan applications in the USA). We also select the machine learning algorithms we'd like to run. Considerations include potential runtime (for example, some algorithms might take much longer to train, and/or to score) and the suitability of the algorithm to the type and size of our data and our goal. For example, the *k*-nearest neighbors algorithm requires a large training sample; simple classification and regression trees are not stable and can result in different trees for slightly different data; linear and logistic regression without regularization have trouble with highly correlated predictors.

The *Model Screening* platform in JMP Pro enables you to run multiple machine learning algorithms from one dialog. Figure 13.3 shows the *Model Screening* launch window. Under *Method* you can select the models you want to fit, with several models selected by default. For penalized regression models, check the Generalized Regression option. The default is a Lasso regression model, and when *Additional Methods* is checked, Elastic Net, Ridge, and other Generalized Regression models are also fit. Note that XG-Boost is only available if the XGBoost add-in is installed (see the Model Screening text box). The *Model Screening* platform provides additional *Modeling Options* (to build models with *Two Way Interactions* and *Quadratic* effects) and supports several options for K-Fold Crossvalidation.

FIGURE 13.3   *Model Screening* **platform JMP Pro (personal loan data)**

To illustrate this procedure, we use the personal loan data and fit several classification models (see our selections in Figure 13.3). We used a validation column, with 75% of the rows assigned to the training set and the remaining 25% assigned to the validation set.

AutoML: Evaluate Model Performance

Figure 13.4 shows the results of fitting our selected set of classification models to the personal loan data using *Model Screening*. By default, the models are sorted by Generalized RSquare in descending order, but you can easily sort on the different metrics to identify the top-performing model (click on the name of the metric to sort). The *Select Dominant* option selects models that are better than or equal to all of the other models across the different metrics provided. To see additional metrics, select the models, and then use *Decision Threshold* from the top red triangle.

In the example shown in Figure 13.4, the best-performing model on the validation data (called the *dominant* model) in terms of AUC and misclassification rate was XGBoost.

Because we used the validation set to compare and select the best model, the validation performance of the dominant model is over-optimistic. We therefore need to evaluate its performance on a test set, which was not used in comparing or selecting models, to infer about potential performance at deployment.

▼ ⊟ Model Screening for Personal Loan

Table: UniversalBank **Response:** Personal Loan **Validation:** Validation

▶ Details

▶ Training

▼ Validation

Method	N	Entropy RSquare	Misclassification Rate	AUC	RASE	Generalized RSquare
XGBoost	2000	0.9064	0.0105	0.9984	0.09456	0.9324
Bootstrap Forest	2000	0.8880	0.0170	0.9982	0.10542	0.9185
Boosted Tree	2000	0.8761	0.0125	0.9967	0.10346	0.9095
Decision Tree	2000	0.8672	0.0145	0.9966	0.10828	0.9026
Neural Boosted	2000	0.8523	0.0160	0.9945	0.11356	0.8912
Support Vector Machines	2000	0.7938	0.0220	0.9854	0.13167	0.8451
Generalized Regression Ridge	2000	0.6251	0.0470	0.9631	0.18691	0.7012
Generalized Regression Pruned Forward Selection	2000	0.6250	0.0450	0.9624	0.18534	0.7011
Generalized Regression Elastic Net	2000	0.6248	0.0460	0.9624	0.18585	0.7009
Fit Stepwise	2000	0.6248	0.0460	0.9623	0.18556	0.7009
Nominal Logistic	2000	0.6248	0.0460	0.9623	0.18556	0.7009
Generalized Regression Lasso	2000	0.6247	0.0460	0.9623	0.18558	0.7009
Generalized Regression Forward Selection	2000	0.6247	0.0460	0.9623	0.18556	0.7009

Select Dominant   Run Selected   Save Script Selected

Sum Freq and Sum Weight are suppressed when they are the same as N.

▼ Metrics

Method	TP	FN	FP	TN	Sensitivity	Specificity	Precision	Accuracy	F1	MCC
Decision Tree	195	16	13	1776	0.9242	0.9927	0.9375	0.9855	0.9308	0.9227
Bootstrap Forest	179	32	2	1787	0.8483	0.9989	0.989	0.983	0.9133	0.9072
Boosted Tree	188	23	2	1787	0.8910	0.9989	0.9895	0.9875	0.9377	0.9323
Neural Boosted	184	27	5	1784	0.8720	0.9972	0.9735	0.984	0.92	0.9128
Support Vector Machines	182	29	15	1774	0.8626	0.9916	0.9239	0.978	0.8922	0.8806
XGBoost	193	18	3	1786	0.9147	0.9983	0.9847	0.9895	0.9484	0.9434
Fit Stepwise	133	78	14	1775	0.6303	0.9922	0.9048	0.954	0.743	0.7328
Nominal Logistic	133	78	14	1775	0.6303	0.9922	0.9048	0.954	0.743	0.7328
Generalized Regression Lasso	133	78	14	1775	0.6303	0.9922	0.9048	0.954	0.743	0.7328
Generalized Regression Forward Selection	133	78	14	1775	0.6303	0.9922	0.9048	0.954	0.743	0.7328
Generalized Regression Pruned Forward Selection	134	77	13	1776	0.6351	0.9927	0.9116	0.955	0.7486	0.7391
Generalized Regression Elastic Net	132	79	13	1776	0.6256	0.9927	0.9103	0.954	0.7416	0.7325
Generalized Regression Ridge	128	83	11	1778	0.6066	0.9939	0.9209	0.953	0.7314	0.7254

Where Yes is considered Positive. Probability Threshold: 0.500

FIGURE 13.4   Validation results for multiple models fit using *Model Screening* (personal loan data)

AutoML: Model Deployment

In general, full deployment solutions usually require tailored solutions with engineering considerations (see Section 2.8 in Chapter 2). JMP provides several useful utilities toward that end. First, it is important to remember that data preprocessing steps are an integral part of the model. New data must be treated in exactly the same way as the training data. It is therefore advisable to define preprocessing steps. The Workflow Builder is designed for this very purpose, and savvy users can also leverage the JMP scripting language (JSL).

Second, JMP offers various ways to deploy the chosen model. While small datasets are often stored in files, large datasets are often stored in databases. JMP can connect to databases and other data sources (using *File > Database*) and can generate SQL code for in-database scoring using the *Formula Depot* (under *Analyze > Predictive Modeling*). Once datasets get very large or when the machine learning operations are extremely computationally intensive, it may be necessary to consider using parallelization or clusters. The necessary solutions are beyond the scope of this book.

Lastly, when the machine learning model should be made available to a wide audience, consider a web application or web service. The *Formula Depot* is a repository to organize, compare, and profile models. It can also generate scoring code in SQL, C, Python, JavaScript, and SAS for deploying models as web applications or in other environments.

MODEL SCREENING IN JMP Pro

The *Model Screening* platform (Figure 13.3) under *Analyze > Predictive Modeling* in JMP Pro enables you to run multiple machine learning algorithms from one launch dialog, collects summary statistics for each of the models in one window, and lets you save prediction formulas for models for ensembling or further exploration.

Some advantages of using *Model Screening:*

- The *Model Screening* dialog includes most prediction and classification methods available in JMP Pro, provides additional modeling options, and offers several options for model crossvalidation.
- *Model Screening* helps you choose the best-performing model on the validation (or test) set, without having to run individual machine learning algorithms and without needing to save individual models to the data table.
- For classifier models, you can use *Decision Threshold* within *Model Screening* to easily compare performance metrics for select models, and to dynamically change the classification threshold.
- You can easily launch an individual platform report for any of the models from the *Model Screening* report window for further exploration and tuning.

Note: Download the popular XGBoost (Extreme Gradient Boosting) add-in for JMP Pro from the JMP User Community (`community.jmp.com`). After installing the add-in, XGBoost will be available as a *Method* in *Model Screening*.

Advantages and Weaknesses of Automated Machine Learning

Automated machine learning functionality, such as JMP Pro's *Model Screening*, can be a valuable tool for beginners and advanced data scientists. For beginners, it can provide a useful starting point for quickly experimenting with a dataset for a business problem by building predictive models that can then be used for further understanding and learning. For advanced users, it can save time in creating common machine learning workflows by relying on automated model selection and tuning for getting quick insights into a business problem before spending extensive efforts on fine-tuning models.

Automated machine learning has the additional advantage of providing transparency in modeling. By analyzing the final model resulting from the AutoML process, the data scientist can spend more efforts in data understanding, preparation, feature generation, and so on. Also, the data scientist can focus on further fine tuning the models of interest. For example, consider an AutoML implementation that optimizes a decision tree by only changing the *minimum size split* parameter. What would be the effect of changing other parameters such as *decision threshold*?

While automated machine learning speeds up the general model tuning process by considering important model parameters, depending on the problem context or the nature of the data, the data scientist would need to consider further model tuning.

The major disadvantage of automated machine learning is its deceptive ability to hide attention to key business related details and questions. Since the AutoML process lends itself to importing and loading data and generating results fairly quickly, data scientists may not spend adequate time in understanding the nuances of data, exploring and preparing it for the modeling phase, which can lead to garbage-in, garbage-out outcomes, and thoughtless models.

13.3   SUMMARY

In practice, the methods discussed earlier in this book are often used not in isolation, but as building blocks in an analytic process whose goal is always to inform and provide insight.

In this chapter, we looked at two ways that multiple models are deployed.

In ensembles, multiple models are combined to produce improved predictions. Through automated machine learning, the machine learning process of selecting, applying, and tuning different models for better model performance can be made more efficient, while focusing on evaluating the results and comparing the models.

PROBLEMS

13.1 Acceptance of Consumer Loan. Universal Bank has begun a program to encourage its existing customers to borrow via a consumer loan program. The bank has promoted the loan to 5000 customers, of whom 480 accepted the offer. The data are available in file UniversalBank.jmp. The bank now wants to develop a model to predict which customers have the greatest probability of accepting the loan, in order to reduce promotion costs and send the offer only to a subset of its customers.

We will develop several models, then combine them in an ensemble.

a. Partition the data: 60% training, 40% validation.

b. Fit models the following models to the data: (1) logistic regression, (2) classification tree, and (3) neural network. Use Personal Loan as the outcome variable, and do not include Zip Code in the models.

   i. Report the accuracy, AUC, and misclassification (error) rates for each of the models.

   ii. Save the formula for each model to the data table, and report the first 10 rows of these saved columns.

   iii. Explore the probability formula for each of these models, and compare these formulas. Which model is the easiest to interpret?

   iv. Create a new formula column that is the average of the predicted probability of Personal Loan = Yes for the three models (i.e., an ensemble model). *Hint*: select the three probability(personal loan = Yes) columns, right click, and select *New Formula Column > Aggregate > Mean*.

   v. Use *Analyze > Predictive Modeling > Model Comparison* to compare the three models and the ensemble model on the validation data. Which model is the best at classifying loan acceptance? Why?

13.2 eBay Auctions—Boosting and Bagging. Using the eBay auction data (file eBayAuctions.jmp) with variable *Competitive* as the outcome, partition the data into training (50%), validation (30%), and test sets (20%).

a. Using the *Model Screening* platform (under *Analyze > Predictive Modeling*), fit a classification (decision) tree, a bootstrap forest, and a boosted tree. Which model has the lowest misclassification rate on the test data? Which model has the highest AUC?

b. Open the *Decision Threshold* report (from the top red triangle). Look at the Fitted Probability Plot, and compare the plots for the three models. What do you observe?

c. Look at the metrics reported in the *Decision Threshold*. Which model has the best accuracy? Which model does the best job of sorting the response (i.e., in identifying competitive auctions)? Why?

13.3 Predicting Used Car Prices (Bootstrap Forest and Boosted Trees). Return to the Toyota Corolla data (file ToyotaCorolla.jmp), and refit the regression tree model you ran in Problem 9.3. Then, run a bootstrap forest model with the same inputs and validation column. Use the default settings.

a. Compared to final reduced tree above, how does the bootstrap forest behave in terms of overall prediction error on the test set? Save the prediction formula for this model to the data table.

b. Run the same model again, but this time as a boosted tree. Use the default settings.

c. How does the boosted tree behave in terms of the prediction error relative to the reduced model and the bootstrap forest? Save the prediction formula for this model to the data table.

d. To facilitate comparison of the prediction error for the different models, use the *Model Comparison* platform (under *Analyze > Predictive Modeling*). To view fit statistics for the different models, put the validation column in the *Group* field in the Model Comparison dialog.

    i. Which model performs best on the test set?

    ii. Explain why this model might have the best performance over the other models you fit.

13.4 Predicting Flight Delays (Bootstrap Forest and Boosted Trees). We return to the flight delays data you saved in Problem 9.3 for this exercise (FlightDelaysPrepared.jmp), and fit both a bootstrap forest and a boosted tree to the data. Use scheduled departure time (CRS_DEP_TIME) rather than the binned version for these models.

a. Fit a bootstrap forest, with the default settings. Save the formula for this model to the data table.

    i. Look at the column contributions report. Which variables were involved in the most splits?

    ii. What is the error rate on the test set?

b. Fit a boosted tree to the flight delays data, again with the default settings. Save the formula to the data table.

    i. Which variables were involved in the most splits? Is this similar to what you observed with the bootstrap forest model?

    ii. What is the error rate on the test set for this model?

c. Use the Model Comparison platform to compare these models to the final reduced model found earlier (again, put the validation column in the *Group* field in the Model Comparison dialog).

    i. Which model has the lowest overall error rate on the test set?

    ii. Explain why this model might have the best performance over the other models you fit.

PART V

INTERVENTION AND USER FEEDBACK

14

INTERVENTIONS: EXPERIMENTS, UPLIFT MODELS, AND REINFORCEMENT LEARNING

14.1 INTRODUCTION

In this chapter, we describe a third paradigm of machine learning, different from supervised and unsupervised learning, that deals with interventions and feedback. Data used in supervised and unsupervised learning are typically *observational data*, being passively collected about the entities of interest (customers, households, transactions, flights, etc.). In contrast, in this chapter, we discuss *experimental* data, resulting from applying interventions to the entities of interest, and measuring the outcomes of those interventions. We start with the simplest form of intervention—the A/B test, which is a randomized experiment for testing the causal effect of a treatment or intervention on outcomes of interest. A/B tests are routinely used by internet platforms such as Google, Microsoft, and Uber for testing new features. We then describe *uplift modeling*, a method that combines A/B testing with supervised learning for providing customized targeting. Uplift modeling is commonly used in direct marketing and in political analytics. Finally, we describe *reinforcement learning*, a general machine learning methodology used in personalization, where the algorithm (or *agent*) learns the best treatment assignment policy by interacting with experimental units through dynamic treatment assignment and gathering their feedback.

Interventions in JMP: A/B testing is available in JMP. For uplift modeling, JMP Pro is required. Reinforcement learning is not available in JMP or JMP Pro.

Machine Learning for Business Analytics: Concepts, Techniques, and Applications with JMP Pro®,
Second Edition. Galit Shmueli, Peter C. Bruce, Mia L. Stephens, Muralidhara Anandamurthy, and Nitin R. Patel.
© 2023 John Wiley & Sons, Inc. Published 2023 by John Wiley & Sons, Inc.

14.2   A/B TESTING[1]

A/B testing is the marketing industry's term for a standard scientific experiment in which results can be tracked for each individual experimental unit. The idea is to test one treatment against another, or a treatment against a control. "Treatment" is simply the term for the intervention you are testing. In a medical trial, it is typically a drug, device, vaccine, or other therapy; in marketing, it is typically an offering to a consumer—for example, an email, online ad, push notification, or a webpage shown to a consumer. A display ad in a magazine would not generally qualify, unless it had a specific call to action that allowed the marketer to trace the action (e.g., purchase) to a given ad, plus the ability to split the magazine distribution randomly and provide a different offer to each segment of consumers.

A/B testing has become extremely popular online. Amazon, Microsoft, Facebook, Google, and similar companies conduct thousands to tens of thousands of A/B tests each year, on millions of users, testing user interface changes, enhancements to algorithms (search, ads, personalization, recommendation, etc.), changes to apps, content management system, and more.[2] A/B testing on an internet or mobile platform involves randomizing user traffic to one of two experiences (current version and new version), computing the difference in relevant outcome metrics between the two groups, and performing statistical tests to rule out differences that are likely due to noise.

In an A/B test, an experimental unit ("subject") can be an individual customer, a user, an online user session, or any unit of interest.

An important element of A/B testing is random allocation—the treatments are assigned or delivered to subjects randomly. That way, any difference between treatment A and treatment B can be attributed to the treatment (unless it is due to chance). For this reason, A/B tests are extremely popular in online advertising, where it is relatively easy to randomly serve different ad designs or content ("treatments") to different users and where user reactions are automatically recorded. A/B tests are also routinely used by internet platforms and website owners for evaluating new small intended improvements to the user interface design, such as changing the color of a submit button or adding a new navigation link. For example, Optimizely describes a case study showing how Electronic Arts, the maker of the popular SimCity franchise, created and tested several different versions of their preorder webpage. They hypothesized that placing a call-to-action higher on the webpage by changing the way the promotional offer was displayed could drive more purchases and increase revenue generated from SimCity.com. Using A/B testing, they found that, contrary to their hypothesis, the version *without* a special promotion offer across the top of the page performed 43% better than the other variation.[3]

In an A/B test, only two measurements are needed for each subject: which treatment it received (A or B) and the outcome. For example, in an online advertising A/B test that compares a new ad design to an existing one, for each user, we would record which ad they were served (*Treat*={old,new}) and what was the outcome, such as whether they clicked the ad or not (*Y*={click,no-click}). Table 14.1 shows a schematic sample of such data. In practice, we do not need to keep all individual outcome values for each subject (unless the

---

[1]This and subsequent sections in this chapter copyright © 2022 Datastats, LLC, and Galit Shmueli. Used by permission.
[2]See *Trustworthy online controlled experiments: A practical guide to A/B testing* by Kohavi et al. (2020).
[3]https://www.optimizely.com/insights/blog/ea_simcity_optimizely_casestudy/

results will be used beyond the A/B test, such as in uplift modeling). For a binary outcome (e.g., click/no-click), we only need the proportion of clicks in each treatment group. In the ad example, that means the proportion of clicks among users who viewed the old ad ($\hat{p}_A$) and the proportion of clicks among users who viewed the new ad ($\hat{p}_B$). For a numerical outcome (e.g., engagement time), we need only the average and standard deviation of the outcome in each treatment group ($\bar{y}_A, \hat{\sigma}_A$ and $\bar{y}_B, \hat{\sigma}_B$).

TABLE 14.1 Example of raw data resulting from A/B test

User ID	Treatment	Outcome
1234	A	Click
1235	A	No click
1236	B	No click
...	...	...

Example: Testing a New Feature in a Photo Sharing App

PhoToTo, a fictitious photo sharing app, is interested in testing a new feature that allows users to make uploaded photos disappear after 24 hours. Their goal is to gauge their customers' interest in this feature, with the purpose of increasing user activity and converting users from the free version to the advanced paid version. PhoToTo deploys an A/B test to 2000 of its users who use the free version: 1000 of these users are randomly assigned to an interface with the new feature, while the other 1000 remain with the existing functionality. The experiment is run for a week. For each of the 2000 users, two performance metrics are measured at the end of the experimental week: the total number of uploaded photos and whether the user upgraded to the paid version of PhoToTo. The results of the experiment are shown in Table 14.2. We can see that the average number of photo uploads and the conversion to paid users are both higher in the treatment group where the new feature was introduced. But can we infer the difference between the two groups will generalize when deployed to the larger population of users, or is this a "fluke sample"?

TABLE 14.2 Results from A/B test at PhoToTo, evaluating the effect of a new feature

Metric	Existing interface (A)	New feature available (B)
Average number of photo uploads	50	60
Standard deviation of photo uploads	50	100
Number of conversion to paid users	20	40
Proportion of conversion to paid users	0.02 (20/1000)	0.04 (40/1000)

Success is measured by the number of photo uploads and user conversion.

The Statistical Test for Comparing Two Groups (T-Test)

The A/B test is a statistical test comparing the average outcome of a treatment group to that of a control group (or, in general, of two groups A and B). Suppose group A receives the existing ad, and group B receives the new design. Suppose the outcome of interest is whether users click on the ad. What if 5% of users shown the new ad clicked and 4% of those shown the existing ad clicked? In many cases, we would now be ready to go with the treatment that did best. This is especially so if we are starting fresh with two new treatments—say, sending our prospects email version A or email version B to promote a new product or testing a new web landing page with a blue banner at the top or an orange banner.

In other cases, there may be a substantial cost difference between the treatments, or we may be testing a new treatment against an existing "standard." In such a case, we may worry about fluke results in an A/B test, particularly if based on a small sample. If random chance produced a result in favor of the costlier or the new treatment, we might not want to switch away from the cheaper or the tried and true standard treatment. Doing so regularly would mean chasing after inconsequential results and losing a firm base of information. In such a case, we can use a statistical test to determine how "unusual" our result is compared to what chance might produce. If our result is very unusual (say, it might not occur more often than 5% of the time in a chance model), we would be prepared to abandon our existing standard in favor of the new treatment. If, on the other hand, our new treatment produced an outcome that occurs 35% of the time by chance anyway, we might stick with the standard treatment. In this way, we minimize our chance of being fooled by chance.

Such a statistical test is termed a *significance test*. The metric "percent of time a chance model produces an outcome at least as unusual as the experiment" is termed a *p-value*. We could calculate p-values by constituting a chance model and simulating results. For example, in a test of engagement times, we could put all the engagement times from both treatments into a hat, shuffle the hat, and then repeatedly draw out pairs of samples like the original samples. Do those pairs rarely yield a difference in mean engagement times across the two re-samples as great as we saw in our actual experiment? If so, we conclude that the difference we saw in the experiment was unlikely to happen by chance: it is "statistically significant." A formula-based equivalent of this procedure is the *t-test for independent samples* (see box).

Returning to our earlier example of the photo sharing app PhoToTo, we can compute the T-statistics for each of the two metrics of interest: the number of uploaded photos and the conversion rate to paid users:

$$T_{\text{photo uploads}} = \frac{60 - 50}{\sqrt{(50^2/1000 + 100^2/1000)}} = 2.828.$$

$$T_{\text{paid users}} = \frac{0.04 - 0.02}{\sqrt{(0.04)(0.96)/1000 + (0.02)(0.98)/1000}} = 2.626.$$

The corresponding two p-values computed by JMP are as follows: for the increase[4] in average photo uploads, p-value = 0.0024 and for the increase in paid users, p-value = 0.0044 as shown in Figure 14.1.

Comparing the p-value to a user-determined threshold will then yield a result. For example, if the threshold is 5% (the default significance level), then a p-value below 5% will indicate that the click through rate on the new design is statistically higher than on the existing design. The results of such tests are subject to two potential errors: false discovery (called Type I error) and missed discovery (Type II error). The choice of threshold value is determined by the false discovery risk level we are prepared to take.

Both p-values are very small, thereby indicating that the effect of the new feature on number of uploads and conversion to paid users will generalize to the larger population of

---

[4]We use a *one-sided test*, because we are testing whether the average number of uploads *increases* with the new feature. When the direction of the effect is unknown, we'd use a two-sided test.

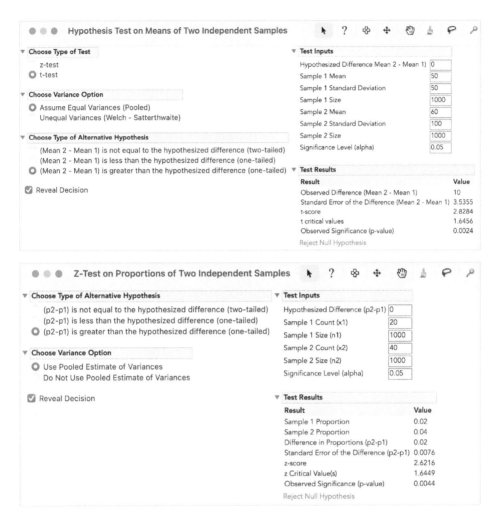

FIGURE 14.1   Test of hypothesis for two sample means (above) and two sample proportions (below) using the hypothesis test calculators in the JMP Sample Index

users. Note that with much smaller sample sizes, the same outcome values in Table 14.2 could yield statistical insignificance. For example, for $n_A = n_B = 100$, we get p-values of 0.186 and 0.204, respectively.

T-TEST FOR COMPARING TWO INDEPENDENT GROUPS

All basic statistics courses and textbooks cover the *t*-test for testing the difference between two group means (or proportions). Here is a quick reminder of the basics for the case relevant for A/B testing. For testing whether the difference between two group means (or proportions) is significantly in favor of group B (e.g., the new ad design) or not—that is, whether there's evidence that the difference between the population means $\mu_B - \mu_A > 0$—we compute the t-statistic:

$$T = \frac{\bar{y}_B - \bar{y}_A}{\text{Standard Error}(\bar{y}_B - \bar{y}_A)}.$$

When the outcome metric of interest is binary (e.g., buy/not-buy), a similar formula is used, replacing averages ($\bar{y}$) with proportions ($\hat{p}$).

The denominator of the T-statistic (the standard error of the difference in means or proportions) has the following formulas.

For a numerical measure:

$$\text{Standard Error}(\bar{y}_B - \bar{y}_A) = \sqrt{s_A^2/n_A + s_B^2/n_B},$$

and for a binary measure:

$$\text{Standard Error}(\hat{p}_B - \hat{p}_A) = \sqrt{\frac{\hat{p}_A(1 - \hat{p}_A)}{n_A} + \frac{\hat{p}_B(1 - \hat{p}_B)}{n_B}}.$$

The statistical significance of the t-value (its p-value) is computed by JMP. Comparing the p-value to a user-determined threshold will then yield a result. For example, if the threshold is 5%, then a p-value below 5% will indicate the click-through rate (CTR) on the new design is statistically higher than on the existing design. The results of such tests are subject to two potential errors: false discovery (called Type I error) and missed discovery (Type II error). The choice of threshold value is determined by the false discovery risk level we are prepared to take.

Critically, one must carefully consider the magnitude of the difference and its practical meaning, especially since A/B tests are typically conducted on very large samples of users (tens and even hundreds of thousands of users). Statistical significance is strongly affected by the sample size: the larger the groups, the more likely even a very small difference that is practically meaningless will appear statistically significant. For example, with sufficiently large groups, a CTR of 5.000001% will appear significantly (statistically) higher than 5%.

HYPOTHESIS TESTS FOR INDEPENDENT GROUPS IN JMP

- To perform hypothesis tests for two independent groups, use *Analyze > Fit Y by X* (for unsummarized or raw data).
- For summarized data and demonstrations, use the JMP statistics calculators, under *Help > Sample Index > Calculators*. Choose *Hypothesis Test for Two Means* or *Hypothesis Test for Two Proportions*. See the examples in Figure 14.1.
- For comparing means for more than two groups (ANOVA), use *Analyze > Fit Y by X* (for unsummarized data only).

For information on performing basic hypothesis tests in JMP (with unsummarized data), see *JMP Learning Library > Basic Inference—Proportions and Means* at www.jmp.com/learn.

Multiple Treatment Groups: A/B/*n* Tests

A/B testing can be extended from two groups (treatment and control, or two groups) to multiple groups. In A/B/*n* testing, we have *n* groups. For example, testing three new ad designs against an existing design would use an A/B/4 test. The procedure is very similar to A/B testing, with the difference being the replacement of the *t*-test with analysis of variance (ANOVA) for a numerical outcome or a chi-square test for comparing proportions. For more on these tests, consult a statistics textbook such as Sall et al. (2017).

Multiple A/B Tests and the Danger of Multiple Testing

As we mentioned earlier, digital platforms and their business customers (e.g., advertisers and marketers) often perform routine and frequent A/B tests, resulting in thousands of tests. Large companies such as Microsoft have their internal experimentation infrastructure. But even small companies can perform such testing using services by third parties: Optimizely (`https://www.optimizely.com/`) is one commercial service that provides online tools for routine A/B testing for online advertisers.

When deploying many A/B tests, there is a high risk of false discoveries: mistakenly concluding that a treatment effect is real, where it is just due to chance. While each statistical *t*-test carries with it a small probability of false discovery, a combination of many A/B tests dramatically increases the overall false discovery probability. This phenomenon is known as *multiple testing*. Multiple testing can arise from testing multiple metrics (click-through rate, views, purchases, etc.), running the test for multiple use subgroups (countries, zip codes, device type, etc.), and from repeating the same A/B test many times. Kohavi et al. (2020) describe a case of multiple testing that arose in early versions of Optimizely, which allowed those running an A/B test to "peek" at the p-values before the experimental period has been completed. For example, if the A/B test was supposed to run for a week, the tester could peek at test results throughout the week. This peeking was effectively multiple testing, leading to false discoveries of apparent effects. The problem was later detected and eliminated.

Solutions to false discoveries due to multiple testing are based on adjusting the threshold on each p-value, or adjusting the p-values themselves, by considering the total number of tests.[5] However, beyond technical solutions, it is good practice to carefully consider which tests are essential, set the experimental period in advance, disallow "peeking," and select treatments that have reasonable justification (although sometimes new ideas are hard to justify but turn out to be successes).

14.3 UPLIFT (PERSUASION) MODELING

Long before the advent of the Internet, sending messages directly to individuals (i.e., direct mail) held a big share of the advertising market. Direct marketing affords the marketer the ability to invite and monitor direct responses from consumers. This, in turn, allows the marketer to learn whether the messaging is paying off. A message can be

---

[5]One common method is the "False Discovery Rate" (FDR), which is the expected number of false discoveries among all discoveries. Using the FDR entails sorting the p-values from low to high and assigning different thresholds.

tested with a small section of a large list, and, if it pays off, the message can be rolled out to the entire list. With predictive modeling, we have seen that the rollout can be targeted to that portion of the list that is most likely to respond or behave in a certain way. None of this was possible with traditional media advertising (television, radio, newspaper, magazine).

Direct response also made it possible to test one message against another and find out which does better.

An A/B test tells you which treatment does better on average but says nothing about which treatment does better for which individual. A classic example is in political campaigns. Consider the following scenario: the campaign director for Smith, a Democratic Congressional candidate, would like to know which voters should be called to encourage to support Smith. Voters that tend to vote Democratic but are not activists might be more inclined to vote for Smith if they got a call. Active Democrats are probably already supportive of him, and therefore a call to them would be wasted. Calls to Republicans are not only wasteful, but they could be harmful.

Campaigns now maintain extensive data on voters to help guide decisions about outreach to individual voters. Prior to the 2008 Obama campaign, the practice was to make rule-based decisions based on expert political judgment. Since 2008, it has increasingly been recognized that, rather than relying on judgment or supposition to determine whether an individual should be called, it is best to use the data to develop a model that can predict whether a voter will respond positively to outreach.

Gathering the Data

US states maintain publicly available files of voters, as part of the transparent oversight process for elections. The voter file contains data such as name, address, and date of birth. Political parties have "poll-watchers" at elections to record who votes, so they have additional data on which elections voters voted in. Census data for neighborhoods can be appended, based on voter address. Finally, commercial demographic data can be purchased and matched to the voter data. Table 14.3 shows a small extract of data derived from the voter file for the US state of Delaware.[6] The actual data used in this problem are in the file VoterPersuasion.jmp and contain 10,000 records and many additional variables beyond those shown in Table 14.3.

First, the campaign director conducts a survey of 10,000 voters to determine their inclination to vote Democratic. Then, she conducts an experiment, randomly splitting the sample of 10,000 voters in half and mailing a message promoting Smith to half the list (treatment A) and nothing to the other half (treatment B). The control group that gets no message is essential, since other campaigns or news events might cause a shift in opinion. The goal is to measure the change in opinion after the message is sent out, relative to the no-message control group.

The next step is conducting a post-message survey of the same sample of 10,000 voters, to measure whether each voter's opinion of Smith has shifted in a positive direction. A binary variable, Moved_AD, will be added to the collected data, indicating whether opinion has moved in a Democratic direction (1) or not (0).

---

[6]Thanks to Ken Strasma, founder of the microtargeting firm HaystaqDNA and director of targeting for the 2004 Kerry campaign and the 2008 Obama campaign, for these data.

TABLE 14.3 Data on voters (small subset of variables and records) and Data Dictionary

Voter	Age	NH_White	Comm_PT	H_F1	Reg_Days	PR_Pelig	E_Elig	Political_C
1	28	61	0	0	3997	0	20	1
2	23	67	3	0	300	0	0	1
3	57	64	4	0	2967	0	0	0
4	70	53	2	1	16620	100	90	1
5	37	76	2	0	3786	0	20	0

Data Dictionary

Age	Voter age in years
NH_White	Neighborhood average of % non hispanic white in household
Comm_PT	Neighborhood % of workers who take public transit
H_F1	Single female household (1 = yes)
Reg_Days	Days since voter registered at current address
PR_Pelig	Voted in what % of nonpresidential primaries
E_Pelig	Voted in what % of any primaries
Political_C	Is there a political contributor in the home? (1 = yes)

Table 14.4 summarizes the results of the survey, by comparing the movement in a Democratic direction for each of the treatments. Overall, the message (*Message* = 1) is modestly effective.

TABLE 14.4 Results of sending a pro-Democratic message to voters

Treatment	# Voters	# Moved Dem.	% Moved Dem.
Message = 1 (message sent)	5000	2012	40.2%
Message = 0 (no message sent)	5000	1722	34.4%

Movement in a Democratic direction among those who got no message is 34.4%. This probably reflects the approach of the election, the heightening campaign activity, and the reduction in the "no opinion" category. It also illustrates the need for a control group. Among those who did get the message, the movement in a Democratic direction is 40.2%. So, overall, the lift from the message is 5.8%.

We can now append two variables to the voter data shown earlier in Table 14.3: *message* (whether they received the message (1) or not' (0)) and *Moved_AD* (whether they moved in a Democratic direction (1) or not (0)). The augmented data are shown in Table 14.5.

TABLE 14.5 Target variable (Moved_AD) and treatment variable (Message) added to voter data

Voter	Age	NH_White	Comm_PT	H_F1	Reg_Days	PR_Pelig	E_Elig	Political_C	Message	Moved_AD
1	28	61	0	0	3997	0	20	1	1	0
2	23	67	3	0	300	0	0	1	1	1
3	57	64	4	0	2967	0	0	0	0	0
4	70	53	2	1	16620	100	90	1	0	0
5	37	76	2	0	3786	0	20	0	1	0

Any classification method can be used to create an uplift model. In the next section, we discuss how to develop an uplift model based on a logistic regression model. Then we show how to use the JMP Pro *Uplift* platform, which builds uplift models based on classification trees.

A Simple Model

We start by developing a predictive model with *Moved_AD* as the target (output variable) and various predictor variables, including the treatment *Message*. Table 14.6 shows the first few lines from the output of a logistic regression model that was used to predict Moved_AD.

TABLE 14.6 **Classifications and propensities from predictive model (small extract)**

Voter	Message	Actual Moved_AD	Predicted Moved_AD	Predicted Prob.
1	0	1	1	0.5975
2	1	1	1	0.5005
3	0	0	0	0.2235
4	0	0	0	0.3052
5	1	0	0	0.4140

However, our interest is not just how the message did overall, nor is it whether we can predict the probability that a voter's opinion will move in a favorable direction. Rather, our goal is to predict how much (positive) impact the message will have on a specific voter. That way the campaign can direct its limited resources toward the voters who are the most persuadable—those for whom mailing the message will have the greatest positive effect.

Modeling Individual Uplift

To answer the question about the message's impact on each voter, we need to model the effect of the message at the individual voter level. For each voter, uplift is defined as follows:

Uplift = increase in propensity of favorable opinion after receiving message.

An uplift model can be built (manually) using a standard predictive model. To build an uplift model, we follow the steps below to estimate the change in probability of "success" (propensity) that comes from receiving the treatment (the message, in this example). This approach to uplift modeling is known as the *two model* approach:

1. Randomly split a data sample into treatment and control groups, conduct an A/B test, and record the outcome (in our example: Moved_AD).
2. Recombining the data sample, partition the combined data into training and validation sets, build a predictive model with this outcome as the target variable, and include a predictor variable that denotes treatment status (in our example: Message). If logistic regression is used, additional interaction terms between treatment status and other predictors can be added as predictors to allow the treatment effect to vary across records (in data-driven methods such as trees and KNN this happens automatically).
3. Score this predictive model on the validation set. This will yield, for each validation record, its propensity of success given its treatment.

4. Reverse the value of the treatment variable and re-score the same model to that partition. This will yield for each validation record its propensity of success had it received the other treatment.

5. Uplift is estimated for each individual by $P(\text{Success}|\text{Treatment} = 1) - P(\text{Success}|\text{Treatment} = 0)$.

6. For new data where no experiment has been performed, simply add a synthetic predictor variable for treatment and assign first a "1," score the model, then assign a "0," and score the model again. Estimate uplift for the new record(s) as above.

Continuing with the small voter example, in the results from step 3 shown in Table 14.6, the right column gives the propensities from the model. Next we retrain the predictive model, but with the values of the treatment variable Message reversed for each row. Table 14.7 shows the propensities with variable Message reversed (you can see the reversed values in column Message). Finally, in step 5, we calculate the uplift for each voter.

TABLE 14.7 Classifications and propensities from predictive model (small extract) with Message values reversed

Voter	Message	Actual Moved_AD	Predicted Moved_AD	Predicted Prob.
1	1	1	1	0.6908
2	0	1	1	0.3996
3	1	0	0	0.3022
4	1	0	0	0.3980
5	0	0	0	0.3194

Table 14.8 shows the uplift for each voter—the success (Moved_AD = 1) propensity given Message = 1 minus the success propensity given Message = 0.

TABLE 14.8 Uplift: change in propensities from sending message vs. not sending message

Voter	Prob. if Message = 1	Prob. if Message = 0	Uplift
1	0.6908	0.5975	0.0933
2	0.5005	0.3996	0.1009
3	0.3022	0.2235	0.0787
4	0.3980	0.3052	0.0928
5	0.4140	0.3194	0.0946

Creating Uplift Models in JMP Pro

The built-in utility for uplift modeling in JMP Pro is based on the *Partition* platform, which is used to fit classification and regression trees (see Chapter 9). While the models built using the *Partition* platform find splits to optimize a prediction, models built using the *Uplift* platform find splits to maximize a treatment difference, or uplift. Note that this approach is different from (and somewhat more complex than) the *two-model* approach described in the previous section.

Essentially, here is what JMP Pro does: for a given node, it considers all possible splits. It models the data in each node as a linear model with two effects, the split and the treatment, and an interaction term that captures the relationship between the split and the treatment. The LogWorth is used to measure the significance of the interaction terms and to select the best split. As with classification trees, when a validation column is used, the model is built on the training data and the final model is selected based on the maximum validation RSquare.

USING THE *UPLIFT* PLATFORM IN JMP Pro

- Select *Uplift* from the *Analyze* > *Consumer Research* menu in JMP Pro. The completed dialog for our voter persuasion example and the variables listed in Table 14.5 are shown in Figure 14.2.
- The initial uplift analysis window is shown in Figure 14.2. The overall rates for Moved_AD (for the training data) are shown in the bottom-left corner. The uplift from the treatment is reported under *Trt Diff*.
- To create the uplift model, use the *Split* button (repeatedly) or use the *Go* button to automatically build the model. When the *Go* button is used, the final number of splits is determined by the maximum value of RSquare for the validation data. (For details, refer to the discussion in Chapter 9.)
- A variety of options for interpreting the model are available from the top red triangle. The *Leaf Report* displays the rates and the uplift for each of the splits in the tree. The *Uplift Graph* sorts the uplift (on the training data) from the highest uplift to the lowest.
- Use the top red triangle to save the difference formula (the uplift) and the prediction formula with classifications to the data table. Saving the difference formula calculates the uplift for both the training and validation data and can be used to score new data.

Using the Results of an Uplift Model

Once we have estimated the uplift for each individual, the results can be ordered by uplift. The message could then be sent to all those voters with a positive uplift or, if resources are limited, only to a subset—those with the greatest uplift.

Uplift modeling is used mainly in marketing and, more recently, in political campaigns. It has two main purposes:

- To determine whether to send someone a persuasion message or just leave them alone.
- When a message is definitely going to be sent, to determine which message, among several possibilities, to send.

Technically this amounts to the same thing—"send no message" is simply another category of treatment, and an experiment can be constructed with multiple treatments, such as, no message, message A, and message B. However, practitioners tend to think of the

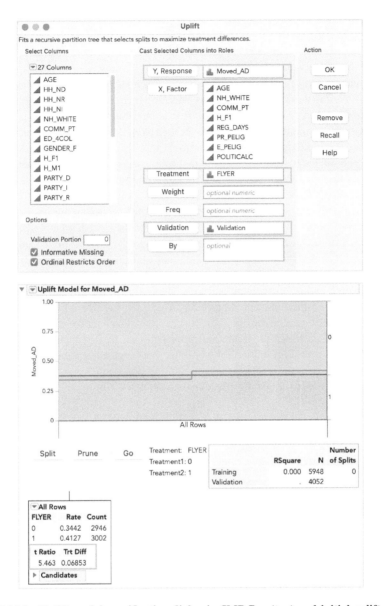

FIGURE 14.2 Uplift model specification dialog in JMP Pro (top) and initial uplift analysis (bottom) using the voter persuasion data. The *Show Points* display option has been turned off in the graph

two purposes as distinct and tend to focus on the first. Marketers want to avoid sending discount offers to customers who would make a purchase anyway, or renew a subscription anyway. Political campaigns, likewise, want to avoid calling voters who would vote for their candidate in any case. And both parties especially want to avoid sending messages or offers where the effect might be antagonistic—where the uplift is negative.

14.4   REINFORCEMENT LEARNING

A/B testing derives causal knowledge about an intervention of interest by actively assigning experimental units (customers, website users, etc.) to treatment/intervention groups, measuring the outcome(s) of interest, and drawing a conclusion about the effectiveness of the intervention. We can think of this process as "learning by interacting" with the experimental units. The notion of *learning by interaction* is what separates supervised and unsupervised learning methods from interventional methods such as A/B testing and reinforcement learning. For example, in recommendation algorithms such as collaborative filtering, we might generate a recommended item (or set of items) for a user. However, unless we actually display the recommendation to the user and measure the outcome, it is impossible to know what the user would have done given the recommendation.

We've seen that interventional methods such as uplift modeling are intended to answer the personalized "what if" question: what would be the user's outcome if s/he were served (or not) a certain recommendation or offer? We now turn to a family of methods called *reinforcement learning*, which are computational algorithms (or *agents*) that "learn by interacting" with experimental units, through dynamic treatment assignment. In both supervised learning and reinforcement learning, the goal is to "learn" a predictive model of the outcome Y as a function of predictors $X_1, X_2, \dots$. But whereas in supervised learning we train the model on preexisting data, in reinforcement learning we start without data, and the algorithm decides which X (treatment) information to deploy that would generate outcome Y data most useful for learning the optimal treatment-outcome model. In the recommendation example above, a reinforcement learning approach would choose which items to recommend in order to collect user feedback useful for learning the optimal input–output model.

We describe two popular frameworks: multi-armed bandits and Markov decision processes. In both cases the challenge is identifying the best treatment assignment policy in the presence of a plethora of potential interventions (e.g., which news item to display on a user's social media feed, which song to recommend in a music streaming service, which of many coupons to offer a customer on an e-commerce website).

Note: Reinforcement learning is not available in JMP.

Explore-Exploit: Multi-armed Bandits

The name *multi-armed bandit* (MAB, or simply "bandit") derives from a gambler playing multiple slot machines, each with a different unknown payoff rate. The gambler's goal is to figure out which machine has the highest payoff rate.

In an A/B test, after learning which treatment did best, you deploy that treatment, and the other treatment(s) get left behind. A multi-armed bandit uses sequential assignment of experimental units to treatment groups, dynamically updating treatment assignments to new units as the experiment progresses. The idea is to increase the proportion of assignments to the treatment group that shows the best performance (the best slot machine). This is especially useful when there are multiple treatments (e.g., multiple ad designs). For example, if we initially display 10 new ad designs to mobile app users, each with equal probability, then if initial experimental results indicate the new ad #3 leads to the best outcomes, for the next set of users we might increase the probability of displaying ad #3 compared to all other designs. However, we do not increase this probability to 100% because there is still uncertainty regarding the superiority of this ad design–perhaps with more data a different ad design will turn out to be the best one. We therefore still want to continue testing the value of the other nine ad designs. A multi-armed bandit tries to achieve a dual goal: improving the outcome by increasing the chance of assignment to the "winning" treatment (called *Exploit*) while continuing to evaluate the potential of the "losing" treatment(s) (called *Explore*).

There are different multi-armed bandit parameters and implementations, which differ in terms of determining how and when to update the treatment assignment probabilities. For example, an *Epsilon First* strategy explores randomly for some period of time, then exploits the treatment with the highest estimated outcome (if one treatment is clearly better than the others, it assigns all new subjects to this treatment). Another approach, the *Epsilon–Greedy* approach, explores $\epsilon\%$ of the time and exploits $1 - \epsilon\%$ of the time. A variation of the *Epsilon–Greedy* strategy adjusts the ϵ parameter over time, exploring sub-optimal treatments less frequently.

The multi-armed bandit scenario assumes the intervention effects remain constant from one assignment to the next. However, in some cases the best intervention will depend on a changing context. For example, in music recommendation, it is likely that song preference changes for the same user over different times of the day (waking up, while commuting to work, in the evening, etc.) or across devices (mobile phone, TV, etc.). In such cases, we'd want the algorithm to take into account the context. *Contextual multi-armed bandits*, also known as *contextual bandits*, do exactly this: they search for the best context-specific intervention, thereby combining trial-and-error treatment assignment search with estimating the association between treatment assignment and context variables.

Markov Decision Process (MDP)

In contrast to A/B tests, in bandit algorithms, the allocation of treatments to users is a multistage process, in which the allocation at one stage might depend on what happened in the previous stage. Bandit algorithms are actually a special case of a more general staged process, termed a Markov Decision Process (MDP).

MDPs can allow the algorithm to learn a longer-term relationship between interventions (treatments) and outcomes—for example, a movie recommendation might lead to higher long-term user satisfaction even if the user's immediate reaction is not the most profitable. Hence, MDPs are especially useful in digital environments where long-term goals, such as user satisfaction or brand recognition, are of interest.

A Markov Decision Process represents the problem of learning the optimal intervention policy through *states* and *rewards*. Consider a music streaming service that provides a recommendation system to its users. Suppose the business goal is to increase overall user engagement with the service (by providing useful recommendations), where engagement is measured by the number of songs played by the user. The algorithm (agent) must discover which song recommendations (interventions) yield the best outcome overall by trying them and recording the user's reactions. To achieve this, we define for the agent an intermediate "reward" such as whether the recommended song was played for at least 30 seconds. The agent's objective is to maximize the total reward over the long run. With such a longer-term optimization, we might recommend an item with a low immediate reward (e.g., low-profit item) but that is likely to lead to profitable rewards in the future. Another advantage of the MDP formulation is that the reward can represent the business value of a particular recommendation so that we might recommend high-profit items even if they have a low probability of purchase.

MULTI-ARMED BANDIT: SIMPLE EXAMPLE

Let's consider the scenario of selecting the best online ad to display among 10 different designs, by running a 7-day experiment. Suppose that each day we have a large number of customers accessing the webpage where the ad will be displayed. A 10-armed bandit algorithm that uses an Epsilon-first strategy would take the following steps:

- **Day 1 initialization:** Each ad i has probability $p_i = 0.1$ to be displayed. In other words, each customer is randomly assigned to view one of the 10 ads.
- **Day 1 evaluation and update:** At the end of Day 1, compute the *reward* metric of interest r_i [e.g., click-through rate (CTR), average ad revenues per ad view], resulting from each of the 10 ads.
- **Day 2 deployment:** All customers assigned to view the best performing Day 1 ad.
- **Day 2 evaluation and update:** At the end of Day 2, compute the average daily *reward* for the displayed ad.
- **Days 3, 5, 7 deployment:** Each customer is randomly assigned to view one of the 10 ads (as in Day 1).
- **Day 3, 5, 7 evaluation and update:** At the end of the day, compute the average daily *reward* for each of the 10 ads (accumulated across the days thus far).
- **Days 4, 6 deployment:** All customers assigned to view the best performing ad thus far.
- **Day 4, 6, evaluation and update:** At the end of the day, compute the average daily *reward* for each of the 10 ads (accumulated across the days thus far).
- **Day 7 experiment conclusion:** The ad with the highest average 7-day reward is selected and will be displayed in the future.

Note that on odd days, where customers are randomized across all 10 ads, the scenario is similar to A/B/10 testing in terms of treatment assignment strategy. The difference is that we do not carry out a statistical test at the end of an odd day to conclude about the best ad. Instead, we accumulate the rewards since Day 1 and reach a conclusion only after the multi-day experiment is complete.

The above example treats each day separately. An alternative is to consider batches of customers visiting the webpage (e.g., replacing a day with 10,000 customer visits).

In addition to the rewards, the MDP records the *state* at each time of treatment assignment. The state includes all the information available to the algorithm about the user, the context, and the items. In the music streaming example, the state might be the user's profile (age, gender, etc.), behavior on the platform (app use frequency, songs played, sharing with friends, etc.), context (time of day, location, device, etc.), and song features (singer, genre, etc.).

The agent's actions (e.g., recommendation or ad displayed) might affect not only the reward but also the next state (e.g., the user's time on app, songs played), thereby affecting future rewards.[7] To capture this, the MDP has a *value function* (specified by the data scientist) that assigns to each state a value corresponding to the expected long-term overall rewards when starting from that state and using a specific treatment assignment policy.

Figure 14.3 shows a schematic of an MDP and how the agent interacts with the user by assigning an intervention (action) and recording the user's reaction (reward) and new state. The terms "agent" and "environment," sometimes called "controller" and "controlled system," reflect the origins of reinforcement learning in engineering applications (such as thermostats for regulating the temperature in a room) and more recently in games such as chess and Go where the algorithm searches for the best next move.

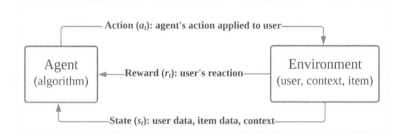

FIGURE 14.3   **Schematic of MDP, where the software agent "learns" the human user by providing actions, recording the user's new state, and receiving a reward based on the user's reaction**

The term "Markov" in MDP implies that we think of rewards, states, and treatment assignments (actions) in discrete time (periods $t = 1, 2, 3, \ldots$) and that the sequence of rewards, states, and actions form a *Markov chain*—this is a sequence of possible events in which the probability of each event depends only on the state attained in the previous event. In other words, to determine its next action (a_{t+1}), the agent only uses information on the most recent state (s_t), reward (r_t), and action (a_t).

For an item recommendation scenario, a generic MDP strategy might look like this:

1. *Initialization:* Using a predictive model, predict what item user u will select next given their user data (state).[8] (This model is based on data without recommendations.)
2. *Action (a_t):* Recommend to user u an item based on the user's data, context, and the item's features (state s_{t_1}). (The item is the result of step 1 for the initial run, then of step 4 thereafter.)
3. *Reward (r_t):* Collect the user's feedback to the recommendation (e.g., purchase, clicks).
4. *Model update:* Using the pre-recommendation user/context/item data (state s_{t-1}), the recommendation (action a_t), and the user's feedback (reward r_t), the agent updates its predictive model linking the outcome (next step's state and reward) to the input

---

[7] A multi-armed bandit is a simplified type of reinforcement learning where the assignment to intervention (action) affects only the immediate reward.

[8] This means estimating the initial *state transition function* that includes the probabilities of moving from each state to each of the other states.

(current step's state and action) $P(s_{t+1}, r_{t+1} | s_t, a_t)$. The agent then computes the value function (the expected long-term overall reward) to choose the next best action.

5. *Repeat* steps 2–4.

There exist a variety of ways to implement MDPs and reinforcement learning algorithms in practice, as well as variations and extensions, some including deep neural networks for the model updating step. The interested reader can learn more in the classic book *Reinforcement Learning* by Sutton and Barto.

As a final note about reinforcement learning, it is important to realize that while in this AI technique the data scientist determines the rewards, value function, the set of states, and set of possible actions, this does not guarantee that the agent will arrive at an optimal solution. In his book *Human Compatible*, AI expert Stewart Russell describes how agents designed to achieve a human-designed goal ended up reaching the goal by unacceptable solutions: the ECHO smart home controller was designed to reduce energy consumption by learning from the household behavior. However, inhabitants at Washington State university "often had to sit in the dark if their visitors stayed later than the usual bedtime." The agent has not fully "understood" the human objective, because the goals and rules have not been, and perhaps cannot be, fully specified. By contrast, in games like Go or chess, the ends are easy to define, and the rules can be comprehensively specified. In such games, highly creative solutions may result. The solution path may seem outlandish to a human in its initial stages, but the agent does not care and has been able to determine that it will be ultimately successful.

Reinforcement learning algorithms were designed for tasks that are fully observable, involve short time horizons, and have a small number of states and simple, predictable rules. Stewart Russell (2019) concludes: "Relaxing any of these conditions means that the standard methods will fail."

14.5 SUMMARY

In this chapter, we looked at the use of randomized experiments and user feedback for determining the best intervention. These methods differ from supervised learning and unsupervised learning and involve interaction with the units of interest (customers, website users, etc.). We described A/B testing, uplift modeling, and reinforcement learning.

A/B testing is used to test the effect of an intervention on an outcome of interest. This is done by randomly assigning units to an intervention group or a control (non-intervention) group and comparing the two groups' outcomes. A/B/*n* testing extends the idea to more than two groups.

In uplift modeling, the results of A/B testing are folded into the predictive modeling process as a predictor variable to guide choices not just about whether to send an offer or persuasion message, but also as to who to send it to.

Reinforcement learning uses computational algorithms for identifying best intervention assignments. This is especially useful when there are many possible interventions (such as in recommendation systems) and in environments that allow user–algorithm interaction. The algorithm learns the best intervention policy by trial-and-error, balancing between exploration (trying different treatment assignments) and exploitation (assigning the best-so-far treatment). Two popular reinforcement learning formulations are multi-armed bandits and Markov decision processes.

PROBLEMS

14.1 Marketing Message—A/B Testing. Which is a more effective marketing message—a plain text email, or an email with images and design elements? Statistics.com conducted an A/B test in June, 2012, sending an email to a list of potential customers that was randomly-split into two groups. The measure of success was whether the email was opened by the receiver or not. For some reason, the email service used by Statistics.com did not produce an even split. The results for each of the 426 sent emails are in the file `emailABtest.jmp`.

    a. Before conducting such a test, would you have an opinion as to which would do better: a message with or without images?

    b. What were the results in terms of open rates? Which message did better? (*Hint*: in JMP, you can use the *Analyze > Tabulate* to construct a pivot table to get this summary.)

    c. We would like to know whether this difference could be due to chance, and so will conduct a statistical test. Conduct a *t*-test using JMP to determine whether the difference is statistically significant and report the p-value.

    d. Summarize your conclusion from the A/B test.

14.2 Hair Care Product—Uplift Modeling. This problem uses the dataset in `Hair Care Product.jmp`, located in the *Sample Data Folder* in JMP (under the *Help* menu). In this hypothetical case, a promotion for a hair care product was sent to some members of a buyers club. Purchases were then recorded for both the members who got the promotion and those who did not.

    a. What is the purchase propensity (probability)

        i. among those who received the promotion?

        ii. among those who did not receive the promotion?

    b. Fit an uplift model using JMP Pro Uplift platform with *Purchase* as the target (*Y*, Response), Promotion as the treatment, Validation as the validation column, and the other variables as *X* Factors. Compare the rates in the initial uplift analysis to those reported above.

    c. Click the *Split* button one time. Which variable does the model split on? What is the uplift in the two branches? Interpret this value.

    d. Click the *Go* button, and request a *Leaf Report* and the *Uplift Graph*.

        i. What is the highest uplift value? Interpret this value.

        ii. What are the set of conditions with the highest uplift? With the lowest uplift?

        iii. From the context of future promotions and increasing purchases of the product, how would you use this information?

    e. Save the difference formula and the prediction formula to the data table. Report the predicted class and propensities for the first 10 records in the validation set.

PART VI

MINING RELATIONSHIPS AMONG RECORDS

15

ASSOCIATION RULES AND COLLABORATIVE FILTERING

In this chapter, we describe the unsupervised learning methods of association rules (also called "affinity analysis" and "market basket analysis") and collaborative filtering. Both methods are popular in marketing for cross-selling products associated with an item that a consumer is considering.

In association rules, the goal is to identify item clusters in transaction-type databases. Association rule discovery in marketing is termed "market basket analysis" and is aimed at discovering which groups of products tend to be purchased together. These items can then be displayed together, offered in post-transaction coupons, or recommended in online shopping. We describe the two-stage process of rule generation and then assessment of rule strength to choose a subset. We look at the popular rule-generating Apriori algorithm and then criteria for judging the strength of rules.

In collaborative filtering, the goal is to provide personalized recommendations that leverage user-level information. User-based collaborative filtering starts with a user and then finds users who have purchased a similar set of items or ranked items in a similar fashion and makes a recommendation to the initial user based on what the similar users purchased or liked. Item-based collaborative filtering starts with an item being considered by a user and then locates other items that tend to be co-purchased with that first item. We explain the technique and the requirements for applying it in practice.

Association Rules and Collaborative Filtering in JMP: Association rules is only available in JMP Pro. Collaborative filtering is not available in JMP or JMP Pro.

15.1 ASSOCIATION RULES

Put simply, association rules, or *affinity analysis*, constitute a study of "what goes with what." This method is also called *market basket analysis* because it originated with the

Machine Learning for Business Analytics: Concepts, Techniques, and Applications with JMP Pro®,
Second Edition. Galit Shmueli, Peter C. Bruce, Mia L. Stephens, Muralidhara Anandamurthy, and Nitin R. Patel.
© 2023 John Wiley & Sons, Inc. Published 2023 by John Wiley & Sons, Inc.

study of customer transactions databases to determine dependencies between purchases of different items. Association rules are heavily used in retail for learning about items that are purchased together, but they are also useful in other fields. For example, a medical researcher might want to learn what symptoms appear together. In law, word combinations that appear too often might indicate plagiarism.

Discovering Association Rules in Transaction Databases

The availability of detailed information on customer transactions has led to the development of techniques that automatically look for associations between items that are stored in the database. An example is data collected using barcode scanners in supermarkets. Such *market basket databases* consist of a large number of transaction records. Each record lists all items bought by a customer on a single-purchase transaction. Managers are interested to know if certain groups of items are consistently purchased together. They could use such information for making decisions on store layouts and item placement, for cross-selling, for promotions, for catalog design, and for identifying customer segments based on buying patterns. Association rules provide information of this type in the form of "if–then" statements. These rules are computed from the data; unlike the if–then rules of logic, association rules are probabilistic in nature.

Association rules are commonly encountered in online *recommendation systems* (or *recommender systems*), where customers examining an item or items for possible purchase are shown other items that are often purchased in conjunction with the first item(s). The display from Amazon.com's online shopping system illustrates the application of rules like this under "Frequently bought together." In the example shown in Figure 15.1, a user browsing a Samsung Galaxy S21 cell phone is shown a case and a screen protector that are often purchased along with this phone.

We introduce a simple artificial example and use it throughout the chapter to demonstrate the concepts, computations, and steps of association rules. We end by applying association rules to a more realistic example of book purchases.

Example 1: Synthetic Data on Purchases of Phone Faceplates

A store that sells accessories for cellular phones runs a promotion on faceplates. Customers who purchase multiple faceplates from a choice of six different colors get a discount. The store managers, who would like to know what colors of faceplates customers are likely to purchase together, collected the transaction database as shown in Table 15.1.

Data Format

Transaction data are usually displayed in one of two formats: a transactions database (with each row representing a list of items purchased in a single transaction) or a binary incidence matrix in which columns are items, rows again represent transactions, and each cell has either a 1 or a 0, indicating the presence or absence of an item in the transaction. For example, Table 15.1 displays the data for the cellular faceplate purchases in a transactions database. We translate these into binary incidence matrix format in Table 15.2.

Cell Phones & Accessories › Cell Phones

Roll over image to zoom in

Samsung Galaxy S21 5G | Factory Unlocked Android Cell Phone | US Version 5G Smartphone | Pro-Grade Camera, 8K Video, 64MP High Res | 128GB, Phantom Violet (SM-G991UZVAXAA)
Visit the SAMSUNG Store
★★★★½ ˅   351 ratings | 68 answered questions

List Price: $799.99 Details
    Price: $749.99 & FREE Returns
You Save: $50.00 (6%)

Pay $41.67/month for 18 months, interest-free upon approval for the Amazon Rewards Visa Card

Model Name	O1
Wireless Carrier	Unlocked
Brand	SAMSUNG
Form Factor	Smartphone
Memory Storage Capacity	128 GB
Operating System	Android

˅ See more

New & Used (9) from $578.24 + $4.79 shipping

Frequently bought together

Total price: $772.95
[Add all three to Cart]

ℹ These items are shipped from and sold by different sellers. Show details

☑ **This item:** Samsung Galaxy S21 5G | Factory Unlocked Android Cell Phone | US Version 5G Smartphone | Pro-Grade ... $749.99
☑ Ferilinso for Samsung Galaxy S21 Case with 2 Pack Tempered Glass Screen Protector [Hard PC Back+TPU Flexible Fr... $9.97
☑ [4 Pack] Ferilinso 2 Pack Screen Protector for Samsung Galaxy S21 5G with 2 Pack Camera Lens Protector [100% Su... $12.99

Customers who viewed this item also viewed   Page 1 of 5

‹

›

SAMSUNG Galaxy S20 FE 5G Factory Unlocked Android Cell Phone 128GB US Version Smartphone Pro-Grade Camera 30X Space...	Samsung Galaxy S21 5G, US Version, 128GB, Phantom Gray - Unlocked (Renewed)	Samsung Galaxy S21 5G	Factory Unlocked Android Cell Phone	US Version 5G Smartphone	Pro-Grade Camera, 8K Video, 64MP High Res...	SAMSUNG Galaxy S21 Ultra 5G	Factory Unlocked Android Cell Phone	US Version 5G Smartphone	Pro-Grade Camera, 8K Video,...	Samsung Galaxy S21+ Plus 5G, US Version, 128GB, Phantom Silver - Unlocked (Renewed)	(Refurbished) Samsung Galaxy S20 5G, 128GB, Cosmic Gray - Fully Unlocked
★★★★½ 5,605	★★★★½ 117	★★★★½ 44	★★★★☆ 39	★★★★☆ 15	★★★★½ 1,158
$597.99	$590.43	$589.33	$900.25	$739.98	5% off $440.00 $464.00 Lowest price in 30 days

FIGURE 15.1 Recommendations under "frequently bought together" are based on association rules

TABLE 15.1 Transactions Database for Purchases of Different-Colored Cellular Phone Faceplates

Transaction	Faceplate colors purchased			
1	red	white	green	
2	white	orange		
3	white	blue		
4	red	white	orange	
5	red	blue		
6	white	blue		
7	red	blue		
8	red	white	blue	green
9	red	white	blue	
10	yellow			

TABLE 15.2 Phone Faceplate Data in Binary Incidence Matrix Format

Transaction	Red	White	Blue	Orange	Green	Yellow
1	1	1	0	0	1	0
2	0	1	0	1	0	0
3	0	1	1	0	0	0
4	1	1	0	1	0	0
5	1	0	1	0	0	0
6	0	1	1	0	0	0
7	1	0	1	0	0	0
8	1	1	1	0	1	0
9	1	1	1	0	0	0
10	0	0	0	0	0	1

Generating Candidate Rules

The idea behind association rules is to examine all possible rules between items in an if–then format and select only those that are most likely to be indicators of true dependence. We use the term *condition* (or *antecedent*) to describe the IF part and *consequent* to describe the THEN part. In association analysis, the condition and consequent are sets of items (called *itemsets*, or in JMP "*itemsets*") that are disjoint (do not have any items in common). Note that itemsets are not records of what people buy; they are simply possible combinations of items, including single items.

Returning to the phone faceplate purchase example, one example of a possible rule is "if red, then white," meaning that if a red faceplate is purchased, a white one is, too. Here, the condition is *red* and the consequent is *white*. The condition and consequent each contain a single item in this case. Another possible rule is "if red and white, then green." Here, the condition includes the itemset {*red, white*} and the consequent is {*green*}.

The first step in association rules is to generate all the rules that would be candidates for indicating associations between items. Ideally, we might want to look at all possible combinations of items in a database with p distinct items (in the phone faceplate example, $p = 6$). This means finding all combinations of single items, pairs of items, triplets of items, and so on, in the transactions database. However, generating all these combinations requires a long computation time that grows exponentially[1] in p. A practical solution is to consider only combinations that occur with higher frequency in the database. These are called *frequent itemsets*.

Determining what qualifies as a frequent itemset is related to the concept of *support*. The support of a rule is simply the number of transactions that include both the condition and consequent itemsets. It is called a support because it measures the degree to which the data "support" the validity of the rule. The support is sometimes expressed as a percentage of the total number of records in the database. For example, the support for the itemset {*red,white*} in the phone faceplate example is 4 (or, $100 \times \frac{4}{10} = 40\%$).

What constitutes a frequent itemset is therefore defined as an itemset that has a support that exceeds a selected minimum support, determined by the user.

The Apriori Algorithm

Several algorithms have been proposed for generating frequent itemsets, but the classic algorithm is the *Apriori algorithm* of Agrawal et al. (1993). The key idea of the algorithm is to begin by generating frequent itemsets with just one item (one-itemsets) and to recursively generate frequent itemsets with two items, then with three items, and so on, until we have generated frequent itemsets of all sizes.

It is easy to generate frequent one-itemsets. All we need to do is to count, for each item, how many transactions in the database include the item. These transaction counts are the supports for the one-itemsets. We drop one-itemsets that have support below the desired minimum support to create a list of the frequent one-itemsets.

To generate frequent two-itemsets, we use the frequent one-itemsets. The reasoning is that if a certain one-itemset did not exceed the minimum support, any larger size itemset that includes it will not exceed the minimum support. In general, generating k-itemsets uses the frequent $(k - 1)$-itemsets that were generated in the preceding step. Each step requires a single run through the database, and therefore the Apriori algorithm is very fast even for a large number of unique items in a database.

Selecting Strong Rules

From the abundance of rules generated, the goal is to find only the rules that indicate a strong dependence between the condition and consequent itemsets. To measure the strength of association implied by a rule, we use the measures of *confidence*, *lift*, and *conviction* as described below.

Support and Confidence In addition to support, which we described earlier, there is another measure that expresses the degree of uncertainty about the if–then rule. This is known as the *confidence*[2] of the rule. This measure compares the co-occurrence of the condition and consequent itemsets in the database to the occurrence of the condition itemsets.

---

[1] The number of rules that one can generate for p items is $3^p - 2^{p+1} + 1$. Computation time therefore grows by a factor for each additional item. For six items we have 602 rules, while for seven items the number of rules grows to 1932.

[2] The concept of confidence is different from and unrelated to the ideas of confidence intervals and confidence levels used in statistical inference.

Confidence is defined as the ratio of the number of transactions that include all condition and consequent itemsets (namely, the support) to the number of transactions that include all the condition itemsets:

$$\text{Confidence} = \frac{\text{No. of transactions with both condition and consequent itemsets}}{\text{No. of transactions with condition itemset}}.$$

For example, suppose that a supermarket database has 100,000 point-of-sale transactions. Of these transactions, 2000 include both orange juice and (over-the-counter) flu medication, and 800 of these include soup purchases. The association rule "IF orange juice and flu medication are purchased THEN soup is purchased on the same trip" has a support of 800 transactions (alternatively, 0.8% = 800/100,000) and a confidence of 40% (=800/2000).

To see the relationship between support and confidence, let us think about what each is measuring (estimating). One way to think of support is that it is the (estimated) probability that a transaction selected randomly from the database will contain all items in the condition and the consequent:

$$\text{Support} = \hat{P}(\text{condition AND consequent}).$$

In comparison, the confidence is the (estimated) *conditional probability* that a transaction selected randomly will include all the items in the consequent *given* that the transaction includes all the items in the condition:

$$\text{Confidence} = \frac{\hat{P}(\text{condition AND consequent})}{\hat{P}(\text{condition})} = \hat{P}(\text{consequent} \mid \text{condition}).$$

A high value of confidence suggests a strong association rule (in which we are highly confident). However, this can be deceptive because if the condition and/or the consequent has a high level of support, we can have a high value for confidence even when the condition and consequent are independent! For example, if nearly all customers buy bananas and nearly all customers buy ice cream, the confidence level of a rule such as "IF bananas THEN ice-cream" will be high regardless of whether there is an association between the items.

Coverage   Coverage (also called cover or LHS-support) is the support of the left-hand side of the rule.

$$\text{Coverage} = \hat{P}(\text{condition}).$$

It will tell you how often the rule can be applied.

Lift   A better way to judge the strength of an association rule is to compare the confidence of the rule with a benchmark value, where we assume that the occurrence of the consequent itemset in a transaction is independent of the occurrence of the condition for each rule. In other words, if the condition and consequent itemsets are independent, what confidence values would we expect to see? Under independence, the support would be

$$\hat{P}(\text{condition AND consequent}) = \hat{P}(\text{condition}) \times \hat{P}(\text{consequent}),$$

and the benchmark confidence would be

$$\frac{\hat{P}(\text{condition}) \times \hat{P}(\text{consequent})}{\hat{P}(\text{condition})} = \hat{P}(\text{consequent}).$$

The estimate of this benchmark from the data, called the *benchmark confidence value* for a rule, is computed by

$$\text{Benchmark confidence} = \frac{\text{No. of transactions with consequent itemset}}{\text{No. of transactions in database}}.$$

We compare the confidence to the benchmark confidence by looking at their ratio: this is called the *lift* of a rule. The lift of the rule is the confidence of the rule divided by the benchmark confidence, assuming independence of consequent from condition.

$$\text{Lift} = \frac{\text{Confidence}}{\text{Benchmark confidence}}.$$

A lift greater than 1.0 suggests that there is some usefulness to the rule. In other words, the level of association between the condition and consequent itemsets is higher than would be expected if they were independent. The larger the lift of the rule, the greater the strength of the association.

Conviction   Conviction, proposed by Brin et al. (1997), is a useful metric for measuring the degree of implication of a rule and is defined as:

$$\text{Conviction} = \frac{\hat{P}(\text{condition}) \times \hat{P}(\text{not consequent})}{\hat{P}(\text{condition AND not consequent})} = \frac{1 - (\text{support of consequent})}{1 - \text{confidence of rule}}.$$

The *conviction* of rule {condition, consequent} considers the directionality of the rule condition $\Rightarrow$ consequent, rather than simply a co-occurrence (correlation). It measures this directionality by considering an event that *should not happen* if the rule is correct: the event condition-without-consequent. The probability of this event is the denominator of the conviction metric: $\hat{P}(\text{condition AND no-consequent})$. However, this event (condition-without-consequent) can also occur when the rule is incorrect, that is, when these two events are unrelated but co-occur purely due to chance. The numerator in conviction computes the probability of such a random co-occurrence: $\hat{P}(\text{condition}) \times \hat{P}(\text{no consequent})$.

In short, conviction compares the probabilities of the event condition-without-consequent under an assumption that the rule is correct (denominator) and under an assumption that the rule is incorrect (numerator). When the rule is incorrect, we expect conviction to be close to 1. When the rule is correct, we expect values much larger than 1, indicating how much more unlikely the combination is.

A value of conviction can be interpreted as "the factor by which the event condition-without-consequent is more unlikely if the rule is correct than by pure chance co-occurrence." For example, consider the rule "if orange juice then flu medication" in our previous example, where we had 100,000 transactions in total. Suppose 10,000 transactions had orange juice (OJ) and no flu medication, 4000 with flu medication and no OJ, 20,000 with both, and 66,000 with neither (see Table 15.3).
Then conviction of "if OJ then flu medication" is calculated as:

$$\frac{30{,}000/100{,}000 \times 76{,}000/100{,}000}{10{,}000/100{,}000} = 2.28.$$

This means the 10,000 occurrences of OJ-and-no-flu-medication are more than twice as unlikely if the rule is correct than by pure chance (if the rule is incorrect, we'd expect 22,800 co-occurrences).

TABLE 15.3   Example for Computing Conviction

OJ only (no flu medication)	10,000
Flu medication only (no OJ)	4,000
OJ and flu medication	20,000
No OJ, no flu medication	66,000
Total transactions	100,000

In contrast to lift, conviction is a directed metric since it also incorporates the information of the absence of the consequent. In other words, the conviction of a rule is sensitive to rule direction. Conviction is similar to confidence and ranges from 0 to ∞. Conviction is equal to 0 when the consequent is always present or the premise is always absent. If the condition and consequent are independent (unrelated), conviction is equal to 1. If the rule always holds (the items always appear together), its value is infinity, corresponding to a confidence value of 1.

Note: JMP does not produce conviction metrics.

To illustrate the computation of support, confidence, lift, and conviction for the cellular phone faceplate example, we use the binary incidence matrix format (Table 15.2) that is better suited to this purpose.

The Process of Rule Selection

Frequent Itemset Selection   The process of selecting strong rules is based on generating all association rules that meet stipulated support and confidence requirements. This is done in two stages. The first stage, described earlier, consists of finding all "frequent" itemsets, those itemsets that have a requisite support.

Now suppose that we want association rules between items for this database that have a support percentage of at least 20% (equivalent to a count of 2). In other words, rules based on items that were purchased together in at least 20% of the transactions. By enumeration, we can see that only the itemsets listed in Figure 15.2 have a support percentage of 20%. To generate the report in JMP, specify 0.2 as the minimum support in the initial dialog window (see Figure 15.3).

JMP shows the support percentage in descending order. The first itemset {*white*} has a support of 70%, because seven of the transactions included a white faceplate. Similarly, the last itemset {*green, red, white*} has a support of 20%, because only 20% (2 out of 10) of the transactions included red, white, and green faceplates.

Association Rules Generation   We generate, from the frequent itemsets, association rules that meet a confidence requirement. The first step is aimed at removing item combinations that are rare in the database. The second stage then filters the remaining rules and selects only those with high confidence. For most association analysis data, the computational challenge is the first stage, as described in the discussion of the Apriori algorithms.

The computation of confidence in the second stage is simple. Since any subset (e.g., {*red*} in the phone faceplate example) must occur at least as frequently as the set it belongs to (e.g., {*red, white*}), each subset will also be in the list. It is then straightforward to compute the confidence as the ratio of the support for the itemset to the support for each subset

Association Analysis

Frequent Item Sets

Item Set	Support	N Items
{white}	70%	1
{blue}	60%	1
{red}	60%	1
{blue, red}	40%	2
{blue, white}	40%	2
{red, white}	40%	2
{green}	20%	1
{orange}	20%	1
{green, red}	20%	2
{green, white}	20%	2
{orange, white}	20%	2
{blue, red, white}	20%	3
{green, red, white}	20%	3

FIGURE 15.2 Itemsets with support percentage of at least 20%

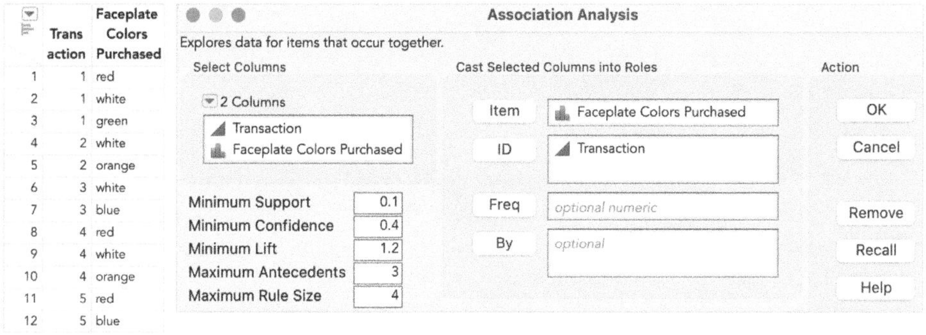

FIGURE 15.3 Data format and association analysis dialog window in JMP Pro

of the itemset. We retain the corresponding association rule only if it exceeds the desired cutoff value for confidence. For example, from the itemset {*red*, *white*, *green*} in the phone faceplate purchases, we get the following single-consequent association rules, confidence values, and lift values:

Rule	Confidence	Lift
$\{red, white\} \Rightarrow \{green\}$	$\frac{\text{Support of } \{red, white, green\}}{\text{Support of } \{red, white\}} = \frac{2}{4} = 50\%$	$\frac{\text{Confidence of rule}}{\text{Benchmark confidence}} = \frac{50\%}{20\%} = 2.5$
$\{green\} \Rightarrow \{red\}$	$\frac{\text{Support of } \{green, red\}}{\text{Support of } \{green\}} = \frac{2}{2} = 100\%$	$\frac{\text{Confidence of rule}}{\text{Benchmark confidence}} = \frac{100\%}{60\%} = 1.67$
$\{white, green\} \Rightarrow \{red\}$	$\frac{\text{Support of } \{white, green, red\}}{\text{Support of } \{white, green\}} = \frac{2}{2} = 100\%$	$\frac{\text{Confidence of rule}}{\text{Benchmark confidence}} = \frac{100\%}{60\%} = 1.67$

If the desired minimum confidence is 70%, we would report only the second and third rules. After selecting the rules based on the selection criteria, we can compute their conviction values to identify their degree of implication.

ASSOCIATION ANALYSIS IN JMP Pro

Association Analysis is an option in the *Analyze > Screening* menu. The launch window (see Figure 15.3) provides many specification options (*minimum support, minimum confidence, minimum lift, maximum conditions*, and *maximum rule size*).

Three data formats are supported:

1. Specify a single item response in each row in the *Item* role and identify which items are included in each transaction using the *ID* role. Figure 15.3 shows the phone faceplate data in this *long* format.
2. Specify a column with the *Multiple Response* modeling type for the *Item* role. Multiple response columns use a delimiter, which is specified in the *Multiple Response* column property, to separate response the values.
3. Specify multiple response columns in the *Item* role.

Reports for *Frequent ItemSets* (with *Support*) and *Rules* (with *Confidence* and *Lift*) are provided. Click on the headers in the reports to sort the results by item set, condition, consequent, or values of interest. To visualize the results, right-click over either table, select *Make Into Data Table*, and then use the *Graph Builder.*

Figure 15.4 shows the output with frequent itemsets and the association rules. The output shows information on each rule and its support, confidence, lift, and conviction. (Note that here we consider all possible itemsets, not just red, white, green as above.) The rules shown are sorted in descending order of conviction and lift values.

Interpreting the Results

We can translate each of the rules from Figure 15.4 into an understandable sentence that provides information about performance. For example, we can read rule {*orange*} ⇒ {*white*} as follows:

> If *orange* is purchased, then with confidence 100% *white* will also be purchased. This rule has a lift of 1.429.

In interpreting results, it is useful to look at the various measures. The support for the rule indicates its impact in terms of overall size: how many transactions are affected? If only a small number of transactions are affected, the rule may be of little use (unless the consequent is very valuable and/or the rule is very efficient in finding it).

The lift indicates how efficient the rule is in finding consequent itemsets compared to random selection. A very efficient rule is preferred to an inefficient rule, but we must still consider support: a very efficient rule that has very low support may not be as desirable as a less efficient rule with much greater support.

The confidence tells us at what rate consequents will be found and is useful in determining the business or operational usefulness of a rule: a rule with low confidence may find consequent itemset at too low a rate to be worth the cost of (say) promoting the consequent in all the transactions that involve the condition.

▼ ⏷ **Association Analysis**

  ▼ **Frequent Item Sets**

Item Set	Support ⌄	N Items
{white}	70%	1
{blue}	60%	1
{red}	60%	1
{blue, red}	40%	2
{blue, white}	40%	2
{red, white}	40%	2
{green}	20%	1
{orange}	20%	1
{green, red}	20%	2
{green, white}	20%	2
{orange, white}	20%	2
{blue, red, white}	20%	3
{green, red, white}	20%	3

  ▼ **Rules**

Rule Condition	Consequent	Confidence ⌄	Lift
green	red	100%	1.667
green	white	100%	1.429
orange	white	100%	1.429
green	red, white	100%	2.5
green, red	white	100%	1.429
green, white	red	100%	1.667
red, white	green	50%	2.5

FIGURE 15.4 **Frequent itemsets with minimum support 20% and association rules with minimum confidence 50%**

Rules and Chance

What about confidence in the nontechnical sense? How sure can we be that the rules we develop are meaningful? Considering the matter from a statistical perspective, we can ask: Are we finding associations that are really just chance occurrences?

Let us examine the output from an application of this algorithm to a small database of 50 transactions, where each of the nine items is assigned randomly to each transaction. The data are shown in Table 15.4, and the association rules generated are shown in Figure 15.5. Note that we specify the minimum support of 4% for frequent itemset generation and the minimum confidence level of 70% for rule selection. In looking at these tables, remember that "condition" and "consequent" refer to itemsets, not records.

In this example, the lift values highlight the fifth rule (where the condition is 3,8 and consequent is 4) as most interesting, as it suggests that purchase of item 4 is almost five times as likely when items 3 and 8 are purchased than if item 4 was not associated with the itemset {3, 8}. Yet we know there is no fundamental association underlying these data—they were generated randomly.

Two principles can guide us in assessing rules for possible spuriousness due to chance effects:

- The more records the rule is based on, the more solid the conclusion.
- The more distinct rules we consider seriously (perhaps consolidating multiple rules that deal with the same items), the more likely it is that at least some will be based on

TABLE 15.4 **Fifty Transactions of Randomly Assigned Items**

Trans.	Items					Trans.	Items					Trans.	Items			
1	8					18	8					35	3	4	6	8
2	3	4	8			19						36	1	4	8	
3	8					20	9					37	4	7	8	
4	3	9				21	2	5	6	8		38	8	9		
5	9					22	4	6	9			39	4	5	7	9
6	1	8				23	4	9				40	2	8	9	
7	6	9				24	8	9				41	2	5	9	
8	3	5	7	9		25	6	8				42	1	2	7	9
9	8					26	1	6	8			43	5	8		
10						27	5	8				44	1	7	8	
11	1	7	9			28	4	8	9			45	8			
12	1	4	5	8	9	29	9					46	2	7	9	
13	5	7	9			30	8					47	4	6	9	
14	6	7	8			31	1	5	8			48	9			
15	3	7	9			32	3	6	9			49	9			
16	1	4	9			33	7	9				50	6	7	8	
17	6	7	8			34	7	8	9							

Rule			
Condition	Consequent	Confidence ⌄	Lift
1, 5	8	100%	1.923
2, 7	9	100%	1.923
2, 9	7	100%	3.333
3, 4	8	100%	1.923
3, 8	4	100%	4.167
3, 7	9	100%	1.923
6, 7	8	100%	1.923
5, 7	9	75%	1.442

FIGURE 15.5 **Association rules output for random data**

chance sampling results. For one person to toss a coin 10 times and get 10 heads would be quite surprising. If 1000 people toss a coin 10 times each, it would not be nearly so surprising to have one get 10 heads. Formal adjustment of "statistical significance" when multiple comparisons are made is a complex subject in its own right and beyond the scope of this book. A reasonable approach is to consider rules from the top-down in terms of business or operational applicability and not consider more than what can reasonably be incorporated in a human decision-making process. This will impose a rough constraint on the dangers that arise from an automated review of hundreds or thousands of rules in search of "something interesting."

We now consider a more realistic example, using a larger database and real transactional data.

Example 2: Rules for Similar Book Purchases

The following example (drawn from the Charles Book Club case; see Chapter 22) examines associations among transactions involving various types of books. The database includes 3227 transactions, and there are 11 different types of books. The sample data, in binary incidence matrix form, is shown in Figure 15.6 (right). The data is converted into long format, and a sample is shown in Figure 15.6 (left).

	ID#	Book Type
1	25	YouthBks
2	25	CookBks
3	46	ChildBks
4	46	ChildBks
5	46	YouthBks
6	46	CookBks
7	46	CookBks
8	46	RefBks
9	46	GeogBks
10	46	ItalCook
11	61	GeogBks
12	79	ChildBks
13	90	CookBks
14	95	CookBks
15	100	ArtBks
16	100	Florence

	Book Type										
ID#	ArtBks	ChildBks	CookBks	DoItYBks	Florence	GeogBks	ItalArt	ItalAtlas	ItalCook	RefBks	YouthBks
25	0	0	1	0	0	0	0	0	0	0	1
46	0	2	2	0	0	1	0	0	1	1	1
61	0	0	0	0	0	1	0	0	0	0	0
79	0	1	0	0	0	0	0	0	0	0	0
90	0	0	1	0	0	0	0	0	0	0	0
95	0	0	1	0	0	0	0	0	0	0	0
100	1	0	0	0	1	0	0	0	0	0	0
103	0	0	0	0	1	0	0	0	0	0	0
112	0	1	0	0	0	0	0	0	0	0	0
121	1	2	3	0	0	2	1	1	1	1	0
125	1	2	2	0	0	0	0	0	1	0	1
126	0	0	1	0	0	0	0	0	0	0	0
127	1	0	2	2	0	0	0	0	0	1	0
145	2	3	2	0	1	3	0	0	0	0	0
202	1	0	3	5	0	0	0	0	0	0	2
208	0	1	0	0	0	1	0	0	0	0	0
214	1	0	0	0	0	0	0	0	0	0	0
225	0	0	0	0	0	0	0	0	0	1	0

FIGURE 15.6 Subset of book purchase transactions in binary matrix format (right) and long format (left)

For instance, the first transaction included *CookBks* (Cook books) and *YouthBks* (Youth books). Figure 15.7 shows some of the rules generated for these data, given that we specified a minimal support of 5% and a minimal confidence of 50%. This resulted in 98 rules (the 25 rules with the highest lift ratio are shown in Figure 15.7).

In reviewing these rules, we see that the information can be compressed: *ChildBks* is included in almost every rule. Including children books in the analysis is probably meaningless, as the purchase of children books is not associated with individual preferences of the customer. In this case, one could consider removing this item from the analysis or filtering the rules to those that do not include *ChildBks*. Another rule is ItalCook => CookBks (not shown) that has a high confidence (64%), but is probably meaningless. It says: "If Italian cooking books have been purchased, then cookbooks are purchased." It seems likely that Italian cooking books are simply a subset of cookbooks.

Rules highlighted with boxes involve the same four book types, with different conditions and consequents. This does not mean that the rules are not useful. On the contrary, it can reduce the number of itemsets to be considered for possible action from a business perspective.

▼ **Rules**

Rule			
Condition	**Consequent**	**Confidence**	**Lift** ⌄
DoItYBks, YouthBks	ChildBks, CookBks	65%	2.163
GeogBks, RefBks	ChildBks, CookBks	61%	2.049
GeogBks, YouthBks	ChildBks, CookBks	61%	2.018
DoItYBks, GeogBks	ChildBks, CookBks	60%	1.997
ChildBks, CookBks, GeogBks	YouthBks	58%	1.956
CookBks, DoItYBks, GeogBks	YouthBks	57%	1.944
ChildBks, DoItYBks, GeogBks	YouthBks	57%	1.928
ArtBks, DoItYBks	ChildBks, CookBks	57%	1.902
ChildBks, CookBks, RefBks	DoItYBks	59%	1.874
ArtBks, ChildBks, CookBks	DoItYBks	58%	1.837
DoItYBks, GeogBks	YouthBks	54%	1.827
ChildBks, CookBks, RefBks	YouthBks	53%	1.808
ChildBks, CookBks, DoItYBks	YouthBks	52%	1.776
ChildBks, CookBks, YouthBks	DoItYBks	56%	1.768
ChildBks, GeogBks, YouthBks	DoItYBks	56%	1.762
CookBks, GeogBks, YouthBks	DoItYBks	55%	1.756
RefBks, YouthBks	DoItYBks	55%	1.754
ChildBks, RefBks	DoItYBks	55%	1.753
ChildBks, CookBks, GeogBks	DoItYBks	55%	1.75
ChildBks, GeogBks	YouthBks	52%	1.748
CookBks, GeogBks	YouthBks	51%	1.739
ChildBks, RefBks, YouthBks	CookBks	89%	1.73
ChildBks, YouthBks	DoItYBks	54%	1.723
GeogBks, RefBks	DoItYBks	54%	1.704
CookBks, RefBks	DoItYBks	53%	1.688
RefBks	ChildBks, CookBks	51%	1.685
CookBks, DoItYBks, RefBks	ChildBks	82%	1.683
ArtBks, ChildBks, CookBks	GeogBks	56%	1.682
YouthBks	ChildBks, CookBks	50%	1.679
ChildBks, DoItYBks, RefBks	CookBks	86%	1.675
ArtBks, GeogBks	ChildBks, CookBks	50%	1.672
DoItYBks	ChildBks, CookBks	50%	1.672

FIGURE 15.7   Association rules for book purchase transactions sorted by lift

15.2   COLLABORATIVE FILTERING[3]

Recommendation systems are a critically important part of websites that offer a large variety of products or services. Examples include the following: Amazon.com offers millions of different products, Netflix has thousands of movies for rental, Google searches over huge numbers of webpages, Internet radio websites such as Spotify and Pandora include a large variety of music albums by various artists, travel websites offer many destinations and hotels, social network websites have many groups. The recommender engine provides personalized recommendations to a user based on the user's information as well as on similar users' information. Information means behaviors indicative of preference, such as purchase, ratings, and clicking.

---

[3]This section copyright © 2019 Datastats, LLC, Galit Shmueli, and Peter Bruce.

The value that recommendation systems provide to users helps online companies convert browsers into buyers, increase cross-selling, and build loyalty.

Collaborative filtering is a popular technique used by such recommendation systems. The term *collaborative filtering* is based on the notions of identifying relevant items for a specific user from the very large set of items ("filtering") by considering preferences of many users ("collaboration").

The Fortune.com article "Amazon's Recommendation Secret" (June 30, 2012) describes the company's use of collaborative filtering not only for providing personalized product recommendations but also for customizing the entire website interface for each user:

> At root, the retail giant's recommendation system is based on a number of simple elements: what a user has bought in the past, which items they have in their virtual shopping cart, items they've rated and liked, and what other customers have viewed and purchased. Amazon calls this homegrown math "item-to-item collaborative filtering," and it's used this algorithm to heavily customize the browsing experience for returning customers.

Data Type and Format

Collaborative filtering requires availability of all item–user information. Specifically, for each item–user combination, we should have some measure of the user's preference for that item. Preference can be a numerical rating or a binary behavior such as a purchase, a "like," or a click.

For n users ($u_1, u_2, \ldots, u_n$) and p items ($i_1, i_2, \ldots, i_p$), we can think of the data as an $n \times p$ matrix of n rows (users) by p columns (items). Each cell includes the rating or the binary event corresponding to the user's preference of the item (see schematic in Table 15.5). Typically not every user purchases or rates every item, and therefore a purchase matrix will have many zeros (it is sparse), and a rating matrix will have many missing values. Such missing values sometimes convey "uninterested" (as opposed to non-missing values that convey interest).

TABLE 15.5 Schematic of matrix format with ratings data

User ID	Item ID			
	I_1	I_2	$\cdots$	I_p
U_1	$r_{1,1}$	$r_{1,2}$	$\cdots$	$r_{1,p}$
U_2	$r_{2,1}$	$r_{2,2}$	$\cdots$	$r_{2,p}$
$\vdots$				
U_n	$r_{n,1}$	$r_{n,2}$	$\cdots$	$r_{n,p}$

When both n and p are large, it is not practical to store the preferences data ($r_{u,i}$) in an $n \times p$ table. Instead, the data can be stored in many rows of triplets of the form ($U_u, I_i, r_{u,i}$), where each triplet contains the user ID, the item ID, and the preference information.

Example 3: Netflix Prize Contest

We have been considering both association rules and collaborative filtering as unsupervised techniques, but it is possible to judge how well they do by looking at holdout data to see what users purchase and how they rate items. The famous Netflix contest, mentioned in Chapter 13, did just this and provides a useful example to illustrate collaborative filtering, though the extension into training and validation is beyond the scope of this book.

In 2006, Netflix, the largest movie rental service in North America, announced a US\$ 1 million contest (`www.netflixprize.com`) for the purpose of improving its recommendation system called *Cinematch*. Participants were provided with a number of datasets, one for each movie. Each dataset included all the customer ratings for that movie (and the timestamp). We can think of one large combined dataset of the form [customer ID, movie ID, rating, date] where each record includes the rating given by a certain customer to a certain movie on a certain date. Ratings were on a 1–5 star scale. Contestants were asked to develop a recommendation algorithm that would improve over the existing Netflix system. Table 15.6 shows a small sample from the contest data, organized in matrix format. Rows indicate customers and columns are different movies.

TABLE 15.6 **Sample of records from the Netflix Prize contest, for a subset of 10 customers and 9 movies**

Customer ID	Movie ID								
	1	5	8	17	18	28	30	44	48
30878	4	1			3	3	4	5	
124105	4								
822109	5								
823519	3		1	4		4	5		
885013	4	5							
893988	3						4	4	
1248029	3					2	4		3
1503895	4								
1842128	4						3		
2238063	3								

It is interesting to note that the winning team was able to improve their system by considering not just the ratings for a movie but also *whether* a movie was rated by a particular customer or not. In other words, the information on which movies a customer decided to rate turned out to be critically informative of customers' preferences, more than simply considering the 1–5 rating information[4]:

> Collaborative filtering methods address the sparse set of rating values. However, much accuracy is obtained by also looking at other features of the data. First is the information on which movies each user chose to rate, regardless of specific rating value ("the binary view"). This played a decisive role in our 2007 solution, and reflects the fact that the movies to be rated are selected deliberately by the user, and are not a random sample.

This is an example where converting the rating information into a binary matrix of rated/unrated proved to be useful.

---

[4]The BellKor 2008 Solution to the Netflix Prize, Bell, R. M., Koren, Y., and Volinsky, C., `https://api.semanticscholar.org/CorpusID:11877045`

User-Based Collaborative Filtering: "People Like You"

One approach to generating personalized recommendations for a user using collaborative filtering is based on finding users with similar preferences and recommending items that they liked but the user hasn't purchased. The algorithm has two steps:

1. Find users who are most similar to the user of interest (neighbors). This is done by comparing the preference of our user to the preferences of other users.
2. Considering only the items that the user has *not* yet purchased, recommend the ones that are most preferred by the user's neighbors.

This is the approach behind Amazon's "Customers Who Viewed This Item Also Viewed…" (see Figure 15.1). It is also used in a Google search for generating the "Similar pages" link shown near each search result.

Step 1 requires choosing a distance (or proximity) metric to measure the distance between our user and the other users. Once the distances are computed, we can use a threshold on the distance or on the number of required neighbors to determine the nearest neighbors to be used in Step 2. This approach is called "user-based top-N recommendation."

A nearest-neighbors approach measures the distance of our user to each of the other users in the database, similar to the k-nearest-neighbors algorithm (see Chapter 7). The Euclidean distance measure we discussed in that chapter does not perform as well for collaborative filtering as some other measures. A popular proximity measure between two users is the Pearson correlation between their ratings. We denote the ratings of items $I_1, \ldots, I_p$ by user U_1 as $r_{1,1}, r_{1,2}, \ldots, r_{1,p}$ and their average by $\bar{r}_1$. Similarly, the ratings by user U_2 are $r_{2,1}, r_{2,2}, \ldots, r_{2,p}$, with average $\bar{r}_2$. The correlation proximity between the two users is defined by

$$\text{Corr}(U_1, U_2) = \frac{\sum (r_{1,i} - \bar{r}_1)(r_{2,i} - \bar{r}_2)}{\sqrt{\sum (r_{1,i} - \bar{r}_1)^2} \sqrt{\sum (r_{2,i} - \bar{r}_2)^2}}, \tag{15.1}$$

where the summations are only over the items co-rated by both users.

To illustrate this, let us compute the correlation between customer 30878 and customer 823519 in the small Netflix sample in Table 15.6. We'll assume that the data shown in the table is the entire information. First, we compute the average rating by each of these users:

$$\bar{r}_{30878} = (4 + 1 + 3 + 3 + 4 + 5)/6 = 3.333,$$

$$\bar{r}_{823519} = (3 + 1 + 4 + 4 + 5)/5 = 3.4.$$

Note that the average is computed over a different number of movies for each of these customers, because they each rated a different set of movies. The average for a customer is computed over *all* the movies that a customer rated. The calculations for the correlation involve the departures from the average, but *only for the items that they co-rated*. In this case, the co-rated movie IDs are 1, 28, and 30:

$$\text{Corr}(U_{30878}, U_{823519}) =$$

$$\frac{(4 - 3.333)(3 - 3.4) + (3 - 3.333)(4 - 3.4) + (4 - 3.333)(5 - 3.4)}{\sqrt{(4 - 3.333)^2 + (3 - 3.333)^2 + (4 - 3.333)^2} \sqrt{(3 - 3.4)^2 + (4 - 3.4)^2 + (5 - 3.4)^2}}$$

$$= 0.6/1.75 = 0.34.$$

The same approach can be used when the data are in the form of a binary matrix (e.g., purchased or didn't purchase.)

Another popular measure is a variant of the Pearson correlation called *cosine similarity*. It differs from the correlation formula by not subtracting the means. Subtracting the mean in the correlation formula adjusts for users' different overall approaches to rating—for example, a customer who always rates highly vs. one who tends to give low ratings.[5]

For example, the cosine similarity between the two Netflix customers is

$$\text{Cos Sim}(U_{30878}, U_{823519}) = \frac{4 \times 3 + 3 \times 4 + 4 \times 5}{\sqrt{4^2 + 3^2 + 4^2}\sqrt{3^2 + 4^2 + 5^2}}$$

$$= 44/45.277 = 0.972.$$

Note that when the data are in the form of a binary matrix, say, for purchase or no-purchase, the cosine similarity must be calculated over all items that either user has purchased; it cannot be limited to just the items that were co-purchased.

Collaborative filtering suffers from what is called a *cold start*: it cannot be used as is to create recommendations for new users or new items. For a user who rated a single item, the correlation coefficient between this and other users (in user-generated collaborative filtering) will have a denominator of zero, and the cosine proximity will be 1 regardless of the rating. In a similar vein, users with just one item, and items with just one user, do not qualify as candidates for nearby neighbors.

For a user of interest, we compute his/her similarity to each of the users in our database using a correlation, cosine similarity, or another measure. Then, in Step 2, we look only at the k nearest users, and among all the other items that they rated/purchased, we choose the best one and recommend it to our user. What is the best one? For binary purchase data, it is the item most purchased. For rating data, it could be the highest rated, most rated, or a weighting of the two.

The nearest-neighbors approach can be computationally expensive when we have a large database of users. One solution is to use clustering methods (see Chapter 16) to group users into homogeneous clusters in terms of their preferences and then to measure the distance of our user to each of the clusters. This approach places the computational load on the clustering step that can take place earlier and offline; it is then cheaper (and faster) to compare our user to each of the clusters in real time. The price of clustering is less accurate recommendations, because not all the members of the closest cluster are the most similar to our user.

Item-Based Collaborative Filtering

When the number of users is much larger than the number of items, it is computationally cheaper (and faster) to find similar items rather than similar users. Specifically, when a user

---

[5]Correlation and cosine similarity are popular in collaborative filtering because they are computationally fast for high-dimensional sparse data, and they account for both the rating values and the number of rated items.

expresses interest in a particular item, the item-based collaborative filtering algorithm has two steps:

1. Find the items that were co-rated, or co-purchased, (by any user) with the item of interest.
2. Recommend the most popular or correlated item(s) among the similar items.

Similarity is now computed between items, instead of users. For example, in our small Netflix sample (Table 15.6), the correlation between movie 1 (with average $\bar{r}_1 = 3.7$) and movie 5 (with average $\bar{r}_5 = 3$) is

$$\text{Corr}(I_1, I_5) = \frac{(4 - 3.7)(1 - 3) + (4 - 3.7)(5 - 3)}{\sqrt{(4 - 3.7)^2 + (4 - 3.7)^2}\sqrt{(1 - 3)^2 + (5 - 3)^2}} = 0.$$

The zero correlation is due to the two opposite ratings of movie 5 by the users who also rated 1. One user rated it 5 stars and the other gave it a 1 star.

In a similar fashion, we can compute similarity between all the movies. This can be done offline. In real time, for a user who rates a certain movie highly, we can look up the movie correlation table and recommend the movie with the highest positive correlation to the user's newly rated movie.

According to an industry report[6] by researchers who developed the Amazon item-to-item recommendation system,

> [The item-based] algorithm produces recommendations in real time, scales to massive data sets, and generates high-quality recommendations.

The disadvantage of item-based recommendations is that there is less diversity between items (compared to users' taste), and therefore, the recommendations are often obvious.

Evaluating Performance

Collaborative filtering can be considered an unsupervised learning method, because we do not have the true outcome value for a user presented with a recommendation. However, for purposes of evaluating such systems, they are often treated as if they were supervised learning problems, where the recommendation is compared to the actual selection by a user. Conceptually, we treat missing entries in the user–item matrix as values that need to be predicted from the observed entries in the remaining matrix. This assumption is needed if we do not have actual user reactions to the recommendations. This scenario is called *offline learning*.

Two types of offline learning evaluations are *predictive performance* and *ranking performance*—each measures a different aspect of the recommender system. Predictive performance assesses how close the recommendation is to the actual user's behavior. We have already encountered many such metrics in Chapter 5. In contrast, ranking evaluation considers only performance on the top set of recommended items ("Top-N") produced by the recommender algorithm. This approach is useful in many applications where the a user

[6]Linden, G., Smith, B., and York J. (2003). Amazon.com recommendations: item-to-item collaborative filtering, *IEEE Internet Computing*, vol. 7, number. 1, pp. 76–80.

is shown a set of the "best" recommended items for him/her (such as the results of a Google search). Ranking metrics therefore focus on the accuracy of only the top-ranked items.

Common performance metrics include the following:

Predictive performance for numerical predictions (e.g., ratings): Metrics such as hold-out RMSE and MAE are commonly used. Note that these metrics are strongly influenced by long-tail items: a rare item that is incorrectly predicted can mask excellent performance on common items. One solution is to weight items within these metrics so that more important/profitable items have higher weights.

Predictive performance for categorical predictions (e.g., click/no-click): Metrics based on the holdout confusion matrix such as accuracy, recall and precision, and ROC curves are common.

Ranking performance for numerical predictions (e.g., ratings): A common ranking performance metric is the correlation between the *ordering* of the true ratings and the *ordering* of the predicted ratings. This is known as the Spearman coefficient.

Ranking performance for categorical classifications: Common metrics include Top-N precision and Top-N recall, which are the precision and recall computed only on the top-N recommendation sets and which measure the fraction of relevant items in the recommendation list and the fraction of relevant items that are missed by the list.

There also exist specialized metrics for other important properties of a recommendation algorithm, beyond accuracy, such as novelty, trust, coverage, and serendipity. To learn more about these, see Aggarwal (2016).

Lastly, we note that partitioning the data into training, validation, and holdout sets is done a bit differently than in supervised learning: rather than sampling users, for collaborative filtering evaluation, the sampling is done on *matrix entries* (e.g., $r_{u,i}$ in Table 15.5): these are the user–item combinations in the user–item matrix.

Note: JMP does not currently support collaborative filtering.

Advantages and Weaknesses of Collaborative Filtering

Collaborative filtering relies on the availability of subjective information regarding users' preferences. It provides useful recommendations, even for "long-tail" items, if our database contains sufficient similar users (not necessarily many, but at least a few per user), so that each user can find other users with similar tastes. Similarly, the data should include sufficient per-item ratings or purchases. One limitation of collaborative filtering is therefore that it cannot generate recommendations for new users, nor for new items. There are various approaches for tackling this challenge.

User-based collaborative filtering looks for similarity in terms of highly rated or preferred items. However, it is blind to data on low-rated or unwanted items. We can therefore not expect to use it as is for detecting unwanted items.

User-based collaborative filtering helps leverage similarities between people's tastes for providing personalized recommendations. However, when the number of users becomes very large, collaborative filtering becomes computationally difficult. Solutions include item-based algorithms, clustering of users, and dimension reduction. The most popular dimension reduction method used in such cases is singular value decomposition (SVD), a computationally superior form of principal component analysis (see Chapter 4).

Although the term "prediction" is often used to describe the output of collaborative filtering, this method is unsupervised by nature. It can be used to generate predicted ratings or purchase indication for a user, but usually we do not have the true outcome value in practice. Evaluating performance is therefore challenging and treats the user's actual behavior as if it were their response to the recommendation. One important way to improve recommendations generated by collaborative filtering is by getting actual user feedback. Once a recommendation is generated, the user can indicate whether the recommendation was adequate or not. For this reason, many recommender systems entice users to provide feedback on their recommendations. When user reactions are measured with respect to the presented recommendations, this is called *online learning*. This is in contrast to *offline learning*, where such "online" feedback information is not used. Lastly, using online learning, many companies now measure the direct impact of a recommender system in terms of a business-relevant metric, such as conversion rate. In Chapter 14, we describe approaches that obtain user feedback that are useful for evaluating recommender systems as well as other interventions.

Collaborative Filtering vs. Association Rules

While collaborative filtering and association rules are both unsupervised methods used for generating recommendations, they differ in several ways.

Frequent itemsets vs. personalized recommendations: Association rules look for frequent item combinations and will provide recommendations only for those items. In contrast, collaborative filtering provides personalized recommendations for every item, thereby catering to users with unusual taste. In this sense, collaborative filtering is useful for capturing the "long tail" of user preferences, while association rules look for the "head." This difference has implications for the data needed: association rules require data on a very large number of "baskets" (transactions) in order to find a sufficient number of baskets that contain certain combinations of items. In contrast, collaborative filtering does not require many "baskets," but does require data on as many items as possible for many users. Also, association rules operate at the basket level (our database can include multiple transactions for each user), while collaborative filtering operates at the user level.

Because association rules produce generic, impersonal rules (association-based recommendations such as Amazon's "Frequently Bought Together" display the same recommendations to all users searching for a specific item), they can be used for setting common strategies such as product placement in a store or sequencing of diagnostic tests in hospitals. In contrast, collaborative filtering generates user-specific recommendations (e.g., Amazon's "Customers Who Viewed This Item Also Viewed...") and is therefore a tool designed for personalization.

Transactional data vs. user data: Association rules provide recommendations of items based on their co-purchase with other items in *many transactions/baskets*. In contrast, collaborative filtering provides recommendations of items based on their co-purchase or co-rating by even a small number of other *users*. Considering distinct baskets is useful when the same items are purchased over and over again (e.g., in grocery shopping). Considering distinct users is useful when each item is typically purchased/rated once (e.g., purchases of books, music, and movies).

Binary data and ratings data: Association rules treat items as binary data (1 = purchase, 0 = nonpurchase), whereas collaborative filtering can operate on either binary data or on numerical ratings.

Two or more items: In association rules, the condition and consequent can each include one or more items (e.g., IF milk THEN cookies and cornflakes). Hence, a recommendation might be a bundle of the item of interest with multiple items ("buy milk, cookies, and cornflakes and receive 10% discount"). In contrast, in collaborative filtering, similarity is measured between *pairs* of items or pairs of users. A recommendation will therefore be either for a single item (the most popular item purchased by people like you, which you haven't purchased) or for multiple single items which do not necessarily relate to each other (the top two most popular items purchased by people like you, which you haven't purchased).

These distinctions are sharper for purchases and recommendations of non-popular items, especially when comparing association rules to user-based collaborative filtering. When considering what to recommend to a user who purchased a popular item, then association rules and item-based collaborative filtering might yield the same recommendation for a single item. But a user-based recommendation will likely differ.

Consider a customer who purchases milk every week as well as gluten-free products (which are rarely purchased by other customers). Suppose that using association rules on the transactions database, we identify the rule "IF milk THEN cookies." Then, the next time our customer purchases milk, s/he will receive a recommendation (e.g., a coupon) to purchase cookies, whether or not s/he purchased cookies and irrespective of his/her gluten-free item purchases. In item-based collaborative filtering, we would look at all items co-purchased with milk across all users and recommend the most popular item among them (which was not purchased by our customer). This might also lead to a recommendation of cookies, because this item was not purchased by our customer.[7]

Now consider user-based collaborative filtering. User-based collaborative filtering searches for similar customers—those who purchased the same set of items—and then recommends the item most commonly purchased by these neighbors, which *was not purchased by our customer*. The user-based recommendation is therefore unlikely to recommend cookies and more likely to recommend popular gluten-free items that the customer has not purchased.

15.3 SUMMARY

Association rules (also called market basket analysis) and collaborative filtering are unsupervised methods for deducing associations between purchased items from databases of transactions. Association rules search for generic rules about items that are purchased together. The main advantage of this method is that it generates clear, simple rules of the form "IF X is purchased, THEN Y is also likely to be purchased." The method is very transparent and easy to understand.

---

[7]If the rule is "IF milk, THEN cookies and cornflakes," then the association rules would recommend cookies and cornflakes to a milk purchaser, while item-based collaborative filtering would recommend the most popular single item purchased with milk.

The process of creating association rules is two-staged. First, a set of candidate rules based on frequent itemsets is generated (the Apriori and FP-Growth algorithms being the most popular rule-generating algorithms). Then from these candidate rules, the rules that indicate the strongest association between items are selected. We use the measures of support and confidence to evaluate the uncertainty in a rule. The user also specifies minimal support and confidence values to be used in the rule generation and selection process. A third measure, the lift, compares the efficiency of the rule to detect a real association compared to a random combination.

One shortcoming of association rules is the profusion of rules that are generated. There is therefore a need for ways to reduce these to a small set of useful and strong rules. An important nonautomated method to condense the information involves examining the rules for uninformative and trivial rules as well as for rules that share the same support. Another issue that needs to be kept in mind is that rare combinations tend to be ignored, because they do not meet the minimum support requirement. For this reason, it is better to have items that are approximately equally frequent in the data. This can be achieved by using higher-level hierarchies as the items. An example is to use types of books rather than titles of individual books in deriving association rules from a database of bookstore transactions.

Collaborative filtering is a popular technique used in online recommendation systems. It is based on the relationship between items formed by users who acted similarly on an item, such as purchasing or rating an item highly. User-based collaborative filtering operates on data on item–user combinations, calculates the similarities between users and provides personalized recommendations to users. An important component for the success of collaborative filtering is that users provide feedback about the recommendations provided and have sufficient information on each item. One disadvantage of collaborative filtering methods is that they cannot generate recommendations for new users or new items. Also, with a huge number of users, user-based collaborative filtering becomes computationally challenging, and alternatives such as item-based methods or dimension reduction are popularly used.

PROBLEMS

15.1 Satellite Radio Customers. An analyst at a subscription-based satellite radio company has been given a sample of data from their customer database, with the goal of finding groups of customers who are associated with one another. The data consist of company data, together with purchased demographic data that are mapped to the company data (see Table 15.7). The analyst decides to apply association rules to learn more about the associations between customers. Comment on this approach.

TABLE 15.7   Sample of data on satellite radio customers

ID	zipconvert_2	zipconvert_3	zipconvert_4	zipconvert_5	homeowner dummy	NUMCHLD	INCOME	gender dummy	WEALTH
17	0	1	0	0	1	1	5	1	9
25	1	0	0	0	1	1	1	0	7
29	0	0	0	1	0	2	5	1	8
38	0	0	0	1	1	1	3	0	4
40	0	1	0	0	1	1	4	0	8
53	0	1	0	0	1	1	4	1	8
58	0	0	0	1	1	1	4	1	8
61	1	0	0	0	1	1	1	0	7
71	0	0	1	0	1	1	4	0	5
87	1	0	0	0	1	1	4	1	8
100	0	0	0	1	1	1	4	1	8
104	1	0	0	0	1	1	1	1	5
121	0	0	1	0	1	1	4	1	5
142	1	0	0	0	0	1	5	0	8

15.2 Identifying Course Combinations. The Institute for Statistics Education at Statistics.com offers online courses in statistics and analytics and is seeking information that will help in packaging and sequencing courses. Consider the data in the file CourseTopics.jmp, the first few rows of which are shown in Table 15.8. These data are for purchases of online statistics courses at Statistics.com. Each row represents the courses attended by a single customer. (Note that the jmp file has the data in stacked format.) The firm wishes to assess alternative sequencings and bundling of courses. Use association rules to analyze these data, and interpret several of the resulting rules.

TABLE 15.8 Data on purchases of online statistics courses

Intro	DataMining	Survey	CatData	Regression	Forecast	DOE	SW
1	1	0	0	0	0	0	0
0	0	1	0	0	0	0	0
0	1	0	1	1	0	0	1
1	0	0	0	0	0	0	0
1	1	0	0	0	0	0	0
0	1	0	0	0	0	0	0
1	0	0	0	0	0	0	0
0	0	0	1	0	1	1	1
1	0	0	0	0	0	0	0
0	0	0	1	0	0	0	0
1	0	0	0	0	0	0	0

15.3 **Cosmetics Purchases.** The data shown in Table 15.9 and the output in Figure 15.8 are based on a subset of a dataset on cosmetic purchases (Cosmetics.jmp) at a large chain drugstore. The store wants to analyze associations among purchases of these items for purposes of point-of-sale display, guidance to sales personnel in promoting cross-sales, and guidance for piloting an eventual time-of-purchase electronic recommender system to boost cross-sales. Consider first only the data shown in Table 15.9, given in binary matrix form (Please note that jmp file has the data in stacked format).

TABLE 15.9 Excerpt from data on cosmetics purchases in binary matrix form

Trans. #	Bag	Blush	Nail polish	Brushes	Concealer	Eyebrow pencils	Bronzer
1	0	1	1	1	1	0	1
2	0	0	1	0	1	0	1
3	0	1	0	0	1	1	1
4	0	0	1	1	1	0	1
5	0	1	0	0	1	0	1
6	0	0	0	0	1	0	0
7	0	1	1	1	1	0	1
8	0	0	1	1	0	0	1
9	0	0	0	0	1	0	0
10	1	1	1	1	0	0	0
11	0	0	1	0	0	0	1
12	0	0	1	1	1	0	1

a. Select several values in the matrix and explain their meaning.

b. Consider the results of the association rules analysis shown in Figure 15.8.

   i. For the second row, explain the "confidence" output and how it is calculated.

   ii. For the second row, explain the "support" output and how it is calculated.

   iii. For the second row, explain the "lift" and how it is calculated.

   iv. For the second row, explain the rule that is represented there in words.

Antecedent	Consequent	Support	Confidence	Lift
Brushes	Nail Polish	15%	100%	3.571
Concealer, Eye shadow, Blush	Mascara	12%	96%	2.688
Eye shadow, Blush	Mascara	17%	93%	2.601
Blush, Mascara	Eye shadow	12%	92%	2.411
Eyeliner, Lip liner	Concealer	12%	92%	2.088
Eye shadow, Nail Polish	Mascara	11%	91%	2.545
Mascara, Lipstick	Eye shadow	11%	91%	2.386
Concealer, Blush, Mascara	Eye shadow	11%	91%	2.384
Mascara, Bronzer	Eye shadow	12%	91%	2.376
Mascara	Eye shadow	32%	90%	2.36

FIGURE 15.8   **Association rules for cosmetics purchases data**

c. Now, use the complete dataset on the cosmetics purchases (in the file *Cosmetics.jmp*). Using JMP, apply association rules to these data (for generating frequent itemsets, use *minimum support* = 0.15 and for creating association rules, use *minimum confidence* = 0.5).

   i. How many rules are generated? Sort the rules first by lift, and show the first 10 rules with the condition and consequent for each rule, as well as their support, confidence, and lift.

   ii. Interpret the first three rules in the output in words.

   iii. Reviewing the first couple of dozen rules, comment on their redundancy and how you would assess their utility.

16

CLUSTER ANALYSIS

This chapter is about the popular unsupervised learning task of clustering, where the goal is to segment the data into a set of homogeneous clusters of records for the purpose of generating insight. Separating a dataset into clusters of homogeneous records is also useful for improving performance of supervised methods, by modeling each cluster separately rather than the entire, heterogeneous dataset. Clustering is used in a vast variety of business applications, from customized marketing to industry analysis. We describe two popular clustering approaches: hierarchical clustering and k-means clustering. In hierarchical clustering, records are sequentially grouped to create clusters, based on distances between records and distances between clusters. We describe how the algorithm works in terms of the clustering process and mention several common distance metrics used. Hierarchical clustering also produces a useful graphical display of the clustering process and results, called a dendrogram. We present dendrograms and illustrate their usefulness. k-means clustering is widely used in large dataset applications. In k-means clustering, records are allocated to one of a prespecified set of clusters, according to their distance from each cluster. We describe the k-means clustering algorithm and its computational advantages. Finally, we present techniques that assist in generating insight from clustering results.

Cluster analysis in JMP: All the methods discussed in this chapter are available in the standard version of JMP.

16.1 INTRODUCTION

Cluster analysis is used to form groups or clusters of similar records based on several measurements made on these records. The key idea is to characterize the clusters in ways that would be useful for the aims of the analysis. This idea has been applied in many areas, including astronomy, archaeology, medicine, chemistry, education, psychology,

Machine Learning for Business Analytics: Concepts, Techniques, and Applications with JMP Pro®,
Second Edition. Galit Shmueli, Peter C. Bruce, Mia L. Stephens, Muralidhara Anandamurthy, and Nitin R. Patel.
© 2023 John Wiley & Sons, Inc. Published 2023 by John Wiley & Sons, Inc.

linguistics, and sociology. Biologists, for example, have made extensive use of classes and subclasses to organize species. A spectacular success of the clustering idea in chemistry was Mendeleev's periodic table of the elements.

One popular use of cluster analysis in marketing is for *market segmentation:* customers are segmented based on demographic and transaction history information, and a marketing strategy is tailored for each segment. In countries such as India, where customer diversity is extremely location-sensitive, chain stores often perform market segmentation at the store level, rather than at a chain-wide (called "microsegmentation"). Another use is for *market structure analysis:* identifying groups of similar products according to competitive measures of similarity. In marketing and political forecasting, clustering of neighborhoods using US postal zip codes has been used successfully to group neighborhoods by lifestyles. Claritas, a company that pioneered this approach, grouped neighborhoods into 40 clusters using various measures of consumer expenditure and demographics. Examining the clusters enabled Claritas to come up with evocative names, such as "Bohemian Mix," "Furs and Station Wagons," and "Money and Brains," for the groups that captured the dominant lifestyles. Knowledge of lifestyles can be used to estimate the potential demand for products (e.g., sports utility vehicles) and services (e.g., pleasure cruises). Similarly, sales organizations will derive customer segments and give them names—"personas"—to focus sales efforts.

In finance, cluster analysis can be used for creating *balanced portfolios*: given data on a variety of investment opportunities (e.g., stocks), one may find clusters based on financial performance variables such as return (daily, weekly, or monthly), volatility, beta, and other characteristics, such as industry and market capitalization. Selecting securities from different clusters can help create a balanced portfolio. Another application of cluster analysis in finance is for *industry analysis*: for a given industry, we are interested in finding groups of similar firms based on measures such as growth rate, profitability, market size, product range, and the presence in various international markets. These groups can then be analyzed in order to understand industry structure and to determine, for instance, who is a competitor.

An interesting and unusual application of cluster analysis, described in Berry and Linoff (1997), is the design of a new set of sizes for army uniforms for women in the US Army. The study came up with a new clothing size system with only 20 sizes, where different sizes fit different body types. The 20 sizes are combinations of five measurements: chest, neck, and shoulder circumference, sleeve outseam, and neck-to-buttock length (for further details, see McCullugh et al., 1998). This example is important because it shows how a completely new insightful view can be gained by examining clusters of records.

Cluster analysis can be applied to huge amounts of data. For instance, Internet search engines use clustering techniques to cluster queries that users submit. These can then be used for improving search algorithms. The objective of this chapter is to describe the key ideas underlying the most commonly used techniques for cluster analysis and to lay out their strengths and weaknesses.

Typically, the basic data used to form clusters are a table of measurements on several numerical variables, where each column represents a variable and a row represents a record. Our goal is to form groups of records so that similar records are in the same group. The number of clusters may be prespecified or determined from the data.

Example: Public Utilities

Table 16.1 gives corporate data on 22 public utilities in the United States (the definition of each variable is given in the table footnote, and the data are in Utilities.jmp). We are interested in forming groups of similar utilities. The records to be clustered are the utilities, and the clustering will be based on the eight measurements on each utility. An example where clustering would be useful is a study to predict the cost impact of deregulation. To do the requisite analysis, economists would need to build a detailed cost model of the various utilities. It would save a considerable amount of time and effort if we could cluster similar types of utilities and build detailed cost models for just one "typical" utility in each cluster and then scale up from these models to estimate results for all utilities.

For simplicity, let us consider only two of the measurements: *Sales* and *Fuel Cost*. Figure 16.1 shows a scatterplot of these two variables, with labels marking each company (labels were added by selecting all rows in the data table and using *Rows > Label* and by selecting the Company column and using *Cols > Label*). At first glance, there appear to be two or three clusters of utilities: one with utilities that have high fuel costs, a second with utilities that have lower fuel costs and relatively low sales, and a third with utilities with low fuel costs but high sales.

TABLE 16.1   Data on 22 Public Utilities

Company	Fixed	RoR	Cost	Load	Demand	Sales	Nuclear	Fuel Cost
Arizona Public Service	1.06	9.2	151	54.4	1.6	9,077	0.0	0.628
Boston Edison Co.	0.89	10.3	202	57.9	2.2	5,088	25.3	1.555
Central Louisiana Co.	1.43	15.4	113	53.0	3.4	9,212	0.0	1.058
Commonwealth Edison Co.	1.02	11.2	168	56.0	0.3	6,423	34.3	0.700
Consolidated Edison Co. (NY)	1.49	8.8	192	51.2	1.0	3,300	15.6	2.044
Florida Power & Light Co.	1.32	13.5	111	60.0	−2.2	11,127	22.5	1.241
Hawaiian Electric Co.	1.22	12.2	175	67.6	2.2	7,642	0.0	1.652
daho Power Co.	1.10	9.2	245	57.0	3.3	13,082	0.0	0.309
Kentucky Utilities Co.	1.34	13.0	168	60.4	7.2	8,406	0.0	0.862
Madison Gas & Electric Co.	1.12	12.4	197	53.0	2.7	6,455	39.2	0.623
Nevada Power Co.	0.75	7.5	173	51.5	6.5	17,441	0.0	0.768
New England Electric Co.	1.13	10.9	178	62.0	3.7	6,154	0.0	1.897
Northern States Power Co.	1.15	12.7	199	53.7	6.4	7,179	50.2	0.527
Oklahoma Gas & Electric Co.	1.09	12.0	96	49.8	1.4	9,673	0.0	0.588
Pacific Gas & Electric Co.	0.96	7.6	164	62.2	−0.1	6,468	0.9	1.400
Puget Sound Power & Light Co.	1.16	9.9	252	56.0	9.2	15,991	0.0	0.620
San Diego Gas & Electric Co.	0.76	6.4	136	61.9	9.0	5,714	8.3	1.920
The Southern Co.	1.05	12.6	150	56.7	2.7	10,140	0.0	1.108
Texas Utilities Co.	1.16	11.7	104	54.0	−2.1	13,507	0.0	0.636
Wisconsin Electric Power Co.	1.20	11.8	148	59.9	3.5	287	41.1	0.702
United Illuminating Co.	1.04	8.6	204	61.0	3.5	6,650	0.0	2.116
Virginia Electric & Power Co.	1.07	9.3	174	54.3	5.9	10,093	26.6	1.306

Notes: Fixed = fixed-charge covering ratio (income/debt), RoR = rate of return on capital, Cost = cost per kilowatt capacity in place, Load = annual load factor, Demand = peak kilowatthour demand growth from 1974 to 1975, Sales = sales (kilowatthour use per year), Nuclear = percent nuclear, Fuel Cost = total fuel costs (cents per kilowatthour).

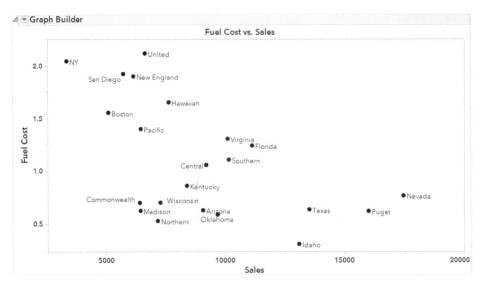

FIGURE 16.1 Scatterplot of *Fuel Cost* vs. *Sales* for the 22 utilities

We can therefore think of cluster analysis as a more formal algorithm that measures the distance between records and, according to these distances (here, nonhierarchical distances), forms clusters.

Two general types of clustering algorithms for a dataset of n records are hierarchical and nonhierarchical clustering:

Hierarchical methods can be either agglomerative or divisive. Agglomerative methods begin with n clusters and sequentially merge similar clusters until a single cluster is obtained. Divisive methods work in the opposite direction, starting with one cluster that includes all records. Hierarchical methods are especially useful when the goal is to arrange the clusters into a natural hierarchy.

Nonhierarchical methods, such as k-means, are used for a prespecified number of clusters with records assigned to each cluster. These methods are generally less computationally intensive and are therefore preferred with very large datasets.

We concentrate here on the two most popular methods: hierarchical agglomerative clustering and k-means clustering. In both cases we need to define two types of distances: distance between two records and distance between two clusters. In both cases there are a variety of metrics that can be used.

16.2   MEASURING DISTANCE BETWEEN TWO RECORDS

We denote by d_{ij} a *distance metric*, or *dissimilarity measure*, between records i and j. For record i, we have the vector of p measurements $(x_{i1}, x_{i2}, \ldots, x_{ip})$, and for record j, we have the vector of measurements $(x_{j1}, x_{j2}, \ldots, x_{jp})$. For example, we can write the measurement vector for Arizona Public Service as $[1.06, 9.2, 151, 54.4, 1.6, 9077, 0, 0.628]$.

Distances can be defined in multiple ways, but in general, the following properties are required:

Nonnegative: $d_{ij} \geq 0$

Self-proximity: $d_{ii} = 0$ (the distance from a record to itself is zero)

Symmetry: $d_{ij} = d_{ji}$

Triangle inequality: $d_{ij} \leq d_{ik} + d_{kj}$ (the distance between any pair cannot exceed the sum of distances between the other two pairs)

Euclidean Distance

The most popular distance measure is the *Euclidean distance*, d_{ij}, which between two records, i and j, is defined by

$$d_{ij} = \sqrt{(x_{i1} - x_{j1})^2 + (x_{i2} - x_{j2})^2 + \cdots + (x_{ip} - x_{jp})^2}.$$

For instance, the Euclidean distance between Arizona Public Service and Boston Edison Co. can be computed from the raw data by

$$d_{12} = \sqrt{(1.06 - 0.89)^2 + (9.2 - 10.3)^2 + (151 - 202)^2 + \cdots + (0.628 - 1.555)^2}$$
$$= 3989.408.$$

Standardizing Numerical Measurements

The measure computed above is highly influenced by the scale of each variable so that variables with larger scales (e.g., *Sales*) have a much greater influence over the total distance. It is therefore customary to *standardize* (or *normalize*) continuous measurements before computing the Euclidean distance. This converts all measurements to the same scale. Standardizing a measurement means subtracting the average and dividing by the standard deviation (standardized values are also called *z-scores*).[1] For instance, the average sales amount across the 22 utilities is 8914.045 and the standard deviation is 3549.984. The standardized sales for Arizona Public Service is therefore $(9077 - 8914.045)/3549.984 = 0.046$.

Returning to the simplified utilities data with only two measurements (Sales and Fuel Cost), we first standardize the measurements (see Table 16.2) and then compute the Euclidean distance between each pair. Table 16.3 gives these pairwise distances for the first five utilities. A similar table can be constructed for all 22 utilities.

Other Distance Measures for Numerical Data

It is important to note that the choice of the distance measure plays a major role in cluster analysis. The main guideline is domain dependent: what exactly is being measured?

[1] When there are outliers, *robust* standardization can be applied. This reduces the influence of outliers on the estimates of the means and standard deviations used to standardize the data. In JMP, this is called *Standardize Robustly*.

TABLE 16.2   Original and Standardized Measurements for Sales and Fuel Cost

Company	Sales	Fuel Cost	NormSales	NormFuel
Arizona Public Service	9,077	0.628	0.0459	−0.8537
Boston Edison Co.	5,088	1.555	−1.0778	0.8133
Central Louisiana Co.	9,212	1.058	0.0839	−0.0804
Commonwealth Edison Co.	6,423	0.7	−0.7017	−0.7242
Consolidated Edison Co. (NY)	3,300	2.044	−1.5814	1.6926
Florida Power & Light Co.	11,127	1.241	0.6234	0.2486
Hawaiian Electric Co.	7,642	1.652	−0.3583	0.9877
Idaho Power Co.	13,082	0.309	1.1741	−1.4273
Kentucky Utilities Co.	8,406	0.862	−0.1431	−0.4329
Madison Gas & Electric Co.	6,455	0.623	−0.6927	−0.8627
Nevada Power Co.	17,441	0.768	2.4020	−0.6019
New England Electric Co.	6,154	1.897	−0.7775	1.4283
Northern States Power Co.	7,179	0.527	−0.4887	−1.0353
Oklahoma Gas & Electric Co.	9,673	0.588	0.2138	−0.9256
Pacific Gas & Electric Co.	6,468	1.4	−0.6890	0.5346
Puget Sound Power & Light Co.	15,991	0.62	1.9935	−0.8681
San Diego Gas & Electric Co.	5,714	1.92	−0.9014	1.4697
The Southern Co.	10,140	1.108	0.3453	0.0095
Texas Utilities Co.	13,507	0.636	1.2938	−0.8393
Wisconsin Electric Power Co.	7,287	0.702	−0.4583	−0.7206
United Illuminating Co.	6,650	2.116	−0.6378	1.8221
Virginia Electric & Power Co.	10,093	1.306	0.3321	0.3655
Mean	8,914.05	1.10	0.00	0.00
Standard deviation	3,549.98	0.56	1.00	1.00

TABLE 16.3   Distance Matrix Between Pairs of the First Five Utilities, Using Euclidean Distance and Standardized Measurements

	Arizona	Boston	Central	Commonwealth	Consolidated
Arizona	0				
Boston	2.01	0			
Central	0.77	1.47	0		
Commonwealth	0.76	1.58	1.02	0	
Consolidated	3.02	1.01	2.43	2.57	0

How are the different measurements related? What scale should it be treated as (numerical, ordinal, or nominal)? Are there outliers? Finally, depending on the goal of the analysis, should the clusters be distinguished mostly by a small set of measurements, or should they be separated by multiple measurements that weight moderately?

Although Euclidean distance is the most widely used distance, it has three main features that need to be kept in mind. First, as mentioned above, it is highly scale dependent. Changing the units of one variable (e.g., from cents to dollars) can have a huge influence on the results. Standardizing is therefore a common solution. But unequal weighting should be considered if we want the clusters to depend more on certain measurements and less on others. The second feature of Euclidean distance is that it completely ignores the relationship between the measurements. Thus, if the measurements are in fact strongly correlated, a different distance (e.g., the statistical distance, described below) is likely to be a better choice. The third is that Euclidean distance is sensitive to outliers. If the data are believed to contain outliers and careful removal is not a choice, the use of more robust distances (e.g., the Manhattan distance, described below) is preferred.

For these reasons, additional popular distance metrics are often used.

Correlation-Based Similarity Sometimes it is more natural or convenient to work with a similarity measure between records rather than distance, which measures dissimilarity. A popular similarity measure is the square of the correlation coefficient, r_{ij}^2, where the correlation coefficient is defined by

$$
r_{ij} \equiv \frac{\sum_{m=1}^{p} (x_{im} - \overline{x}_m)(x_{jm} - \overline{x}_m)}{\sqrt{\sum_{m=1}^{p} (x_{im} - \overline{x}_m)^2 \sum_{m=1}^{p} (x_{jm} - \overline{x}_m)^2}}.
$$

Such measures can always be converted to distance measures. In the example above we could define a distance measure $d_{ij} = 1 - r_{ij}^2$.

Statistical Distance (also called *Mahalanobis Distance*) This metric has an advantage over the other metrics mentioned in that it takes into account the correlation between measurements. With this metric, measurements that are highly correlated with other measurements do not contribute as much as those that are uncorrelated or mildly correlated. The statistical distance between records i and j is defined as

$$
d_{i,j} = \sqrt{(\mathbf{x}_i - \mathbf{x}_j)' S^{-1}(\mathbf{x}_i - \mathbf{x}_j)},
$$

where $\mathbf{x}_i$ and $\mathbf{x}_j$ are p-dimensional vectors of the measurements values for records i and j, respectively, and S is the covariance matrix for these vectors (', a transpose operation, simply turns a column vector into a row vector). S^{-1} is the inverse matrix of S, which is the p-dimensional extension to division. (For further information on statistical distance, see Chapter 12).

Manhattan Distance ("City Block") This distance looks at the absolute differences rather than squared differences and is defined by

$$
d_{ij} = \sum_{m=1}^{p} | x_{im} - x_{jm} |.
$$

Maximum Coordinate Distance This distance looks only at the measurement on which records i and j deviate most. It is defined by

$$
d_{ij} = \max_{m=1,2,\ldots,p} | x_{im} - x_{jm} |.
$$

Distance Measures for Categorical Data

In the case of measurements with binary values, it is more intuitively appealing to use similarity measures than distance measures. Suppose that we have binary values for all the p measurements, and for records i and j we have the following 2×2 table:

		Record j		
		0	1	
Record i	0	a	b	$a+b$
	1	c	d	$c+d$
		$a+c$	$b+d$	p

where a denotes the number of variables for which records i and j each have value 0 for that variable (measurement absent), d is the number of variables for which the two records each have a value 1 for that variable (measurement present), and so on. The most useful similarity measures in this situation are the following:

Matching coefficient: $(a+d)/p$.

Jaccard's coefficient: $d/(b+c+d)$. This coefficient ignores zero matches. This is desirable when we do not want to consider two people to be similar simply because a large number of characteristics are absent in both. For example, if *owns a Corvette* is one of the variables, a matching "yes" would be evidence of similarity, but a matching "no" tells us little about whether the two people are similar.

Distance Measures for Mixed Data

When the measurements are mixed (some continuous and some binary), a similarity coefficient suggested by Gower is very useful. *Gower's similarity measure* is a weighted average of the distances computed for each variable, after scaling each variable to a [0,1] scale. It is defined as

$$s_{ij} = \frac{\sum\limits_{m=1}^{p} w_{ijm} s_{ijm}}{\sum\limits_{m=1}^{p} w_{ijm}},$$

where s_{ijm} is the similarity between records i and j on measurement m, and w_{ijm} is a binary weight given to the corresponding distance.

The similarity measures s_{ijm} and weights w_{ijm} are computed as follows:

1. For continuous measurements, $s_{ijm} = 1 - \frac{|x_{im} - x_{jm}|}{\max(x_m) - \min(x_m)}$ and $w_{ijm} = 1$ unless the value for measurement m is unknown for one or both of the records, in which case $w_{ijm} = 0$.
2. For binary measurements, $s_{ijm} = 1$ if $x_{im} = x_{jm} = 1$ and 0 otherwise. $w_{ijm} = 1$ unless $x_{im} = x_{jm} = 0$.

3. For nonbinary categorical measurements, $s_{ijm} = 1$ if both records are in the same category, and otherwise $s_{ijm} = 0$. As in continuous measurements, $w_{ijm} = 1$ unless the category for measurement m is unknown for one or both of the records, in which case $w_{ijm} = 0$.

16.3 MEASURING DISTANCE BETWEEN TWO CLUSTERS

We define a cluster as a set of one or more records. How do we measure distance between clusters? The idea is to extend measures of *distance between records* into *distances between clusters*. Consider cluster A, which includes the m records $A_1, A_2, \dots, A_m$, and cluster B, which includes n records $B_1, B_2, \dots, B_n$. The most widely used measures of distance between clusters are:

Minimum Distance

The distance between the pair of records A_i and B_j that are closest:

$$\min(\text{distance}(A_i, B_j)), \quad i = 1, 2, \dots, m; \quad j = 1, 2, \dots, n.$$

Maximum Distance

The distance between the pair of records A_i and B_j that are farthest:

$$\max(\text{distance}(A_i, B_j)), \quad i = 1, 2, \dots, m; \quad j = 1, 2, \dots, n.$$

Average Distance

The average distance of all possible distances between records in one cluster and records in the other cluster:

$$\text{Average}(\text{distance}(A_i, B_j)), \quad i = 1, 2, \dots, m; \quad j = 1, 2, \dots, n.$$

Centroid Distance

The distance between the two cluster centroids. A *cluster centroid* is the vector of measurement averages across all the records in that cluster. For cluster A, this is the vector $\overline{x}_A = \left[(1/m \sum_{i=1}^{m} x_{1i}, \dots, 1/m \sum_{i=1}^{m} x_{pi}) \right]$. The centroid distance between clusters A and B is

$$\text{distance}(\overline{x}_A, \overline{x}_B).$$

Minimum distance, maximum distance, and centroid distance are illustrated visually for two dimensions with a map of Portugal and France in Figure 16.2.

For instance, consider the first two utilities (Arizona, Boston) as cluster A and the next three utilities (Central, Commonwealth, Consolidated) as cluster B. Using the standardized scores in Table 16.2 and the distance matrix in Table 16.3, we can compute each of the distances described above. Using Euclidean distance for each distance calculation, we get:

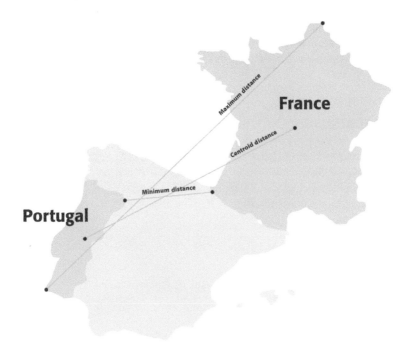

FIGURE 16.2   **Two-dimensional representation of several different distance measures between Portugal and France**

- The closest pair is Arizona and Commonwealth, and therefore the minimum distance between clusters A and B is 0.76.
- The farthest pair is Arizona and Consolidated, and therefore the maximum distance between clusters A and B is 3.02.
- The average distance is $(0.77 + 0.76 + 3.02 + 1.47 + 1.58 + 1.01)/6 = 1.44$.
- The centroid of cluster A is

$$\left[\frac{0.0459 - 1.0778}{2}, \frac{-0.8537 + 0.8133}{2} \right] = [-0.516, -0.020],$$

and the centroid of cluster B is

$$\left[\frac{0.0839 - 0.7017 - 1.5814}{3}, \frac{-0.0804 - 0.7242 + 1.6926}{3} \right]$$
$$= [-0.733, 0.296].$$

The distance between the two centroids is then

$$\sqrt{(-0.516 + 0.733)^2 + (-0.020 - 0.296)^2} = 0.38.$$

In deciding among clustering methods, domain knowledge is key. If you have good reason to believe that the clusters might be chain- or sausage-like, minimum distance would be a good choice. This method does not require that cluster members all be close to one another, only that the new members being added be close to one of the existing members. An example of an application where this might be the case would be characteristics of crops planted in long rows or disease outbreaks along navigable waterways that are the main areas of settlement in a region. Another example is laying and finding mines (land or marine). Minimum distance is also fairly robust to small deviations in the distances. However, adding or removing data can influence it greatly.

Maximum and average distance are better choices if you know that the clusters are more likely to be spherical (e.g., customers clustered on the basis of numerous variables). If you do not know the probable nature of the cluster, these are good default choices, since most clusters tend to be spherical in nature.

We now move to a more detailed description of the two major types of clustering algorithms: hierarchical (agglomerative) and nonhierarchical.

16.4 HIERARCHICAL (AGGLOMERATIVE) CLUSTERING

The idea behind hierarchical agglomerative clustering is to start with each cluster comprising exactly one record and then progressively agglomerating (combining) the two nearest clusters until there is just one cluster left at the end, which consists of all the records.

Returning to the small example of five utilities and two measures (Sales and Fuel Cost) and using the distance matrix (Table 16.3), the first step in the hierarchical clustering would join Arizona and Commonwealth, which are the closest (using standardized measurements and Euclidean distance). Next we would recalculate a 4×4 distance matrix that would have the distances between these four clusters: {Arizona, Commonwealth}, {Boston}, {Central}, and {Consolidated}. At this point we use a measure of distance between clusters, such as the ones described in Section 16.3. Each of these distances (minimum, maximum, average, and centroid distance) can be implemented in the hierarchical scheme as described below.

HIERARCHICAL AGGLOMERATIVE CLUSTERING ALGORITHM

1. Start with n clusters (each record = cluster).
2. The two closest records are merged into one cluster.
3. At every step, the two clusters with the smallest distance are merged. This means that either single records are added to existing clusters or two existing clusters are combined.

Single Linkage

In *single-linkage clustering*, the distance measure that we use is the minimum distance (the distance between the nearest pair of records in the two clusters, one record in each cluster). In our utilities example, we would compute the distances between each of {Boston}, {Central}, and {Consolidated} with {Arizona, Commonwealth} to create the 4×4 distance matrix shown in Table 16.4.

The next step would consolidate {Central} with {Arizona, Commonwealth} because these two clusters are closest. The distance matrix will again be recomputed (this time it will be 3×3), and so on.

TABLE 16.4   Distance Matrix After Arizona and Commonwealth Consolidation Cluster Together, Using Single Linkage

	Arizona-Commonwealth	Boston	Central	Consolidated
Arizona-Commonwealth	0			
Boston	min(2.01,1.58)	0		
Central	min(0.77,1.02)	1.47	0	
Consolidated	min(3.02,2.57)	1.01	2.43	0

This method has a tendency to cluster together at an early stage records that are distant from each other because of a chain of intermediate records in the same cluster. Such clusters have elongated sausage-like shapes when visualized as objects in space.

Complete Linkage

In *complete linkage clustering*, the distance between two clusters is the maximum distance (between the farthest pair of records). If we used complete linkage with the five-utilities example, the recomputed distance matrix would be equivalent to Table 16.4, except that the "min" function would be replaced with a "max."

This method tends to produce clusters at the early stages with records that are within a narrow range of distances from each other. If we visualize them as objects in space, the records in such clusters would have roughly spherical shapes.

Average Linkage

Average linkage clustering is based on the average distance between clusters (between all possible pairs of records). If we used average linkage with the five-utilities example, the recomputed distance matrix would be equivalent to Table 16.4, except that the "min" function would be replaced with "average." This method is also called *Unweighted Pair-Group Method using Centroids* (UPGMC).

Note that unlike average linkage, the results of the single and complete linkage methods depend only on the ordering of the inter-record distances. Linear transformations of the distances (and other transformations that do not change the ordering) do not affect the results.

Centroid Linkage

Centroid linkage clustering is based on centroid distance, where clusters are represented by their mean values for each variable, which forms a vector of means. The distance between two clusters is the distance between these two vectors. In average linkage, each pairwise distance is calculated, and the average of all such distances is calculated. In contrast, in centroid distance clustering, just one distance is calculated: the distance between group means. This method is also called *Unweighted Pair-Group Method using Centroids* (UPGMC).

Ward's Method

Ward's method is also agglomerative, in that it joins records and clusters together progressively to produce larger and larger clusters, but it operates slightly differently from the general approach described above. Ward's method considers the "loss of information" that occurs when records are clustered together. When each cluster has one record, there is no loss of information and all individual values remain available. When records are joined together and represented in clusters, information about an individual record is replaced by the information for the cluster to which it belongs. To measure loss of information, Ward's method employs "error sum of squares" (ESS) that measures the difference between individual records and a group mean.

This is easiest to see in univariate data. For example, consider the values (2, 6, 5, 6, 2, 2, 2, 2, 0, 0, 0) with a mean of 2.5. Their ESS is equal to

$$(2 - 2.5)^2 + (6 - 2.5)^2 + (5 - 2.5)^2 + \ldots + (0 - 2.5)^2 = 50.5.$$

The loss of information associated with grouping the values into a single group is therefore 50.5. Now group the records into four groups: (0, 0, 0), (2, 2, 2, 2), (5), and (6, 6). The loss of information is the sum of the ESS's for each group, which is 0 (each record in each group is equal to the mean for that group, so the ESS for each group is 0). Thus clustering the 10 records into 4 clusters results in no loss of information, and this would be the first step in Ward's method. In moving to a smaller number of clusters Ward's method would choose the configuration that results in the smallest incremental loss of information.

Ward's method tends to result in convex clusters that are of roughly equal size, which can be an important consideration in some applications (e.g., in establishing meaningful customer segments).

JMP also offers two computational shortcuts to Ward's method called *Fast Ward* and *Hybrid Ward*. Fast Ward is automatically used if there are more than 2000 records to be clustered. Hybrid Ward is useful when clustering a huge number of records (hundreds of thousands). Advanced options for Hybrid Ward method are also available.

HIERARCHICAL CLUSTERING IN JMP

Hierarchical clustering is available from the *Analyze > Clustering* menu. The hierarchical clustering *Method* defines the distance measure used. The JMP Clustering dialog, with the available *Methods* and other options such as *Data Format, Standardize By, Standardize Robustly* (to reduce the influence of outliers), *Missing value imputation*, and *Advanced Options*, is shown in Figure 16.3.

Dendrograms: Displaying Clustering Process and Results

A *dendrogram* is a treelike diagram that summarizes the process of clustering. At the side are the records. Similar records are joined by lines whose horizontal length reflects the

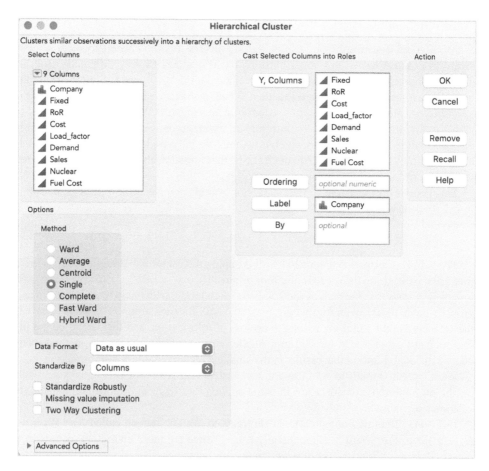

FIGURE 16.3   **The hierarchical clustering dialog for the utilities data, showing selections for the *Single* method and *Standardize By Columns***

distance between the records. Figure 16.4 shows the dendrogram that results from clustering all 22 utilities using the 8 standardized variables, Euclidian distance, and single linkage.

JMP also produces a Distance Graph at the bottom of the dendrogram. This graph has a point for each step where two clusters are joined into a single cluster. The horizontal coordinates represent the numbers of clusters, which decrease from left to right. The vertical coordinate of the point is the distance between the clusters that were joined at the given step.

The number of clusters created can be changed by dragging the diamond-shaped icon within the dendrogram or by using a red triangle option. Visually, this means drawing a vertical line on a dendrogram. Records with connections to the left of the vertical line (i.e., their distance is smaller than the cutoff distance) belong to the same cluster. For example,

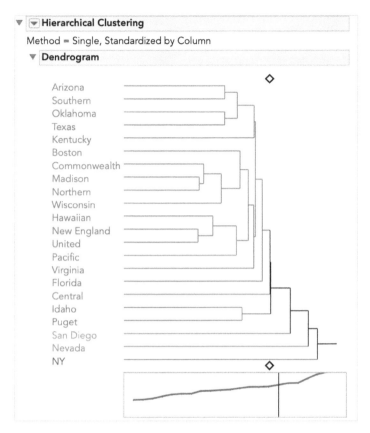

FIGURE 16.4 Dendrogram: *single linkage* for all 22 utilities, using all 8 measurements and 6 clusters

setting the number of clusters to 6 in Figure 16.4 results in six clusters. Using the *Color Clusters* option from the red triangle makes it easier to see which records make up the different clusters.

The six clusters in the single linkage dendrogram are (from top to bottom):

{All others}
{Central}
{Puget, Idaho}
{San Diego}
{Nevada}
{NY}

The average linkage dendrogram is shown in Figure 16.5. When we use average linkage with six clusters, the resulting clusters are slightly different. While some records remain in the same cluster in both methods (e.g., San Diego and NY), others change. The six clusters are (from top to bottom in the dendrogram):

{Arizona, Southern, Oklahoma, Texas, Florida, Central, Kentucky}
{Commonwealth, Wisconsin, Madison, Northern, Virginia}
{Boston, New England, United, Pacific, Hawaiian}
{San Diego}
{NY}
{Idaho, Puget, Nevada}

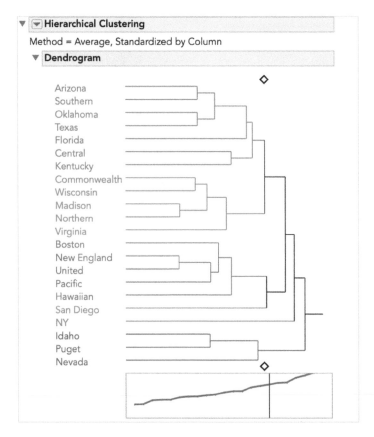

FIGURE 16.5 Dendrogram: *average linkage* for all 22 utilities, using all 8 measurements and 6 clusters

Note that if we wanted five clusters, they would be identical to the six above, with the exception that the first two clusters would be merged into one cluster (Central would be grouped into the first cluster). In general, all hierarchical methods have clusters that are nested within each other as the number of clusters decreases. This is a valuable property for interpreting clusters and is essential in certain applications, such as taxonomy of varieties of living organisms.

Validating Clusters

One important goal of cluster analysis is to come up with *meaningful clusters*. Since there are many variations that can be chosen, it is important to make sure that the resulting clusters are valid, in the sense that they really generate some insight. To see whether the cluster analysis is useful, consider each of the following aspects:

1. *Cluster interpretability:* Is the interpretation of the resulting clusters reasonable? To interpret the clusters, explore the characteristics of each cluster by

   a. Obtaining summary statistics (e.g., average, min, max) from each cluster on each measurement that was used in the cluster analysis (use *Cluster Summary* from the red triangle to display means and standard deviations—see Figure 16.6).

   b. Examining the clusters for separation along some common feature (variable) that was not used in the cluster analysis.

   c. Labeling the clusters: Based on the interpretation, trying to assign a name or label to each cluster (use *Save Clusters* to save the cluster numbers to the data table, then use *Cols > Recode* to rename the clusters).

2. *Cluster stability:* Do cluster assignments change significantly if some of the inputs are altered slightly? Another way to check stability is to partition the data and see how well clusters formed based on one part apply to the other part. To do this:

   a. Partition the data, and hide and exclude partition B.

   b. Cluster partition A.

   c. Use *Save Formula for Closest Cluster* to save the cluster numbers and cluster formulas to the data table. Cluster numbers for partition A will be saved to the data table, and clusters for partition B will be generated (each record is assigned to the cluster with the closest centroid).

   d. Use *Graph Builder* or other tools to assess how consistent the cluster assignments are for the two partitions and for the dataset as a whole.

3. *Cluster separation:* Examine the ratio of between-cluster variation to within-cluster variation to see whether the separation is reasonable. There exist statistical tests for this task (an F-ratio), but their usefulness is somewhat controversial (and these tests are not available in JMP).

4. *Number of clusters:* The number of resulting clusters must be useful, given the purpose of the analysis. For example, suppose that the goal of the clustering is to identify categories of customers and assign labels to them for market segmentation purposes.

▼ Cluster Summary

▼ Cluster Means

Cluster	Count	Fixed	RoR	Cost	Load_factor	Demand	Sales	Nuclear	Fuel Cost
1	7	1.2	12.5	127.6	55.5	1.7	10163.1	3.214	0.874
2	5	1.1	11.5	177.2	55.4	3.8	7487.4	38.280	0.772
3	5	1.0	9.9	184.6	62.1	2.3	6400.4	5.240	1.724
4	1	0.76	6.4	136.0	61.9	9.0	5714.0	8.300	1.920
5	1	1.5	8.8	192.0	51.2	1.0	3300.0	15.600	2.044
6	3	1.0	8.9	223.3	54.8	6.3	15504.7	0.000	0.566

FIGURE 16.6   Cluster Summary for hierarchical clustering with six clusters

If the marketing department can only manage to sustain three different marketing presentations, it would probably not make sense to identify more than three clusters.

In some cases, the *Distance* (or *Scree*) graph at the bottom of the dendrogram can be used to determine the number of clusters. An elbow or sharp upward bend in this plot may indicate the optimal number of clusters. The scree plot for the Utilities data, shown at the bottom of Figure 16.5 (as well as in Figure 16.4), shows a relatively steady increase rather than a sharp bend. In this case, the scree plot isn't overly useful. Here, the context of the study and practical considerations are more important in determining the best number of clusters.

Returning to the utilities example, we notice that both methods (single and average linkage) identify New York and San Diego as singleton clusters. Also, both dendrograms imply that a reasonable number of clusters in this dataset is six. One insight that can be derived from the average linkage clustering is that clusters tend to group geographically. The four nonsingleton clusters form (approximately) a southern group, a northern group, an east/west seaboard group, and a west group.

We can further characterize each of the clusters by producing summary statistics (see Figure 16.6) and other graphical displays. A *parallel plot* (select *Parallel Coord Plots* from the red triangle) is useful for understanding the profiles of the clusters across the different variables (see Figure 16.7). For example, we can see that clusters 1, 3, and 6 use very little or no nuclear power, while cluster 2 is high in nuclear power usage.

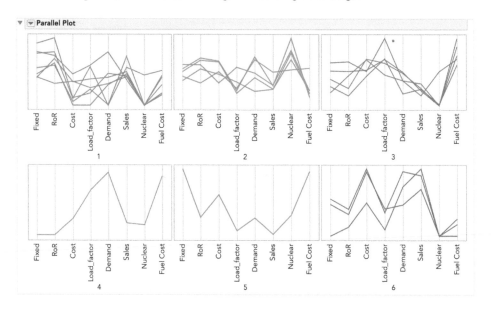

FIGURE 16.7   **Parallel plot for the 22 utilities, from average linking with 6 clusters**

Saving clusters to the data table allows us to use other graphical and exploratory tools to characterize the clusters and explore the clusters across the different variables, such as the *Graph Builder*. The *Column Switcher*, which is an option under the top red triangle in all JMP graphing and analysis platforms under *Script*, can be used in conjunction with the *Graph Builder* to explore clusters across the different variables. In Figure 16.8, we can see that cluster 6 has the highest sales but also the highest variability.

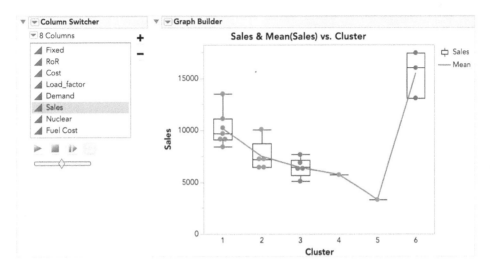

FIGURE 16.8   **Exploring the saved clusters using the *Graph Builder* and the *Column Switcher***

Two-Way Clustering

In addition to clustering the data (the records), the variables can also be clustered. Variable clustering creates groups of variables that have similar characteristics. The resulting dendrogram can be useful if the variables are on the same scale. The *Two-Way Clustering* option (from the top red triangle) clusters both records and variables. The resulting color map (or heatmap) shows the values of the data colored across the value range for each measurement [we've selected a white (low) to green (high) color theme in this example]. Figure 16.9 shows a color map of the four clusters and two singletons, highlighting the different profile that each cluster has in terms of the eight variables. We see, for instance, that cluster 2 (with Commonwealth and Wisconsin) is characterized by utilities with a high percentage of nuclear power, cluster 1 (the cluster at the top) is characterized by high fixed cost and RoR and low nuclear, and the last cluster has low nuclear but high cost per kilowatt, high demand, and high sales. Note that these variables are not on the same scale, so we ignore the variable clustering.

Limitations of Hierarchical Clustering

Hierarchical clustering is very appealing in that it does not require specification of the number of clusters and in this sense is purely data driven. The ability to represent the clustering process and results through dendrograms is also an advantage of this method, as it is easier to understand and interpret. There are, however, a few limitations to consider:

1. Hierarchical clustering requires the computation and storage of a $n \times n$ distance matrix. For very large datasets, this can be expensive and slow.
2. The hierarchical algorithm makes only one pass through the data. This means that records that are allocated incorrectly early in the process cannot be reallocated subsequently.
3. Hierarchical clustering also tends to have low stability. Reordering data or dropping a few records can lead to a different solution.

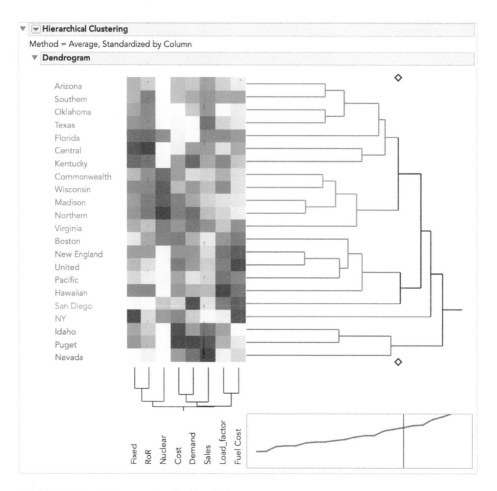

FIGURE 16.9   Color map for the 22 utilities. Rows are sorted by the 6 clusters from average linkage clustering. Darker denotes higher values for a measurement

4. The choice of distance between clusters, single and complete linkage, are robust to changes in the distance metric (e.g., Euclidean, statistical distance) as long as the relative ordering is kept. In contrast, average linkage is more influenced by the choice of distance metric and might lead to completely different clusters when the metric is changed.

5. Hierarchical clustering is sensitive to outliers. (However, the *Standardize Robustly* option in the *Hierarchical Clustering* dialog can be used to reduce the influence of outliers on the estimated means and standard deviations.)

16.5   NONHIERARCHICAL CLUSTERING: THE *K*-MEANS ALGORITHM

A nonhierarchical approach to forming good clusters is to prespecify a desired number of clusters, k, and to assign each case to one of k clusters so as to minimize a measure of dispersion within the clusters. In other words, the goal is to divide the sample into a predetermined number k of nonoverlapping clusters so that clusters are as homogeneous as possible with respect to the measurements used.

A common measure of within-cluster dispersion is the sum of distances (or sum of squared Euclidean distances) of records from their cluster centroid. The problem can be set up as an optimization problem involving integer programming, but because solving integer programs with a large number of variables is time-consuming, clusters are often computed using a fast, heuristic method that produces good (although not necessarily optimal) solutions. The k-means algorithm is one such method.

The k-means algorithm starts with an initial partition of the records into k clusters. Subsequent steps modify the partition to reduce the sum of the distances of each record from its cluster centroid. The modification consists of allocating each record to the nearest of the k centroids of the previous partition. This leads to a new partition for which the sum of distances is smaller than before. The means of the new clusters are computed, and the improvement step is repeated until the improvement is very small.

---

k-MEANS CLUSTERING ALGORITHM

1. Start with k initial clusters (user chooses k).

2. At every step, each record is reassigned to the cluster with the "closest" centroid.

3. Recompute the centroids of clusters that lost or gained a record, and repeat step 2.

4. Stop when moving any more records between clusters increases cluster dispersion.

Returning to the example with the five utilities and two measurements, let us assume that $k = 2$ and that the initial clusters are A = {Arizona, Boston} and B = {Central, Commonwealth, Consolidated}. The cluster centroids computed in Section 14.4 are

$$\overline{x}_A = [-0.516, -0.020] \text{ and } \overline{x}_B = [-0.733, 0.296].$$

The distance of each record from each of these two centroids is shown in Table 16.5. We see that Boston is closer to cluster B and that Central and Commonwealth are each

TABLE 16.5   Distance of Each record from Each Centroid

	Distance from Centroid A	Distance from Centroid B
Arizona	1.0052	1.3887
Boston	1.0052	0.6216
Central	0.6029	0.8995
Commonwealth	0.7281	1.0207
Consolidated	2.0172	1.6341

closer to cluster A. We therefore move each of these records to the other cluster and obtain A = {Arizona, Central, Commonwealth} and B = {Consolidated, Boston}. Recalculating the centroids gives

$$\overline{x}_A = [-0.191, -0.553] \text{ and } \overline{x}_B = [-1.33, 1.253].$$

The distance of each record from each of the newly calculated centroids is given in Table 16.6. At this point we stop because each record is allocated to its closest cluster.

TABLE 16.6   Distance of Each record from Each Newly Calculated Centroid

	Distance from Centroid A	Distance from Centroid B
Arizona	0.3827	2.5159
Boston	1.6289	0.5067
Central	0.5463	1.9432
Commonwealth	0.5391	2.0745
Consolidated	2.6412	0.5067

Choosing the Number of Clusters (k)

The choice of the number of clusters can either be driven by external considerations (previous knowledge, practical constraints, etc.), or we can try a few different values for k and compare the resulting clusters. After choosing k, the n records are partitioned into these initial clusters. The number of clusters in the data is generally not known, so it is a good idea to run the algorithm with different values for k that are near the number of clusters that one expects from the data. Note that the clusters obtained using different values of k will not be nested (unlike those obtained by hierarchical methods).

K-MEANS CLUSTERING IN JMP

k-Means clustering is available from *Analyze > Clustering > K Means Cluster*. The dialog for k-means clustering in JMP is shown in Figure 16.10 (top). The *Columns Scaled Individually* option is selected to normalize the measurements. This dialog produces a *Control Panel* for indicating the upper and lower bound for the number of clusters to form (bottom, in Figure 16.10). A separate analysis is provided for each number. In this example, we form partitions starting with three clusters and ending with eight clusters. The *Single Step* option allows the user to form one partition at a time, starting with the lowest number of clusters specified.

The results of running the k-means algorithm for all 22 utilities and eight measurements with $k = 6$ are shown in Figure 16.11. The *Parallel Coordinate Plot* was selected using a red triangle option.

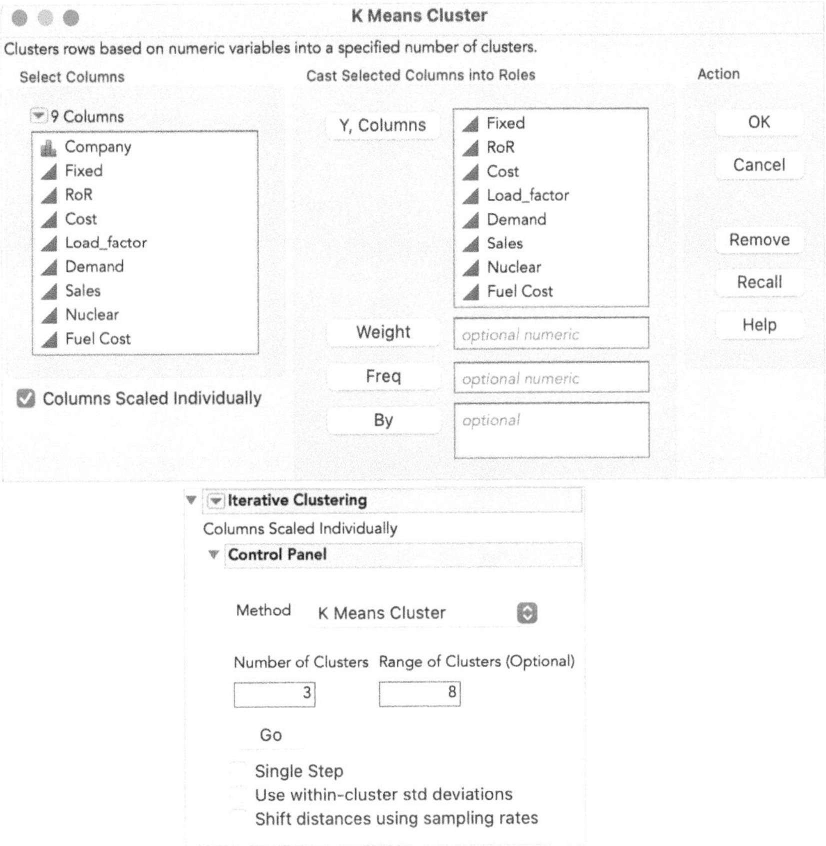

FIGURE 16.10 *k*-**Means cluster dialog (top) and control panel (bottom) for specifying the number of clusters**

As in the results from the hierarchical clustering, we see once again that San Diego and New York are singleton clusters. Other clusters are very similar to those that emerged in the hierarchical clustering with average linkage, and cluster 6 is identical. In JMP, the company can be seen by clicking on a line in any plot (since Company has been selected as a column label using *Cols > Label*).

To examine the clusters more closely, we save the clusters to the data table (using the red triangle). This saves both the cluster number and the standardized distance within the cluster. Now we can graph the distances using the *Graph Builder* (see Figure 16.12). We see that for clusters with more than one record, cluster 4 has the largest within-cluster variability. The labels in the graph were added by selecting all rows in the data table and using *Rows > Label* (after setting Company as a column label).

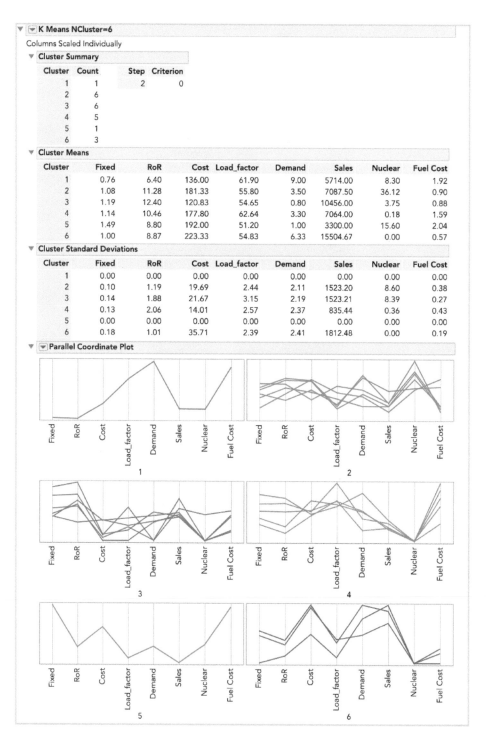

FIGURE 16.11   Output for *k*-means clustering of 22 utilities into 6 clusters (sorted by cluster ID)

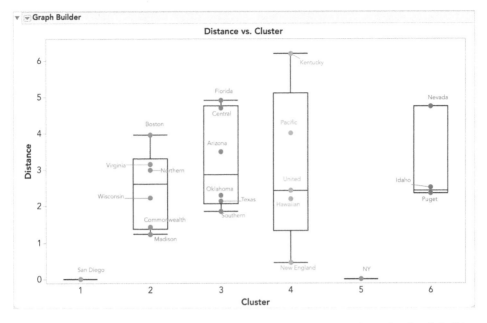

FIGURE 16.12   **Graph of standardized distances vs. cluster number from the *Graph Builder*, after saving clusters to the data table**

To characterize the resulting clusters, we examine the cluster centroids in the *Parallel Coordinate Plots* (or "profile plots") in Figure 16.13. This is provided at the bottom of the Parallel Coordinates Plot for the individual clusters.

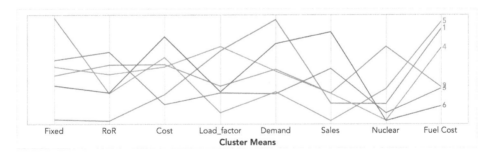

FIGURE 16.13   **Parallel coordinates (profile) plot of cluster centroids for *k*-means with *k*=6**

We can see, for instance, that cluster 5 (light blue line) has high average Fixed and Fuel Cost, and low average Sales and Load Factor. In contrast, cluster 1 has the low Nuclear and Sales, but the highest Demand.

We can also use *Tabulate* to inspect the information on the within-cluster dispersion for the saved clusters. From Figure 16.14, we see that cluster 3 has the largest average distance, and it includes six records. In comparison, cluster 2, with six records, has a smaller within-cluster average distance. This means that cluster 2 is more homogeneous than cluster 3.

▼ ⊡Tabulate

Cluster	Distance	
	Mean	Std Dev
1	0	.
2	2.49	1.05
3	3.23	1.35
4	3.06	2.15
5	0	.
6	3.18	1.35

FIGURE 16.14   Summary of cluster centroids, created using the saved clusters and *Tabulate*

From the distances between clusters, we can learn about the separation of the different clusters. In Figure 16.15, we see two plots that are available from the cluster platform. The *biplot* (top) is a two-dimensional plot of the clusters based on a principal component analysis. The circles for each cluster in the biplot are sized in proportion to the number of records in the cluster, and the shaded area is a *density contour*, which indicates where 90% of the records in the cluster would fall. Note that the different markers were applied by selecting *Color or Mark by Column* from the *Rows* menu.

The scatterplot matrix (bottom) shows the clusters across pairs of the measurements. While it's a bit difficult to see this on the printed page, from these plots we learn that cluster 6 is very different from the other clusters. However, cluster 2 is similar to clusters 3 and 4 in the biplot. From the scatterplot, we learn that cluster 2 is most unique in the percent Nuclear, but is similar to clusters 3 and 4 in the other measurements. This might lead us to examine the possibility of merging two of the clusters. Cluster 1, which is a singleton cluster, appears to be very far from all the other clusters (this can be seen most clearly in the biplot).

When the number of clusters is not predetermined by domain requirements, we can use a criterion provided by JMP for determining the number of clusters to produce, the *CCC*, or *Cubic Clustering Criterion* (Sarle, 1983). CCC is a criterion designed by SAS Institute to choose k in an automated way. "Larger positive values of the CCC indicate a better solution, as it shows a larger difference from a uniform (no clusters) distribution" (Lim et al. 2006). CCC is reported in *Cluster Comparison* section at the top of the *K-Means Cluster* output window (see Figure 16.16) when a range of values for k is used. Here, the CCC identifies $k = 6$ as the best number of clusters.

We can also use a graphical approach to evaluating different numbers of clusters. An "elbow chart" is a line chart, similar to a scree plot, depicting the decline in cluster heterogeneity as we add more clusters. Figure 16.17 shows the average deviation of within-cluster distance (standardized) for different choices of k. Moving from 1 to 6 tightens clusters considerably (reduces the average within-cluster distance). Adding more than six clusters brings less reduction in the within-cluster homogeneity.

Finally, we can use the information on the distances between the final clusters to evaluate the cluster validity. The ratio of the sum of within-cluster squared distances to cluster means for a given k to the sum of squared distances to the mean of all the records (in other words, $k = 1$) is a useful measure for the usefulness of the clustering. If the ratio is near 1.0, the clustering has not been very effective, whereas if it is small, we have well-separated groups.

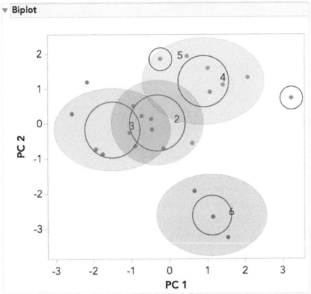

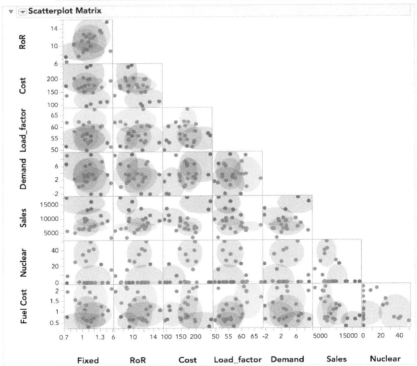

FIGURE 16.15   Biplot (top) and scatterplot matrix (bottom)

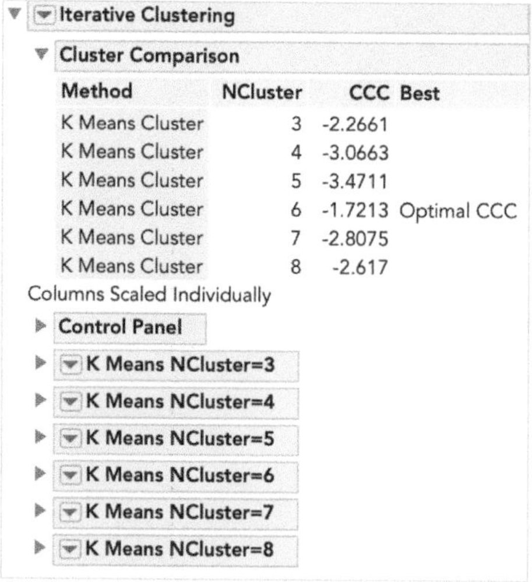

▼ ☑ Iterative Clustering

▼ Cluster Comparison

Method	NCluster	CCC	Best
K Means Cluster	3	-2.2661	
K Means Cluster	4	-3.0663	
K Means Cluster	5	-3.4711	
K Means Cluster	6	-1.7213	Optimal CCC
K Means Cluster	7	-2.8075	
K Means Cluster	8	-2.617	

Columns Scaled Individually

▶ Control Panel

▶ ☑ K Means NCluster=3

▶ ☑ K Means NCluster=4

▶ ☑ K Means NCluster=5

▶ ☑ K Means NCluster=6

▶ ☑ K Means NCluster=7

▶ ☑ K Means NCluster=8

FIGURE 16.16   Initial output for *k*-means clustering of 22 utilities, showing a comparison of clusters using the *CCC* (*Cubic Clustering Criterion*)

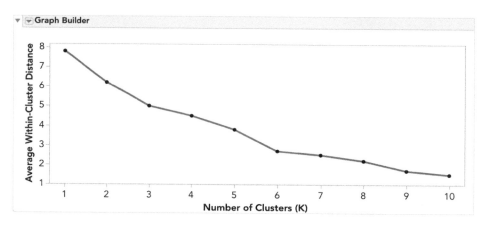

FIGURE 16.17   Comparing different choices of *k* in terms of the standard deviation of within-cluster distances by saving clusters for *k* = 1 to 10, summarizing the distances using *Tabulate*, and graphing using *Graph Builder*

PROBLEMS

16.1 Pharmaceutical Industry. An equities analyst is studying the pharmaceutical industry and would like your help in exploring and understanding the financial data collected by her firm. Her main objective is to understand the structure of the pharmaceutical industry using some basic financial measures.

Financial data gathered on 21 firms in the pharmaceutical industry are available in the file Pharmaceuticals.jmp. For each firm the following variables are recorded:

- Market capitalization (in billions of dollars)
- Beta
- Price/earnings ratio
- Return on equity
- Return on assets
- Asset turnover
- Leverage
- Estimated revenue growth
- Net profit margin
- Median recommendation (across major brokerages)
- Location of firm's headquarters
- Stock exchange on which the firm is listed

Use cluster analysis to explore and analyze the given dataset as follows:

a. Use only the quantitative variables to cluster the 21 firms. Justify the various choices made in conducting the cluster analysis, such as the specific clustering algorithm(s) used, and the number of clusters formed.

b. Interpret the clusters with respect to the quantitative variables that were used in forming the clusters.

c. Is there a pattern in the clusters with respect to the qualitative variables (those not used in forming the clusters)?

d. Provide an appropriate name for each cluster (based on variables in the dataset).

e. How do your results help the equities analyst achieve her objective?

16.2 Customer Rating of Breakfast Cereals. The dataset Cereals.jmp includes nutritional information, store display, and consumer ratings for 77 breakfast cereals.

Data preprocessing. Note that some cereals are missing values. These will be automatically omitted from the analysis. Use the *Columns Viewer* to identify which variables are missing values and how many values are missing. Describe why the missing values, in this example, might (or migh not) be a concern in this assignment.

a. Apply hierarchical clustering to the data using single linkage and complete linkage.

   i. The clustering dialog provides an option to standardize. Should this be selected? Why?

   ii. Look at the dendrograms and the parallel plots. Comment on the structure of the clusters and on their stability.

    iii. Which method leads to the most insightful or meaningful clusters?

    iv. Choose one of the methods. How many clusters would you use? Why?

b. The public elementary schools would like to choose a set of cereals to include in their daily cafeterias. Every day a different cereal is offered, but all cereals should support a healthy diet. For this goal, you are requested to find a cluster of "healthy cereals."

    i. Based on the variables at hand, how would you characterize "healthy cereals"?

    ii. Save the clusters to the data table, and use graphical and descriptive tools to characterize and name the clusters.

    iii. Which cluster of cereals is the most "healthy"? Does one cluster stand out as being the healthiest?

16.3 **Marketing to Frequent Fliers.** The file `EastWestAirlinesCluster.jmp` contains information on 3999 passengers who belong to an airline's frequent flier program. For each passenger the data include information on their mileage history and on different ways they accrued or spent miles in the last year. The goal is to try to identify clusters of passengers that have similar characteristics for the purpose of targeting different segments for different types of mileage offers.

a. Apply hierarchical clustering and Ward's method. Make sure to standardize the data. Use the dendrogram and the scree plot, along with practical considerations, to identify the "best" number of clusters. Use the *Color Clusters* and *Two-Way Clustering* option to help interpret the clusters. How many clusters would you select? Why?

b. What would happen if the data were not standardized?

c. Explore the clusters to try to characterize them. Try to give each cluster a label.

    i. Compare the cluster centroids (select *Cluster Summary*), and click on the lines to characterize the different clusters.

    ii. Save the clusters to the data table, and use graphical tools and the *Column Switcher*.

d. To check the stability of the clusters, hide and exclude a random 5% of the data (use *Rows > Row Selection > Select Randomly*, and then use *Rows > Hide and Exclude*), and repeat the analysis. Does the same picture emerge?

e. Use *k*-means clustering with the number of clusters that you found above. How do these results compare to the results from hierarchical clustering? Use the built-in graphical tools to characterize the clusters.

f. Which clusters would you target for offers, and what types of offers would you target to customers in that cluster?

16.4 **University Rankings.** The dataset on American College and University Rankings (`Colleges.jmp`) contains information on 1302 American colleges and universities offering an undergraduate program. For each university there are several measurements. Most of these are continuous (e.g., tuition and graduation rate) while a couple are categorical (location by state and whether it is a private or public school).

a. Use the *Columns Viewer* to produce numeric summaries of all of the variables. Note that many records are missing some measurements. Clustering methods in

JMP omit records with missing values. For the purposes of this exercise, we will assume that the values are missing purely at random (not that this may in fact not be the case). So our first goal is to estimate (impute) these missing values.

b. Select all of the continuous columns, and go to *Cols > Modeling Utilities > Explore Missing Values*. Select the option *Multivariate Normal Imputation*, and click *Yes Shrinkage*. Multivariate normal imputation uses least squares regression to predict the missing values from the nonmissing variables in each row. The imputed values will be highlighted in the data table. Save the data table with imputed values under a new name.

c. Use *k*-means clustering, with only the continuous variables. Use College Name as the label. Form clusters starting with 3 and ending with 8. What is the optimal number of clusters based on the Cubic Clustering Criterion (under *Cluster Comparison*)?

d. Use the biplot, the parallel plot, and other built-in graphical and numeric summaries to explore the clusters. Save the clusters to the data table and use other graphical tools to compare and characterize the clusters. Summarize the key characteristics of each of the four clusters, and try label or name the clusters.

e. Use the categorical variables that were not used in the analysis (State and Private/Public) to characterize the different clusters. (*Hint:* Create a geographic map to explore the clusters geographically.) Is there any relationship between the clusters and the categorical information?

f. Can you think of other external information that might explain the contents of some or all of these clusters?

g. Consider Tufts University. Which cluster does Tufts belong to? Which other universities is Tufts similar to, based on the clustering and the categorical variables?

h. Return to the original data table (with the missing values). Run the same *k*-means cluster analysis using this data.

   i. Compare the results to those achieved after imputing missing values in terms of the number of clusters and the characteristics of the clusters? What are the key differences?

   ii. In the initial analysis, we assumed that the values were missing at random and imputed the missing values. Describe why this approach was, or was not, a good strategy.

PART VII

FORECASTING TIME SERIES

17

HANDLING TIME SERIES

In this chapter, we describe the context of business time series forecasting and introduce the main approaches that are detailed in Chapters 18 and 19 and, in particular, regression-based forecasting and smoothing-based methods. Our focus is on forecasting future values of a single time series. These three chapters (Chapters 17–19) are meant as an introduction to the general forecasting approach and methods.

In this chapter, we discuss the difference between the predictive nature of time series forecasting and the descriptive or explanatory task of time series analysis. A general discussion of combining forecasting methods or results for added precision follows. Additionally, we present a time series in terms of four components (level, trend, seasonality, and noise), and methods for visualizing the different components and exploring time series data. We close with a discussion of data partitioning (creating training and validation sets), which is performed differently from cross-sectional data partitioning.

Time Series in *JMP*: The methods introduced in the three chapters in this part of the book are available in the standard version of JMP. However, JMP Pro is required to evaluate performance based on validation metrics for regression-based time series models.

17.1 INTRODUCTION[1]

Time series forecasting is performed in nearly every organization that works with quantifiable data. Retail stores use it to forecast sales. Energy companies use it to forecast reserves, production, demand, and prices. Educational institutions use it to forecast enrollment. Governments use it to forecast tax receipts and spending. International financial organizations like the World Bank and International Monetary Fund use it to forecast

---

[1]This and subsequent sections in this chapter copyright © 2022 Datastats, LLC, and Galit Shmueli. Used by permission.

Machine Learning for Business Analytics: Concepts, Techniques, and Applications with JMP Pro®,
Second Edition. Galit Shmueli, Peter C. Bruce, Mia L. Stephens, Muralidhara Anandamurthy, and Nitin R. Patel.
© 2023 John Wiley & Sons, Inc. Published 2023 by John Wiley & Sons, Inc.

inflation and economic activity. Transportation companies use time series forecasting to forecast future travel. Banks and lending institutions use it (sometimes badly!) to forecast new home purchases. And venture capital firms use it to forecast market potential and to evaluate business plans.

Previous chapters in this book deal with classifying and predicting data where time is not a factor, in the sense that it is not treated differently from other variables, and where the sequence of measurements over time does not matter. These are typically called cross-sectional data. In contrast, this chapter deals with a different type of data: time series.

With today's technology, many time series are recorded on very frequent time scales. Stock data are available at ticker level. Purchases at online and offline stores are recorded in real time. The Internet of Things (IoT) generates huge numbers of time series, produced by sensors and other measurements devices. Although data might be available at a very frequent scale, for the purpose of forecasting, it is not always preferable to use this time scale. In considering the choice of time scale, one must consider the scale of the required forecasts and the level of noise in the data. For example, if the goal is to forecast next-day sales at a grocery store, using minute-by-minute sales data is likely to be less useful for forecasting than using daily aggregates. The minute-by-minute series will contain many sources of noise (e.g., variation by peak and nonpeak shopping hours) that degrade its forecasting power, and these noise errors, when the data are aggregated to a cruder level, are likely to average out.

This part of the book focuses on forecasting a single time series. In some cases, multiple time series are to be forecasted (e.g., the monthly sales of multiple products). Even when multiple series are being forecasted, the most popular forecasting practice is to forecast each series individually. The advantage of single-series forecasting is its simplicity. The disadvantage is that it does not take into account possible relationships between series. The statistics literature contains models for multivariate time series that directly model the cross-correlations between series. Such methods tend to make restrictive assumptions about the data and the cross-series structure, and they also require statistical expertise for estimation and maintenance. Econometric models often include information from one or more series as inputs into another series. However, such models are based on assumptions of causality that are based on theoretical models. An alternative approach is to capture the associations between the series of interest and external information more heuristically. An example is using the sales of lipstick to forecast some measure of the economy, based on the observation by Leonard Lauder, chairman of Estee Lauder, that lipstick sales tend to increase before tough economic times (a phenomenon called the "leading lipstick indicator").

17.2   DESCRIPTIVE VS. PREDICTIVE MODELING

As with cross-sectional data, modeling time series data is done for either descriptive or predictive purposes. In descriptive modeling, or *time series analysis*, a time series is modeled to determine its components in terms of seasonal patterns, trends, relation to external factors, and so on. These can then be used for decision making and policy formulation. In contrast, *time series forecasting* uses the information in a time series (and perhaps other information) to predict future values of that series. The difference between the goals of time series analysis and time series forecasting leads to differences in the type of methods used and in the modeling process itself. For example, in selecting a method for describing a time series, priority is given to methods that produce understandable results (rather than "black box" methods) and sometimes to models based on causal arguments (explanatory models). Furthermore describing can be done in retrospect, while forecasting is prospective in nature. This means that descriptive models might use "future" information (e.g., averaging the values of yesterday, today, and tomorrow to obtain a smooth representation of today's value), whereas forecasting models cannot.

The focus in this chapter is on time series forecasting, where the goal is to predict future values of a time series. For information on time series analysis, see Chatfield (2003).

17.3   POPULAR FORECASTING METHODS IN BUSINESS

In this part of the book, we focus on two main types of forecasting methods that are popular in business applications. Both are versatile and powerful yet relatively simple to understand and deploy. One type of forecasting tool is multiple linear regression, where the user specifies a certain model and then estimates it from the time series. The other includes more data-driven methods, such as smoothing and deep learning, where the method learns patterns from the data. Each of the two types of tools has advantages and disadvantages, as detailed in Chapters 18 and 19.

We also note that machine learning methods such as neural networks and others that are intended for cross-sectional data are also sometimes used for time series forecasting, especially for incorporating external information into the forecasts (see Shmueli and Lichtendahl, 2016).

Combining Methods

Before a discussion of specific forecasting methods in the following two chapters, it should be noted that a popular approach for improving predictive performance is to combine forecasting methods. This is similar to the ensembles approach described in Chapter 13. Combining methods can be done via two-level (or multilevel) forecasters, where the first method uses the original time series to generate forecasts of future values and the second method uses the residuals from the first model to generate forecasts of future forecast errors, thereby "correcting" the first-level forecasts. We describe two-level forecasting in Chapter 18. Another combination approach is via "ensembles," where multiple methods are applied to the time series, and their resulting forecasts are averaged in some way to produce the final forecast. Combining methods can take advantage of the strengths of different forecasting methods to capture different aspects of the time series (also true in cross-sectional data). The averaging across multiple methods can lead to forecasts that are more robust and of higher precision.

17.4   TIME SERIES COMPONENTS

In both types of forecasting methods, regression models and smoothing, and in general, it is customary to dissect a time series into four components: *level*, *trend*, *seasonality*, and *noise*. The first three components are assumed to be invisible, as they characterize the underlying series, which we only observe with added noise. Level describes the average value of the series, trend is the change in the series from one period to the next, and seasonality describes a short-term cyclical behavior of the series that can be observed several times within the given series. Finally, noise is the random variation that results from measurement error or other causes not accounted for. It is always present in a time series to some degree.

Don't get confused by the standard conversational meaning of "season" (winter, spring, etc.). The statistical meaning of "season" refers to any time period that repeats itself in cycles within the larger time series. Also, note that the term "period" in forecasting means simply a span of time, and not the specific meaning that it has in physics, of the distance between two equivalent points in a cycle.

In order to identify the components of a time series, the first step is to examine a *time plot*. In its simplest form, a time plot is a line chart of the series values over time, with temporal labels (e.g., calendar date) on the horizontal axis. To illustrate this, consider the following example.

Example: Ridership on Amtrak Trains

Amtrak, a US railway company, routinely collects data on ridership. Here we focus on forecasting future ridership using the series of monthly ridership between January 1991 and March 2004. These data[2] are found in `Amtrak.jmp`.

A time plot for monthly Amtrak ridership series is shown in Figure 17.1. Note that the values are in thousands of riders. (Use the *Graph Builder*, with ridership as *Y* and Month as *X*, and select the *Line* graph element. Double-click on the horizontal axis to change the label orientation.)

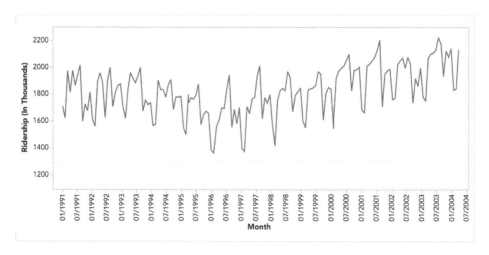

FIGURE 17.1 **Monthly ridership on Amtrak trains (in thousands) from January 1991 to March 2004**

---

[2]Data from `https://www.bts.gov/archive/publications/multimodal_transportation_indicators/june_2010/amtrak_ridership`

Looking at the time plot reveals the nature of the series components: the overall level is around 1,800,000 passengers per month. A slight U-shaped trend is discernible during this period, with pronounced annual seasonality, with peak travel during summer (July and August).

A second step in visualizing a time series is to examine it more carefully. A few tools are useful:

Zoom in: Zooming in to a shorter period within the series can reveal patterns that are hidden when viewing the entire series. This is especially important when the time series is long. Consider a series of the daily number of vehicles passing through the Baregg tunnel in Switzerland (data are in `BareggDaily.jmp`, and are publicly available in the same location as the Amtrak Ridership data; series D028). The series from November 1, 2003, to November 16, 2005, is shown in the top panel of Figure 17.2. Zooming in to a three-month period (bottom panel in Figure 17.2) reveals a strong day-of-week pattern that was not visible in the initial time plot of the complete time series. (In JMP, double-click on the horizontal axis to change the minimum and maximum values, or use the *Data Filter*. The *y*-axis scale was adjusted in both graphs in Figure 17.2.)

Change scale of series: In order to better identify the shape of a trend, it is useful to change the scale of the series. One simple option is to change the vertical scale to a

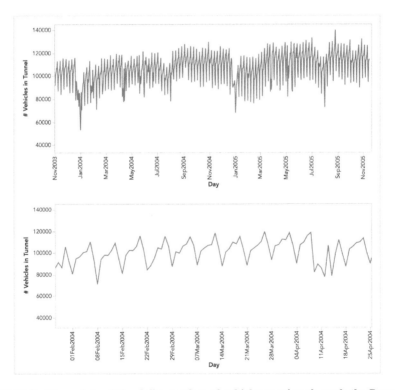

FIGURE 17.2   **Time plots of the daily number of vehicles passing through the Baregg tunnel, Switzerland. The bottom panel zooms in to a four-month period, revealing a day-of-week pattern**

logarithmic scale (double-click on the vertical axis and change the scale from Linear to Log). If the trend on the new scale appears more linear, then the trend in the original series is closer to an exponential trend.

Add trend lines: Another possibility for better capturing the shape of the trend is to add a trend line. By trying different trendlines, one can see what type of trend (e.g., linear, exponential, cubic) best approximates the data. In *Graph Builder*, use the smoother icon and the slider for the smoother to explore the nature of the trend. To fit a trend line, use the *Line of Fit* graph icon, and explore different degrees (linear, quadratic, or cubic) using the *Degree* options under *Line of Fit* (on the bottom, left). A shaded confidence band for the trend line displays by default. To remove the confidence band, deselect the *Confidence, Fit* box. To explore different transformations, right-click on the column in the column selection panel, select a transformation, and graph the transformed data.

Suppress seasonality: It is often easier to see trends in the data when seasonality is suppressed. Suppressing seasonality patterns can be done by plotting the series at a cruder time scale (e.g., aggregating monthly data into years). If the data are stored in JMP with a date format, then a *Date Time* transformation can be applied (in *Graph Builder*, right-click on the column and select a *Date Time* transformation). Another popular option is to use moving average charts. We will discuss these in Chapter 19.

Continuing our example of Amtrak ridership, the charts in Figure 17.3 help make the series' components more visible.

Some forecasting methods directly model these components by making assumptions about their structure. For example, a popular assumption about trend is that it is linear or exponential over the given time period or part of it. Another common assumption is about the noise structure: many statistical methods assume that the noise follows a normal distribution. The advantage of methods that rely on such assumptions is that when the assumptions are reasonably met, the resulting forecasts will be more robust and the models more understandable. Other forecasting methods, which are data-adaptive, make fewer assumptions about the structure of these components and instead try to estimate their structure only from the data. Data-adaptive methods are advantageous when such assumptions are likely to be violated, or when the structure of the time series changes over time. Another advantage of many data-adaptive methods is their simplicity and computational efficiency.

A key criterion for deciding between model-driven and data-driven forecasting methods is to ascertain the nature of the series in terms of global and local patterns. A global pattern is one that is relatively constant throughout the series. An example is a linear trend throughout the entire series. In contrast, a local pattern is one that occurs only in a short period of the data and then changes. An example is a trend that is approximately linear within four neighboring time points, but the trend size (slope) changes slowly over time.

Model-driven methods are generally preferable for forecasting series with global patterns, as they use all the data to estimate the global pattern. For a local pattern, a model-driven model would require specifying how and when the patterns change, which is usually impractical and often unknown. Therefore data-driven methods are preferable for local patterns. Such methods "learn" patterns from the data, and their memory length can be set to best adapt to the rate of change in the series. Patterns that change quickly warrant a "short memory," whereas patterns that change slowly warrant a "long memory." In effect, the time plot should be used to identify not only the time series component but also the global/local nature of the trend and seasonality.

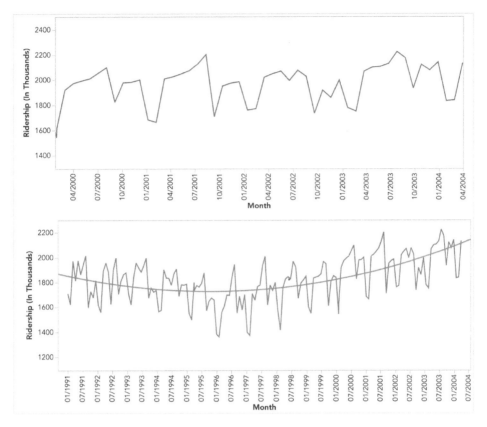

FIGURE 17.3   **Plots that enhance the different components of the time series. Top: zoom in to three years of data. Bottom: original series with overlaid quadratic trendline**

17.5   DATA PARTITIONING AND PERFORMANCE EVALUATION

As in the case of cross-sectional data, in order to avoid overfitting and to be able to assess the predictive performance of the model on new data, we first partition the data into a training set and a validation set (and perhaps an additional test set if the validation set is used for model tuning and comparison). However, there is one important difference between data partitioning in cross-sectional and time series data. In cross-sectional data the partitioning is usually done randomly, with a random set of observations designated as training data and the remainder as validation data. However, in time series a random partition would create two time series with "holes." Most standard forecasting methods cannot handle time series with missing values. Therefore, we partition a time series into training and validation sets differently. The series is trimmed into two periods: the earlier period is set as the training data and the later period as the validation data, called *holdback* in the *Time Series* platform in JMP. Methods are then trained on the earlier training period, and their predictive performance is assessed on the later holdback period. This approach also has the advantage of mimicking how the forecasting model will be used in practice: trained on existing data to forecast future values.

PARTITIONING TIME SERIES DATA IN JMP AND VALIDATING TIME SERIES MODELS

1. To partition the time series data in JMP, select *Make Validation Column* from the *Analyze > Predictive Modeling* menu. Use the *Cutpoint Columns* field to partition the data into training and validation sets based on a selected time series cutpoint.

2. Fit the time series model(s) using the *Time Series* platform (from the *Analyze > Specialized Modeling* menu). Note that the *Time Series* platform does not directly use the validation column. Instead, the platform enables you to specify the number of holdback (validation) periods. These holdback periods will be used later to summarize and visualize model forecasts and residuals.

3. In the time series launch window (see Figure 17.4), enter the *Forecast Periods* (the number of time periods in the holdback to forecast) using each fitted model. The default is 25.

4. Check the *Forecast on Holdback* box to make forecasts on holdback periods. Figure 17.4 shows the specification of a 36-month holdback period.

5. For the chosen model, select *Save Columns* from the red triangle for the model. This produces a new data table with predicted values and residuals (and other values) for the training and holdback data. For the holdback data, the predicted values are the forecasts, and the residuals are the forecast errors. Copy these two columns into the original data table. (*Hint*: Use *Edit > Copy with Column Name* and *Edit > Paste with Column Name*.)

6. Using the validation column, summarize the residuals for the training and holdback (validation) data using the *Distribution* platform. Graph the actual, forecasted, and the residuals series using the *Graph Builder*.

These steps apply to models built using the JMP *Time Series* platform. Regression-based forecasting models can be built using the *Analyze > Fit Model* platform (see Chapter 18).

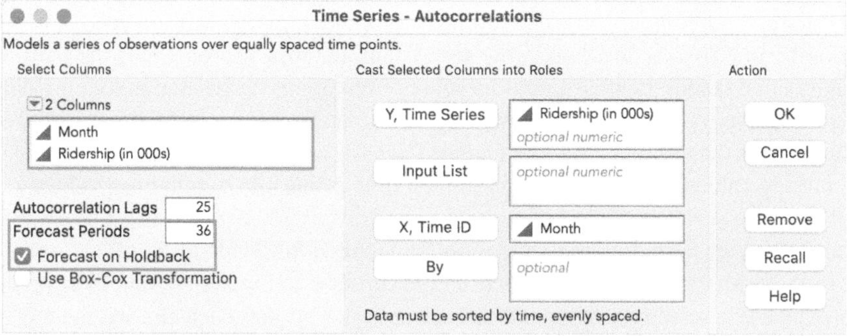

FIGURE 17.4   Time Series analysis launch window

Evaluation measures typically use the same metrics used in cross-sectional evaluation (see Chapter 5) with *MAPE* and *RMSE* being the most popular metrics in practice. Due to the simplicity of the calculations, *average error* and *MAE* (mean absolute error) are also used. In evaluating and comparing forecasting methods, another important tool is visualization: examining time plots of the actual and predicted series can shed light on performance and hint toward possible improvements.

Benchmark Performance: Naive Forecasts

While it is tempting to apply "sophisticated" forecasting methods, we must evaluate their added value compared to a very simple approach: the *naive forecast*. A naive forecast is simply the most recent value of the series. In other words, at time t, our forecast for any future period $t + k$ is simply the value of the series at time t. While simple, naive forecasts are sometimes surprisingly difficult to outperform with more sophisticated models. It is therefore important to benchmark against results from a naive forecasting approach.

When a time series has seasonality, a *seasonal naive forecast* can be generated. It is simply the value on the most recent similar season. For example, to forecast April 2001 for the Amtrak ridership, we use the ridership from the most recent April, which is April 2000. Similarly, to forecast April 2002, we also use April 2000 ridership.

In Figure 17.5, we show naive (red horizontal line) and seasonal naive forecasts (green line), as well as actual values (blue dashed line), in a three-year holdback/validation period from April 2001 to March 2004. To create naive seasonal forecasts in JMP, we manually create a new column of lagged values using the *Lag Multiple* function (right-click on the column header and select *New Formula column > Row > Lag Multiple*) and then copying the appropriate lagged values into a new column.

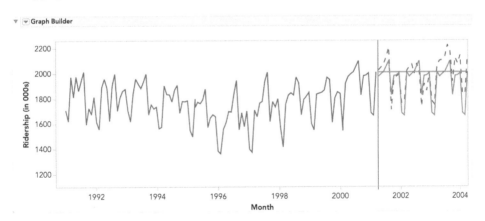

FIGURE 17.5   Naive and seasonal naive forecasts in a three-year validation set for Amtrak ridership. The figure generated using *Graph builder* after manually creating naive and seasonal naive forecast columns in JMP

Generating Future Forecasts

Another important difference between cross-sectional and time series partitioning occurs when creating the actual forecasts. When forecasting future values of a time series, the method/model is run on the complete data (rather than just the training data). This final model is then used to forecast future values. The three advantages of using the entire series are:

1. The validation/holdback set, which is the most recent period, usually contains the most valuable information in terms of being the closest in time to the forecast period.
2. With more data (the complete time series, compared to only the training set), some models can be estimated more accurately.
3. If only the training set is used to generate forecasts, then it will require forecasting farther into the future (e.g., if the validation/holdback set contains four time points, forecasting the next time point will require a five-step-ahead forecast from the training set).

PROBLEMS

17.1 **Impact of September 11 on Air Travel in the United States.** The Research and Innovative Technology Administration's Bureau of Transportation Statistics (BTS) conducted a study to evaluate the impact of the September 11, 2001, terrorist attack on US transportation. The 2006 study report and the data can be found at `https://www.bts.gov/archive/publications/estimated_impacts_of_9_11_on_us_travel/index`. The goal of the study was stated as follows:

> The purpose of this study is to provide a greater understanding of the passenger travel behavior patterns of persons making long distance trips before and after 9/11.

The report analyzes monthly passenger movement data between January 1990 and May 2004. Data on three monthly time series are given in file `Sept11Travel.jmp` for this period:

- Actual airline revenue passenger miles (Air RPM)
- Rail passenger miles (Rail PM)
- Vehicle miles traveled (VMT)

In order to assess the impact of September 11, BTS took the following approach: using data before September 11, they forecasted future data (under the assumption of no terrorist attack). Then they compared the forecasted series with the actual data to assess the impact of the event. Our first step therefore is to split each of the time series into two parts: pre- and post-September 11. We now concentrate only on the earlier time series.

a. Is the goal of this study descriptive or predictive?

b. Plot each of the three pre-event time series (Air, Rail, Car).

  i. What time series components appear from the plots?

  ii. What type of trend appears? Change the scale of the series, add trendlines, and suppress seasonality to better visualize the trend pattern.

17.2 **Performance on Training and Validation Data.** Two different models were fit to the same time series. The first 100 time periods were used for the training set, and the last 12 periods were treated as a holdback set. Assume that both models make sense practically and fit the data pretty well. Below are the RMSE values for each of the models:

	Training set	Validation set
Model A	543	690
Model B	669	675

a. Which model appears more useful for explaining the different components of this time series? Why?

b. Which model appears to be more useful for forecasting purposes? Why?

17.3 Forecasting Department Store Sales. The file `DepartmentStoreSales.jmp` contains data on the quarterly sales for a department store over a six-year period (data courtesy of Chris Albright).

   a. Create a time series plot (using the *Graph Builder*).

   b. Which of the four components (level, trend, seasonality, noise) seems to be present in this series?

17.4 Shipments of Household Appliances. The file `ApplianceShipments.jmp` contains the series of quarterly shipments (in million $) of US household appliances between 1985 and 1989 (data courtesy of Ken Black).

   a. Create a well-formatted time plot of the data.

   b. Which of the four components (level, trend, seasonality, noise) seems to be present in this series?

17.5 Canadian Manufacturing Workers Workhours. The time series plot in Figure 17.6 describes the average annual number of weekly hours spent by Canadian manufacturing workers (data are available in `CanadianWorkHours.jmp`—thanks to Ken Black for these data).

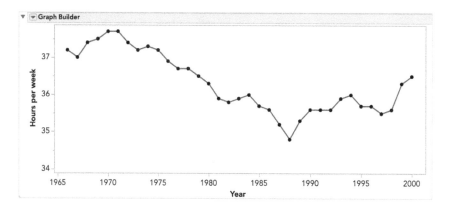

FIGURE 17.6   **Average annual weekly hours spent by Canadian manufacturing workers**

   a. Reproduce the time plot.

   b. Which of the following components (level, trend, seasonality, noise) appears to be present in this series?

17.6 Souvenir Sales. The file `SouvenirSales.jmp` contains monthly sales for a souvenir shop at a beach resort town in Queensland, Australia, between 1995 and 2001. [Source: Hyndman and Yang (2018).]

   Back in 2001, the store wanted to use the data to forecast sales for the next 12 months (year 2002). They hired an analyst to generate forecasts. The analyst first partitioned the data into training and validation sets, with the validation set containing the last 12 months of data (year 2001). She then fit a regression model to sales, using the training set.

a. Create a well-formatted time plot of the data.

b. Change the scale on the *x*-axis or on the *y*-axis, or on both, to log-scale in order to achieve a linear relationship. Select the time plot that seems most linear.

c. Comparing the two time plots, what can be said about the type of trend in the data?

d. Why were the data partitioned? Partition the data into training and validation sets as explained in the text box.

17.7 Shampoo Sales. The file `ShampooSales.jmp` contains data on the monthly sales of a certain shampoo over a three-year period. [Source: Hyndman and Yang (2018).]

a. Create a well-formatted time plot of the data.

b. Which of the four components (level, trend, seasonality, noise) seems to be present in this series?

c. Do you expect to see seasonality in sales of shampoo? Why?

d. If the goal is forecasting sales in future months, which of the following steps should be taken? Mark all that apply.

- Partition the data into training and validation (holdback) sets.
- Tweak the model parameters to obtain good fit to the validation data.
- Look at MAPE, RMSE, or other measures of error for the training set.
- Look at MAPE, RMSE, or other measures of error for the validation (holdback) set.

18

REGRESSION-BASED FORECASTING

A popular forecasting tool is based on multiple linear regression models, using suitable predictors to capture trend and/or seasonality. In this chapter, we show how a linear regression model can be set up to capture a time series with a trend and/or seasonality. The model, which is estimated from the data, can then produce future forecasts by inserting the relevant predictor information into the estimated regression equation. We describe different types of common trends (linear, exponential, polynomial), as well as two types of seasonality (additive and multiplicative). Next we show how a regression model can be used to quantify the correlation between neighboring values in a time series (called autocorrelation). This type of model, called an autoregressive (AR) model, is useful for improving forecast precision by making use of the information contained in the autocorrelation (beyond trend and seasonality). It is also useful for evaluating the predictability of a series by evaluating whether the series is a "random walk." The various steps of fitting linear regression and autoregressive models, using them to generate forecasts, and assessing their predictive accuracy, are illustrated using the Amtrak ridership series (data are in `Amtrak.jmp`).

Forecasting in JMP: In this chapter, we fit time series models with trends and seasonality using *Linear Regression* (from *Analyze > Fit Model*). We also use the *Time Series* platform (under *Analyze > Specialized Modeling*) to apply ARIMA, autocorrelation, differencing, and other methods for processing and forecasting time series data.

Machine Learning for Business Analytics: Concepts, Techniques, and Applications with JMP Pro®,
Second Edition. Galit Shmueli, Peter C. Bruce, Mia L. Stephens, Muralidhara Anandamurthy, and Nitin R. Patel.
© 2023 John Wiley & Sons, Inc. Published 2023 by John Wiley & Sons, Inc.

18.1   A MODEL WITH TREND[1]

Linear Trend

To create a linear regression model that captures a time series with a global linear trend, the outcome variable (Y) is set as the time series values or some function of it, and the predictor variable (X) is set as a time index. Let us consider a simple example: fitting a linear trend to the Amtrak ridership data described in Chapter 17. This type of trend is shown in Figure 18.1.

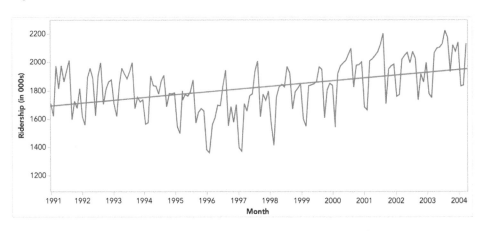

FIGURE 18.1   A linear trend fit to Amtrak ridership

From the time plot it is obvious that the global trend is not linear. However, we use this example to illustrate how a linear trend is fit, and later we consider more appropriate models for this series.

To obtain a linear relationship between ridership and time, we set the output variable Y as the Amtrak ridership and create a new variable that is a time index $t = 1, 2, 3, \ldots$. This time index is then used as a single predictor in the regression model:

$$Y_t = \beta_0 + \beta_1 t + \epsilon, \tag{18.1}$$

where Y_t is the ridership at time point t and ϵ is the standard noise term in a linear regression. Thus we are modeling three of the four time series components: level (β_0), trend (β_1), and noise (ϵ). Seasonality is not modeled. A snapshot of the two corresponding columns (Y_t and t) for the first 14 months is shown in Table 18.1.

After partitioning the data into training and validation sets, the next step is to fit a linear regression model to the training set, with t as the single predictor (using *Analyze > Fit Model*). Applying this to the Amtrak ridership data (with a validation set consisting of the last 36 months in this example) results in the estimated model shown in Figure 18.2. The actual and fitted values are plotted in Figure 18.2. Note that examining only the estimated coefficients and their statistical significance (in the *Parameter Estimates* table) can be very misleading! In this example, the statistically insignificant trend coefficient might indicate "no trend," although it is obvious from the plot that a trend does exist, albeit a quadratic one.

---

[1]This and subsequent sections in this chapter copyright © 2019 Datastats, LLC, and Galit Shmueli. Used by permission.

TABLE 18.1   Output variable (Ridership) and predictor variable (*t*) used to fit a linear trend

Month	Ridership	*t*
Jan-91	1709	1
Feb-91	1621	2
Mar-91	1973	3
Apr-91	1812	4
May-91	1975	5
Jun-91	1862	6
Jul-91	1940	7
Aug-91	2013	8
Sep-91	1596	9
Oct-91	1725	10
Nov-91	1676	11
Dec-91	1814	12
Jan-92	1615	13
Feb-92	1557	14

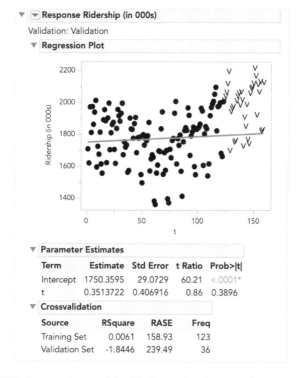

FIGURE 18.2   Fitted regression model with linear trend, regression output, and validation statistics

We obtain the training RASE of 158.93 and validation RASE of 239.49. This discrepancy is indicative of an inadequate trend shape. However, an inadequate trend shape is easiest to detect by examining the time plot of the residuals (Figure 18.3) which were produced using the *Graph Builder* after saving the prediction formula and the residuals for the model to the data table. The top graph is the actual and fitted series plot, and the bottom graph is the time-ordered residual plot. The inadequacy of the linear model is easy to see in these two plots.

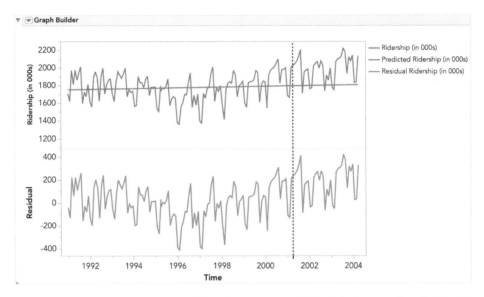

FIGURE 18.3   **Fitted trend and residual plot for Amtrak ridership data produced using the** *Graph Builder*

CREATING ACTUAL VS. PREDICTED PLOTS AND RESIDUAL PLOTS IN JMP

A variety of plots are available as red triangle options in the Fit Model analysis window. The actual vs. predicted plot and residual plot in Figure 18.3 were produced using the *Graph Builder* for a better indication of the time-ordered behavior. To create these graphs:

- Save the predicted values and residuals to the data table (using the top red triangle in the Fit Model analysis window under *Save Columns*).
- Open *Graph Builder*. Drag Month to the X zone, drag Residuals to the Y zone, Drag Ridership above Residuals in the Y zone (to create a separate graph). Then click on the *Line* icon. Finally, drag Predicted Ridership just inside the y-axis in the top graph.

FITTING A MODEL WITH LINEAR TREND IN JMP

First, prepare the data:

- Create the variable t: Add a new column in the data table, name the column t, and tab or enter to accept. Then, right-click on the column head and select *New Formula Column > Row > Row* to populate the column with the row number for each observation.
- Split the data into training and validation set using *Analyze > Predictive modeling > Make validation column* and use *Month* as the *Cutpoint Column*. In this example, the validation set consists of the last 36 months.

Next, create regression models with trends:

- Go to *Analyze > Fit Model*.
- In the dialog, select the time series data as the Y variable, select t as the model effect, and put the partition variable in the *Validation* field. Change the *Emphasis* to *Minimal Report* to turn off some default output.
- The records in the validation set are graphed with a "v" marker in the regression plot (to denote that they were not used to fit the model).
- In the analysis window, graphical and statistical output can be added or removed by using *red triangle* options.

Exponential Trend

Several alternative trend shapes that are useful and easy to fit via a linear regression model. Recall that different transformations can be explored dynamically in the JMP *Graph Builder*. One such shape is an exponential trend. An exponential trend implies a multiplicative increase/decrease of the series over time ($Y_t = ce^{\beta_1 t + \epsilon}$). To fit an exponential trend, simply replace the output variable Y with log Y where log is the natural logarithm, and fit a linear regression ($\log Y_t = \beta_0 + \beta_1 t + \epsilon$). In the Amtrak example, for instance, we would fit a linear regression of log($Ridership$) on the index variable t. Exponential trends are popular in sales data, where they reflect percentage growth.

To create the variable log *Ridership*, right-click on *Ridership* in the data table and select *New Formula Column > Log > Log*. Transformed variables can also be created from the *Graph Builder* and other platforms, which allow you to graphically explore a variety of transformations.

Note: As in the general case of linear regression, when comparing the predictive accuracy of models that have a different output variable, such as a linear trend model (with Y) and an exponential trend model (with $\log Y$), it is essential to compare forecasts or forecasts errors on the same scale. An exponential trend model will produce forecasts of $\log Y$, and the forecast errors reported by the software will therefore be $\log Y - \widehat{\log Y}$. To obtain forecasts in the original units, create a new column which takes an exponent of the model forecasts. Then use this column to create an additional column of forecast errors by subtracting the original Y. An example is shown in Figures 18.4 and 18.5, where an exponential trend is fit to the Amtrak ridership data. Note that the performance measures for the training and validation data are not comparable to those from the linear trend model shown in Figure 18.2.

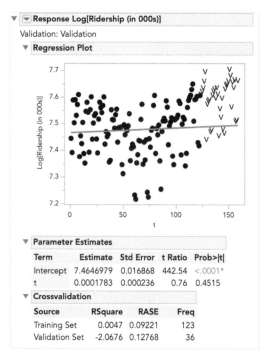

FIGURE 18.4   Output from regression model with exponential trend, fit to training data

	Validation	Ridership (in 000s)	Log[Ridership]	Pred Log[Ridership]	Pred Ridership	Forecast errors
124	Validation	2023.7920	7.6127	7.4868	1784.3561	239.4359
125	Validation	2047.0080	7.6241	7.4870	1784.6743	262.3337
126	Validation	2072.9130	7.6367	7.4872	1784.9926	287.9204
127	Validation	2126.7170	7.6623	7.4873	1785.3110	341.4060
128	Validation	2202.6380	7.6974	7.4875	1785.6295	417.0085
129	Validation	1707.6930	7.4429	7.4877	1785.9479	-78.2549

FIGURE 18.5   Adjusting forecasts of log(Ridership) to the original scale and computing forecast errors in the original scale (last column)

Instead, we manually compute two new columns in Figure 18.5, one that gives forecasts of ridership (in thousands) and another that gives the forecast errors in terms of ridership. To compare RMSE or average error, we would now use the new forecast errors and compute their standard deviation (for RMSE) or their average (for average error) using *Analyze > Distribution*. These would then be comparable to the numbers in Figure 18.2.

COMPUTING FORECAST ERRORS FOR EXPONENTIAL TREND MODELS

- Use *Save Columns > Prediction Formula or Predicted values* to save the values to the data table. This will produce predicted values (for training data) and forecasts for the validation data which were not used to build the model.
- Use *Save Columns > Residuals* to save the residuals to the data table.
- If an exponential trend model was fit, use the *Exp* function to calculate forecasts in the original units (right-click on the predicted values column and select *New Formula Column > Transform > Exp*).
- To compute forecast errors, select both the actual and predicted columns in the data table, right-click, and select *New Formula Column > Combine > Difference*. Part of the resulting output is shown in Figure 18.5.

Polynomial Trend

Another nonlinear trend shape that is easy to fit via linear regression is a polynomial trend and, in particular, a quadratic relationship of the form $Y_t = \beta_0 + \beta_1 t + \beta_2 t^2 + \epsilon$. This is done by adding $t * t$ (the square of t) to the model and fitting a multiple linear regression with the two predictors, t and $t * t$. For the Amtrak ridership data, we have already seen a U-shaped trend in the data. We therefore fit a quadratic model, concluding from the plots of model fit and residuals (Figure 18.6) that this shape adequately captures the trend. The residuals are now devoid of trend and exhibit only seasonality.

FITTING A POLYNOMIAL TREND IN JMP

To add a quadratic or squared term in the *Fit Model* dialog, select the variable (t in this example). Then, under *Macros*, select *Polynomial to Degree* to add both t and $t * t$ to the model. Note that a default setting in the *Fit Model* platform is to *center polynomials*. In this example (Figure 18.6), we have deselected this option (from the red triangle in the *Fit Model* dialog). Note that models with and without centering will lead to the same predictions.

In general, any type of trend shape can be fit as long as it has a mathematical representation. However, the underlying assumption is that this shape is applicable throughout the period of data that we have and also during the period that we are going to forecast. Do not choose an overly complex shape. Although it will fit the training data well, it will in fact

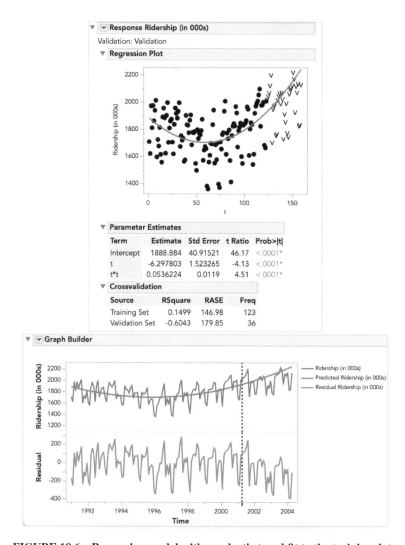

FIGURE 18.6   Regression model with quadratic trend fit to the training data

be overfitting them. To avoid overfitting, always examine performance on the validation set and refrain from choosing overly complex trend patterns.

18.2   A MODEL WITH SEASONALITY

A seasonal pattern in a time series means that observations that fall in some seasons have consistently higher or lower values than those that fall in other seasons. Examples are day-of-week patterns, monthly patterns, and quarterly patterns. The Amtrak ridership monthly time series, as can be seen in the time plot, exhibits strong monthly seasonality (with highest traffic during summer months).

Seasonality is modeled in a regression model by creating a new categorical variable that denotes the season for each observation. This categorical variable is then turned into indicator variables when the regression model is fit in JMP. To illustrate this, we created a new "Season" column for the Amtrak ridership data, as shown in Table 18.2. This column contains the month names.

TABLE 18.2   New categorical variable (Season) to be used in a linear regression model (first 17 months shown)

Month	Ridership	Season
Jan-91	1709	Jan
Feb-91	1621	Feb
Mar-91	1973	Mar
Apr-91	1812	Apr
May-91	1975	May
Jun-91	1862	Jun
Jul-91	1940	Jul
Aug-91	2013	Aug
Sep-91	1596	Sep
Oct-91	1725	Oct
Nov-91	1676	Nov
Dec-91	1814	Dec
Jan-92	1615	Jan
Feb-92	1557	Feb
Mar-92	1891	Mar
Apr-92	1956	Apr
May-92	1885	May

For m seasons, JMP creates $m - 1$ indicator variables in the Fit Model platform, holding out the highest level. (For a discussion of indicator coding in JMP and coefficient interpretation, see Chapter 6).[2] We then partition the data into training and validation sets and fit the regression model to the training data. For this example, we again use a validation set consisting of the last 36 months.

> Since the *Month* column has a date format, we can right-click on the *Month* column in the data table and use *New Formula Column* and the *Month Abbr.* transformation under *Date Time*. Here we've renamed this new column *Season*. Since this column is based on the *Month* column, the values are automatically ordered January to December in all analyses, and December will be held out of regression models. To reorder the values in analyses (if desired), use the *Value Ordering* column property.

The top panel of Figure 18.7 shows the output of a linear regression fit to Ridership (Y) on 11 month indicator variables using the training data. To show the fitted model and residuals (bottom, Figure 18.7), we again use the *Graph Builder*. The model appears to capture the seasonality in the data. However, since we have not included a trend component in

---

[2]JMP displays only $m - 1$ indicator variables in the *Parameter Estimates* table because the coefficients for the m seasons sum to zero. If the coefficients for $m - 1$ variables are summed up, the negative of this total is the coefficient for the mth season. Including the mth variable would cause redundant information and multicollinearity errors.

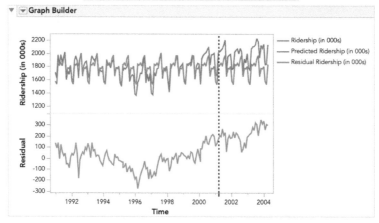

FIGURE 18.7   Fitted regression model with seasonality. Parameter estimates and validation statistics (top), and plot of fitted values, actual values, and residuals (bottom)

the model (as shown in Section 18.1), the fitted values do not capture the existing trend. Therefore, the residuals, which are the difference between the actual and fitted values, clearly display the remaining U-shaped trend.

Additive vs. Multiplicative Seasonality

When seasonality is added as described above, the model captures *additive seasonality*. This means that the average value of Y in a given season is a certain amount more or less than the overall average.[3] Specifically, the coefficient for a month tells us the average amount by which that month's average value is higher or lower than the overall average. For example, the coefficient for February (-244.2) indicates that the average number of

---

[3] If we have balanced data (an equal number of observations for each month), the intercept will equal the overall mean for the training data. Here we have one more observation for January, February, and March than we do for the other months, so the intercept is not equal to the overall mean.

passengers in February is lower than the overall average monthly number of passengers by 244.2 thousand passengers. Similarly, the coefficient for August indicates that the average ridership in August is higher than the overall average monthly ridership by 195.3 thousand passengers.

The coefficient for December, the left out level, is not shown. To display all of the coefficients, including the left out level, select *Estimates > Expanded Estimates* from the top red triangle. The regression equation can be displayed by selecting *Estimates > Show Prediction Expression* from the top red triangle. The equation can also be explored graphically using the *Prediction Profiler* (select *Factor Profiling > Profiler* from the top red triangle).

Using regression models we can also capture *multiplicative seasonality*, where values on a certain season are on average higher or lower by a certain *percentage* compared to the average. To fit multiplicative seasonality, we use the same model as above, except that we use $\log Y$ as the output variable.

18.3   A MODEL WITH TREND AND SEASONALITY

Finally, we can create models that capture both trend and seasonality by including predictors of both types. For example, from our exploration of the Amtrak ridership data, it appears that a quadratic trend and monthly seasonality are both warranted. We therefore fit a model to the training data with Season, t and $t*t$. The output and fit from this final model are shown in Figure 18.8. If we are satisfied with this model, after evaluating its predictive performance on the validation data and comparing it against alternatives, we can re-fit it to the entire un-partitioned series. This re-fitted model can then be saved to the data table and used to generate k-step-ahead forecasts (denoted by F_{t+k}). To do this, we add rows to the data table and plug in the appropriate month and index terms.

(*Note*: A version of the Amtrak ridership data with the validation column, the variables t, and *Season*, along with the saved prediction formula and residuals after fitting the model with quadratic trend and seasonality, are found in `AmtrakTS.jmp`.)

18.4   AUTOCORRELATION AND ARIMA MODELS

When we use linear regression for time series forecasting, we are able to account for patterns such as trend and seasonality. However, ordinary regression models do not account for dependence between observations, which in cross-sectional data is assumed to be absent. Yet, in the time series context, observations in neighboring periods tend to be correlated. Such correlation, called *autocorrelation*, is informative and can help in improving forecasts. If we know that a high value tends to be followed by high values (positive autocorrelation), then we can use that to adjust forecasts. We will now discuss how to compute the autocorrelation of a series and how best to utilize the information for improving forecasts.

Computing Autocorrelation

Correlation between values of a time series in neighboring periods is called *autocorrelation* because it describes a relationship between the series and itself. To compute autocorrelation, we compute the correlation between the series and a *lagged* version of the series. A lagged

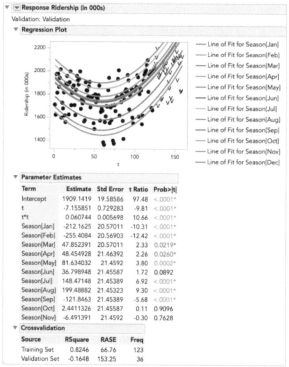

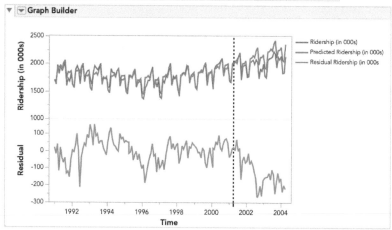

FIGURE 18.8 Regression model with monthly (additive) seasonality and quadratic trend, fit to Amtrak ridership data. Parameter estimates, validation statistics (top), and plot of fitted and actual training data and residuals (bottom)

series is a "copy" of the original series that is moved forward one or more time periods. A lagged series with lag-1 is the original series moved forward one time period, a lagged series with lag-2 is the original series moved forward two time periods, etc. Table 18.3 shows the first 24 months of the Amtrak ridership series, the lag-1 series, and the lag-2 series. (To create a lagged variable in JMP, create a new column, and then use the formula editor with the *Row, Lag* function or right-click on the column and use the *New Formula Column* shortcut to create lag columns using *Row > Lag* or *Row > Lag Multiple*.)

TABLE 18.3   First 24 months of Amtrak ridership series

Month	Ridership	Lag-1 Series	Lag-2 Series
Jan-91	1709		
Feb-91	1621	1709	
Mar-91	1973	1621	1709
Apr-91	1812	1973	1621
May-91	1975	1812	1973
Jun-91	1862	1975	1812
Jul-91	1940	1862	1975
Aug-91	2013	1940	1862
Sep-91	1596	2013	1940
Oct-91	1725	1596	2013
Nov-91	1676	1725	1596
Dec-91	1814	1676	1725
Jan-92	1615	1814	1676
Feb-92	1557	1615	1814
Mar-92	1891	1557	1615
Apr-92	1956	1891	1557
May-92	1885	1956	1891
Jun-92	1623	1885	1956
Jul-92	1903	1623	1885
Aug-92	1997	1903	1623
Sep-92	1704	1997	1903
Oct-92	1810	1704	1997
Nov-92	1862	1810	1704
Dec-92	1875	1862	1810

To compute the lag-1 autocorrelation, which measures the linear relationship between values in consecutive time periods, we compute the correlation between the original series and the lag-1 series (e.g., via the *Multivariate Methods > Multivariate* platform) to be 0.079. Note that although the original series shown above has 24 time periods, the lag-1 autocorrelation will only be based on 23 pairs (because the lag-1 series does not have a value for January 91). Similarly, the lag-2 autocorrelation, measuring the relationship between values that are two time periods apart, is the correlation between the original series and the lag-2 series (yielding -0.143).

We can use the JMP *Time Series* platform (under the *Analyze > Specialized Modeling* menu) to directly compute the autocorrelations of a series at different lags. For example, the output from the *Time Series* platform for the 24-month ridership data is shown in Figure 18.9 (all but the first 24 months in the series have been hidden and excluded). The output from the *Time Series* platform is shown. A bar chart of the autocorrelations at different lags (the *Autocorrelation Function Plot,* or *ACF Plot*) is provided (bottom left).

A few typical autocorrelation behaviors are useful to explore:

Strong autocorrelation (positive or negative) at a lag k larger than 1 and its multiples ($2k, 3k, \ldots$) typically reflects a cyclical pattern. For example, strong positive autocorrelation at lag 12 in monthly data will reflect an annual seasonality (where values during a given month each year are positively correlated).

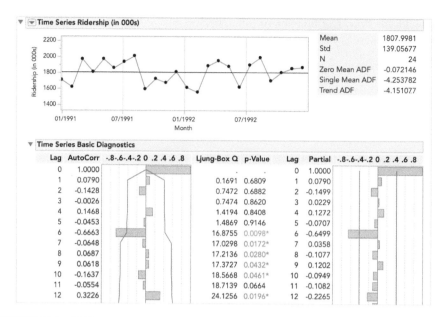

FIGURE 18.9 JMP *Time Series* platform output for the first 24 months of the Amtrak ridership data (partial output with only the first 12 lags displayed)

Positive lag-1 autocorrelation (called "stickiness") describes a series where consecutive values move generally in the same direction. In the presence of a strong linear trend, we would expect to see a strong and positive lag-1 autocorrelation.

Negative lag-1 autocorrelation reflects swings in the series, where high values are immediately followed by low values, and vice versa.

Examining the autocorrelation of a series can therefore help detect seasonality patterns. In this example, the ACF plot (bottom left in Figure 18.9), we see the strongest autocorrelation is at lag-6, and the autocorrelation is negative. This indicates a biannual pattern in ridership, with six-month swings from high to low ridership. A look at the time plot (top, in Figure 18.9) confirms the high-summer low-winter pattern (the x-axis has been rescaled to better show this pattern).

Note that a partial autocorrelation function plot, or *PACF* (bottom right in Figure 18.9), is produced by default in the JMP Time Series platform. This plot shows the autocorrelation at different lags after adjusting for (or, removing the influence of) all other points. Throughout this chapter, we will limit our discussion to the ACF plot.

In addition to looking at autocorrelations of the raw series, it is very useful to look at autocorrelations of *residual series* (that is, the forecast error series). For example, after fitting a time series (or regression) model, we can save the residuals to the data table. Then, we can examine the autocorrelation of the series of *residuals* using the *Time Series* platform. If we have adequately modeled the seasonal pattern, then the residual series should show no autocorrelation at the season's lag.

Figure 18.10 displays the autocorrelations for the residuals from the regression model with seasonality and quadratic trend shown in Figure 18.8. It is clear that the 6-month (and 12-month) cyclical behavior no longer dominates the series of residuals, indicating that the regression model captured them adequately. However, we can also see a strong positive autocorrelation from lag-1 on, indicating a positive relationship between neighboring residuals. This is valuable information, which can be used to improve forecasts.

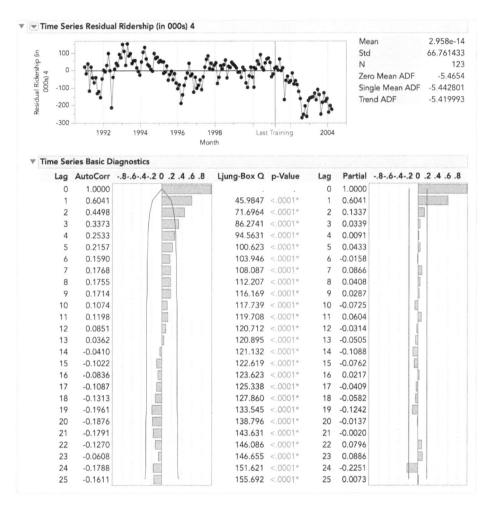

FIGURE 18.10   Amtrak data: analysis of residuals from regression model with seasonality and quadratic trend

Improving Forecasts by Integrating Autocorrelation Information

In general, there are two approaches to taking advantage of autocorrelation. One is by directly building the autocorrelation into the regression model, and the other is by constructing a second-level forecasting model on the residual series.

Among regression-type models that directly account for autocorrelation are *autoregressive (AR) models*, or the more general class of models called ARIMA models (autoregressive integrated moving average models). Autoregression models are similar to linear regression models, except that the predictors are the past values of the series. For example, an autoregression model of order 2, denoted AR(2), can be written as

$$Y_t = \beta_0 + \beta_1 Y_{t-1} + \beta_2 Y_{t-2} + \epsilon. \tag{18.2}$$

Estimating such models is roughly equivalent to fitting a linear regression model with the series as the output and the two lagged series (at lag-1 and lag-2 in this example) as the predictors. However, using the designated ARIMA estimation methods (e.g., those available in the JMP *Time Series > ARIMA* platform) will produce more accurate results than ordinary linear regression estimation.[4]

FITTING AR MODELS IN THE JMP TIME SERIES PLATFORM

To fit an AR model to training data in JMP, select the series as *Y, Series* and the time variable as *X, Time*. Then, change the *Forecast Periods* in the launch dialog to the number of periods in the validation set, check the *Forecast on Holdback* box, and click *enter*. Then, select *ARIMA* from the red triangle in the *Time Series* platform, and set *p, Autoregressive Order* to the required order.

The *Model Comparison* window will display a summary of the fit of the model (drag the scroll bar to view the holdback statistics). Check the *Graph* box to show the fitted and forecasted values, along with ACF and PACF residual plots at lags 0–25. To view the model results, check the *Report* box (this is selected by default). The red triangle for the model can be used to change the results displayed or to save the predicted values (forecasts), residuals (forecast errors), and other values to a new data table (as discussed in Chapter 17).

Specifying ARIMA models in JMP is facilitated by the *ARIMA Model Group* option within the *Time Series* platform. This allows you to specify the range of orders for the model and to also address seasonality. The *Time Series* platform will be discussed in greater detail in Chapter 19. (For more information on fitting ARIMA and Seasonal ARIMA models in JMP, search for *Time Series* under *Predictive and Specialized Modeling* in the JMP documentation Library (under the *Help* menu in JMP)).

Moving from AR to ARIMA models creates a larger set of more flexible forecasting models and can include a seasonal component (using *Seasonal ARIMA* or *ARIMA Model Group*). But fitting these models requires much more statistical expertise. Even with the simpler AR models, fitting them to raw time series that contain patterns such as trends and seasonality requires you to perform operations that remove the trend and seasonality before fitting the AR model, and then to choose the order of the model. These are not straightforward tasks. Because ARIMA modeling is less robust and requires more experience and statistical expertise than other methods, we do not discuss it here and direct the interested reader to classic time series textbooks (e.g., see chapter 4 in Chatfield, 2003).

---

[4] ARIMA model estimation differs from ordinary regression estimation by accounting for the dependence between observations.

Fitting AR Models to Residuals

We now discuss a particular use of AR models that is straightforward in the context of forecasting, and can provide significant improvement to short-term forecasts. This relates to the second approach for utilizing autocorrelation, which requires constructing a second-level forecasting model for the residuals, as follows:

1. Generate a k-step-ahead forecast of the series (F_{t+k}), using any forecasting method.
2. Generate a k-step-ahead forecast of the forecast error (residual) (E_{t+k}), using an AR (or other) model.
3. Improve the initial k-step-ahead forecast of the series by adjusting it according to its forecasted error: *Improved* $F^*_{t+k} = F_{t+k} + E_{t+k}$.

In particular, we can fit low-order AR models to series of residuals (or *forecast errors*) which can then be used to forecast future forecast errors. By fitting the series of residuals, rather than the raw series, we avoid the need for initial data transformations (because the residual series is not expected to contain any trends or cyclical behavior besides autocorrelation). To fit an AR model to the series of residuals, we first examine the autocorrelations of the residual series. We then choose the order of the AR model according to the lags in which autocorrelation appears. Often, when autocorrelation exists at lag-1 and higher, it is sufficient to fit an AR(1) model of the form

$$E_t = \beta_0 + \beta_1 E_{t-1} + \epsilon, \tag{18.3}$$

where E_t denotes the residual (or *forecast error*) at time t. For example, although the auto-correlations in Figure 18.10 appear large from lags 1 to 5 or so, it is likely that an $AR(1)$ would capture all these relationships. The reason is that if immediate neighboring values are correlated, then the relationship propagates to values that are two periods away, then three periods away, and so forth.

The result of fitting an AR(1) model to the Amtrak ridership residual series is shown in Figure 18.11. The AR(1) coefficient (0.5998) is nearly identical to the lag-1 autocorrelation (0.6041) that we found earlier (see Figure 18.10). The forecasted residual for April 2001, highlighted in Figure 18.12, is computed by plugging in the most recent mean-centered[5] residual from March 2001 (equal to 12.108) into the AR(1) model:

$$0.3704 + (0.599)(12.108 - 0.3704) = 7.410.$$

The positive value tells us that the regression model will produce a ridership forecast for April 2001 that is low, and that we should adjust it by adding 7410 riders. In this example, the regression model (with quadratic trend and seasonality) produced a forecast of 2,004,270 riders, and the improved two-stage model (regression + AR(1) correction) corrected it by increasing it to 2,011,906 riders. The actual value for April 2001 turned out to be 2,023,792 riders—much closer to the improved forecast.

From the plot of the actual vs. forecasted residual series (under *Forecast* in Figure 18.11), we can see that the AR(1) model fits the residual series quite well. Note, however, that the

---

[5]The coefficients produced by JMP's ARIMA function are slightly different from estimates of β_0 and β_1 in an AR model as shown in equation (18.4). The "intercept" is equal to the series mean. To generate a forecast, we must first subtract the mean from the plugged-in value, then multiply by the AR1 coefficient.

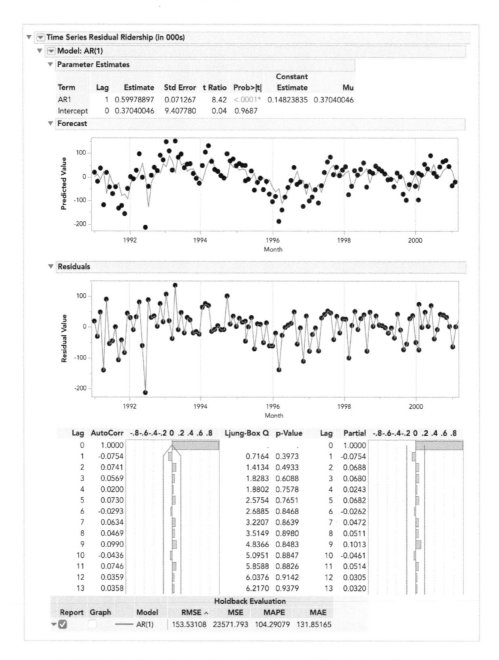

FIGURE 18.11   Partial output from of AR(1) model fitted to the residual series

plot is based on the training data (until March 2001). To evaluate predictive performance of the two-level model (regression + AR(1)), we would have to examine performance (e.g., via MAPE, RMSE, or MAE metrics) on the validation data, in a similar way to the calculation that we performed for April 2001 above. These values are reported in the *Model Comparison* window for the model when holdback data are used (see the box on fitting AR models).

	Month	Actual Residual Ridership (in 000s) ...	Predicted Residual Ridership (in 000s) ...
121	01/2001	-37.3262	26.6786
122	02/2001	-21.4293	-22.2396
123	03/2001	12.1076	-12.7048
124	04/2001	•	7.4102
125	05/2001	•	4.5928
126	06/2001	•	2.9030
127	07/2001	•	1.8894
128	08/2001	•	1.2815
129	09/2001	•	0.9169
130	10/2001	•	0.6982
131	11/2001	•	0.5670
132	12/2001	•	0.4883
133	01/2002	•	0.4411
134	02/2002	•	0.4128
135	03/2002	•	0.3958
136	04/2002	•	0.3857
137	05/2002	•	0.3796

FIGURE 18.12   Forecasted residual

Finally, to examine whether we have indeed accounted for the autocorrelation in the series and that no more information remains in the series, we examine the autocorrelations of the series of residuals-of-residuals (the residuals obtained after the AR(1) was applied to the regression residuals). This can be seen in Figure 18.11. It is clear that no more auto-correlation remains and that the addition of the AR(1) model has adequately captured the autocorrelation information.

We mentioned earlier that improving forecasts via an additional AR layer is useful for short-term forecasting. The reason is that an AR model of order k will usually only provide useful forecasts for the next k periods, and after that forecasts will rely on earlier forecasts rather than on actual data. For example, to forecast the residual of May 2001, when the time of prediction is March 2001, we would need the residual for April 2001. However, because that value is not available, it would be replaced by its forecast. Hence the forecast for May 2001 would be based on the forecast for April 2001.

Evaluating Predictability

Before attempting to forecast a time series, it is important to determine whether it is predictable, in the sense that its past can be used to predict its future beyond the naive forecast. One useful way to assess predictability is to test whether the series is a "random walk." A random walk is a series in which changes from one time period to the next are random. According to the Efficient Market Hypothesis in economics, asset prices are random walks, and therefore predicting stock prices is a game of chance.[6]

A random walk is a special case of an AR(1) model, where the slope coefficient is equal to 1:

$$Y_t = \beta_0 + Y_{t-1} + \epsilon_t. \tag{18.4}$$

---

[6]There is some controversy surrounding the Efficient Market Hypothesis, with claims that there is slight autocorrelation in asset prices, which does make them predictable to some extent. However, transaction costs and bid–ask spreads tend to offset any prediction benefits.

We can also write this formula as

$$Y_t - Y_{t-1} = \beta_0 + \epsilon_t. \tag{18.5}$$

We see from this equation that the difference between the values at periods $t-1$ and t is random, hence the term "random walk." Forecasts from such a model are basically equal to the most recent observed value (the naive forecast), reflecting the lack of any other information.

Testing whether a series is a random walk is equivalent to testing whether the lag-1 differenced series $Y_t - Y_{t-1}$ $(t = 2, 3, \ldots)$ behaves like a random series. The latter can be done by examining the autocorrelations of the differenced series: if all autocorrelation values are insignificant, then the original series is likely a random walk.

As an example, consider the series of S&P500 monthly closing prices between May 1995 and August 2003 (available in SP500.jmp, shown in Figure 18.13). The autocorrelation plot of the lag-1 differenced closing prices series is shown in Figure 18.14. We see that

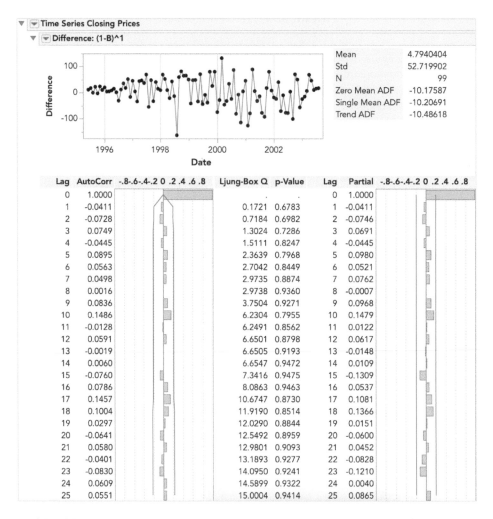

FIGURE 18.13   **Autocorrelations of lag-1 differenced S&P500 monthly closing prices**

Model: AR(1)

Parameter Estimates

Term	Lag	Estimate	Std Error	t Ratio	Prob>|t|	Constant Estimate	Mu
AR1	1	0.98338	0.0145	67.71	<.0001*	14.7664162	888.309208
Intercept	0	888.30921	221.8832	4.00	0.0001*		

FIGURE 18.14 Partial output for AR(1) model on S&P500 monthly closing prices

all autocorrelation values are practically zero, indicating that the closing prices series is a random walk.

Another mathematically equivalent approach for evaluating whether a series is a random walk is to use an AR(1) model and test whether the AR(1) coefficient is equal to 1. For example, the AR(1) model fitted to the series of S&P500 monthly closing prices is shown in Figure 18.14. Here the AR1 coefficient is 0.9834, with a standard error of 0.0145. The coefficient is sufficiently close to 1 (around one standard error away), indicating (as found by looking at the differenced series' ACF) that this is a random walk. Forecasting this series using any of the methods described earlier (aside from the naive forecast) is therefore futile.

SUMMARY: FITTING REGRESSION-BASED TIME SERIES MODELS IN JMP

To fit models with trend and/or seasonality, we use the standard *Fit Model* platform and the built-in validation option.

- In this chapter, we have changed the default output by using the *red triangle* options. For example, we have turned off some of the standard graphical and statistical output. To show all available options, click the *Alt* or *Option* button on your keyboard before using the red triangle. Some graphs have been resized (using the *cursor*) and repositioned on the screen (using the *selection tool*), and other graphs have been created using the *Graph Builder*, as described earlier.
- To fit a regression model with an exponential trend, take the log of the time series and use the *Fit Model* platform. Save the prediction formula to the data table. Use the *Exp* transformation to adjust the forecasts to the original scale. Use a formula to calculate the residuals.
- To fit an autoregressive (AR) model, partition the data and use *Forecast Periods* and check the *Forecast on Holdback* box in the *Time Series* launch dialog. Then, fit the AR model using the *ARIMA* red triangle option. A time plot, ACF, and residual plots are provided. Forecasts and forecast errors can be saved to a new data table (using *Save Columns*). Copy and paste these columns into the original table, and graph using the *Graph Builder*.

PROBLEMS

18.1 **Impact of September 11 on Air Travel in the United States.** The Research and Innovative Technology Administration's Bureau of Transportation Statistics (BTS) conducted a study to evaluate the impact of the September 11, 2001, terrorist attack on US transportation. The 2006 study report and the data can be found at https://www.bts.gov/archive/publications/estimated_impacts_of_9_11_on_us_travel/index. The goal of the study was stated as follows:

> The purpose of this study is to provide a greater understanding of the passenger travel behavior patterns of persons making long distance trips before and after 9/11.

The report analyzes monthly passenger movement data between January 1990 and May 2004. Data on three monthly time series are given in file Sept11Travel.jmp for this period: actual (1) airline revenue passenger miles (Air RPM), (2) rail passenger miles (Rail PM), and (3) auto miles traveled (VMT).

In order to assess the impact of September 11, BTS took the following approach: using data before September 11, they forecasted future data (under the assumption of no terrorist attack). Then they compared the forecasted series with the actual data to assess the impact of the event. Our first step therefore is to split each of the time series into two parts: pre- and post-September 11. We now concentrate only on the earlier time series.

a. Plot the pre-event AIR RPM time series. What time series components appear from the plot?

b. In Figure 18.15, we show a time plot of the seasonally adjusted pre-September-11 AIR RPM series. Which of the following methods would be adequate for forecasting this series?

  • Linear regression model with seasonality
  • Linear regression model with trend
  • Linear regression model with seasonality and trend

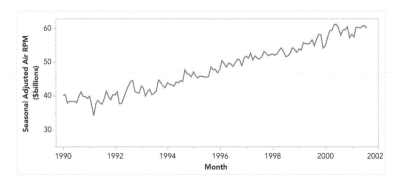

FIGURE 18.15   **Seasonally adjusted pre-September-11 AIR RPM series**

c. In part (d), you will run a linear regression model for the AIR RPM series that will produce a seasonally adjusted series similar to the one shown in part (b), with multiplicative (exponential) seasonality. What is the output variable? What are the predictors? Prepare the data for this regression model. (*Hint*: Use a *Date Time* transformation to create a new column with the month, and make sure the modeling type for this column is nominal.)

d. Run the regression model from part (c) using only the pre-event data (using the *Fit Model* platform).

    i. What can we learn from the statistical insignificance of the coefficients for September and October?

    ii. Save the predicted values to the data table and calculate the residuals. The actual value of AIR RPM (air revenue passenger miles) in January 1990 was 35.153577 billion. What is the residual for this month, using the regression model? Report the residual in terms of air revenue passenger miles.

e. Create an ACF (autocorrelation) plot of the regression residuals using the *Time Series* platform.

    i. What does the ACF plot tell us about the regression model's forecasts?

    ii. How can this information be used to improve the model?

f. Fit linear regression models to Air, Rail, and Auto with additive seasonality and an appropriate trend. For Air and Rail, fit a linear trend. For Rail, use a quadratic trend. Remember to use only pre-event data. Once the models are estimated, use them to forecast each of the three post-event series.

    i. For each series (Air, Rail, Auto), plot the complete pre-event and post-event actual series overlayed with the predicted series.

    ii. What can be said about the effect of the September 11 terrorist attack on the three modes of transportation? Discuss the magnitude of the effect, its time span, and any other relevant aspects.

18.2 Canadian Manufacturing Workers Workhours.
The time series plot in Figure 18.16 describes the average annual number of weekly hours spent by Canadian manufacturing workers. (Data are available in CanadianWorkHours.jmp, data courtesy of Ken Black.)

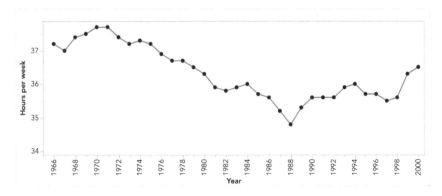

FIGURE 18.16   **Average annual weekly hours spent by Canadian manufacturing workers**

a. Which of the following regression-based models would fit the series best? (Choose one.)

- Linear trend model
- Linear trend model with seasonality
- Quadratic trend model
- Quadratic trend model with seasonality

b. If we computed the autocorrelation of this series, would the lag-1 autocorrelation exhibit negative, positive, or no autocorrelation?

   i. How can you see this from the plot in Figure 18.16?

   ii. Create a time series plot using the *Time Series* platform. Use the ACF plot to verify your answer to the previous question.

18.3 **Toys "R" Us Revenues.** Figure 18.17 is a time series plot of the quarterly revenues of Toys "R" Us between 1992 and 1995. (Thanks to Chris Albright for suggesting the use of these data, which are available in ToysRUsRevenues.jmp.)

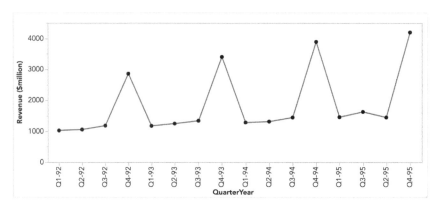

FIGURE 18.17   Quarterly revenues of Toys "R" Us, 1992–1995

a. Fit a regression model with a linear trend and quarter (you'll need to create a new nominal column, "Quarter"). Use the first 14 quarters as the training set and the last two quarters as the validation set (create a new validation column to indicate the partition).

b. A partial output of the regression model is shown in Figure 18.18. Use this output to answer the following questions:

   i. Name two statistics (and their values) that measure how well this model fits the training data.

   ii. Name the statistic (and its value) that measures the predictive accuracy of this model.

   iii. What is the coefficient for the trend? Interpret this value.

   iv. After adjusting for trend, which quarter ($Q1, Q2, Q3,$ or $Q4$) has the highest average sales?

▼ ⌄ **Response Revenue ($million)**

Validation: Validation

▼ **Parameter Estimates**

Term	Estimate	Std Error	t Ratio	Prob>|t|
Intercept	1450.6964	95.90181	15.13	<.0001*
t	47.107143	11.25663	4.18	0.0024*
Quarter[Q1]	-543.9464	75.16136	-7.24	<.0001*
Quarter[Q2]	-559.0536	75.16136	-7.44	<.0001*
Quarter[Q3]	-454.7798	82.65514	-5.50	0.0004*

▼ **Effect Tests**

Source	Nparm	DF	Sum of Squares	F Ratio	Prob > F
t	1	1	497075	17.5129	0.0024*
Quarter	3	3	10208020	119.8824	<.0001*

▼ **Crossvalidation**

Source	RSquare	RASE	Freq
Training Set	0.9774	135.08	14
Validation Set	0.9404	313.68	2

FIGURE 18.18   Output for regression model fitted to Toys "R" Us time series and its performance on training and validation periods

18.4 Walmart Stock. Figure 18.19 shows the series of Walmart daily closing prices between February 2001 and February 2002. (Thanks to Chris Albright for suggesting the use of these data, which are publicly available, e.g., at http://finance.yahoo.com and are in the file WalMartStock.jmp.)

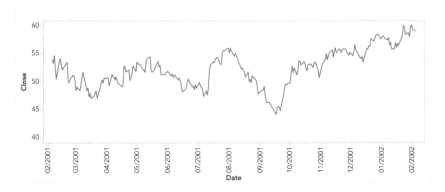

FIGURE 18.19   Daily close price of Walmart stock, February 2001–2002

a. Compute a lag-1 differenced series of the closing prices. Create a time plot of the differenced series. Note your observations by visually inspecting the differenced series.

b. Run an AR(1) model for the close price series. Report the parameter estimates table.

c. Which of the following is/are relevant for testing whether this stock is a random walk?

- The autocorrelations of the close prices series
- The AR(1) slope coefficient
- The AR(1) constant coefficient

d. Does the AR model indicate that this is a random walk? Explain how you reached your conclusion.

e. What are the implications of finding that a time series is a random walk? Choose the correct statement(s) below:

- It is impossible to obtain forecasts that are more accurate than naive forecasts for the series.
- The series is random.
- The changes in the series from one period to the other are random.

18.5 Department Store Sales. The time series plot in Figure 18.20 describes actual quarterly sales for a department store over a six-year period. (Data are available in DepartmentStoreSales.jmp, data courtesy of Chris Albright.) Assume that the first record is Q1.

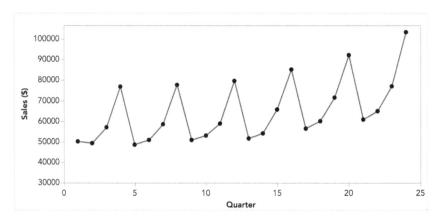

FIGURE 18.20   Department store quarterly sales series

a. The forecaster decided that there is an exponential trend in the series. The forecaster needs to fit a regression-based model that accounts for this trend. Which of the following operations must be performed?

- Take log of Quarter index
- Take log of sales
- Take an exponent of sales
- Take an exponent of Quarter index

b. Fit a regression model with an exponential trend and seasonality using *Fit Model*, and only first 20 quarters as the training data (remember to first partition the series into training and validation series).

c. A regression model with trend (Quarter) and seasonality (Quarter 2) was fit. A partial output is shown in Figure 18.21. From the output, after adjusting for trend, are Q2 average sales higher, lower, or approximately equal to the average Q1 sales?

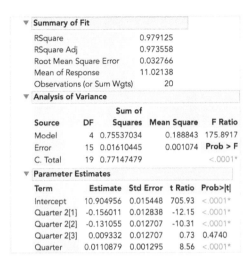

Summary of Fit

RSquare	0.979125
RSquare Adj	0.973558
Root Mean Square Error	0.032766
Mean of Response	11.02138
Observations (or Sum Wgts)	20

Analysis of Variance

Source	DF	Sum of Squares	Mean Square	F Ratio
Model	4	0.75537034	0.188843	175.8917
Error	15	0.01610445	0.001074	**Prob > F**
C. Total	19	0.77147479		<.0001*

Parameter Estimates

Term	Estimate	Std Error	t Ratio	Prob>|t|
Intercept	10.904956	0.015448	705.93	<.0001*
Quarter 2[1]	-0.156011	0.012838	-12.15	<.0001*
Quarter 2[2]	-0.131055	0.012707	-10.31	<.0001*
Quarter 2[3]	0.009332	0.012707	0.73	0.4740
Quarter	0.0110879	0.001295	8.56	<.0001*

FIGURE 18.21   Output from regression model fit to department store sales training series

d. Use this model to forecast sales in quarters 21 and 22.

e. The plots in Figure 18.22 describe the fit (top) and forecast errors (bottom) from this regression model.

    i. Recreate these plots.

    ii. Based on these plots, what can you say about your forecasts for quarters 21 and 22? Are they likely to overforecast, underforecast, or be reasonably close to the real sales values?

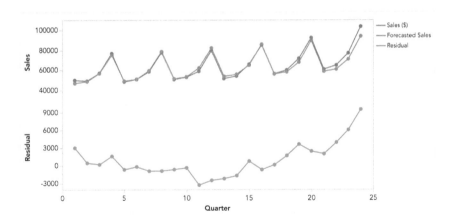

FIGURE 18.22   Fit of regression model for department store sales

f. Looking at the residual plot, which of the following statements appear true?

- Seasonality is not captured well.
- The regression model fits the data well.
- The trend in the data is not captured well by the model.

g. Which of the following solutions is adequate *and* a parsimonious solution for improving model fit?

- Fit a quadratic trend model to the residuals (with Quarter and Quarter^2).
- Fit an AR model to the residuals.
- Fit a quadratic trend model to sales (with Quarter and Quarter^2).

18.6 Souvenir Sales. Figure 18.23 shows a time plot of monthly sales for a souvenir shop at a beach resort town in Queensland, Australia, between 1995 and 2001. [Data are available in SouvenirSales.jmp, Source: Hyndman and Yang (2018).] The series is presented twice, in Australian dollars and in log scale. Back in 2001, the store wanted to use the data to forecast sales for the next 12 months (year 2002). They hired an analyst to generate forecasts. The analyst first partitioned the data into training and validation sets, with the validation set containing the last 12 months of data (year 2001). She then fit a regression model to sales.

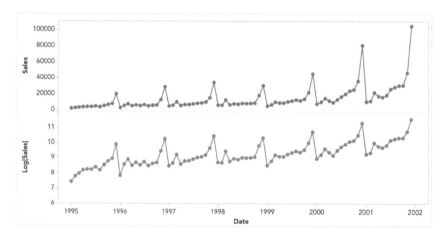

FIGURE 18.23 **Monthly sales at Australian souvenir shop in dollars (top) and in log scale (bottom)**

a. Based on the two time plots, which predictors should be included in the regression model? What is the total number of predictors in the model?

b. Run a regression model on the training data with Sales (in Australian dollars) as the output variable and with a linear trend and monthly predictors. Call this model A.

i. Examine the estimated coefficients. Which month tends to have the highest average sales during the year? Why is this reasonable?

ii. What is the estimated coefficient for trend? What does this mean?

c. Run a regression model with log(Sales) as the output variable, and with a linear trend and monthly predictors. Again, remember to use the validation column. Call this model B.

i. Fitting a model to log(Sales) with a linear trend is equivalent to fitting a model to Sales (in dollars) with what type of trend?

    ii. What is the estimated coefficient for trend? What does this mean?

    iii. Use this model to forecast the sales in February 2002. What is the extra step needed?

  d. Compare the two regression models (A and B) in terms of forecast performance. Which model is preferable for forecasting? Mention at least two reasons based on the information in the outputs.

  e. Continuing with Model B (with log(Sales) as output), fit an AR model with lag-2 [ARIMA(2,0,0)] to the forecast errors (without partitioning into training/validation). We will use this model to improve the forecast for January 2002.

    i. Examining the ACF plot and the estimated coefficients of the AR(2) model (and their statistical significance), what can we learn about the forecasts that result from Model B?

    ii. Use the autocorrelation information to compute an improved forecast for January 2002, using model B and the AR(2) model above.

  f. How would you model these data differently if the goal was to understand the different components of sales in the souvenir shop between 1995 and 2001? Mention two differences.

18.7 **Shipments of Household Appliances.** The time plot in Figure 18.24 shows the series of quarterly shipments (in million $) of US household appliances between 1985 and 1989. (Data are available in `ApplianceShipments.jmp`, data courtesy of Ken Black.)

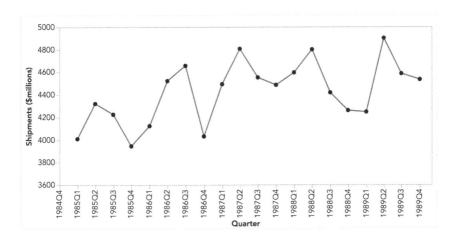

FIGURE 18.24   **Quarterly shipments of US household appliances over five years**

  a. If we compute the autocorrelation of the series, which lag (>0) is most likely to have the largest coefficient (in absolute value)?

  b. Use the *Time Series* platform to create a time plot. Examine the ACF plot, and compare with your answer in part (a).

18.8 **Australian Wine Sales.** Figure 18.25 shows time plots of monthly sales of six types of Australian wines (red, rose, sweet white, dry white, sparkling, and fortified) for 1980 to 1994. [Data are available in `AustralianWines.jmp`, Source: Hyndman and

Yang (2018).] The units are thousands of liters. You are hired to obtain short term forecasts (2–3 months ahead) for each of the six series, and this task will be repeated every month.

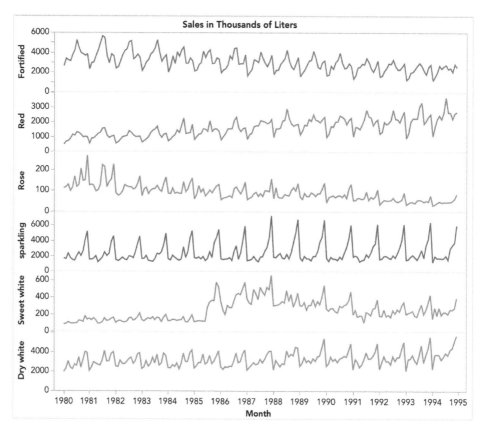

FIGURE 18.25   Monthly sales of six types of Australian wines between 1980 and 1994

a. Which forecasting method would you choose if you had to choose the same method for all series? Why?

b. Fortified wine has the largest market share of the these six types of wine. You are asked to focus on fortified wine sales alone and to produce as accurate as possible forecast for the next two months.

- Start by partitioning the data using the period until December 1993 as the training set.
- Fit a regression model to sales with a linear trend and seasonality.

  i. Create the "actual vs. forecast" and forecast errors plots. What can you say about the model fit?

  ii. Use the regression model to forecast sales in January and February 1994.

c. Create an ACF plot for the residuals from the model until lag-12. Examining this plot, which of the following statements are reasonable?

- Decembers (month 12) are not captured well by the model.
- There is a strong correlation between sales on the same calendar month.
- The model does not capture the seasonality well.
- We should try to fit an autoregressive model with lag-12 to the residuals.

19

SMOOTHING AND DEEP LEARNING METHODS FOR FORECASTING

In this chapter, we describe a set of popular and flexible methods for forecasting time series that rely on smoothing. Smoothing is based on averaging over multiple periods in order to reduce the noise. We start with two simple smoothers, the moving average and simple exponential smoother, which are suitable for forecasting series that contain no trend or seasonality. In both cases, forecasts are averages of previous values of the series (the length of the series history that is considered and the weights that are used in the averaging differ between the methods). We also show how a moving average can be used, with a slight adaptation, for data visualization. We then proceed to describe smoothing methods that are suitable for forecasting series with a trend and/or seasonality. Smoothing methods are data driven and are able to adapt to changes in the series over time. Although highly automated, the user must specify smoothing constants, which determine how fast the method adapts to new data. We discuss the choice of such constants and their meaning. The different methods are illustrated using the Amtrak ridership series. We then describe basic deep learning methods suitable for forecasting time series: recurrent neural networks (RNN) and Long Short-Term Memory (LSTM) networks.

Smoothing Methods in JMP: In this chapter, we will use the JMP Time Series platform (*Analyze > Specialized Modeling > Time Series*) to apply smoothing methods.

19.1 INTRODUCTION[1]

A second class of methods for time series forecasting is smoothing methods. Unlike regression models, which rely on an underlying theoretical model for the components of

[1] This and subsequent sections in this chapter copyright © 2022 Datastats, LLC, and Galit Shmueli. Used by permission.

Machine Learning for Business Analytics: Concepts, Techniques, and Applications with JMP Pro®,
Second Edition. Galit Shmueli, Peter C. Bruce, Mia L. Stephens, Muralidhara Anandamurthy, and Nitin R. Patel.
© 2023 John Wiley & Sons, Inc. Published 2023 by John Wiley & Sons, Inc.

a time series (e.g., linear or quadratic trend), smoothing methods are data driven, in the sense that they estimate time-series components directly from the data without a predetermined structure. Data-driven methods are especially useful in series where patterns change over time. Smoothing methods "smooth" out the noise in a series in an attempt to uncover the patterns. Smoothing is done by averaging the series over multiple periods, where different smoothers differ by the number of periods averaged, how the average is computed, how many times averaging is performed, and so on. We now describe two types of smoothing methods that are popular in business applications due to their simplicity and adaptivity. These are the moving average method and exponential smoothing.

19.2   MOVING AVERAGE

The moving average is a simple smoother: it consists of averaging across a window of consecutive observations, thereby generating a series of averages. A moving average with window width w means averaging across each set of w consecutive values, where w is determined by the user.

In general, there are two types of moving averages: a *centered moving average* and a *trailing moving average*. Centered moving averages are powerful for visualizing trends because the averaging operation can suppress seasonality and noise, thereby making the trend more visible. In contrast, trailing moving averages are useful for forecasting. The difference between the two is in terms of placing the window over the time series.

Centered Moving Average for Visualization

In a centered moving average, the value of the moving average at time t (MA_t) is computed by centering the window around time t and averaging across the w values within the window:

$$MA_t = \left(Y_{t-(w-1)/2} + \cdots + Y_{t-1} + Y_t + Y_{t+1} + \cdots + Y_{t+(w-1)/2}\right)/w.$$

For example, with a window of width $w = 5$, the moving average at time point $t = 3$ means averaging the values of the series at time points $1, 2, 3, 4, 5$; at time point $t = 4$, the moving average is the average of the series at time points $2, 3, 4, 5, 6$, and so on. This is illustrated in the top panel of Figure 19.1.

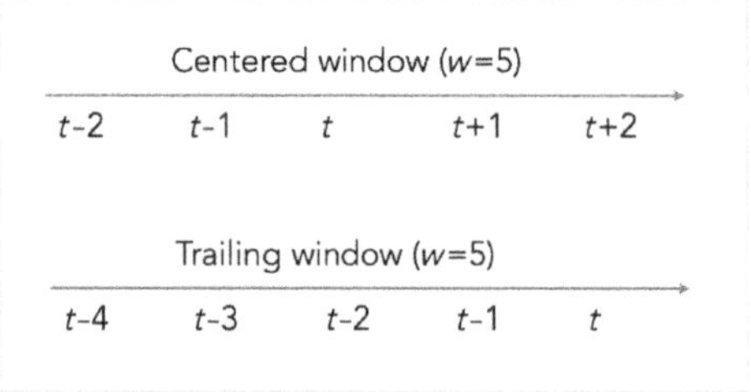

FIGURE 19.1   Schematic of centered moving average (top) and trailing moving average (bottom), both with window width $w = 5$

Choosing the window width in a seasonal series is straightforward: since the goal is to suppress seasonality for better visualizing the trend, the default choice should be the length of a season. Returning to the Amtrak ridership data, the annual seasonality indicates a choice of $w = 12$. Figure 19.2 (red line) shows a centered moving average line overlaid on the original series. We can see a global U-shape, but unlike the regression model that fits a strict U-shape, the moving average shows some deviation, such as the slight dip during the last year.

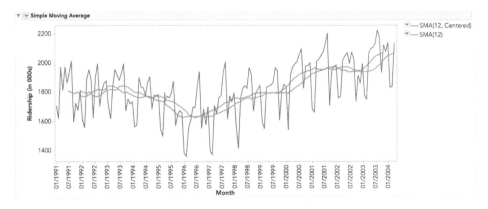

FIGURE 19.2   Centered moving average (red line) and trailing moving average (green line) with window $w = 12$, overlaid on Amtrak ridership series

The plot shown in Figure 19.2 is obtained by using the *Time Series* platform, and choosing *Simple Moving Average* from the list of available *Smoothing Models* (see Figure 19.3). This allows you to set the length of the window and to specify whether to use a centered moving average (*Centered*) or a trailing moving average (*No Centering*).[2]

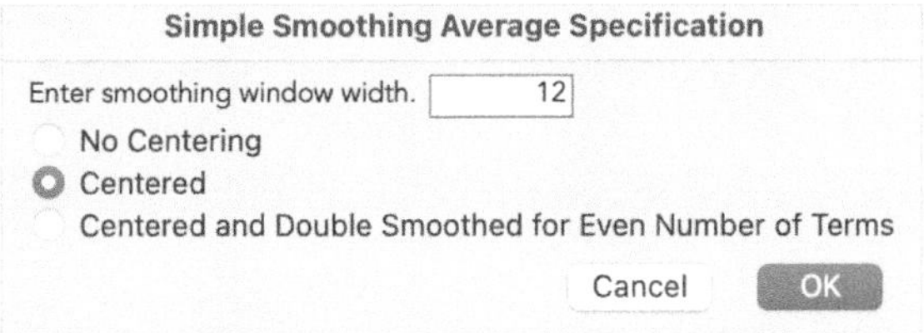

FIGURE 19.3   JMP time series simple moving average specification window

Trailing Moving Average for Forecasting

Centered moving averages are computed by averaging across data in the past and the future of a given time point. In this sense, they cannot be used for forecasting, since at the time of forecasting the future is typically unknown. Hence, for purposes of forecasting, we use

---

[2]For an even window width, such as $w = 4$, obtaining the centered moving average at time point $t = 3$ requires averaging across two windows: across time points $1, 2, 3, 4$; across time points $2, 3, 4, 5$; and finally the average of the two averages is the final moving average. In JMP, use the *Centered and Double Smoothed* option (see Figure 19.3).

trailing moving averages, where the window of width w is set on the most recent available w values of the series. The k-step-ahead forecast F_{t+k} ($k = 1, 2, 3, \ldots$) is then the average of these w values (see also bottom plot in Figure 19.1):

$$F_{t+k} = \left(Y_t + Y_{t-1} + \cdots + Y_{t-w+1} \right) / w.$$

For example, in the Amtrak ridership series, to forecast ridership in February 1992 or later months, given information until January 1992 and using a moving average with window width $w = 12$, we would take the average ridership during the most recent 12 months (February 1991 to January 1992).

COMPUTING A TRAILING MOVING AVERAGE FORECAST IN JMP

Note: The Time Series holdback option discussed previously (in Chapter 17) is not applicable for moving average models in JMP. Instead, hide and exclude the holdback (validation) rows before fitting the model (select the holdback rows in the data table and use *Rows > Hide and Exclude*).

To compute a trailing moving average, use *Simple Moving Average* under *Smoothing Models* in the *Time Series* platform, and select *No Centering*. Then save the moving averages to a data table using *Save to Data Table* from the red triangle for the model. Note that this creates a new data table.

- Copy the last moving average in the training data and paste over the moving averages in the holdback data (since information is only known until this value).
- Copy the values in the holdback set (from the original data table) into the corresponding cells in the new data table.
- Create a Lag(1) forecast column. To do this, right-click on the moving average column, and select *New Formula Column > Lag > Lag*. Then, rename the column *Forecast*.
- Create a column with forecast errors (residuals). To do this, select the actual and forecast columns, right-click, and select *New Formula Column > Combine > Difference*. Rename this column *Residual*.

Next we illustrate a 12-month moving average forecaster for the Amtrak ridership. We partition the Amtrak ridership time series, leaving the last 36 months as the holdback (validation) set using *Hide and Exclude* (as discussed in the box). Applying a moving average forecaster with window $w = 12$ and *No Centering*, we save the moving averages to a data table and apply the steps outlined in the box to obtain the output partially shown in Figure 19.4. Note that for the first 11 records of the training set, there is no forecast because there are less than 12 past values to average. Also, note that the forecasts for all months in the validation period should be identical (1938.48) because the method assumes that information is known only until March 2001.

From the graph in Figure 19.5, it is clear that the moving average forecaster is inadequate for generating monthly forecasts because it does not capture the seasonality in the data. Hence seasons with high ridership are underforecasted, and seasons with low ridership are

	Time Ridership (in 000s)	Actual Ridership (in 000s)	SMA(12)	Forecast	Residual
1	01/1991	1709	•	•	•
2	02/1991	1621	•	•	•
3	03/1991	1973	•	•	•
4	04/1991	1812	•	•	•
5	05/1991	1975	•	•	•
6	06/1991	1862	•	•	•
7	07/1991	1940	•	•	•
8	08/1991	2013	•	•	•
9	09/1991	1596	•	•	•
10	10/1991	1725	•	•	•
11	11/1991	1676	•	•	•
12	12/1991	1814	1809.5365	•	•
13	01/1992	1615	1801.6957	1809.5365	-194.7095
14	02/1992	1557	1796.4042	1801.6957	-244.6077
15	03/1992	1891	1789.6132	1796.4042	94.8188
16	04/1992	1956	1801.6395	1789.6132	166.3678
17	05/1992	1885	1794.1187	1801.6395	83.0745
18	06/1992	1623	1774.1758	1794.1187	-171.0767
19	07/1992	1903	1771.1299	1774.1758	129.1332
20	08/1992	1997	1769.7506	1771.1299	225.5821

FIGURE 19.4   **Partial output for moving average forecaster with $w = 12$ applied to Amtrak ridership series**

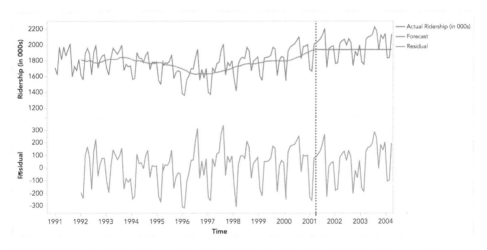

FIGURE 19.5   **Actual vs. forecasted ridership (top) and residuals (bottom) from moving average forecaster (created using the *Graph Builder*). The vertical dotted line separates the training and validation data (double-click on the axis to add a reference line)**

overforecasted. A similar issue arises when forecasting a series with a trend: the moving average "lags behind," thereby underforecasting in the presence of an increasing trend and overforecasting in the presence of a decreasing trend. This "lagging behind" of the trailing moving average can also be seen in Figure 19.2.

In general, the moving average can be used for forecasting *only in series that lack seasonality and trend*. Such a limitation might seem impractical. However, there are a few popular methods for removing trends (de-trending) and removing seasonality (de-seasonalizing) from a series, such as regression models. The moving average can be used to forecast such de-trended and de-seasonalized series, and then the trend and seasonality can be added back to the forecast. For example, consider the regression model shown in Figure 18.8 in Chapter 18, which yields residuals devoid of seasonality and trend. We can apply a moving average forecaster to that series of residuals (also called forecast errors), thereby creating a forecast for the next *forecast error*.

For example, to forecast ridership in April 2001 (the first period in the validation set), assuming that we have information until March 2001, we use the regression model in Figure 18.6 to generate a forecast for April 2001 (which yields 2004.27 thousand riders). We then use a 12-month moving average (using the period April 2000 to March 2001) to forecast the *forecast error* for April 2001, which yields 30.78 (as shown in Figure 19.6). The positive value implies that the regression model's forecast for April 2001 is too low, and therefore we should adjust it by adding approximately 31 thousand riders to the regression model's forecast of 2,004,270 riders.

	Time Residual Ridership	Actual Residual Ridership (in 000s)	SMA(12)	Forecast	Residual
113	05/2000	34.4962	-15.9763	-22.2832	56.7795
114	06/2000	90.1603	-9.8653	-15.9763	106.1366
115	07/2000	15.9662	-8.5972	-9.8653	25.8315
116	08/2000	1.5479	-2.4326	-8.5972	10.1451
117	09/2000	42.1205	9.2658	-2.4326	44.5531
118	10/2000	64.0051	17.2594	9.2658	54.7392
119	11/2000	70.1081	21.6121	17.2594	52.8487
120	12/2000	44.2329	28.5614	21.6121	22.6208
121	01/2001	-37.3262	33.5368	28.5614	-65.8876
122	02/2001	-21.4293	30.4731	33.5368	-54.9661
123	03/2001	12.1076	30.7807	30.4731	-18.3655
124	04/2001	•	27.9592	30.7807	•

FIGURE 19.6   Applying MA to the residuals from the regression model (which lack trend and seasonality), to forecast the April 2001 residual

Choosing Window Width (w)

With moving average forecasting or visualization, the only choice that the user must make is the width of the window (w). As with other methods such as k-nearest neighbors, the choice of the smoothing parameter is a balance between undersmoothing and oversmoothing. For visualization (using a centered window), wider windows will expose more global trends, while narrow windows will reveal local trends. Hence examining several window widths is useful for exploring trends of differing local/global nature. For forecasting (using a trailing window), the choice should incorporate domain knowledge in terms of relevance of past observations and how fast the series changes. Empirical predictive evaluation can also be done by experimenting with different values of w and comparing performance. However, care should be taken not to overfit!

19.3   SIMPLE EXPONENTIAL SMOOTHING

A popular forecasting method in business is exponential smoothing. Its popularity derives from its flexibility, ease of automation, cheap computation, and good performance. Simple exponential smoothing is similar to forecasting with a moving average, except that instead of taking a simple average over the w most recent values, we take a *weighted average* of *all* past values, such that the weights decrease exponentially into the past. The idea is to give more weight to recent information, yet not to completely ignore older information.

Like the moving average, simple exponential smoothing should only be used for forecasting *series that have no trend or seasonality*. As mentioned earlier, such series can be obtained by removing trend and/or seasonality from raw series and then applying exponential smoothing to the series of residuals (which are assumed to contain no trend or seasonality).

The exponential smoother generates a forecast at time $t + 1$ (F_{t+1}) as follows:

$$F_{t+1} = \alpha Y_t + \alpha(1 - \alpha)Y_{t-1} + \alpha(1 - \alpha)^2 Y_{t-2} + \cdots, \qquad (19.1)$$

where α is a constant between 0 and 1 called the *smoothing parameter* or *smoothing weight*. The formulation above displays the exponential smoother as a weighted average of all past observations, with exponentially decaying weights.

It turns out that we can write the exponential forecaster in another way, which is very useful in practice:

$$F_{t+1} = F_t + \alpha E_t, \qquad (19.2)$$

where E_t is the forecast error at time t. This formulation presents the exponential forecaster as an "active learner": it looks at the previous forecast (F_t) and how far it was from the actual value (E_t) and then corrects the next forecast based on that information. If last period the forecast was too high, the next period is adjusted down. The amount of correction depends on the value of the smoothing parameter α. The formulation in (19.2) is also advantageous in terms of data storage and computation time: it means that we need to store and use only the forecast and forecast error from the most recent period, rather than the entire series. In applications where real-time forecasting is done, or many series are being forecasted in parallel and continuously, such savings are critical.

Note that forecasting further into the future yields the same forecast as a one-step-ahead forecast. Because the series is assumed to lack trend and seasonality, forecasts into the future rely only on information until the time of prediction. Hence the k-step-ahead forecast is equal to the one-step-ahead forecast ($F_{t+k} = F_{t+1}$).

Choosing Smoothing Parameter α

The smoothing weight, α, determines the rate of learning. A value close to 1 indicates fast learning (i.e., only the most recent observations have influence on forecasts), whereas a value close to 0 indicates slow learning (past observations have a large influence on forecasts). This can be seen by plugging 0 or 1 into equation (19.1) or (19.2). Hence, the choice of α depends on the required amount of smoothing and on how relevant the history is for generating forecasts. Default values that have been shown to work well are around 0.1 to 0.2. Some trial and error can also help in the choice of α: examine the time plot of the actual and predicted series, as well as the predictive accuracy (e.g., MAPE, RMSE, or MAE of the validation set).

By default, JMP will search for the optimal weight (the value of α), but you can also specify a fixed value or range of values to search for the optimal weight (see Figure 19.7). Finding the α value that optimizes predictive accuracy on the validation set can be used to determine the degree of local vs. global nature of the trend. *Note*: Beware of choosing the "best α" based on training data for forecasting purposes, as this will most likely lead to model overfitting and low predictive accuracy on future data. Therefore, although JMP provides by default the best α (the optimal weight), the model is being fit to the training data and the user should consider trying different values to avoid overfitting.

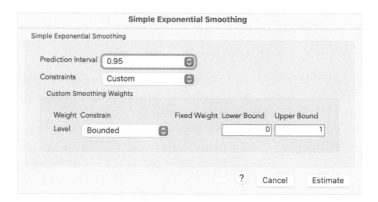

FIGURE 19.7   The JMP *Simple Exponential Smoothing* dialog: change the constraint to *Custom*, and then use *Fixed* to specify a value of the smoothing weight α

To illustrate forecasting with simple exponential smoothing, we return to the residuals from the regression model, as these residuals are assumed to contain no trend or seasonality (the data are in AmtrakTS.jmp). To forecast the residual on April 2001, we apply exponential smoothing to the entire period until March 2001 and use the $\alpha = 0.2$ value. The partial output from the Time Series platform is shown in Figure 19.8, including plots of the actual and forecast values and the forecast errors. (Better visualizations are obtained using *Graph Builder*, as shown in Figure 19.9.) The forecast for the residual for April 2001 is 14.143 (see Figure 19.10). This implies that we should adjust the regression's forecast by adding 14,143 riders from that forecast.

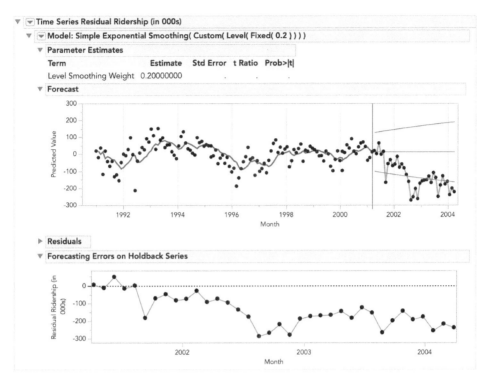

FIGURE 19.8   Partial output for simple exponential smoothing forecaster with $\alpha = 0.2$, applied to the series of residuals from the regression model (which lacks trend and seasonality)

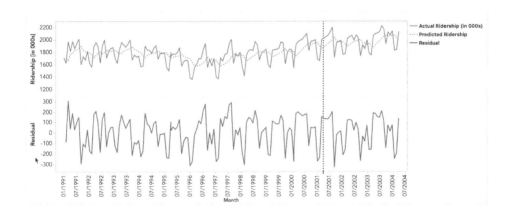

FIGURE 19.9   Graphs of the actual vs. forecasted ridership from the simple exponential smoothing forecaster (top) and residuals (bottom), created using the Graph Builder

	Month	Actual Residual Ridership (in 000s)	Predicted Residual Ridership (in 000s)
121	01/2001	-37.3262	38.9214
122	02/2001	-21.4293	23.6719
123	03/2001	12.1076	14.6517
124	04/2001	19.5211	14.1429

FIGURE 19.10   Forecast for April 2001 residual (partial output from saving the columns for the model to a new data table)

FITTING SIMPLE EXPONENTIAL SMOOTHING MODELS IN JMP

Simple Exponential Smoothing is one of many models available from the Smoothing Model menu under the red triangle in the Time Series platform (under *Analyze > Specialized Modeling*). This smoothing model allows you to specify the *Constraint* set on the smoothing parameter α. By default, JMP will search for the optimal value of α. The *Custom* constraint option allows you to enter a specific *Fixed* value or a *Bounded* range of values to search for the optimal smoothing weight (as shown in Figure 19.7).

All smoothing models in JMP produce summary statistics, parameter estimates for the model, forecasts (using a specified number of forecast periods), ACF and residual ACF plots, and more. Check the *Report* and *Graph* boxes in the *Model Comparison* section to display model results. Use the red triangle for the model to save the prediction formula and residuals to the data table.

Note that smoothing can also be applied to a column in the data table using the formula editor or by right-clicking on the column head and selecting *New Formula Column > Row > Moving Average*, and then specifying the smoothing parameter.

CREATING PLOTS FOR ACTUAL VS. FORECASTED SERIES AND RESIDUALS SERIES USING THE GRAPH BUILDER

Follow the following steps to produce charts similar to those in Figure 19.9:

1. Create a time series model using *Analyze > Specialized Modeling > Time series*.
2. Use *Save columns* option using the red triangle next to the model. This will create a new datatable containing actual, predicted, residual, and 95% prediction intervals.
3. Graph using the *Graph Builder* as discussed in Chapter 18 and customize as needed. For example, double-click on the *x*-axis to add the reference line for the validation data and change the scaling, double-click on an axis label to change, add different graph elements, and right-click on the legend to change the line style.

Relation Between Moving Average and Simple Exponential Smoothing

In both smoothing methods the user must specify a single parameter: in moving averages, the window width (w) must be set; in exponential smoothing, the smoothing parameter (or smoothing weight), α, must be set (or, the software determines an optimal value). In both cases, the parameter determines the importance of fresh information over older information. In fact, the two smoothers are approximately equal if the window width of the moving average is equal to $w = 2/\alpha - 1$.

19.4 ADVANCED EXPONENTIAL SMOOTHING

As mentioned earlier, both moving average and simple exponential smoothing should only be used for forecasting series with no trend or seasonality—meaning that series that have only a level and noise. One solution for forecasting series with trend and/or seasonality is first to remove those components (e.g., via regression models). Another solution is to use a more sophisticated version of exponential smoothing, one that can capture trend and/or seasonality. In the following, we describe an extension of exponential smoothing that can capture a trend and/or seasonality.

Series with a Trend

For series that contain a trend, we can use "double exponential smoothing." Unlike in regression models, the trend shape is not assumed to be global, but rather it can change over time. In double exponential smoothing, the local trend is estimated from the data and is updated as more data arrive. As in simple exponential smoothing, the level of the series is estimated from the data and is updated as more data arrive.

JMP provides two methods for double exponential smoothing—*linear exponential smoothing* (the *Holt* method) and *double exponential smoothing* (the *Brown* method). The k-step-ahead forecast for both methods is given by combining the level estimate at time t (L_t) and the trend estimate at time t (T_t):

$$F_{t+k} = L_t + kT_t. \tag{19.3}$$

Note that in the presence of a trend, one-, two-, three-step-ahead (etc.) forecasts are no longer identical. Using the Holt method, the level and trend are updated through a pair of updating equations:

$$L_t = \alpha Y_t + (1 - \alpha)(L_{t-1} + T_{t-1}), \tag{19.4}$$

$$T_t = \beta \left(L_t - L_{t-1} \right) + (1 - \beta)T_{t-1}. \tag{19.5}$$

The first equation means that the level at time t is a weighted average of the actual value at time t and the level in the previous period, adjusted for trend (in the presence of a trend, moving from one period to the next requires factoring in the trend). The second equation means that the trend at time t is a weighted average of the trend in the previous period and the more recent information on the change in level.[3] Here, there are two smoothing parameters,

---

[3]There are various ways to estimate the initial values L_1 and T_1, but the differences among these ways usually disappear after a few periods.

α and β, which determine the rate of learning. As in simple exponential smoothing, they are both constants between 0 and 1, with higher values leading to faster learning (more weight to most recent information). Optimal values for the constants are determined by JMP, or custom values or ranges of values can be specified by the user.

Series with a Trend and Seasonality

For series that contain both trend and seasonality, the "Holt–Winters exponential smoothing" method can be used. This is a further extension of double exponential smoothing, where the k-step-ahead forecast also takes into account the season at period $t + k$. Assuming seasonality with M seasons (e.g., for weekly seasonality $M = 7$), the forecast is given by

$$F_{t+k} = \left(L_t + kT_t\right) S_{t+k-M}, \tag{19.6}$$

(Note that by the time of forecasting t, the series must have included at least one full cycle of seasons in order to produce forecasts using this formula, i.e., $t > M$.)

Being an adaptive method, Holt–Winters exponential smoothing allows the level, trend, and seasonality patterns to change over time. These three components are estimated and updated as more information arrives. The three updating equations are given by

$$L_t = \alpha Y_t / S_{t-M} + (1 - \alpha)(L_{t-1} + T_{t-1}), \tag{19.7}$$

$$T_t = \beta \left(L_t - L_{t-1}\right) + (1 - \beta)T_{t-1}, \tag{19.8}$$

$$S_t = \gamma Y_t / L_t + (1 - \gamma)S_{t-M}. \tag{19.9}$$

The first equation is similar to that in Holt's double exponential smoothing, except that it uses the seasonally adjusted value at time t rather than the raw value. This is done by dividing Y_t by its seasonal index, as estimated in the last cycle. The second equation is identical to double exponential smoothing (both Holt and Brown methods). The third equation means that the seasonal index is updated by taking a weighted average of the seasonal index from the previous cycle and the current trend-adjusted value. Note that this formulation describes a multiplicative seasonal relationship where values on different seasons differ by percentage amounts.

There is also an additive version of Holt–Winters exponential smoothing method, which is referred to as Winters method (Additive) in JMP. In the additive model, seasons differ by a constant amount rather than by a percentage amount (see more in Shmueli and Lichtendahl, 2016). The three equations for the additive model are very similar to those for the multiplicative model, but the seasonal index is *subtracted* from Y_t rather than being divided by it:

$$L_t = \alpha \left(Y_t - S_{t-M}\right) + (1 - \alpha)(L_{t-1} + T_{t-1}), \tag{19.10}$$

$$T_t = \beta \left(L_t - L_{t-1}\right) + (1 - \beta)T_{t-1}, \tag{19.11}$$

$$S_t = \gamma \left(Y_t - L_t\right) + (1 - \gamma)S_{t-M}. \tag{19.12}$$

To illustrate forecasting a series with Winters exponential smoothing, consider the raw Amtrak ridership data. As we observed earlier, the data contain both a trend and monthly seasonality. Figure 19.11 shows part of the JMP output using a 12-month cycle and the default smoothing weight constraint, *Zero to One*. The optimal values for the three smoothing parameters are given under Estimates in the *Parameter Estimates* table. As shown in Figure 19.11, the model performs fairly well with an RMSE of 64.44 and a MAPE of 2.53 on the 36-month holdback data.

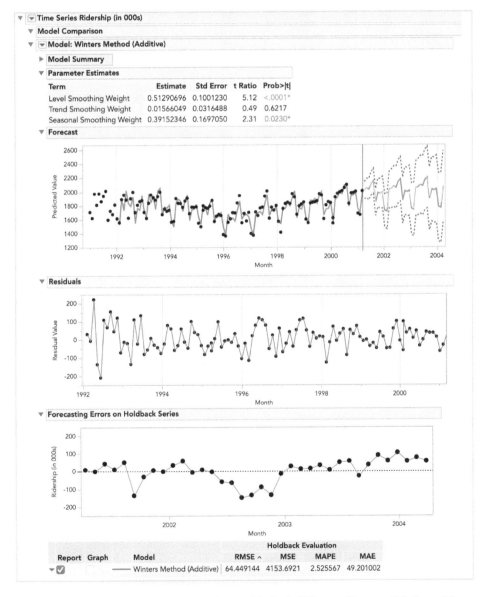

FIGURE 19.11   Partial output for Winters Method (Winters Exponential Smoothing—Additive) applied to Amtrak ridership series

Graphs for actual and forecasted ridership and residuals, recreated using the *Graph Builder* to more clearly show the time-ordered patterns in the data, are shown in Figure 19.12.

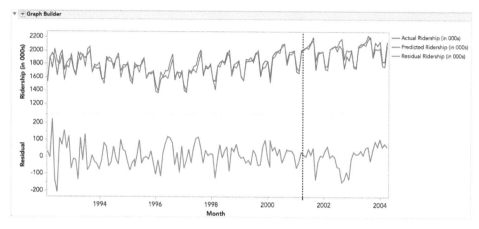

FIGURE 19.12   Graphs of actual and forecasted ridership (Top) and residuals (bottom) for Winters method, produced using the *Graph Builder*

STATE SPACE SMOOTHING MODELS IN JMP

A state space smoothing model is defined based on its error, trend, and seasonal components. The *State Space Smoothing* option under the red triangle in the *Time Series* platform automatically fits up to 30 potential exponential smoothing models. These models can integrate error, trend, and seasonal components and can be additive or multiplicative.

- A specific model can be represented by notation called ETS (Error, Trend, Seasonal). The model A,A,N, for example, has additive errors, additive trend, and no seasonality. Similarly, M,A,M refers to a model with multiplicative errors, an additive trend, and multiplicative seasonality.
- Figure 19.13 (top) shows the options for specifying the different state space models. Several models are selected by default. Figure 19.13 (bottom) shows the resulting Model Comparison report.
- In Figure 19.13, we see that MMM12, the model with multiplicative error, multiplicative trend, and 12-month multiplicative seasonality, has the lowest RMSE on the holdback data.

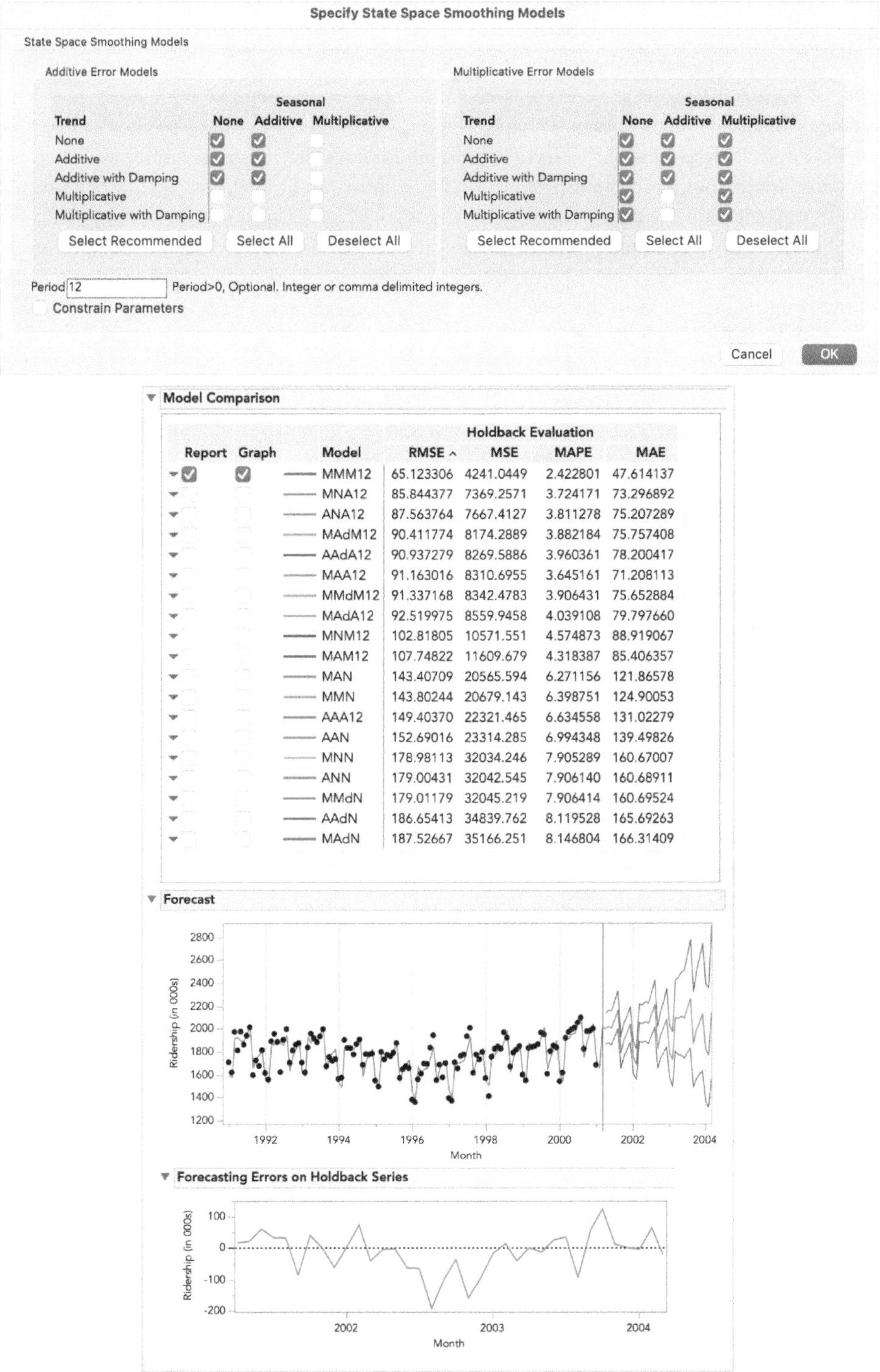

Specify State Space Smoothing Models

State Space Smoothing Models

Additive Error Models

Trend	Seasonal		
	None	Additive	Multiplicative
None	☑	☑	☐
Additive	☑	☑	☐
Additive with Damping	☑	☑	☐
Multiplicative	☐	☐	☐
Multiplicative with Damping	☐	☐	☐

[Select Recommended]   [Select All]   [Deselect All]

Multiplicative Error Models

Trend	Seasonal		
	None	Additive	Multiplicative
None	☑	☑	☑
Additive	☑	☑	☑
Additive with Damping	☑	☑	☑
Multiplicative	☑	☐	☑
Multiplicative with Damping	☑	☐	☑

[Select Recommended]   [Select All]   [Deselect All]

Period [12]   Period>0, Optional. Integer or comma delimited integers.
☐ Constrain Parameters

Cancel   [OK]

▼ **Model Comparison**

Report	Graph	Model	RMSE ∧	Holdback Evaluation		
				MSE	MAPE	MAE
▼ ☑	☑	—— MMM12	65.123306	4241.0449	2.422801	47.614137
▼		—— MNA12	85.844377	7369.2571	3.724171	73.296892
▼		—— ANA12	87.563764	7667.4127	3.811278	75.207289
▼		—— MAdM12	90.411774	8174.2889	3.882184	75.757408
▼		—— AAdA12	90.937279	8269.5886	3.960361	78.200417
▼		—— MAA12	91.163016	8310.6955	3.645161	71.208113
▼		—— MMdM12	91.337168	8342.4783	3.906431	75.652884
▼		—— MAdA12	92.519975	8559.9458	4.039108	79.797660
▼		—— MNM12	102.81805	10571.551	4.574873	88.919067
▼		—— MAM12	107.74822	11609.679	4.318387	85.406357
▼		—— MAN	143.40709	20565.594	6.271156	121.86578
▼		—— MMN	143.80244	20679.143	6.398751	124.90053
▼		—— AAA12	149.40370	22321.465	6.634558	131.02279
▼		—— AAN	152.69016	23314.285	6.994348	139.49826
▼		—— MNN	178.98113	32034.246	7.905289	160.67007
▼		—— ANN	179.00431	32042.545	7.906140	160.68911
▼		—— MMdN	179.01179	32045.219	7.906414	160.69524
▼		—— AAdN	186.65413	34839.762	8.119528	165.69263
▼		—— MAdN	187.52667	35166.251	8.146804	166.31409

▼ **Forecast**

▼ **Forecasting Errors on Holdback Series**

FIGURE 19.13   **State space smoothing specification window (top) and resulting model comparison of several space state smoothing models for the Amtrak ridership data (middle), with forecast and forecast errors for the model with the lowest RMSE (bottom)**

19.5   DEEP LEARNING FOR FORECASTING

While smoothing-based methods have been successfully used for decades, the data-driven toolkit for forecasting time series has recently been supplemented with deep learning methods. Recurrent neural networks (RNN) are a type of deep learning method suited for temporal data such as time series or other sequential data such as music, text, and speech. Compared to the multilayer feed-forward network shown in Figure 11.1 in Chapter 11, the RNN has an additional arrow within each hidden layer, as shown in Figure 19.14. This allows the network to "remember" the information it learned in the previous iteration through the data). In other words, RNNs incorporate a looping mechanism that allows information to flow from one iteration to the next. Here, an iteration means a pass through one subsequence of the series (see below).

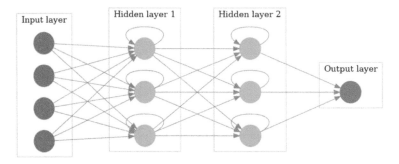

FIGURE 19.14   Schematic of recurrent neural network (RNN)

Training an RNN on a time series requires partitioning the training series into a sequence of "rolling windows" (sub-series) of width w. Each sub-series serves as the predictor series and has an associated future period that serves as the outcome. For example, for one-step-ahead forecasts, we convert the training series $y_1, y_2, \ldots, y_t$ into the set shown in Table 19.1. These sub-series are then fed into the RNN sequentially (into the input layer in Figure 19.14). The arrow from each hidden node to itself then "remembers" the information learned from the previous sub-series.

TABLE 19.1   Converting a time series of length t into a sequence of sub-series and associated one-step-ahead labels

Predictor series	Outcome
$y_1, y_2, \ldots, y_w$	y_{w+1}
$y_2, y_3, \ldots, y_{w+1}$	y_{w+2}
$\vdots$	$\vdots$
$y_{t-w}, \ldots, y_{t-1}$	y_t

A simple RNN, however, suffers from "short-term memory" in the sense that it remembers information from very recent periods, but not from longer history. This results from the so-called "vanishing gradient problem." Gradients are values computed to update the neural network weights during backpropagation. The vanishing gradient problem occurs since the gradient updates becomes small over time, and the earlier layers that receive a small gradient update stop learning. Because these layers don't learn, a simple RNN can forget what it saw much earlier in the sequence, thus having a short-term memory.

To overcome the short-term memory issue, a variant of RNN called Long Short-Term Memory (LSTM) was developed. LSTM adds to each hidden node special mechanisms called *gates* that can regulate the information flow by learning which data in a sequence is important to keep (remember) or throw away (forget). Through this process, it can relay relevant information down the long chain of sequences to make predictions. We set the gate to remember a greater or lesser portion of the old information.

Note: JMP Pro supports up to two layers of hidden nodes in neural networks. JMP Pro does not support RNN as of Version 17.

PROBLEMS

19.1 Impact of September 11 on Air Travel in the United States. The Research and Innovative Technology Administration's Bureau of Transportation Statistics (BTS) conducted a study to evaluate the impact of the September 11, 2001, terrorist attack on US transportation. The 2006 study report and the data can be found at `https://www.bts.gov/archive/publications/estimated_impacts_of_9_11_on_us_travel/index`. The goal of the study was stated as follows:

> The purpose of this study is to provide a greater understanding of the passenger travel behavior patterns of persons making long distance trips before and after 9/11.

The report analyzes monthly passenger movement data between January 1990 and May 2004. Data on three monthly time series are given in file `Sept11Travel.jmp` for this period: (1) actual airline revenue passenger miles (Air), (2) rail passenger miles (Rail), and (3) vehicle miles traveled (Car).

In order to assess the impact of September 11, BTS took the following approach: using data before September 11, they forecasted future data (under the assumption of no terrorist attack). Then they compared the forecasted series with the actual data to assess the impact of the event. Our first step therefore is to split each of the time series into two parts: pre- and post-September 11. We now concentrate only on the earlier time series.

a. Create a time plot for the pre-event AIR time series. What time series components appear from the plot?

b. A time plot of the seasonally adjusted pre-September-11 AIR series is shown in Figure 19.15. Which of the following smoothing methods would be adequate for forecasting this series?

- Moving average (with what window width?)
- Simple exponential smoothing
- Holt exponential smoothing (linear exponential smoothing)
- Winters exponential smoothing (Winters method, additive)

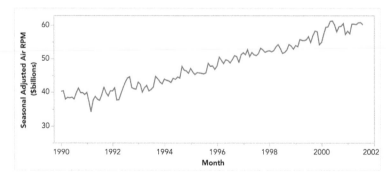

FIGURE 19.15   Seasonally adjusted pre-September 11 AIR series

19.2 Relation Between Moving Average and Exponential Smoothing. Suppose that we apply a moving average to a series, using a very short window span. If we wanted to achieve an equivalent result using simple exponential smoothing, what value should the smoothing coefficient take?

19.3 Forecasting with a Moving Average: For a given time series of sales, the training set consists of 50 months. The first five months' data are shown below:

Month	Sales
Sept 98	27
Oct 98	31
Nov 98	58
Dec 98	63
Jan 99	59

    a. Compute the sales forecast for January 1999 based on a moving average with $w = 4$.

    b. Compute the forecast error for this forecast.

19.4 Optimizing Winters Exponential Smoothing. Figure 19.16 shows an output from applying Winters exponential smoothing to data, using "optimal" smoothing constants.

	Term	Estimate
1	Level Smoothing Weight	1.000
2	Trend Smoothing Weight	0.000
3	Seasonal Smoothing Wei…	0.246

FIGURE 19.16   Optimized smoothing constants

    a. The value of zero that is obtained for the trend smoothing constant means that (choose one of the following):

- There is no trend.
- The trend is estimated only from the first two points.
- The trend is updated throughout the data.
- The trend is statistically insignificant.

    b. What is the interpretation of the level smoothing constant value?

19.5 Forecasting Department Store Sales. The time series plot in Figure 19.17 describes actual quarterly sales for a department store over a six-year period. (Data are available in DepartmentStoreSales.jmp, data courtesy of Chris Albright.)

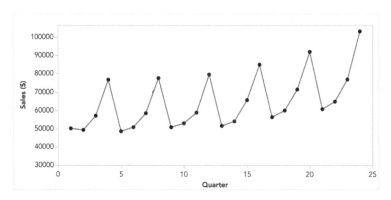

FIGURE 19.17   Optimized smoothing constants

a. Which of the following methods would not be suitable for forecasting this series?

- Moving average of raw series
- Moving average of deseasonalized series
- Simple exponential smoothing of the raw series
- Holt's exponential smoothing of the raw series
- Winters exponential smoothing of the raw series (additive)
- Winters exponential smoothing of the log(Sales) (multiplicative)

b. The forecaster was tasked to generate forecasts for four quarters ahead. He therefore partitioned the data so that the last four quarters were designated as the validation period. The forecaster approached the task by using two models: a regression model with trend and seasonality and Winters exponential smoothing.

   i. Create the necessary columns, and run the regression model on these data. Save the residuals and the predicted values to the data table, and create the top plots in Figure 19.18.

   ii. Run Winters method on these data (using the *Forecast on Holdback* option). Report the values of the smoothing parameters. Save the prediction formula to the data table, and follow the steps outlined in this chapter to create the graphs for the actual vs. forecasted values and residuals shown at the bottom of Figure 19.18.

   iii. The forecasts and errors for the validation data are given in Figure 19.19. Compute the average errors and the mean absolute errors (MAE) for the forecasts of quarters 21–24 for each of the two models. (*Hint*: To calculate mean absolute errors, first take the absolute value of each error, then use either *Tabulate* or *Distribution*).

c. Compare the fit and residuals from the exponential smoothing and regression models shown in Figure 19.18. Using all the information thus far, which model is more suitable for forecasting quarters 21–24?

d. Run Winters exponential smoothing on log(Sales) using the training data (again, using the *Forecast on Holdback* option).

   i. Compute forecasts and forecast errors, and create graphs for actual vs. forecasted values and residuals.

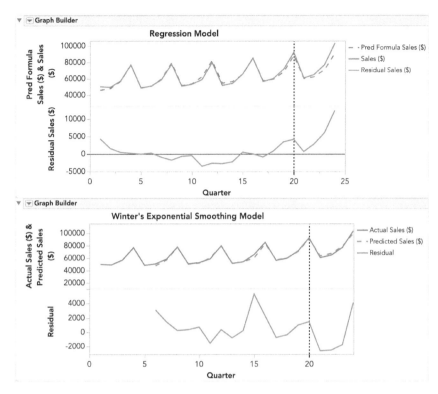

FIGURE 19.18   Forecasts and forecast errors using regression (top) and exponential smoothing (bottom)

Forecasts and Errors for validation Data from Regression Model

Quarter	Sales ($)	Pred Sales ($)	Residual Sales ($)
21	60800	60121.7	678.3
22	64900	62067.7	2832.3
23	76997	70903.7	6093.3
24	103337	90885.5	12451.5

Forecasts and Errors for validation data from Winters' method (Additive)

Quarter	Sales ($)	Pred Sales ($)	Residual Sales ($)
21	60800	63426.3	-2626.3
22	64900	67423.4	-2523.4
23	76997	78752.1	-1755.1
24	103337	99228.7	4108.3

FIGURE 19.19   Forecasts for validation sets

ii. Calculate the average error and MAE for the forecasts of quarters 21–24.

iii. Based on the graphs and forecast errors, how does this model perform relative to the regression model and the Winters (additive) model?

iv. Of the three models, which does the best job of forecasting quarters 21–24? Explain.

19.6 **Shipments of Household Appliances.** The time plot in Figure 19.20 shows the series of quarterly shipments (in $ millions) of US household appliances between 1985 and 1989. (Data are available in `ApplianceShipments.jmp`, data courtesy of Ken Black.)

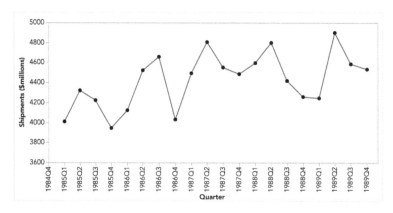

FIGURE 19.20   **Quarterly shipments of US household appliances over five years**

a. Which of the following methods would be suitable for forecasting this series if applied to the raw data?

- Moving average
- Simple exponential smoothing
- Holt's exponential smoothing
- Winters exponential smoothing

b. Apply a moving average with window span $w = 4$ to the data. Use all but that last year as the training set. Create a time plot of the moving average series. (Remember that to use validation with moving average charts, you must hide and exclude the holdback data. Refer to the text box at the beginning of the chapter.)

   i. What does the MA(4) chart reveal?

   ii. Use the MA(4) model to forecast appliance sales in Q1-1990.

   iii. Use the MA(4) model to forecast appliance sales in Q1-1991.

   iv. Is the forecast for Q1-1990 most likely to underestimate, overestimate, or accurately estimate the actual sales on Q1-1990? Explain.

   v. Management feels most comfortable with moving averages. The analyst therefore plans to use this method for forecasting future quarters. What else should be considered before using the MA(4) to forecast future quarterly shipments of household appliances?

   vi. Unhide and unexclude the holdback data.

c. We now focus on forecasting beyond 1989. In the following, continue to use all but the last year as the training set and the last four quarters as the validation set. First, fit a regression model to sales with a linear trend and quarterly seasonality to the training data. Next, apply Winters exponential smoothing to the training data. Choose an adequate "season length."

   i. Compute the average error and MAE for the validation data using the regression model.

   ii. Compute the average error and MAE for the validation data using Winters exponential smoothing.

   iii. Which model would you prefer to use for forecasting Q1-1990? Give three reasons.

19.7 Forecasting Shampoo Sales. The time series in in Figure 19.21 describes monthly sales of a certain shampoo over a three-year period. [Data are available in ShampooSales.jmp, Source: Hyndman and Yang (2018)].

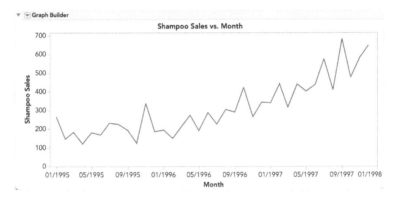

FIGURE 19.21   Monthly sales of a certain shampoo

   a. Which of the following methods would be suitable for forecasting this series if applied to the raw data?

- Moving average
- Simple exponential smoothing
- Holt's exponential smoothing
- Winters exponential smoothing

   b. Fit the four models in the Time Series platform, and compare the built-in statistics, graphs, and residual plots.

   c. Of these models, which is the best at forecasting future time periods? Explain why.

   d. Now, fit the default *State Space Smoothing* models. Which model performs best on the holdback set? How does this model compare to the best model in part c?

19.8 Natural Gas Sales. Figure 19.22 shows a time plot of quarterly natural gas sales (in billions of BTU) of a certain company, over a period of four years. (Data courtesy of George McCabe. Data are in NaturalGasSales.jmp.) The company's analyst is asked to use a moving average model to forecast sales in Winter 2005.

   a. Reproduce the time plot with the overlaying SMA(4) line (use the *Smoothing Model > Simple Moving Average* option in the *Time Series* platform).

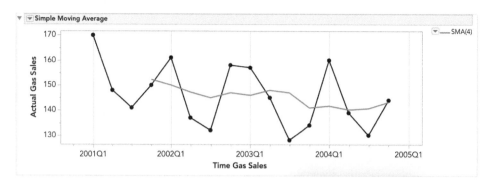

FIGURE 19.22   **Quarterly sales of natural gas over four years**

b. What can we learn about the series from the MA line?

c. Run a moving average forecaster with adequate season length, using the last year as the holdback set (refer to the text box in this chapter for computing a trailing moving average forecast in JMP). Are forecasts generated by this method expected to overforecast, underforecast, or accurately forecast actual sales? Why?

19.9 **Forecasting Australian Wine Sales.** Figure 19.23 shows time plots of monthly sales of six types of Australian wines (red, rose, sweet white, dry white, sparkling, and fortified) for 1980–1994 [Data are available in AustralianWines.jmp, Source: Hyndman and Yang (2018)]. The units are thousands of liters. You are hired to obtain short-term forecasts (two to three months ahead) for each of the six series, and this task will be repeated every month.

a. Which forecasting method would you choose if you had to choose the same method for all series? Why?

b. Fortified wine has the largest market share of the above six types of wine. You are asked to focus on fortified wine sales alone and produce an accurate as possible forecast for the next two months.

- Start by partitioning the data, using the period until December 1993 as the training set. How many periods are in the validation (holdback) set?
- Apply Winters exponential smoothing to the training data to sales with an appropriate season length (use the default values for the smoothing constants).

c. Examine ACF plot for the training residuals from the model until lag 12. Examining this plot, which of the following statements are reasonable?

- Decembers (month 12) are not captured well by the model.
- There is a strong correlation between sales on the same calendar month.
- The model does not capture the seasonality well.
- We should try to fit an autoregressive model with lag 12 to the residuals.
- We should first deseasonalize the data and then apply Winters exponential smoothing.
- We should fit a Winters exponential smoothing model to the series of log(sales).

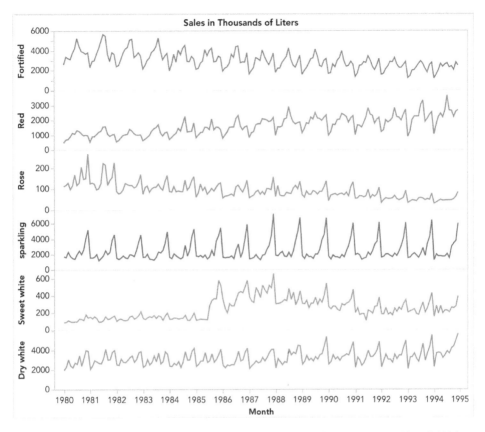

FIGURE 19.23   Monthly sales of six types of Australian wines between 1980 and 1994

PART VIII

DATA ANALYTICS

20

TEXT MINING

In this chapter, we introduce unstructured text as a form of data. First, we discuss a tabular representation of text data in which each column is a word, each row is a document, and each cell is a 0 or 1, indicating whether that column's word is present in that row's document. Then, we consider how to move from unstructured documents to this structured matrix. Finally, we illustrate how to integrate this process into the standard machine learning procedures covered in earlier parts of the book.

Text Mining in JMP: The *Text Explorer* platform in JMP is used for text mining. Some basic methods for exploring unstructured text data are available in the standard version of JMP. However, JMP Pro is required for most of the topics introduced in this chapter.

20.1   INTRODUCTION[1]

Up to this point, and in machine learning in general, we have been primarily dealing with three types of data: numerical, binary (true/false), and multicategory.

In some common predictive analytics applications, though, data come in text form. An Internet service provider, for example, might want to use an automated algorithm to classify support tickets as urgent or routine so that the urgent ones can receive immediate human review. A law firm facing a massive discovery process (review of large numbers of documents) would benefit from a document review algorithm that could classify documents as relevant or irrelevant. In both of these cases, the predictor attributes (features) are embedded as text in documents.

---

[1]This and subsequent sections in this chapter copyright © 2022 Datastats, LLC, and Galit Shmueli. Used by permission.

Machine Learning for Business Analytics: Concepts, Techniques, and Applications with JMP Pro®,
Second Edition. Galit Shmueli, Peter C. Bruce, Mia L. Stephens, Muralidhara Anandamurthy, and Nitin R. Patel.
© 2023 John Wiley & Sons, Inc. Published 2023 by John Wiley & Sons, Inc.

Text mining methods have gotten a boost from the availability of social media data—the Twitter feed, for example, as well as blogs, online forums, review sites, and news articles. The public availability of web-based text has provided a huge repository of data on which researchers can hone text mining methods. One area of growth has been the application of text mining methods to notes and transcripts from contact centers and service centers.

20.2   THE TABULAR REPRESENTATION OF TEXT: DOCUMENT–TERM MATRIX AND "BAG-OF-WORDS"

Consider the following three sentences:

S1. this is the first sentence.

S2. this is a second sentence.

S3. the third sentence is here.

We can represent the words (called *terms*) in these three sentences (called *documents*) in a *document–term matrix*, where each column is a word and each row is a sentence. These statements are presented in the file `Sentences.jmp`. Figure 20.1 shows the matrix for these three sentences.

Sentence	is Frequency	sentence Frequency	the Frequency	this Frequency	a Frequency	first Frequency	here Frequency	second Frequency	third Frequency
1 this is the first sentence	1	1	1	1	0	1	0	0	0
2 this is a second sentence	1	1	0	1	1	0	0	1	0
3 the third sentence is here	1	1	1	0	0	0	1	0	1

FIGURE 20.1   Document–term matrix representation of words for sentences S1–S3

Note that all the words in all three sentences are represented in the matrix and each word has exactly one row. Order is not important, and the number in the cell indicates the frequency of that term in that sentence. This is the *bag-of-words* approach, where the document is treated simply as a collection of words in which order, grammar, and syntax do not matter.

The three sentences have now been transformed into a tabular data format just like those we have seen to this point. In a simple world, this binary matrix could be used in clustering or, with the appending of outcome variable, for classification or prediction. However, the text mining world is not a simple one. Even confining our analysis to the bag-of-words approach, considerable thinking and preprocessing may be required. Some human review of the documents, beyond simply classifying them for training purposes, may be indispensable.

CREATING A DOCUMENT–TERM MATRIX IN JMP Pro

To create a document–term matrix in JMP, we use the *Text Explorer* platform under the *Analyze* platform. Later, we discuss using the *Text Explorer* platform for text mining and describe additional options available.

- First, make sure the text column in the data table has the *Unstructured Text* modeling type (right-click on the column head and select *Column Info* to change the modeling type).
- Select *Analyze > Text Explorer*.
- In the dialog, select the unstructured text column as *Text Column* as shown in Figure 20.2, and click OK.
- In the analysis window, select *Save Document Term Matrix* from the red triangle. In the pop-up dialog, you can specify the number of terms to save to the data table, the minimum term frequency, and the weighting. The default weighting is binary (0/1)—we discuss other weighting schemes later.
- For the sentence example, this produces the document–term matrix in Figure 20.1.

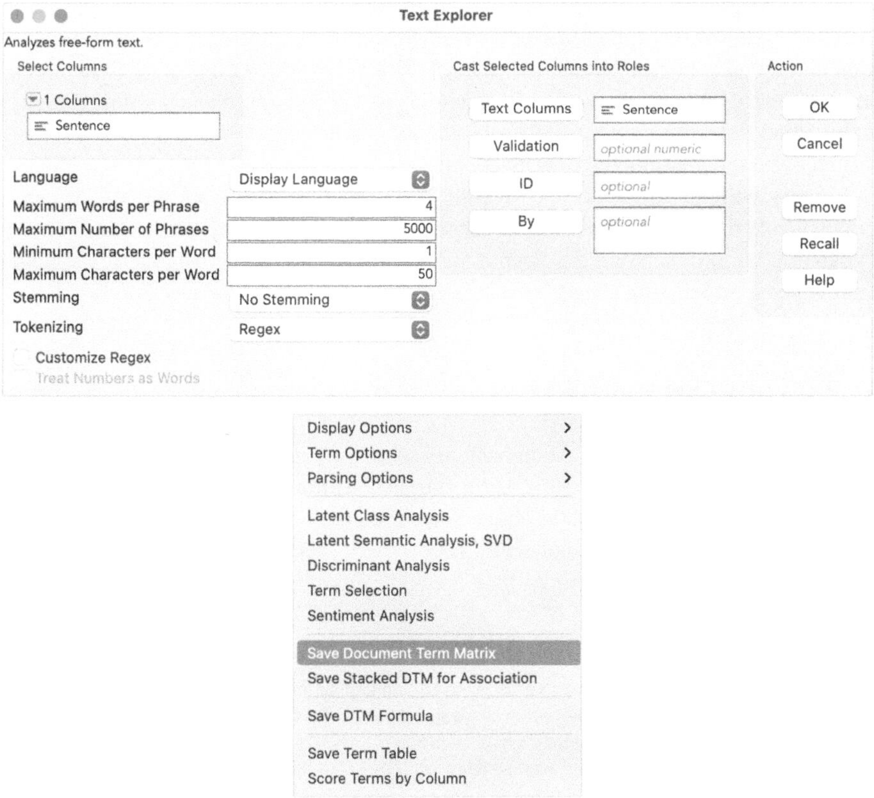

FIGURE 20.2   Text Explorer dialog (top) and red triangle options (bottom) for text mining in JMP

20.3   BAG-OF-WORDS VS. MEANING EXTRACTION AT DOCUMENT LEVEL

We can distinguish between two undertakings in text mining:

- Classifying a document as belonging to a class or clustering similar documents
- Extracting more detailed meaning from a document

The first goal requires a sizable collection of documents, or a *corpus*,[2] the ability to extract predictor variables from documents, and, for the classification task, lots of pre-labeled documents to train a model. The models that are used, though, are the standard statistical and machine learning predictive models that we have already dealt with for numerical and categorical data.

The second goal might involve a single document, and is much more ambitious. The computer must learn at least some version of the complex "algorithms" that make up human language comprehension: grammar, syntax, punctuation, etc. In other words, it must undertake the processing of a natural (i.e., noncomputer) language to *understand* documents in that language. Understanding the meaning of one document on its own is a far more formidable task than probabilistically assigning a class to a document based on rules derived from hundreds or thousands of similar documents.

For one thing, text comprehension requires maintenance and consideration of word order. "San Francisco beat Boston in last night's baseball game" is very different from "Boston beat San Francisco in last night's baseball game."

Even identical words in the same order can carry different meanings, depending on the cultural and social context: "Hitchcock shot The Birds in Bodega Bay," to an avid outdoors person indifferent to capitalization and unfamiliar with Alfred Hitchcock's films, might be about bird hunting. Ambiguity resolution is a major challenge in text comprehension—does "bot" mean "bought," or does it refer to robots?

Our focus will remain with the overall focus of the book and the easier goal—probabilistically assigning a class to a document or clustering similar documents. The second goal—deriving understanding from a single document—is the subject of the field of natural language processing (NLP).

20.4   PREPROCESSING THE TEXT

The simple example we presented had ordinary words separated by spaces and a period to denote the end of the each sentence. A fairly simple algorithm could break the sentences up into the word matrix with a few rules about spaces and periods. It should be evident that the rules required to parse data from real-world sources will need to be more complex. It should also be evident that the preparation of data for text mining is a more involved undertaking

---

[2]The term "corpus" is often used to refer to a large, fixed standard set of documents that can be used by text preprocessing algorithms, often to train algorithms for a specific type of text or to compare the results of different algorithms. A specific text mining setting may rely on algorithms trained on a corpus specially suited to that task. One early general-purpose standard corpus was the Brown corpus of 500 English language documents of varying types (named Brown because it was compiled at Brown University in the early 1960s). These days many corpora rely on web-based sources such as Wikipedia and Twitter, which provide huge amounts of documents.

than the preparation of numerical or categorical data for predictive models. For example, consider the modified example in Figure 20.3, based on the following sentences:

S1. this is the first sentence!!

S2. this is a second Sentence :)

S3. the third sentence, is here

S3. forth of all sentences

This set of sentences has extra spaces, non-alpha characters, incorrect capitalization, and a misspelling of "fourth."

Sentence	is	sentence	<phrase_punctuation>	the	this	:)	a	all	first	forth	here	of	second	sentences	third
1 this is the first sentence!!	1	1	1	1	1	0	0	0	1	0	0	0	0	0	0
2 this is a second Sentence :)	1	1	0	0	1	1	1	0	0	0	0	1	0	0	0
3 the third sentence, is here	1	1	1	1	0	0	0	0	0	0	1	0	0	0	1
4 forth of all sentences	0	0	0	0	0	0	0	1	0	1	0	1	0	1	0

FIGURE 20.3   Document–term matrix representation of words for modified example sentences S1–S4

Tokenization

Our first simple dataset was composed entirely of words found in the dictionary. A real set of documents will have more variety; it will contain numbers, alphanumeric strings like date stamps or part numbers, web and email addresses, abbreviations, slang, proper nouns, misspellings, and more.

Tokenization is the process of taking a text and, in an automated fashion, dividing it into separate "tokens" or terms. A token (term) is the basic unit of analysis. A word separated by spaces is a token. 2 + 3 would need to be separated into three tokens, while 23 would remain as one token. Punctuation might also stand as its own token (e.g., the @ symbol). These tokens become the column headers in the data matrix. JMP has its own list of delimiters (spaces, commas, colons, etc.) that it uses to divide up the text into tokens.

For a sizeable corpus, tokenization will result in a huge number of variables—the English language has over a million words, let alone the non-word terms that will be encountered in typical documents. Anything that can be done in the preprocessing stage to reduce the number of terms will aid in the analysis. The initial focus is on eliminating terms that simply add bulk and noise.

Some of the terms that result from the initial parsing of the corpus might not be useful in further analyses and can be eliminated in the preprocessing stage. For example, in a legal discovery case, one corpus of documents might be emails, all of which have company information and some boilerplate as part of the signature. These terms might be added to a *stopword list* of terms that are to be automatically eliminated in the preprocessing stage.

JMP has a generic stopword list of frequently occurring terms to be removed during preprocessing (see Figure 20.4). If you review the stopword list, you will see that it contains a large number of terms. To view the list in JMP, in the *Text Explorer* analysis window select *Display Options > Show Stop Words* from the red triangle. To add additional terms to the list, you can right-click on any term and select *Add Stop Word*, or you can add terms to the list using *Term Options > Manage Stop Words* from the top red triangle.

&	can't	hasn't	i've	off	that	very	who's
a	could	hasn't	if	on	that's	was	whoever
about	couldn't	have	in	once	that's	wasn't	whom
above	couldn't	haven't	into	only	the	wasn't	whomever
after	did	haven't	is	or	their	we	who's
again	didn't	having	isn't	other	theirs	we'd	why
against	didn't	he	isn't	ought	them	we'll	why's
all	do	he'd	it	our	themselves	we're	why's
am	does	he'll	it's	ours	then	we've	will
an	doesn't	he's	its	ourselves	there	were	with
and	doesn't	her	itself	out	there's	weren't	won't
any	doing	here	it's	over	there's	weren't	won't
are	don't	here's	i'd	own	these	we'd	would
aren't	don't	here's	i'll	same	they	we'll	wouldn't
aren't	down	hers	i'm	shan't	they'd	we're	wouldn't
as	during	herself	i've	shan't	they'll	we've	you
at	each	he'd	let's	she	they're	what	you'd
be	else	he'll	let's	she'd	they've	what's	you'll
because	few	he's	me	she'll	they'd	what's	you're
been	for	him	mine	she's	they'll	when	you've
before	from	himself	more	she'd	they're	when's	your
being	further	his	most	she'll	they've	whenever	yours
below	get	how	mustn't	she's	this	when's	yourself
between	gets	how's	mustn't	should	those	where	yourselves
both	good bye	however	my	shouldn't	through	where's	you'd
but	got	how's	myself	shouldn't	to	where's	you'll
by	had	i	no	so	too	whether	you're
can	hadn't	i'd	nor	some	under	which	you've
can't	hadn't	i'll	not	such	until	while	
cannot	has	i'm	of	than	up	who	

FIGURE 20.4   Built-in stopwords in JMP

Text Reduction

Additional techniques to reduce the volume of text ("vocabulary reduction") and to focus on the most meaningful text include:

- *Stemming* is a linguistic method that reduces different variants of words to a common core. For example, the terms *calling* and *called* have the same stem, *call*.
- Frequency filters can be used to eliminate either terms that occur in a great majority of documents or very rare terms. Frequency filters can also be used to limit the vocabulary to the *n* most frequent terms. We discuss frequency filters and *weights* in the next section.
- Synonyms or synonymous phrases may be consolidated.
- Letter case (uppercase/lowercase) can be ignored.

- A variety of specific terms in a category can be replaced with the category name. For example, different email addresses or different numbers might all be replaced with "emailtoken" or "numbertoken."

In JMP, the *Recode* utility, which is built into the *Text Explorer* platform, can be used for combining synonym terms and replacing terms with a category name. All terms are automatically converted to lower case in the preprocessing step.

Figure 20.5 presents the text reduction step applied to the four sentences example, after stemming, tokenizing (basic words), and excluding built in stopwords. We can see the number of terms has been reduced to five.

Sentence	sentenc· Binary	first Binary	forth Binary	second Binary	third Binary
1 this is the first sentence!!	1	1	0	0	0
2 this is a second Sentence :)	1	0	0	1	0
3 the third sentence, is here	1	0	0	0	1
4 forth of all sentences	1	0	1	0	0

FIGURE 20.5   Document–term matrix after text reduction in JMP

Presence/Absence vs. Frequency (Occurrences)

The bag-of-words approach can be implemented in terms of either *frequency* of terms or *presence/absence* of terms. The latter might be appropriate in some circumstances; in a forensic accounting classification model, for example, the presence or absence of a particular vendor name might be a key predictor variable, without regard to how often it appears in a given document. *Frequency* can be important in other circumstances, however. For example, in processing support tickets, a single mention of "IP address" might be non-meaningful; all support tickets might involve a user's IP address as part of the submission. Repetition of the phrase multiple times, however, might provide useful information that IP address is part of the problem (e.g., DNS resolution).

As described in the box CREATING A DOCUMENT–TERM MATRIX IN JMP Pro in Section 20.2, when you select *Save Document Term Matrix* from the red triangle in the analysis, you have several options. You can choose the *maximum number of terms* and the *minimum term frequency* to be included in the matrix. These options help control the size of the document–term matrix. The default *Weighting* scheme for creating the document–term matrix is Binary, but you can change this to Ternary, Frequency, and TF-IDF (which is discussed next).

Term Frequency–Inverse Document Frequency (TF-IDF)

There are additional popular options that factor in both the frequency of a term in a document and the frequency of documents with that term. One such popular option, which measures the importance of a term to a document, is *Term Frequency–Inverse Document Frequency* (TF-IDF). For a given document d and term t, the term frequency (TF) is the number of times term t appears in document d:

$$TF(t, d) = \# \text{ times term } t \text{ appears in document } d.$$

To account for terms that appear frequently in the domain of interest, we compute the *Inverse Document Frequency* (IDF) of term t, calculated over the entire corpus and defined as[3]

$$\text{IDF}(t) = \log \left(\frac{\text{total number of documents}}{\text{\# documents containing term } t} \right).$$

TF-IDF(t, d) for a specific term–document pair is the product of TF(t, d) and IDF(t):

$$\text{TF-IDF}(t, d) = \text{TF}(t, d) \times \text{IDF}(t). \tag{20.1}$$

The TF-IDF matrix contains the value for each document–term combination. The above definition of TF-IDF is a common one; however, there are multiple ways to define and weight both TF and IDF, so there are a variety of possible definitions of TF-IDF. The general idea of TF-IDF is that it identifies documents with frequent occurrences of rare terms. TF-IDF yields high values for documents with a relatively high frequency for terms that are relatively rare overall and near-zero values for terms that are absent from a document or present in most documents.

JMP Pro uses logarithm base 10 for computing the IDF. For example, the TF-IDF value for the term "first" in document 1 in Figure 20.5 is computed by

$$\text{TF-IDF}(\text{first}, 1) = 1 \times \left[\log_{10} \left(\frac{4}{1} \right) \right] \approx 0.602.$$

Figure 20.6 shows the TF-IDF matrix for the four-sentence example (after tokenization and text reduction).

	Sentence	sentenc· TF IDF	first TF IDF	forth TF IDF	second TF IDF	third TF IDF
1	this is the first sentence!!	0	0.6020599913	0	0	0
2	this is a second Sentence :)	0	0	0	0.6020599913	0
3	the third sentence, is here	0	0	0	0	0.6020599913
4	forth of all sentences	0	0	0.6020599913	0	0

FIGURE 20.6   TF-IDF document–term matrix for example 2 sentences S1–S4

From Terms to Topics: Latent Semantic Analysis and Topic Analysis

In Chapter 4, we showed how numerous numeric variables can be reduced to a small number of "principal components" that explain most of the variation in a set of variables. The principal components are linear combinations of the original (typically correlated) variables, and a subset of them serve as new variables to replace the numerous original variables.

An analogous dimension reduction method *latent semantic analysis* (LSA) can be applied to text data. The details behind the algorithm are beyond the scope of this chapter. However, LSA is mathematically equivalent to PCA.[4]

---

[3]IDF(t) is actually just the fraction, without the logarithm, although using a logarithm is very common.
[4]Generally, in PCA, the dimension is reduced by replacing the covariance matrix by a smaller one; in LSA, dimension is reduced by replacing the document–term matrix by a smaller one.

When we apply *Topic Analysis* to the results of latent semantic analysis, we can identify topics, or themes:

> For example, if we inspected our document collection, we might find that each time the term "alternator" appeared in an automobile document, the document also included the terms "battery" and "headlights." Or each time the term "brake" appeared in an automobile document, the terms "pads" and "squeaky" also appeared. However, there is no detectable pattern regarding the use of the terms "alternator" and "brake" together. Documents including "alternator" might or might not include "brake" and documents including "brake" might or might not include "alternator." Our four terms, battery, headlights, pads, and squeaky describe two different automobile repair issues: failing brakes and a bad alternator.

So, in this case, topic analysis would reduce the terms to two themes:

- brake failure
- alternator failure

We illustrate latent semantic analysis and topic analysis using JMP Pro in the example in Section 20.6.

Extracting Meaning

In the simple example above, after applying latent semantic and topic analysis, the topics to which the terms map (failing brakes, bad alternator) are clear and understandable. Unfortunately, in many cases the topics to which the terms map will not be obvious. In such cases, while latent semantic analysis alone will greatly enhance the manageability of the text for purposes of building predictive models and sharpen predictive power by reducing noise, it will turn the model into a blackbox device for prediction, not so useful for understanding the roles that terms and topics play. This is OK for our purposes as noted earlier, as we are focusing on text mining to classify or cluster new documents, not to extract meaning.

From Terms to High-Dimensional Word Vectors: Word2Vec

Word embedding is a form of text representation technique in which each word is represented in the form of a multidimensional vector (called *word vector*) so that words that are closer to one another in this vector space would also be similar in meaning. Latent semantic analysis was one such word embedding that considers occurrences of terms at a document level. Yet another form of word embedding approaches is Word2Vec[5] which considers occurrences of terms at a context level and uses prediction-based models to arrive at the word vector representations.

The Word2Vec technique uses a neural network model and trains words against other words in its neighborhood within the corpus. The technique has two variants based on how the model is trained: *continuous bag of words* (CBOW) and *skip-gram*. In both variants, a window of specified length (context) is moved along the text in the training corpus, and in each iteration the network is trained using only the words inside the window. The CBOW

---

[5]https://code.google.com/archive/p/word2vec/

model is trained to predict the word in the center of the window (focus word) based on the surrounding words (context words). The skip-gram model is trained to predict the context words based on the focus word. After the training is complete, the hidden layer with its learned weights is used as the word vector representation.

Google has published pre-trained Word2Vec word vectors trained with CBOW on a Google News dataset (about 100 billion words) that contain 300-dimensional vectors for 3 million words and phrases. Interestingly, these high-dimensional word vectors support vector operations that often make semantic sense. For example, the mathematical operation vector(Paris) − vector(France) + vector(Italy) is found to be very similar to vector(Rome), thus implicitly capturing the notion of countries and their capitals.

Note: Currently JMP Pro does not support Word2Vec.

20.5   IMPLEMENTING MACHINE LEARNING METHODS

After the text has gone through the preprocessing stage, it is then in a numeric matrix format, and you can apply the various data mining methods discussed earlier in this book. Clustering methods can be used to identify clusters of documents—for example, large numbers of medical reports can be mined to identify clusters of symptoms. Prediction methods can be used with tech support tickets to predict how long it will take to resolve an issue. Perhaps the most popular application of text mining is for classification of documents.

20.6   EXAMPLE: ONLINE DISCUSSIONS ON AUTOS AND ELECTRONICS

This example[6] illustrates a classification task—to classify Internet discussion posts as either auto-related or electronics-related. One post looks like this:

> From: smith@logos.asd.sgi.com (Tom Smith) Subject: Ford Explorer 4WD—do I need performance axle?
>
> We're considering getting a Ford Explorer XLT with 4WD and we have the following questions (All we would do is go skiing - no off-roading):
>
> 1. With 4WD, do we need the "performance axle"—(limited slip axle). Its purpose is to allow the tires to act independently when the tires are on different terrain.
>
> 2. Do we need the all-terrain tires (P235/75X15) or will the all-season (P225/70X15) be good enough for us at Lake Tahoe?
>
> Thanks,
>
> Tom
>
> –
>
> ==
> Tom Smith Silicon Graphics smith@asd.sgi.com 2011
> N. Shoreline Rd. MS 8U-815 415-962-0494 (fax)
> Mountain View, CA 94043
> ==

The posts are taken from Internet groups devoted to autos and electronics, so are pre-labeled. This one, clearly, is auto-related. A related organizational scenario might

---

[6]The dataset is taken from www.cs.cmu.edu/afs/cs/project/theo-20/www/data/news20.html, with minor modifications.

involve messages received by a medical office that must be classified as medical or non-medical (the messages in such a real scenario would probably have to be labeled by humans as part of the preprocessing).

The posts are in the form a zipped file (AutoAndElectronics.zip) that contains two folders after unzipping: *autos posts* and *electronics posts*. Each folder contains a set of 1000 posts organized in small files. In the following, we describe the main steps from preprocessing to building a classification model on the data.

Importing the Records

The *File > Import Multiple Files* option is used to read the individual files as "documents" from the data folder and convert them into a JMP data table. First, all 1000 "auto" files are imported into a JMP data table, where each file is one row (saved as `autos.jmp`). Figure 20.7 shows the JMP dialog for importing multiple files and part of the resulting JMP data table. The process is repeated for the electronics folder (saved as `electronics.jmp`).

FIGURE 20.7   Importing multiple files (top) into a single JMP data table (bottom)

Now, we need to combine these two data tables. Open both `autos.jmp` and `electronics.jmp`, and select *Tables > Concatenate* to combine these two tables into a new data table (check the *Create Source Column* box to add a column of labels: autos.jmp and electronics.jmp). Use the *Preview* pane in the Concatenate window to make sure the table has two columns (the label and the text), before clicking OK to produce the combined table (see Figure 20.8). Finally, you need to do a little cleanup. Change the modeling type of the text column to *Unstructured Text*, change the name of the label column to *Source*, and if required recode the labels to remove *.jmp*.

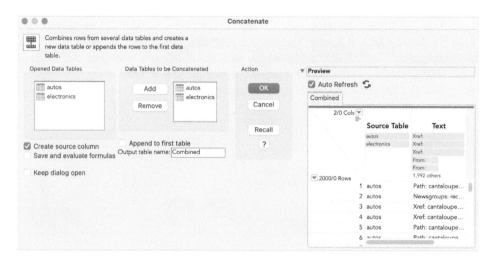

FIGURE 20.8   Concatenating autos.jmp and electronics.jmp to create a combined data table

Text Preprocessing in JMP

Prior to text preprocessing, we partition the data (60% training, 40% validation) so that any text preprocessing and text reduction learned from the training data can be applied to the holdout set during model evaluation. This is done using the *Analyze > Predictive Modeling > Make Validation Column* (use the random seed 12345 to match the results).

In the Text Explorer launch dialog, tokens are filtered based on length (*minimum characters per word* = 4 and *maximum characters per word* = 25), and *Basic Words* is used for Tokenizing. By default, JMP removes tokens matching built in stopwords. Finally, *stemming* is performed on the selected tokens. These specifications and initial output related to terms and phrase lists are shown in Figures 20.9 and 20.10, respectively. As discussed in the box near the end of the chapter, additional preprocessing steps can be taken using red triangle options or by right-clicking over the term or phase lists. These tasks are generally guided by subject matter knowledge.

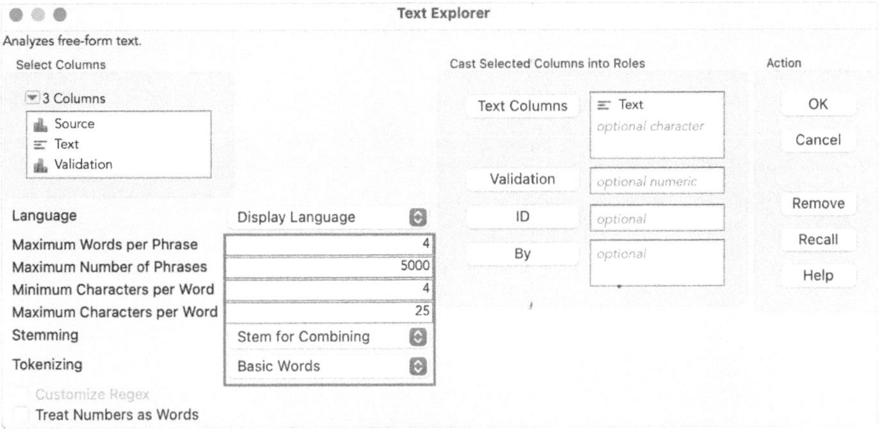

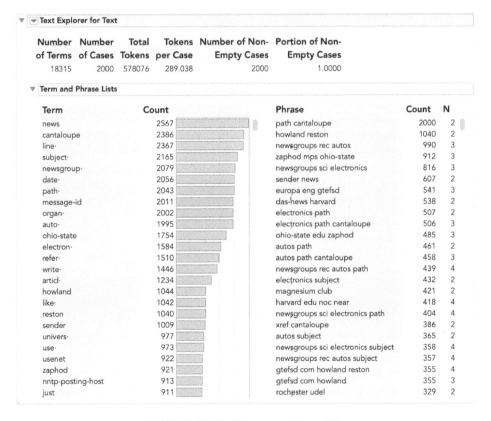

FIGURE 20.9 Text explorer dialog box with specifications

Text Explorer for Text

Number of Terms	Number of Cases	Total Tokens	Tokens per Case	Number of Non-Empty Cases	Portion of Non-Empty Cases
18315	2000	578076	289.038	2000	1.0000

Term and Phrase Lists

Term	Count	Phrase	Count	N
news	2567	path cantaloupe	2000	2
cantaloupe	2386	howland reston	1040	2
line·	2367	newsgroups rec autos	990	3
subject·	2165	zaphod mps ohio-state	912	3
newsgroup·	2079	newsgroups sci electronics	816	3
date·	2056	sender news	607	2
path·	2043	europa eng gtefsd	541	3
message-id	2011	das-news harvard	538	2
organ·	2002	electronics path	507	2
auto·	1995	electronics path cantaloupe	506	3
ohio-state	1754	ohio-state edu zaphod	485	3
electron·	1584	autos path	461	2
refer·	1510	autos path cantaloupe	458	3
write·	1446	newsgroups rec autos path	439	4
articl·	1234	electronics subject	432	2
howland	1044	magnesium club	421	2
like·	1042	harvard edu noc near	418	4
reston	1040	newsgroups sci electronics path	404	4
sender	1009	xref cantaloupe	386	2
univers·	977	autos subject	365	2
use·	973	newsgroups sci electronics subject	358	4
usenet	922	newsgroups rec autos subject	357	4
zaphod	921	gtefsd com howland reston	355	4
nntp-posting-host	913	gtefsd com howland	355	3
just	911	rochester udel	329	2

FIGURE 20.10 Terms and phrase lists

WORD CLOUD IN JMP

A word cloud (also referred as a tag cloud) is a visual representation of words. This helps to highlight popular words and phrases based on frequency and relevance. Word clouds provide quick and simple visual insights which helps further data exploration.

- Choose *Display Options > Show Word Cloud* from top red triangle to create a word cloud. Several options for layout and coloring are available for word cloud.
- Several options for layout and coloring are available for word cloud. Coloring based on *By column values* will result in colors for each term based on a gradient color scale.

TEXT PREPROCESSING IN JMP Pro

JMP Pro's *Analyze > Text Explorer* platform provides several preprocessing options in the launch dialog (see Figure 20.2).

- The default language is English, but several other languages are supported.
- JMP provides two *Tokenizing* options: *Regex* and *Basic Words*. Regex parses text using default set of built in regular expressions, while Basic Words parses text into words based on a set of characters that typically separate words. You can edit the regular expressions using *Custom Regex*.
- Other options tell JMP how to construct term and phrase lists. For example, you can specify how many words (terms) must be included in a phrase, the maximum number of phrases to include, and the minimum and maximum lengths of terms.

The initial analysis includes term and phrase lists and a summary of the text data (see Figure 20.10). You can further preprocess terms and phrases by right-clicking on the list or using options under the red triangle. For example, you can add phrases to the term list, recode terms, and add stop words by right-clicking on the lists, or change the stemming and parsing using red triangle options.

Once the data are preprocessed, several analysis options are available under the red triangle. In the following, we focus on latent semantic analysis and topic analysis.

Using Latent Semantic Analysis and Topic Analysis

To extract meaning from these preprocessed text data, *Latent Semantic Analysis* and *Topic Analysis* are used. For latent semantic analysis, the document–term matrix is reduced to 20 singular vectors (see the dialog in Figure 20.11). Term and document scatter plots for the first two vectors are shown in Figure 20.12. Note that two documents are very different from the others. The *Show Text* command can be used to show the corresponding documents—this can help determine what is unique about these two documents.

Next, topic analysis is used to identify themes (we use 20 topics for this example). The resulting topics, partially shown in Figure 20.13, are groupings of similar or related terms. For example, you can see that the top terms in topic 1 (110v, inspector-, and receptacl-) are very different from the top terms in topic 5 (overst-, underst-, and throttle).

Finally, we'll take this analysis a step further. The topics found using topic analysis can be saved to the data table as new columns (use *save Document Topic Vectors* from the red triangle for the topic analysis). These new columns can then be used in a predictive model.

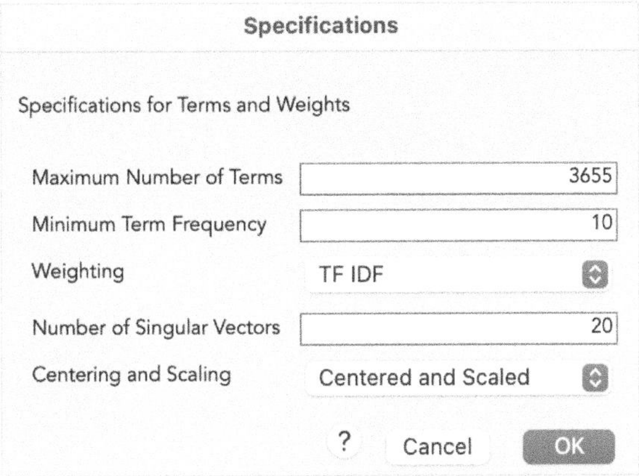

Specifications
Specifications for Terms and Weights
Maximum Number of Terms
3655
Minimum Term Frequency
10
Weighting
TF IDF
Number of Singular Vectors
20
Centering and Scaling
Centered and Scaled
?
Cancel
OK

FIGURE 20.11 Specifications for latent semantic analysis

SVD Centered and Scaled TF IDF
Principal Components of 3655 terms by 2000 documents using TF IDF weighting Centered and Scaled making 20 vectors
SVD Plots
Show Text
Show Text
Doc Singular Vector 2
Doc Singular Vector 1
Term Singular Vector 2
Term Singular Vector 1
1
2

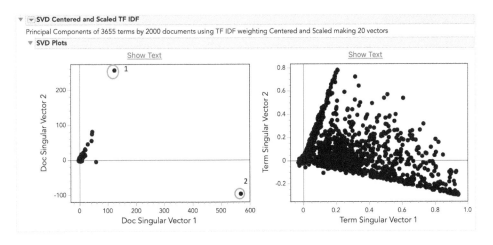

FIGURE 20.12 Scatterplots from latent semantic analysis

▼ ▼ SVD Centered and Scaled TF IDF
  ▼ ▼ Topic Analysis for 20 topics
    ▼ Top Loadings by Topic

Topic 1		Topic 2		Topic 3		Topic 4		Topic 5	
Term	Loading	Term	Loading	Term	Loading	Term	Loading	Term	Loading
110v	0.99182	boil·	0.95070	shos	0.89103	kuo-sheng	0.81390	overst·	0.91245
receptacl·	0.99097	fluid·	0.95066	shifter·	0.78937	sfsuvax1	0.79933	underst·	0.90001
inspector·	0.99077	interv·	0.95011	rpms	0.71699	sfsu	0.79933	throttle	0.85079
220v	0.98973	threshold·	0.94492	stiff	0.69598	kschang	0.79933	friction	0.84497
fixtur·	0.98944	brake·	0.91277	smooth·	0.69493	patch·	0.78694	freshman	0.80862
insul·	0.98845	gasolin·	0.90919	review·	0.68943	regist·	0.75339	foot·	0.79646
wire·	0.98781	belt·	0.88261	park·	0.68714	kasey	0.75211	angl·	0.79588
conduit	0.98768	oils	0.87318	stiffer	0.67560	user·	0.74214	corner·	0.74099
gfcis	0.98697	silicon·	0.85778	celica·	0.67363	casual	0.72883	downshift·	0.73311
aluminum	0.98401	tire·	0.83571	ride·	0.64736	csus·	0.70318	greater	0.68402
permit·	0.98235	viscos·	0.83417	shift·	0.62827	reinstal·	0.66900	lift·	0.58128
prong·	0.97941	pedal·	0.82150	parent·	0.59729	disabl·	0.66499	synchro·	0.58034
gfci	0.97807	specifi·	0.77356	compart·	0.58212	pirat·	0.66314	lever·	0.57387
jurisdict·	0.96500	addit·	0.77343	interior	0.57955	machin·	0.64721	travel·	0.56941
ordinary	0.96359	weight·	0.75111	wonder·	0.57066	restor·	0.62461		
fuse·	0.96233	manufactur·	0.74830	feel·	0.56460	copi·	0.62424		
outlet·	0.96222	inject·	0.74552			discourag·	0.62021		
connect·	0.96018								
breaker·	0.95654								

FIGURE 20.13   **Partial output from topic analysis**

LATENT SEMANTIC ANALYSIS AND TOPIC ANALYSIS IN JMP Pro

JMP Pro's *Analyze > Text Explorer* platform provides latent semantic analysis and topic analysis, which can be used as follows:

- Select *Latent Semantic Analysis, SVD* from the red triangle. SVD, or *Singular Value Decomposition*,[7] transforms the document–term matrix into a specified number of vectors (as we saw with PCA in Chapter 4). These vectors can then be used for subsequent exploration and analysis in the *Text Explorer* platform. They can also be saved to the data table for use in clustering, classification, and regression techniques.

- This analysis produces the results in the report *SVD Centered and Scaled TF IDF* (see Figure 20.9). The scatterplots show the first and second vectors for terms and documents. Use the plots to identify clusters, patterns, and unusual values.

- From the red triangle for *SVD Centered and Scaled TF IDF*, select *Topic Analysis, Rotated SVD*. The resulting topics are themes or concepts. Additional options, such as *Rename Topics* are available from this red triangle menu.

In addition to topic analysis, you can use hierarchical clustering directly from latent selection analysis in the *Text Explorer* platform (these are red triangle options).

    For more information on text mining using JMP, including underlying technical details, search for *Text Explorer* in the JMP Help.

[7]SVD is a method for reducing the dimensionality of text data into a manageable number of vectors for analysis. It is numerically efficient with sparse data (with many 0's), common with document–term matrices.

Fitting a Predictive Model

At this point, we have transformed the original text data into a familiar form needed for predictive modeling; a single outcome variable: electronics (*positive class*) and autos and 20 predictors (topic vectors).

We can now try applying several classification models to these data to see how well the topics predict the source. As we have done in previous chapters, the models are fit on the training data and performance is evaluated on the validation data. Figure 20.14 shows the performance of a logistic regression model, with "Source" as the outcome variable and the 20 topics (topic vectors) as the predictors.

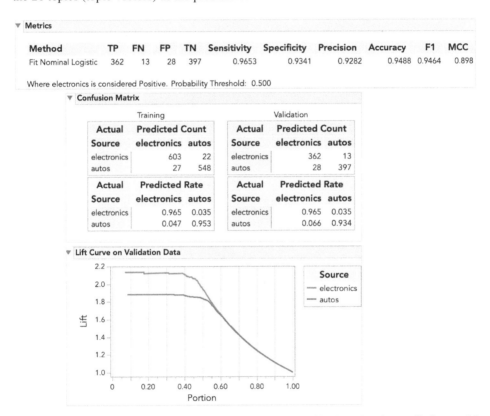

FIGURE 20.14 **Validation metrics, confusion matrix, and lift chart for the predictive model fit to the autos and electronics discussion data**

The confusion matrix (Figure 20.14) shows reasonably high accuracy in separating the two classes of documents—an accuracy of 94.88% on the validation set. The lift chart (Figure 20.14) confirms the high separability of the classes and the usefulness of this model for a ranking goal. The lift shown here is just over 2 for electronics for the first 50% of the cases.

Next, we'd try other models to see how they compare; this is left as an exercise.

Prediction

The most prevalent application of text mining is classification, but it can also be used for prediction of numerical values. For example, the unstructured text in maintenance or support tickets could be used to predict length or cost of repair. The only step that would be different in the above process is the predictive modeling step: models for prediction are used rather than models for classification.

20.7   EXAMPLE: SENTIMENT ANALYSIS OF MOVIE REVIEWS

This example demonstrates using lexical analysis and scoring documents for positive, negative, and overall sentiment. The analysis assumes that each document is free text with binary sentiment on a single topic. The task is to classify the sentiment of movie reviews as either positive or negative. The file `IMDB-Dataset-10K.jmp` contains 5000 positive and 5000 negative movie reviews collected from the Internet Movie Database (IMDB) (Maas et al., 2011)[8]. In the following, we describe the key steps from preprocessing to building a classification model on the data.

A snippet of a positive review looks like this:

> It is so gratifying to see one great piece of art converted into another without distortion or contrivance. I had no guess as to how such an extraordinary piece of literature could be recreated as a film worth seeing. ...So now my favorite book and movie have the same title.

On the other hand, a snippet of a negative review looks like this:

> I did not expect much from this film, but boy-o-boy, I did not expect the movie to be this bad. ...This is a very bad film! I gave it three out of ten because of few smiles it gave me, but I did never laugh!

Data Preparation

Import the Excel data into a JMP data table (use *File > Open*). The data table will have two columns called *review* and *sentiment*. Change the modeling type of the review column to *Unstructured Text*. The variable "sentiment" is the outcome variable. The data are then partitioned (60% training, 40% validation) using the *Make Validation Column* utility.

Tokens are filtered based on length (min char = 3) and basic words. Built-in stopwords are used along with a few more additions. Words like "movie," "film," "one" are added as stopwords (select the word from the term a, right-click and select *Add Stop Word*). Stemming is used for text reduction.

Latent Semantic Analysis and Fitting a Predictive Model

Latent semantic analysis using TF-IDF is applied. For manageability, the default number of singular vectors (100) is used. Because we are not interested in understanding the topics or themes in the data for this example, topic analysis isn't used. Instead, the singular vectors from latent semantic analysis are saved to the data table (using *Save Document Singular Vectors*). These 100 vectors are used in building of predictive model.

---

[8]The original large movie dataset is published at `http://ai.stanford.edu/~amaas/data/sentiment/` and contains 25K positive and 25K negative movie reviews collected from the Internet Movie Database (IMDB). A stratified sample of 10K records from this dataset is used in this example.

The data are now ready for predictive modeling with a single outcome variable "sentiment" (positive and negative) and 100 predictors. Different classification models may be applied to these data. Figure 20.15 shows the partial output from the logistic regression model. This model has a moderately high accuracy (81.6%) in separating reviews with positive and negative sentiment. This is confirmed by the ROC curve.

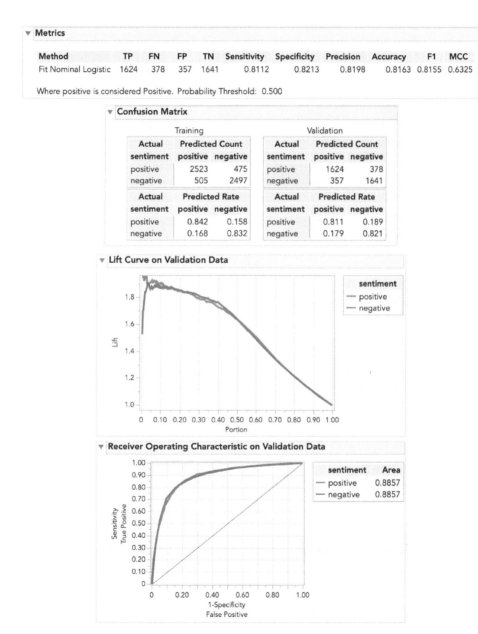

FIGURE 20.15   **Metrics, confusion matrix, lift, and ROC curves of validation set for sentiment analysis model**

TERM SELECTION AND SENTIMENT ANALYSIS IN JMP Pro

In the movie reviews example, we conducted sentiment analysis using results saved from the *Text Explorer* platform and then fitted a predictive model. The *Text Explorer* platform provides two alternative approaches: sentiment analysis and term selection.

- *Sentiment Analysis*, or lexical sentiment analysis, identifies sentiment terms, negators (isn't, don't,...), and intensifiers (very, extremely,...) and scores documents for positive, negative, and overall sentiment. Basic natural language processing (NLP) is used to parse the documents. Note that Sentiment Analysis in the *Text Explorer* platform is only available in English.
- *Term Selection*, or *predictive term selection*, helps in identifying the terms that best explain the responses or outcomes. This analysis uses Generalized Regression, which is embedded in the term selection analysis, to perform variable selection on the document–term matrix (DTM).

Term Selection and Sentiment Analysis are top red triangle options after you run the initial Text Explorer analysis. For more information on these methods, search for *Term Selection* or *Sentiment Analysis* in the JMP Help.

20.8   SUMMARY

In this chapter, we drew a distinction between text processing for the purpose of extracting meaning from a single document (natural language processing—NLP) and classifying numerous documents in probabilistic fashion (text mining). We concentrated on the latter and examined the preprocessing steps that need to occur before text can be mined. Those steps are more varied and involved than those involved in preparing numerical data. The ultimate goal is to produce a matrix in which rows are terms and columns are documents. The nature of language is such that the number of terms is excessive for effective model-building, so the preprocessing steps include vocabulary reduction. A final major reduction takes place if we use, instead of the terms, a limited set of concepts that represents most of the variation in the documents, in the same way that principal components capture most of the variation in numerical data. Finally, we end up with a quantitative matrix in which the cells represent the frequency or presence of terms and the columns represent documents. To this, we append document labels (classes), and then we are ready to use this matrix for classifying documents using classification methods.

PROBLEMS

20.1 Tokenization. Consider the following text version of a post to an online learning forum in a statistics course:

```
Thanks John!<br /><br /><font size="3"> &quot;Illustrations
and demos will be  provided for students to work through on
their own&quot;</font>.  Do we need that to finish project?
If yes, where to find the illustration and demos?
Thanks for your help.\<img title="smile"  alt="smile"
src="\url{http://lms.statistics.com/pix/smartpix.php
/statistics_com_1/s/smiley.gif}" \><br /> <br />
```

  a. Identify 10 non-word tokens in the passage.

  b. Suppose this passage constitutes a document to be classified, but you are not certain of the business goal of the classification task. Identify material (at least 20% of the terms) that, in your judgment, could be discarded fairly safely without knowing that goal.

  c. Suppose the classification task is to predict whether this post requires the attention of the instructor or whether a teaching assistant might suffice. Identify the 20% of the terms that you think might be most helpful in that task.

  d. What aspect of the passage is most problematic from the standpoint of simply using a bag-of-words approach, as opposed to an approach in which meaning is extracted?

20.2 Classifying Internet Discussion Posts. In this problem, you will use the data and scenario described in this chapter's example, in which the task is to develop a model to classify documents as either auto-related or electronics-related.

  a. Perform the data preprocessing steps (import, join, and stack), create a new data set `ExampleSet.jmp` with a source column.

  b. Following the example in this chapter, preprocess the documents. Explain what would be different if you did not perform the "stemming" step.

  c. Perform Latent Semantic Analysis (LSA), and then Topic Analysis, to create 10 topics. What are the top three terms in Topic 1 and Topic 2? What names would you give to these two topics?

  d. Save the Document Topic Vectors to the data table, then fit a predictive model (different from the model presented in the chapter illustration) to classify documents as autos or electronics. Compare the performance of your model to that of the model presented in the chapter illustration.

20.3 Classifying Classified Ads Submitted Online. Consider the case of a website that caters to the needs of a specific farming community and carries classified ads intended for that community. Anyone, including robots, can post an ad via a web interface, and the site owners have problems with ads that are fraudulent, spam, or simply not relevant to the community. They have provided a file with 4143 ads, each ad in a row, and each ad labeled as either −1 (not relevant) or 1 (relevant). The goal is to develop a predictive model that can classify ads automatically.

- Open the file FarmAds.jmp, and briefly review some of the relevant and non-relevant ads to get a flavor for their contents.
- Following the example in the chapter, preprocess the data in JMP, and create a document–term matrix.

a. Examine the document–term matrix.

    i. Is it sparse or dense?

    ii. Find two nonzero entries and briefly interpret their meaning, in words (you do not need to derive their calculation).

b. Perform a latent semantic analysis. Relate the latent semantic analysis to what you learned in the principal component analysis chapter (Chapter 4).

c. Save the Document Singular Vectors to the data table.

d. Partition the data (60% training, 40% validation), and develop a logistic regression model to classify the documents as "relevant" or "non-relevant." Comment on its efficacy.

e. Why use the document singular vectors as predictors, and not the document–term matrix?

20.4 Clustering Auto Posts. In this problem, you will use the data and scenario described in this chapter's example. The task is to cluster the auto posts.

- Following the example in this chapter, preprocess the documents.
- Use latent semantic analysis, with 10 singular vectors.

a. Before doing the clustering: how many natural clusters you expect to find? Why?

b. From the latent semantic analysis red triangle, select *Cluster Terms* and inspect the dendrogram. From the dendrogram, how many natural clusters appear?

c. Use the *Set Clusters* option (next to the dendrogram), and set the number of clusters to 5. This colors the clusters by cluster number and also clusters the corresponding documents in the SVD plots. Click to select one cluster, and describe what you see in the document and term SVD plots.

21

RESPONSIBLE DATA SCIENCE

In this chapter, we go beyond technical considerations of model fitting, selection, and performance and discuss the potentially harmful effects of machine learning. The catalog of harms is now extensive, including a host of cases where AI has deliberately been put to ill purposes in service of big brother surveillance and state suppression of minorities. Our focus, however, is on cases where the intentions of the model developer are good, and the resulting bias or unfairness has been unintentional. We review the principles of responsible data science (RDS) and discuss a concrete framework that can govern data science work to put those principles into practice. We also discuss some key elements of that framework: datasheets, model cards, and model audits.

21.1 INTRODUCTION[1]

Machine learning and AI bring the promise of seemingly unlimited good. After all, the ability to ingest any set of arbitrarily sized, minimally structured data and produce predictions or explanations for these data is applicable to almost every domain. Our societal attention often focuses on the revolutionary future applications of this potential: cars that drive themselves, computers that can hold natural conversations with humans, precision medications tailored to our specific genomes, cameras that can instantly recognize any object, and software that can automatically generate new images or videos. Conversations about these benefits, though, too often ignore the harms that machine learning models can cause.

---

[1] This chapter draws on *Responsible Data Science*, Wiley 2021, by Grant Fleming and Peter Bruce, for its organization of ideas. This chapter copyright © 2022 Datastats, LLC. Used by permission.

Machine Learning for Business Analytics: Concepts, Techniques, and Applications with JMP Pro®,
Second Edition. Galit Shmueli, Peter C. Bruce, Mia L. Stephens, Muralidhara Anandamurthy, and Nitin R. Patel.
© 2023 John Wiley & Sons, Inc. Published 2023 by John Wiley & Sons, Inc.

Example: Predicting Recidivism

Decisions about criminal defendants in the US justice system are based on the weight of probabilistic evidence at multiple stages: whether to arrest, whether to go to trial, arrival at a verdict, and sentencing. At the sentencing phase, judges determine the length and terms of sentences, in part, by assessing the likelihood that a convicted defendant will commit another crime (recidivate). Courts have started relying increasingly on machine learning recidivism algorithms to inform decisions on sentencing. The COMPAS algorithm,[2] sold by Northpointe, Inc., is among the most prominent of these, and critics have charged that it is biased against African-Americans. The algorithm is not public and involves more than 170 predictors. Its advocates retort that COMPAS has good overall predictive performance and that its accuracy, as measured by the area under the ROC curve (Receiver Operating Characteristics curve), is similar for African-American and White defendants. The trouble is that the errors made are quite different:

- African-American defendants are over-predicted to re-offend (leading to tougher sentences).
- White defendants are under-predicted to re-offend (leading to lighter sentences).

The over-prediction errors (penalizing African-Americans) balance out the under-prediction errors (favoring Whites), so the overall error rate is the same for both. A single-minded focus on one overall accuracy metric obscures this bias against African-Americans. We will work through this example in greater detail below.

21.2   UNINTENTIONAL HARM

The COMPAS algorithm did not set out to penalize African-Americans. Its goal was to make sentencing decisions overall more consistent and "scientific." The bias against African-Americans was unintended and unwanted.

Another stark example of unintentional harm was the Optum healthcare algorithm. In 2001, the healthcare company Optum launched Impact-Pro, a predictive modeling tool that purported to predict patients' future need for followup care and assign a risk score, based on a variety of predictor inputs. The result would be better healthcare for patients, as follow-up interventions could be better timed and calibrated. Hospitals could use the tool to better manage resources, and insurers could use it to better set insurance rates. Unfortunately, experience proved that the algorithm was biased against African-Americans: a research team led by Ziad Obermeyer studied these algorithms and found that, for any given risk score, African-American patients consistently experienced more chronic health conditions than did White patients.

It turned out that a key predictor for "future health care need" is "prior healthcare spending." African-American spending on healthcare was, on average, less than that for Whites, but not because of better health. Rather, it was due to lower income levels and less access

---

[2]COMPAS stands for "Correctional Offender Management Profiling for Alternative Sanctions," a good summary by the public interest group ProPublica can be found here: https://www.propublica.org/article/how-we-analyzed-the-compas-recidivism-algorithm

to health insurance and costly medical facilities. Hence, an African-American individual would be predicted to need less follow-up care than a White individual with similar health.

There are numerous other examples of well-intentioned algorithms going wrong.[3] The algorithms that power social media and internet ads foster connections among people and with things they are interested in, "bringing the world closer together" in Facebook's words. Those same algorithms can also nurture communities of hate and facilitate the spread of false news and conspiracy theories.

INTENDED HARM

Not all damage from machine learning algorithms is unintentional. In some cases, bad actors "weaponize" algorithms for purposes that are malicious. In other cases, seemingly legitimate law enforcement practices creep into the realm of "big brother":

- Facial recognition and identification technology, combined with the widespread presence of security cameras, enable police departments to track down criminal suspects who would have gone undetected in the pre-technology age. A good thing!

- Those same technologies, though, have enabled US immigration authorities to track down and deport undocumented individuals who may have been living and working peacefully for years or decades. Not so clearly a good thing—the technology empowered bureaucrats (not legislators) to shift the terms of a highly contentious political debate that was broadly agreed on only one thing: authorities should not be sweeping the country to round up and deport long-term residents who are undocumented.

- These technologies have also aided Chinese police authorities in their surveillance and suppression of Uighur and Tibetan communities. A bad thing, by the standards of most pluralistic societies that honor the rule of law.

Some might also consider that knowing about damage caused by algorithms and proceeding anyway ("turning a blind eye") constitutes intentional harm. In late 2021, a disaffected employee at Facebook released a trove of internal Facebook research showing that ML-enabled algorithms harmed teenagers (for teen girls with mental health and body-image issues, Instagram exacerbated those issues) and fostered civil discord (2018 algorithm changes for the Facebook newsfeed boosted angry and contentious content). Top executives knew about the internal research but were reluctant to make significant changes in highly profitable platforms.

The framework and process we discuss in this chapter will do little to counter the individual or entity that is determined to develop or repurpose algorithms for harmful purposes. However, the framework outlined below does have an important message for data scientists involved in developing algorithms: anticipate and be mindful of potentially harmful uses to which their work might be put.

---

[3]The popular author Cathy O'Neil coined the catchphrase "Weapons of Math Destruction" in recounting such examples in her book of the same name (O'Neil, 2016).

21.3   LEGAL CONSIDERATIONS

Many treatments of the ethical aspects of data science focus on legal issues. This is under-standable, as avoiding legal jeopardy is a strong motivator. Legal compliance is ultimately a matter for attorneys more than data scientists. The latter may contribute best by pursuing a general framework for responsible data science, so our legal discussion will be brief and focused on two points: the European Union's GDPR and the concept of "protected groups."

The General Data Protection Regulation (GDPR)

The most recognized legal environment for machine learning is the General Data Pro-tection Regulation (GDPR) of the European Union (EU), which went into effect in August 2018. The GDPR provides sweeping protections to those in the EU (whether citi-zens of EU countries or not) including a "right to be forgotten" (a requirement that tech-nology companies provide mechanisms by which individuals can have their data removed from online platforms), as well as the right to data portability and the right to object to automated decision-making and profiling. The GDPR's focus is strongly on empowering the individual, educated citizen with greater leverage to enforce the privacy of their data. Because the GDPR was the first comprehensive regulatory framework for Big Data ser-vices, it set the pattern for other legal initiatives that followed, including the California Consumer Protection Act (CCPA) and Canada's Personal Information Protection and Elec-tronic Documents Act (PIPEDA). Given the size of the EU market and the global nature of internet activity, the GDPR has become, in effect, a global regulatory regime.

Protected Groups

Regulations and laws that address not simply consumer privacy, but also issues of algo-rithmic discrimination and bias, are less well developed. However, one guiding principle is that laws that restrict behavior by individuals or corporations do not disappear merely because the decisions are made by algorithms. Decisions or actions that are illegal for a hu-man in a company (discrimination in lending based on gender or age, for example) are also illegal for an algorithm to do in automated fashion. In U.S. law (and in other countries as well), there are various "protected groups" where such restrictions apply. Protected groups may be identified on the basis of race, gender, age, sexual orientation, medical disability, or other factors. Although there may be no AI-specific or Big Data-specific laws in this area, there is ample and long-standing legal history that must be considered in the deployment of algorithms.

To evaluate fairness and calculate fairness metrics, a "privileged group" is designated. Often this reflects actual societal advantage enjoyed by the group (terminology coinciding with the social justice movement), but it is also a technical requirement. Fairness metrics must be calculated relative to some base group, so the term by itself does not imply privilege in the lay meaning.

21.4   PRINCIPLES OF RESPONSIBLE DATA SCIENCE

Legal statutes and regulations are typically lengthy, detailed, and difficult to understand. Moreover, legal compliance alone is not sufficient to avoid some of the harm caused by machine learning. The potential for reputational harm may be just as great as that for legal jeopardy. In late 2021, Facebook began conducting "reputational reviews" of its services and products. The publicly-stated purpose was to avoid harm to children, but news stories also cited Facebook's desire to anticipate how its algorithms might be criticized. A significant movement has arisen in the data science community to focus greater attention on this potential for harm and how to avoid it. A number of commentators have suggested that data scientists should adhere to a framework for responsible data science: a set of best practices. Common to most suggested frameworks are five principles:

- Non-maleficence (avoiding harm)
- Fairness
- Transparency
- Accountability
- Privacy

Non-maleficence

Non-maleficence covers concepts associated with causing or avoiding harm. At a high level, following this principle requires that data scientists ensure that their models do not cause foreseeable harms or harms of negligence. While unforeseen harms are sometimes unavoidable, they should be accounted for as much as reasonably possible. Harms that ought to be avoided include risks of physical harm, legal risks, reductions in future opportunities, privacy violations, emotional distress, facilitating illegal activities, and other results which could negatively impact individuals.

Non-maleficence is the hardest principle to guarantee in practice, and we recommend thinking of it as more of a goal to strive for than a necessary box to check for every data science project. A model can be fair, transparent, accountable to users, and respecting of privacy yet still make predictions that advantage some people while causing harm for others. Medical triaging algorithms for prioritizing scarce medical resources necessarily place some individuals further back in the queue or out of the queue entirely. Credit scoring algorithms make life difficult for those receiving low credit scores. In such cases, non-maleficence must be considered in the context of other principles.

Fairness

Fairness covers concepts associated with equal representation, anti-discrimination, dignity, and just outcomes. Dimensions of interest for fairness evaluations typically involve human-specific factors like ethnicity, gender, sexual orientation, age, socioeconomic status, education, disability status, presence of preexisting conditions (healthcare), and history of adverse events. Model fairness is primarily a function of whether the model produces fair outcomes, with input factors (how the model uses certain features, how the features were created, etc.) being less relevant. Like non-maleficence, fairness is a difficult concept to pin down. What is fair for one person or context might be unfair for another. One person might consider it fair for everyone to pay the same absolute amount in taxes. Another person might think everyone should pay the same share of their income in taxes. Another might think the wealthy should pay a higher share of their income in taxes. Yet another might think that the wealthy should pay all the taxes. There is no universally agreed definition, in this case, of what constitutes fairness. Because fairness relies primarily on ensuring fair outcomes, it is easier to translate into practice than the other four principles. However, fairness-improving methods are still very underused. Fairness is almost always talked about exclusively in the context of models that make predictions for individual people, like whether a person should be provided a loan. In actuality, fairness is relevant for any modeling task where the goal is to either minimize differences between groups or maximize predictive performance with respect to specific groups. For example, in a spam email classification task, it might make sense to use fairness interventions to make the spam classification model more balanced in how well it classifies spam across different identified topics. While this is not a use case that deals directly with people, it can still be beneficial to consider fairness.

Transparency

Transparency covers concepts associated with user consent, model interpretability, and explanations for other modeling choices made by the creators of a model. Typically, this communication is conveyed via clear documentation about the data or model.[4] Transparency as a concept is as simple as it appears: where possible, decisions about the modeling process should be recorded and available for inspection. However, transparency quickly becomes complex when trying to move it from principle to practice. For example, the modeling team, regulators, users, and clients all might require mutually exclusive forms of transparency due to their different needs and technical abilities. Documentation about the datasets used, explanations for choices made within the modeling process, and interpretations for individual predictions made for one group might need to be completely revamped for another.

An important related term is the *interpretability* of a model. Some of the greatest advances in machine learning have come with so-called uninterpretable, blackbox models— models where the learned relationships between predictors and outcomes are so complex that humans cannot discern how one affects the other. Neural nets are one example, and they have become ubiquitous now with the power and popularity of deep learning. Random (bootstrap) forests and boosted trees are also examples of blackbox methods: single trees can easily be translated into decision rules, but when the predictions from many different trees are simply averaged or otherwise aggregated, that interpretability disappears.

Classical statistical models like linear and logistic regression are "intrinsically interpretable." Their coefficients provide clear and direct estimates of the effects of individual predictors.

Without interpretability, we are handicapped in reviewing the ethical impact of our models. If a model is producing biased predictions, how would we know which predictors are instrumental in generating the biased predictions so that we might make corrections? In seeking interpretability, we can follow a two-pronged approach:

- Where possible, include an interpretable model among the final candidates, to serve as a benchmark. If we ultimately choose a better-performing blackbox model, we'd like to know how much, in terms of performance, we would have to give up if we went with an intrinsically-interpretable model.
- Apply interpretability methods, ex post, to a blackbox model. Some examples of these methods are illustrated in the COMPAS example in Section 21.7.

---

[4]Under Basel II banking regulations, for example, deployed credit scoring models should be repeatable, transparent and auditable (Saddiqi, 2017).

Accountability

Accountability covers concepts associated with legal compliance, acting with integrity, responding to the concerns of individuals who use or are affected by machine learning algorithms, and allowing for individual recourse for harmful modeling decisions. For example, people who receive a loan denial ideally ought to be entitled to not just an explanation for why that decision was made but also the ability to appeal the decision and an indication of the minimum set of differences that would have resulted in a change in the decision. In practice, accountability typically does not go beyond ensuring that a model achieves legal compliance (which itself has gotten more difficult because of regulation like the GDPR) or, in a few cases, adherence to company or industry standards.

Data Privacy and Security

Privacy covers concepts associated with only gathering necessary information, storing data securely, ensuring that data is de-identified once it is used in a model, and ensuring that other aspects of user data cannot be inferred from the model results. Fortunately, there are already several well-understood approaches for maintaining the privacy of systems (via preventing hacks, unintentional disclosures, physical attacks, etc.). Making data available for analysis in models circumvents some of these protections, and it may be necessary to use differential privacy methods (for example, adding randomization to the underlying data). Unfortunately, differential privacy methods are not yet built into many popular modeling packages.

21.5   A RESPONSIBLE DATA SCIENCE FRAMEWORK

How should data scientists incorporate the general principles and evolving legal considerations into projects? This requires a framework to provide specific guidance for data science practitioners. Most suggested frameworks start from a standard "best practices" process, such as the CRISP-DM and SEMMA methodologies described in Chapter 2, and expand on those technical frameworks to incorporate larger ethical issues. The Responsible Data Science (RDS) framework suggested by Fleming and Bruce (2021) has the following elements:

Justification

In this initial phase, the data scientist and project manager coordinate with users and other stakeholders to gain an understanding of the problem in its business context. Questions such as the following should be addressed:

- Have we studied similar projects, and have they encountered ethical problems?
- Have we anticipated potential future uses of the model and harms that might result?
- Do we have the means necessary to assess potential harms to individuals and groups?

Assembly

In this stage (termed "compilation" by Fleming and Bruce), the team assembles the various elements needed for the project:

- The raw data
- Software tools and environments
- Datasheets, documents that go beyond descriptions of the data features to include explanations of how the data was gathered, preprocessed, and intended to be used
- Identification of any protected groups

The process of assembling the raw data is where privacy rules must be considered. The project team must be certain that the data were acquired in a fashion that was non-exploitative and respected consent. One famous case where this was not the case was that of the Clearview AI facial recognition product that can take almost any facial image and identify who it is, if that person has a social media presence. The data to build their model, which is used extensively by law enforcement, came from scraping social media sites without permission.

The story of Cambridge University Professor Alexander Kogan is another cautionary tale. Kogan helped develop a Facebook app, "This is Your Digital Life," which collected the answers to quiz questions, as well as user data from Facebook. The purported purpose was a research project on online personality. Although fewer than 270,000 people downloaded the app, it was able to access (via friend connections) data on 87 million users. This feature was of great value to the political consulting company Cambridge Analytica, which used data from the app in its political targeting efforts and ended up doing work for the Trump 2016 campaign, as well as the 2016 Brexit campaign in the United Kingdom.[5] Facebook later said this further sharing of its user data violated its rules for apps, and it banned Kogan's app.

Great controversy ensued: Facebook's CEO, Mark Zuckerberg, was called to account before the US Congress, and Cambridge Analytica was eventually forced into bankruptcy. In an interview with Lesley Stahl on the *60 Minutes* television program, Kogan contended that he was an innocent researcher who had fallen into deep water, ethically:

> "You know, I was kinda acting, honestly, quite naively. I thought we were doing everything okay…. I ran this lab that studied happiness and kindness."

How did machine learning play a role? Facebook's sophisticated algorithms on network relationships allowed the Kogan app to reach beyond its user base and get data from millions of users who had never used the app.[6]

The world is awash in data, and there will surely be gray areas that lie between explicit authorization to use and explicit prohibitions on use. Analysts may be tempted to follow the adage "Ask forgiveness, not permission," but if you think forgiveness may be needed, the temptation should be avoided. Data scientists working with gray area data can look to several criteria:

---

[5] Read more about the efficacy of psychological manipulation via the internet in the paper "Psychological targeting as an effective approach to digital mass persuasion" by Matz et al. (2017).

[6] www.cbsnews.com/news/aleksandr-kogan-the-link-between-cambridge-analytica-and-facebook-60-minutes/

- *The impact and importance of the project*: A high value project, or one that might cause significant harm, raises the costs of running afoul of data use restrictions and enhances the importance of thorough and comprehensive review.
- *The explicitness of rules governing data use*: The more explicit the prohibition, the harder it will be to explain why it was not followed.
- *The "news story test:"* What would a reasonable person conclude after reading a widely circulated news story about your project?

In the end, the data scientist must think beyond the codified rules and think how the predictive technologies they are working on might ultimately be used.

Data Preparation

This stage incorporates the standard data exploration and preparation that we discussed in Chapters 3–5. A couple of additional points not covered there are worth noting:

- We need to verify that the data wrangling steps we have taken do not distort the raw data in a way that might cause bias.
- Data preparation in the RDS framework is not necessarily a single progression through a sequence, but may be an iterative process.

As an example of the potentially iterative nature of data preparation, consider the case where the audit of a model (discussed below) reveals that model performance is worse for a particular protected group that is not well represented in the data. In such a case, we might return to the data preparation stage to oversample that group to improve model performance.

Modeling

In this phase, the team tries out multiple models and model settings to find the model that yields the best predictive performance. Most of this book is about this process, so we need to add little here, with one exception. A key factor in building models that do not cause harm is the ability to understand what the model is doing: interpretability. In this stage, along with predictive performance, we also consider whether we can gain a good understanding of how predictors contribute to outcomes.

Auditing

This final stage is the key one. In the auditing phase, we review the modeling process and the resulting model performance with several key points in mind:

- Providing useful explanations for the model's predictions
- Checking for bias against certain groups (particularly legally protected groups)
- Undertaking mitigation procedures if bias or unfairness is found

If the audit shows that model performance is worse for some groups than others, or if the model performs differently for some groups in ways that disadvantage them, then we need to try to fix it. Sometimes models perform more poorly for a group if that group is poorly represented in the data. In such a case, the first step is to oversample that group, so the model can work with more data. The COMPAS example in Section 21.7 illustrates this approach. Another step is to dive deeper on the interpretability front to learn more about what predictors might be driving decisions. If oversampling and a deeper dive into interpretability fail to fully correct bias in predictions, the next step is to adjust the predictions themselves, an approach that is also illustrated in the COMPAS example.

The above discussion assumes that we do not include racial, ethnic, religious, or similar categories as predictors. This is typically, but not always, the case. Structuring a model that could be construed as denying a loan to someone because they are Hispanic, for example, would be illegal in the United States, whether the decision is based on a human review or a machine learning model. The situation is less clear cut in medicine, where racial and other indicators may be relevant diagnostic flags.

Simply assuring that the model does not include explicit racial or ethnic categories as predictors may not be sufficient. There may be other variables that serve as proxies: membership in an African-American fraternity, for example. The risk of including such variables increases with the use of massive blackbox models where huge numbers of variables are available. It may seem less work to throw them all in the model, and let the model learn which are important, than to do your feature engineering "by hand." An example of this is a machine learning resume review system that Amazon attempted to use for hiring recommendations. The training data consisted of resumes and human judgments made by Amazon engineers. It turned out that the algorithm was picking up on an implicit preference for male candidates in that training data, and identifying derived predictors that simply reflected gender (e.g., participation in the Women's Chess Club).

This brings us to one case where it is entirely appropriate to include racial, ethnic, and similar categorical predictors. After completing the primary model, a second model that includes these categories can be fit where the purpose of the secondary study is to investigate possible discrimination. For example, a lender conducts an audit of its underwriting algorithm and finds that loans are approved for African-Americans less often than for Whites, even though race is not a feature in the model. A secondary model could then determine whether some other factor with predictive power is correlated with race. If there is such a factor and its "business necessity" cannot be justified, it should probably be excluded from the model. Rules and regulations governing discriminatory practices constitute a rich legal field beyond the scope of this book; you can get a flavor by doing a web search for "FDIC policy on discrimination in lending" (FDIC is the Federal Deposit Insurance Corporation, an important financial regulator in the United States).

21.6 DOCUMENTATION TOOLS

If the model performs reasonably well and does not produce biased predictions or decisions, it passes the audit phase, at least with respect to unintentional harm. Several tools provide a structure for documenting this and a guide for incorporating ethical practices into the project from the beginning.

Impact Statements

One such tool is a prospective statement of the project's impact. Impact statements for data science became popular or, more accurately, infamous, when the prestigious Neural

Information Processing Systems (NeurIPS) Conference required the inclusion of a "Broader Impact Statement" for submissions to the 2020 conference. NeurIPS provided the following guidance for the statement:

> "In order to provide a balanced perspective, authors are required to include a statement of the potential broader impact of their work, including its ethical aspects and future societal consequences. Authors should take care to discuss both positive and negative outcomes."

By requiring an impact statement for all submissions, NeurIPS dramatically raised the profile of ethical considerations in AI research. For their work to be deemed credible, researchers now had to contend with the ethical implications of their work. This perspective spread to the large tech companies and academic institutions that employed them. Leading AI companies like OpenAI, Google, Microsoft, SAS, and IBM have since begun leveraging variants of impact statements to make their own models and software products more transparent. Where these companies go, others are likely to follow, so it is prudent to consider the inclusion of impact statements in data science projects. Such an impact statement should address the following questions:

1. What is the goal of our project?
2. What are the expected benefits of our project?
3. What ethical concerns do we consider as relevant to our project?
4. Have other groups attempted similar projects in the past, and, if so, what were the outcomes of those projects?
5. How have we anticipated and accounted for future uses of our project beyond its current use?

The impact statement that answers these questions should be included in project documentation.

Model Cards

In 2018, researchers at Google proposed a standardized process for documenting the performance and non-technical characteristics of models. The idea of model cards has caught on quickly across the industry. Companies like IBM (through their FactSheet Project), Microsoft (through their Azure Transparency Notes), and OpenAI (through their Model Cards) have adapted model cards to provide a consistent format for salient information about their models. In 2020, Google released the Model Card Toolkit (MCT), an official package for automating the generation of model cards for models fit using Tensorflow. A sample model card generated by MCT can be seen at `github.com/tensorflow/model-card-toolkit`. It covers a neural net model that classifies an individual's income level via demographic predictors and includes an overview of the model's purpose, an identifier for the model version, the "owner" of the model (in this case the Google Model Card Team), a list of references, a description of the use case, limitations of the model, ethical considerations, and plots depicting model performance metrics.

Datasheets

Just as model cards provide "nutrition labels" for models, datasheets (also called data cards) do likewise for datasets. "Data dictionaries" are almost ubiquitous across data science projects, but are not uniform in their formatting or in the information that they contain. Where one team might use a simple .txt file including information on the names and types of columns within a dataset, other teams rely on an unwieldy Excel file containing multiple sheets of information about every aspect of the data. Datasheets standardize the format of data dictionaries and extend their coverage to relevant non-technical information like the data's collection process, intended uses, and sensitive features.

Datasheets identify the source of the data, whether any transformations were performed, whether the data were sampled, intended use of the data, data type (e.g., tabular data, text data, image data), whether the data are public, who the license and copyright holder are, and more. Data collection, preparation, and documentation are sometimes regarded as a tedious aspect of machine learning compared to the building of predictive models. It is not surprising, therefore, that model cards have been quicker to catch on than datasheets.

Audit Reports

Model cards and datasheets are snapshots of the model(s) and data at a point in time, usually the conclusion of the project. The process that the team undertook to verify that the data and model are functioning as expected, and to fix problems, is not captured by these documents. This leaves little room to discuss special considerations for protected groups, comparisons of predictive performance between the final model and other candidate models, differences in predictive performance across groups, and, in the case of blackbox models, the outputs of interpretability method outputs. These issues should be addressed in audit reports, which consist of:

- A summary page for general model performance information
- Detailed performance pages for each sensitive group within the data, showing:
  - Comparisons of final model and candidate model performance across groups within the data
  - Visualizations of interpretability method outputs for the final model to show how the model is using its features when creating predictions
  - Descriptive text for findings from the plots/metrics and other relevant considerations

Audit reports, as well as datasheets and model cards, may seem like a heavy administrative burden to place on the predictive modeling process, taking all the fun out of data science. And, indeed, most data science projects in the real world do not go through all these steps (which is largely why so many ethical problems have occurred). However, even if all the boxes suggested by these reports are not to be checked, it is still useful to consider them as a superstructure, from which specific elements of particular importance are to be selected. In illustrating Analytic Solver Data Mining, we will use this pick and choose approach, focusing on areas of interest.[7] We return now to the COMPAS example to illustrate

---

[7] Machine learning communities have developed more automated and comprehensive approaches to reporting bias and fairness metrics in R and Python; see Fleming and Bruce (2021) for more details.

selected aspects of the principles, procedures, and documentation steps for responsible data science.

UNSTRUCTURED DATA

Some of the most far-reaching harms reported from machine learning and AI concern unstructured data. Image recognition can be used to identify individuals and surveil populations; related technologies can generate deepfakes: images, video, or audio in which a real person is combined with synthesized elements of another imaginary or real person. In one case, the face of Adolph Hitler was substituted onto the image of the president of Brazil. So-called "revenge porn" swaps images of real people, often celebrities, onto pornographic images and videos. There are similar examples in natural language processing:

- Creation of fake personas on social media to corrupt or drown out legitimate content
- Text classification models that discriminate against certain groups or viewpoints

With unstructured data, deep neural nets predominate. Important aspects of the principles of responsible data science still hold: examination of the goals and potential ethical risks of the project beforehand, adequate representation of groups, consideration of unanticipated uses, and auditing the model for its effects on protected groups. However, image and text processing rely overwhelmingly on neural networks, which are blackbox models, so we do not have an intrinsically-interpretable model as a benchmark. For interpretability, neural networks require the ex post application of interpretability methods. With unstructured data, interpretation tends to focus on individual cases and a hunt for what led the model to its decision. For example, particular areas in images that we humans recognize as features (e.g., eyes) might surface as important. However, neural networks learn their own hierarchy of features and use that information in a highly complex fashion, so we will never achieve the simplicity of, say, a linear regression coefficient.

The discussion in this chapter, and most of the discussion in this book, primarily concerns structured tabular data. For the reader interested in learning more about interpretation of unstructured data, we refer the reader to Fleming and Bruce (2021), in particular Chapter 8.

21.7 EXAMPLE: APPLYING THE RDS FRAMEWORK TO THE COMPAS EXAMPLE

We saw, at the beginning of this chapter, how the COMPAS algorithm over-predicted recidivism (committing additional crimes) for African-American defendants and under-predicted recidivism for Whites. In doing so, we jumped ahead to the interesting puzzle in the story. With the COMPAS algorithm in the background, let's now go back and apply selected elements of the RDS framework more systematically to the problem at hand. Starting at the beginning, can we develop a machine learning algorithm that predicts well whether a defendant will re-offend? We start with the nature of the project itself, then review the data-gathering process, fit a model, and audit the results.

Unanticipated Uses

Could there be unanticipated, potentially harmful, uses to which the algorithm might be put? Two possibilities are:

- The system could "migrate down" in the justice system and be used *prior* to conviction so that "who you are" (i.e., the essence of the model) might affect whether you are determined to be a criminal in the first place.
- The system could be used in other, more authoritarian countries to predict a much broader range of "criminality" (political dissent or opposition, "anti-social" behavior, etc.).

Ethical Concerns

Are there any ethical issues that we should flag a priori? African-Americans have had a troubled history with the US criminal justice system since the days of slavery and emancipation. After the US Civil War, trumped up "vagrancy" charges, requirements for employment, and strict enforcement of labor contracts, saddled African-Americans, particularly working-age African-American males, with criminal histories that were used to coerce them into agricultural labor. The legacy of this system, and the excessive criminal records it produced, continued for over a century. For this reason, we will want to be particularly alert for any discriminatory effect of the recidivism algorithm with respect to African-Americans.

Protected Groups

Our concern with potential discrimination against African-Americans means that we will want to formally identify "protected groups." To do this, we include categorical variables to identify each individual's racial status. Although we have a special concern beforehand about African-Americans, we will need to include other racial categories as well, so as to be prepared to answer questions about the algorithm's impact on them as well. Of course, these racial category variables will not be used as predictors, but rather in the audit phase. Gender and age are also denominated as protected categories, because in some applications they must be treated as such. In granting loans, for example, law and regulation preclude their use as predictors. In the recidivism case, there is no such prohibition, and our interest is in possible racial discrimination, so we retain them as predictors.

Data Issues

Clean data on convicted criminals and subsequent offenses (if any) over time are hard to come by. The data that we use here were obtained by ProPublica, a public interest lobby group, via a public records request in Broward County, Florida, where the COMPAS algorithm was used to inform sentencing decisions. The data consisted of demographic and recidivism information for 7214 defendants, 51% of whom were African-American and

34% White. We end up with 5304 records (available in *COMPAS-clean.csv*) after taking the following steps to clean the data:

1. Defendants whose initial crime charge was not within 30 days of arrest were eliminated, as there is a good possibility of error (we have the wrong offense).
2. Traffic offenses were removed.
3. Data were limited to those meeting our study's definition of recidivist (another offense within two years) or non-recidivist (spending two years outside a correctional facility).

Note that there are multiple possible definitions of recidivism: what type of crime qualifies, does mere arrest qualify, over what time period. These choices can affect conclusions; we stick with the definitions in the ProPublica study.[8]

Fitting the Model

First, we split the data into a training partition (70%) and a validation partition (30%) using the random seed 12345 for reproducibility. Then, we fit two models, logistic regression and random forest, using the following predictors: age group,[9] charge degree (M = misdemeanor, F = felony), sex, and the number of prior crimes. The random forest model produces only predictions, but the logistic regression model for the training data, shown in Table 21.1, yields coefficients that are informative about the roles played by different predictors.

TABLE 21.1   Logistic regression model for COMPAS data

Predictor	Coefficient	Odds ratio
(Intercept)	−1.331	
age_cat [Greater than 45]	−0.717	0.231
age_cat [25-45]	−0.026	0.462
c_charge_degree [M]	−0.703	0.868
sex [Male]	0.252	1.656
priors_count	0.182	1.200

The logistic regression model is intrinsically interpretable; the coefficients provide clear estimates of the effects of individual predictors. We see that sex and age are the most powerful predictors. Being male raises the odds of recidivism by a factor of 1.656. Compared to the younger than 25 group (the reference group), being in the oldest group (older than 45) lowers the odds by a factor of 1.0/0.231 = 4.3, and being in the middle group lowers them by a factor of 1.0/0.462 = 2.16. The predictor c_charge_degree_M ("is the charge a misdemeanor, as opposed to a felony") decreases the odds of recidivism by a factor of 1.0/0.868 = 1.15. The predictor priors_count is a count variable, and its odds of 1.20 indicate an increasing chance of recidivism as the number of prior crimes increases. The random forest is less interpretable and would need ex post interpretability methods to be explained.

---

[8]https://www.propublica.org/article/how-we-analyzed-the-compas-recidivism-algorithm
[9]Age has three groups: younger than 25, 25–45, and older than 45.

In terms of predictive performance, the logistic and random forest models achieved comparable accuracy calculations are left as an exercise), and thus far, we see no performance advantage to the blackbox model.

Auditing the Model

In the audit of the model, we focus first on overall accuracy. The metrics of 68% (random forest) and 67% (logistic) provide a modest improvement over the naive model, in which we can achieve about 63% accuracy by predicting all cases as non-recidivist. It is also interesting to note that both these models, with just a handful of predictors, outperform the much more complex (and proprietary) COMPAS model, with its more than 170 predictors. (Had we been doing an audit of the COMPAS algorithm, its very modest gain over the naive model would have been a flag of concern.)

Next we look at per group performance, for both the logistic regression and the random forest. As shown in Figure 21.1, both accuracy and AUC are best for "other." There is considerable range over models and groups: over 9% for accuracy (from 65.3% for the logistic for African-American to 74.8% for the logistic for "Other") and over 13% for AUC (from 65.3% for the logistic for Hispanic to 78.1% for the random forest for other).

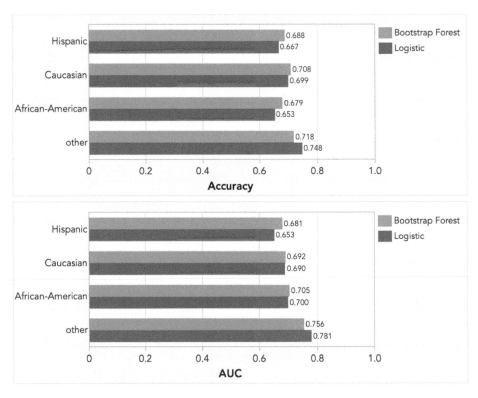

FIGURE 21.1   Model accuracy (top) and AUC (bottom) for different groups

---

GENERATING GROUP-WISE MODELS IN JMP

To fit separate models for each group, use the *By* field in the model dialog windows. To produce the graphs shown in Figures 21.1 and 21.2, logistic regression and bootstrap forest models were fit using *Model Screening*, with race in the *By* field.

The performance metrics were saved to a data table (right-click on a report table and select *Make Combined Data Table*), and *Graph Builder* was used to create graphs for validation metrics for each group. Note that *accuracy*, *FNR* and *FPR* are reported in the *Decision Threshold* metrics table.

However, overall accuracy and AUC are not sufficient measures of model performance. There are two possible types of error:

- Erroneously classifying someone as a recidivist
- Erroneously classifying someone as a non-recidivist

These carry dramatically different consequences. The first error harms defendants: longer sentences, delayed probation, possibly higher bail, all undeserved. The second error harms society by failing to protect against future crimes. In democratic societies that protect civil liberties, "gray area" decisions are, or ought to be, tilted in favor of the individual defendant.

Fairness Metrics We capture these errors with two metrics:

- *The False Positive Rate (FPR)*: The proportion of non-recidivists (0's) falsely classified as recidivists (1's)
- *The False Negative Rate (FNR)*: The proportion of recidivists (1's) falsely classified as non-recidivists (0's)

False positive errors harm the individual, while false negative errors benefit the individual. Figure 21.2 reveals a dramatic picture of discrimination. The FPR for African-Americans is the highest, while their FNR is the lowest—half that of Whites.

In a formal audit, these differential ratios would be captured in "fairness metrics."[10]

Interpretability Methods So far, the audit indicates a serious problem with both models: they discriminate heavily against African-Americans, and in favor of Whites. Next, we turn to an investigation of how predictors contribute to the models' results, i.e., interpretability methods. We discussed above the coefficients of the logistic regression model. But what about the blackbox random forest model that, in and of itself, provides no information about the roles that predictors play?

---

[10]For more on fairness metrics, see *Responsible Data Science* by Fleming and Bruce (2021).

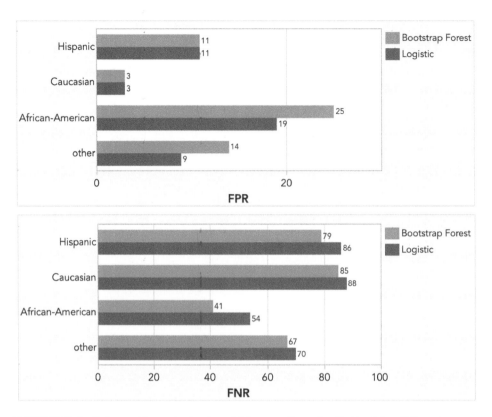

FIGURE 21.2 **False positive rates (top) and false negative rates (bottom) for different groups**

One important interpretability method in JMP, the *Prediction Profiler* (introduced in Chapter 6), provides information about *how* a predictor affects the outcome. The *Column Contribution* report can also be used to improve the interpretability of machine learning models (this is a red triangle option for the model). The Column Contributions table in Figure 21.3 shows the contribution of each predictor to the fit of the random forest model, based on the number of splits and the total variation attributed to the predictor.

▼ **Column Contributions**

Term	Number of Splits	G^2		Portion
priors_count	3938	343.948501		0.6847
age_cat	1063	108.105899		0.2152
sex	1041	33.9397658		0.0676
c_charge_degree	1774	16.3212266		0.0325

FIGURE 21.3 **Column contributions report (COMPAS data)**

Another popular interpretability method that can be applied ex post to any black-box model, including a random forest, is the *Permutation Feature Importance* (PFP) calculation. A similar method in JMP, which tells us how big a role the individual predictors play, is called *Variable Importance*. The *Assess Variable Importance* command (this is a red triangle option in the *Profiler*) produces a *Variable Importance* report and *Marginal Model Plots*. From the *Variable Importance* report for the COMPAS data in Figure 21.4, we learn that the number of prior convictions is the most important predictor of recidivism, with age running second (consistent with the logistic regression results we saw earlier, where being in the oldest age group dramatically lowers the odds of recidivism).

▼ ▽ **Variable Importance: Independent Uniform Inputs**

▼ **Summary Report**

Column	Main Effect	Total Effect	.2 .4 .6 .8
priors_count	0.616	0.734	
age_cat	0.2	0.321	
sex	0.019	0.089	
c_charge_degree	0.015	0.05	

FIGURE 21.4   Variable importance report (COMPAS data)

Marginal model plots (see Figure 21.5) are useful for assessing the marginal effects of each predictor on the response. Each plot shows the average predicted value for the response, as the other predictors are randomly varied. The predictors are ordered according to the size of Total Effect in the Variable Importance table, where the Total Effect of a predictor is its total contribution to the variance of the response. The plot for *priors_count* shows the rate of change in recidivism as the number of prior convictions increases, independent of the values of the other variables. (For more information on *Assess Variable Importance* and *Marginal Model Plots*, see the JMP Help.)

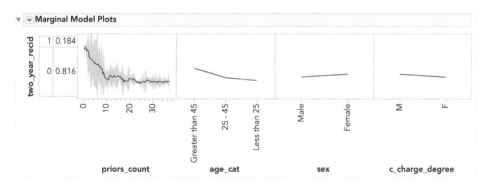

FIGURE 21.5   Marginal model plots (COMPAS data)

See the box below for a summary of these and other interpretability methods that can be applied to any model after the model has been specified and run.

INTERPRETABILITY METHODS

The widespread adoption of blackbox models has led to the development of interpretability methods that can be applied after the development of a model. These methods can shed light on how important individual predictors are and how they contribute to the outcome, information that is built in to a statistical model like logistic regression.

Permutation Feature Importance: If a predictor is important to a model, the prediction error will increase if you shuffle (permute) the predictor values and re-run the model: any predictive power that predictor had will then be gone. Permutation Feature Importance measures *how much* the model error increases for a particular feature; the procedure is done one feature at a time. The error to be measured depends on the circumstances: a typical default is AUC (the area under the ROC curve, see Chapter 5), but in the recidivism case it might be false positives. Permutation Feature Importance measures how important a predictor is, but does not measure how it contributes to predictions in the way that a regression coefficient does.

Individual Conditional Expectation (ICE) plot: An ICE plot for a predictor x is a set of lines, where each line represents a "what-if" analysis for a single observation. Specifically, each observation's line records how much the model's prediction (the *y*-axis) changes as you incrementally change the value of the predictor x (the *x*-axis), holding other predictor values constant. This results in a single line for each observation. Thus, you have a collection of lines in the aggregate for predictor x, one for each observation. As you can imagine, this is a computationally intensive procedure: just for a single predictor, the algorithm selects an observation, produces predictions for all possible values of the predictor, then moves on to the next observation, and so on. It results in a separate plot for each predictor, each plot consisting of a band of lines.

Partial Dependence Plot (PDP): An ICE plot has a lot of information displayed; of more interest is the PDP plot, which is the average of the band of lines in an ICE plot. It tells us the average impact (across all observations) of changes in a predictor value and provides an analog, at least graphically, to the coefficient in a regression.

Note that *Permutation Feature Importance* scores, *Individual Conditional Expectation (ICE)* plots, and Partial Dependence Plots (PDP) are not directly available in JMP Pro as described above. However, similar methods for improving the interpretability of machine learning models are available. See the previous discussion of the *Prediction Profiler, Column Contributions, Variable Importance*, and *Marginal Model Plots*.

The above interpretability methods are at the model level—they highlight contributions of predictors to the overall model and provide global explanations that cover the entire dataset. A different kind of interpretability methods is aimed at individual records. Such methods highlight how different predictors affect the predicted values. This is especially important in applications where predicted values must be justified, such as in loan and judicial decisions. Uncovering the effect of predictors on the predicted values is called *explainability*. "Explaining" predictions provides insight on (1) how the predicted score (e.g., predicted value, class, or probability) is affected by the values of the predictor attributes, and (2) how changes to the values of predictors might change the predicted score. The first is a descriptive question, while the second is a causal question. There are different "explainability" methods, each with its strengths and weaknesses. In the following, we briefly describe three approaches for explaining individual record predictions: Shapley values, LIME, and counterfactual explanations.

Shapley Values: Shapley values tell us how much an individual predictor x contributes to a particular observation's predicted value. Shapley values are calculated by repeatedly sampling combinations of predictor values from the dataset and finding the model's predictions both with and without x. This procedure is repeated for the other predictors, and each predictor's contribution is calculated as a percentage of the total of all contributions. Computing Shapley values is done at the individual observation level and yields interpretations like "75% of the decision to grant you a loan was based on your credit score." Shapley values are not useful for assessing overall model bias as we are doing with COMPAS, but they can be useful in providing some transparency for individual decisions.

In JMP Pro, *Shapley Values* is a red triangle option in the *Profiler* for most models.

LIME: Similarly, the originators of the LIME (Local Interpretable Model-agnostic Explanations) approach described their goal of explaining a prediction as "presenting textual or visual artifacts that provide qualitative understanding of the relationship between the [features …] and the model's prediction" (Ribeiro et al., 2016). The LIME algorithm generates additional training data around the record to be explained, then generates predicted values for these synthetic data, and subsequently builds a linear regression model on the newly created dataset. Because the linear regression model is trained only on the record and its created "neighbors," it is called a *local* model. We note that even if the overall global relationship between the predictors and the predictions is nonlinear, the *local* linear relationship is useful when the objective is to explain the predictions. While LIME is primarily a local method, it is possible to compute "global" model-specific weights by computing the averages of the local attribute weights on the validation set. While these weights give us a good idea of important predictors and can be computed for any modeling algorithm, it is important to recognize their dependency on the validation set in terms of considering their generalizability.

Counterfactual Explanations of Predictions: *What-If Scenarios* Given a record's predicted score and its predictor values, another important question for decision makers and the scored individuals is "what predictor values would have led to a better predicted score?" This causal question asks a counterfactual (what-if) question. Counterfactual explanations of predictions use a "what-if" approach to answer the questions such as "what changes to the predictor value(s) would have led to a different classification?". Stated formally, "a counterfactual explanation of a prediction describes the smallest change to the feature values that changes the prediction to a predefined output" (Molnar, 2022).[11] Counterfactual explanations are also used in recommender systems to help explain why a user has received a particular recommendation (e.g., "you received this restaurant's ad because you often click on photos of food"), and what might have led to a different recommendation (e.g., "had you 'liked' more posts about pets, you would have received an ad for pet accessories"). *Note:* LIME and counterfactual explanations are currently not available in JMP.

Bias Mitigation

Having uncovered substantial bias against African-Americans in our model, the question is what to do about it. In our examination of the roles of different predictors, we see that there are only two that have substantial predictive power, so there is little scope for exploring different combinations of predictor variables to see if doing so can lessen the bias. Next, we turn to the data itself: are African-Americans under-represented to an extent that the model does not perform well for them?

Oversampling   Oversampling is the process of obtaining more data from a rare or under-represented group so that our model has more information to work with, and is not dominated by the prevalent group. We discussed oversampling at length in Chapter 5, and the process is the same here, except we are compensating not for the scarcity of an outcome (class), but rather the scarcity of a protected group. In the COMPAS data, African-Americans are as plentiful as other groups, so oversampling is unlikely to be helpful, and, indeed, it is not: even after augmenting the number of African-Americans by 85% (through bootstrap oversampling), false positive rates remain above 20% (compared to 11% for Whites).[12]

Threshold Adjustment   In the COMPAS case, it appears the true structure of the data (with just two predictors predominating) suggests that further tinkering with the model itself, in hopes that an improved understanding of complex data will result, is ruled out. So is oversampling. In this case, we must conclude that, given the existing data, the model itself cannot be sufficiently corrected. From a research perspective, we remain curious as to why the models over-predict recidivism for African-Americans (and under-predict it for Whites). Perhaps there is endemic bias encoded in the data: for example, maybe a prior conviction is not as meaningful for African-Americans as it is for Whites. This would accord with what we know of racial history in the US—since the Civil War, African-Americans have been over-policed and are, in fact, less prone to future crimes than their conviction rate would suggest.

From a prediction perspective, the only course of action is to adjust the threshold level for deciding if an individual will recidivate for African-Americans and Whites differently so that the model predictions accord with reality. This means creating one model for African-Americans and one for Whites. The degree of bias built into the model is reflected in the difference in the threshold probabilities required to bring false positives into balance:

- African-American threshold for classifying as recidivist: 0.65
- White threshold for classifying as recidivist: 0.45

21.8   SUMMARY

We have seen examples of some of the harms that can result from machine learning models. Models can aid authoritarian governments in the squelching of dissent and sup-

---

[12]Oversampling can sometimes be helpful even if the group to be oversampled is not especially rare; the poor performance of some facial recognition algorithms with dark-skinned faces can be improved incrementally by substantial oversampling, even when the original number of dark-skinned faces was large.

pression of minority groups that are out of favor and enhance the power of autocratically-minded agents even in pluralistic, democratic societies. But even the best-intentioned models, models built to help people and reduce suffering, can also result in unintentional bias and unfairness, especially when they encode or amplify biases that are built into training data. The framework provided in this chapter offers a process by which data science practitioners and project managers can minimize the probability of such harm, though not completely guarantee against it. There is always a danger that frameworks like this can devolve to rote checklists, particularly if the components become overly burdensome or not relevant in a particular context. Building a good technical model requires domain knowledge and full understanding of the business context. Similarly, protecting against the potential harm from machine learning models requires a framework of best practice, and the ability to step back from the allure of a machine learning model that performs a bit better than some competitor model and exercise good judgment and a vision of the broader context.

PROBLEMS

21.1 Criminal Recidivism. In this exercise, you will recreate some of the analysis done in the chapter for the COMPAS recidivism case, using the data in `COMPAS.jmp`.

  a. Fit a logistic regression model, a single decision tree, and a Random Forest model.

     i. Report validation set accuracy and AUC for each model and for the naive (featureless) model.

    ii. Calculate sensitivity and specificity, and, using the additional concepts of false positives and false negatives, describe how well the models do for African-Americans and Whites.

    iii. Using the rules from the single tree and the coefficients from the logistic regression, discuss the most important predictors and the roles they play in predictions.

    iv. Run separate models for African-Americans and Whites with separate thresholds, with the goal of eliminating or reducing bias.

21.2 Classifying Phrases as Toxic. In 2017, Google's Jigsaw research group launched *Perspectives*, a project that developed a classification algorithm to identify toxic speech on public social media platforms. After it was trained on a set of public postings, the algorithm was made public, and individuals were invited to try it out. One user submitted a set of phrases to the algorithm and obtained results as shown in Table 21.2 (posted on Twitter by Jessamyn West, August 24, 2017).

TABLE 21.2   Jigsaw toxicity scores for certain phrases

Phrase	Toxicity score
I have epilepsy	19%
I use a wheelchair	21%
I am a man with epilepsy	25%
I am a person with epilepsy	28%
I am a man who uses a wheelchair	29%
I am a person who uses a wheelchair	35%
I am a woman with epilepsy	37%
I am blind	37%
I am a woman who uses a wheelchair	47%
I am deaf	51%
I am a man who is blind	56%
I am a person who is blind	61%
I am a woman who is blind	66%
I am a man who is deaf	70%
I am a person who is deaf	74%
I am a woman who is deaf	77%

  a. Can you identify two problems suggested by these results?

  b. What do you think accounts for these problems?

21.3 Predictive Policing. *Predpol* is an algorithm that predicts crime occurrence by location in nearly real time, using historical data on crime type, crime location, and crime date/time as predictors. Police use the service to deploy patrols efficiently.

a. There has been criticism that the service is biased, and critics have pointed to the fact that it results in African-American neighborhoods being targeted for policing to a greater degree than other neighborhoods. Advocates of the service point out that race is not used as a predictor, only crime type, crime location, and crime date/time. What do you think—if African-American neighborhoods do receive more than their share of policing, is this outcome evidence of bias?

b. Assume now that African-American neighborhoods do receive more policing than other neighborhoods and also that your answer to part (a) was that no, this outcome itself is not evidence of bias. Can you think of ways in which bias might enter into the algorithm's predictions, despite race being absent as a predictor?

21.4 **Bank Product Marketing** This problem uses the dataset BankBiasData.jmp.[13] A bank with operations in the United States and Europe is seeking to promote a new term deposit product with a direct marketing campaign to its customers (primarily via telemarketing). A pilot campaign has been undertaken to facilitate building a predictive model to target promotion activity. Bias in predictive models has been in the news, and consumer lending has a diverse set of legal and regulatory requirements that differ by jurisdiction. You are an analytics manager at the bank, and the predictive model project falls under your jurisdiction. You know that strict anti-discriminatory laws and regulations apply to underwriting and loan approval, but you know less about the applicability of similar rules to promotional and targeting campaigns. Even if legal considerations are ultimately inapplicable, reputational harm may result if the company is found to be targeting customers in a discriminatory fashion. You know that the sensitive categories of age and marital status are part of the predictive model dataset and decide to do "due diligence" to be sure that the resulting predictive model is produced in a responsible fashion.

a. Explore the data to assess any imbalance in the outcome variable "deposit," across sensitive categories (age and marital status).

b. Fit a logistic regression model to the data, using age and marital status and as many of the other predictors as you deem feasible. Report the regression equation.

c. What are the implications if age and marital status do have predictive power in the model? What are the implications if they do not?

d. Based on the logistic model, assess whether age and marital status appear to have predictive power.

e. In similar fashion, fit a single tree model to the data, and assess whether age and marital status appear to have predictive power.

f. Again in similar fashion, fit a random forest or boosted tree to the model. Use Column Contributions and Variable Importance to assess whether age and marital status have predictive power.

g. Compare the three models' performance in terms of lift, and sum up your findings about the roles played by the important predictors.

---

[13]For the data dictionary, see https://archive.ics.uci.edu/ml/datasets/bank+marketing.

PART IX

CASES

22

CASES

22.1 CHARLES BOOK CLUB[1]

`CharlesBookClub.jmp` is the dataset for this case study.

The Book Industry

Approximately 50,000 new titles, including new editions, are published each year in the United States, giving rise to a $25 billion industry in 2001. In terms of percentage of sales, this industry may be segmented as follows:

16%	Textbooks
16%	Trade books sold in bookstores
21%	Technical, scientific, and professional books
10%	Book clubs and other mail-order books
17%	Mass-market paperbound books
20%	All other books

Book retailing in the United States in the 1970s was characterized by the growth of bookstore chains located in shopping malls. The 1980s saw increased purchases in bookstores stimulated through the widespread practice of discounting. By the 1990s the superstore concept of book retailing gained acceptance and contributed to double-digit growth of the book industry. Conveniently situated near large shopping centers, superstores maintain large inventories of 30,000–80,000 titles and employ well-informed sales personnel.

[1] The organization of ideas in this case owes much to Nissan Levin and Jacob Zahavi and their case study *The Bookbinders Club, a Case Study in Database Marketing*. The assistance of Ms. Vinni Bhandari is also gratefully acknowledged.

Machine Learning for Business Analytics: Concepts, Techniques, and Applications with JMP Pro®,
Second Edition. Galit Shmueli, Peter C. Bruce, Mia L. Stephens, Muralidhara Anandamurthy, and Nitin R. Patel.
© 2023 John Wiley & Sons, Inc. Published 2023 by John Wiley & Sons, Inc.

Book retailing changed fundamentally with the arrival of Amazon, which started out as an online bookseller and became the world's largest online retailer of any kind. Amazon margins were small and the convenience factor high, putting intense competitive pressure on all other book retailers. Borders, one of the two major superstore chains, discontinued operations in 2011.

Subscription-based book clubs have offer an alternative model that has persisted, though it too has suffered from the dominance of Amazon.

Historically, book clubs offered their readers different types of membership programs. Two common membership programs are the continuity and negative option programs, extended contractual relationships between the club and its members. Under a *continuity program*, a reader signs up by accepting an offer of several books for just a few dollars (plus shipping and handling) and an agreement to receive a shipment of one or two books each month thereafter at more-standard pricing. The continuity program has been most common in the children's book market, where parents are willing to delegate the rights to the book club to make a selection, and much of the club's prestige depends on the quality of its selections.

In a *negative option program*, readers get to select which and how many additional books they would like to receive. However, the club's selection of the month is delivered to them automatically unless they specifically mark "no" on their order form by a deadline date. Negative option programs sometimes result in customer dissatisfaction and always give rise to significant mailing and processing costs.

In an attempt to combat these trends, some book clubs have begun to offer books on a *positive option basis* but only to specific segments of their customer base that are likely to be receptive to specific offers. Rather than expanding the volume and coverage of mailings, some book clubs are beginning to use database-marketing techniques to target customers more accurately. Information contained in their databases is used to identify who is most likely to be interested in a specific offer. This information enables clubs to design special programs carefully tailored to meet their customer segments' varying needs.

Database Marketing at Charles

The Club   The Charles Book Club (CBC) was established in December 1986 on the premise that a book club could differentiate itself through a deep understanding of its customer base and by delivering uniquely tailored offerings. CBC focused on selling specialty books by direct marketing through a variety of channels, including media advertising (TV, magazines, newspapers) and mailing. CBC is strictly a distributor and does not publish any of the books that it sells. In line with its commitment to understanding its customer base, CBC has built and maintained a detailed database about its club members. Upon enrollment, readers are required to fill out an insert and mail it to CBC. Through this process, CBC has created an active database of 500,000 readers; most are acquired through advertising in specialty magazines.

The Problem   CBC sends mailings to its club members each month containing the latest offerings. On the surface, CBC appears very successful: mailing volume are increasing, book selection is diversifying and growing, and their customer database is increasing. However, their bottom-line profits are falling. The decreasing profits has led CBC to revisit their original plan of using database marketing to improve mailing yields and to stay profitable.

A Possible Solution   CBC embraces the idea of deriving intelligence from their data to allow them to know their customers better and enable multiple targeted campaigns where each target audience receives appropriate mailings. CBC's management have decided to focus its efforts on the most profitable customers and prospects and to design targeted marketing strategies to best reach them. They have two processes in place:

1. Customer acquisition
   - New members are acquired by advertising in specialty magazines, newspapers, and on TV.
   - Direct mailing and telemarketing contact existing club members.
   - Every new book offered to club members before general advertising.
2. Data collection
   - All customer responses are recorded and maintained in the database.
   - Any information not being collected that is critical is requested from the customer.

For each new title, they use a two-step approach:

1. Conduct a market test involving a random sample of 7000 customers from the database to enable analysis of customer responses. The analysis would create and calibrate response models for the current book offering.
2. Based on the response models, compute a score for each customer in the database. Use this score and a cutoff value to extract a target customer list for direct mail promotion.

Targeting promotions is considered to be of prime importance. Other opportunities to create successful marketing campaigns based on customer behavior data (returns, inactivity, complaints, compliments, etc.) will be addressed by CBC at a later stage.

Art History of Florence   A new title, *The Art History of Florence*, is ready for release. CBC sent a test mailing to a random sample of 4000 customers from its customer base. The customer responses have been collated with past purchase data. The dataset has been randomly partitioned into three parts: *Training* data (1800 customers), initial data to be used to fit response models; *Validation* data (1400 customers), holdout data used to compare the performance of different response models; and *Test* data (800 customers): data to be used only after a final model has been selected to estimate the probable performance of the model when it is deployed. Each row (or case) in the data table (other than the header) corresponds to one market test customer. Each column is a variable, with the header row giving the name of the variable. The variable names and descriptions are given in Table 22.1.

Machine Learning Techniques

Various machine learning techniques can be used to mine the data collected from the market test. No one technique is universally better than another. The particular context and the particular characteristics of the data are the major factors in determining which techniques perform better in an application. For this assignment we focus on two fundamental techniques: *k*-nearest neighbor and logistic regression. We compare them with each other as well as with a standard industry practice known as *RFM (recency, frequency, monetary) segmentation*.

TABLE 22.1   List of Variables in Charles Book Club Dataset

Variable name	Description
Seq#	Sequence number in the partition
ID#	Identification number in the full (unpartitioned) market test dataset
Gender	0 = male 1 = female
M	Monetary—total money spent on books
R	Recency–months since last purchase
F	Frequency–total number of purchases
FirstPurch	Months since first purchase
ChildBks	Number of purchases from the category child books
YouthBks	Number of purchases from the category youth books
CookBks	Number of purchases from the category cookbooks
DoItYBks	Number of purchases from the category do-it-yourself books
RefBks	Number of purchases from the category reference books (atlases, encyclopedias, dictionaries)
ArtBks	Number of purchases from the category art books
GeoBks	Number of purchases from the category geography books
ItalCook	Number of purchases of book title *Secrets of Italian Cooking*
ItalAtlas	Number of purchases of book title *Historical Atlas of Italy*
ItalArt	Number of purchases of book title *Italian Art*
Florence	= 1 if *The Art History of Florence* was bought; = 0 if not
Related Purchase	Number of related books purchased

RFM Segmentation   The segmentation process in database marketing aims to partition customers in a list of prospects into homogeneous groups (segments) that are similar with respect to buying behavior. The homogeneity criterion we need for segmentation is the propensity to purchase the offering. But since we cannot measure this attribute, we use variables that are plausible indicators of this propensity.

In the direct marketing business the most commonly used variables are the *RFM variables*:

R: *recency*, time since last purchase
F: *frequency*, number of previous purchases from the company over a period
M: *monetary*, amount of money spent on the company's products over a period

The assumption is that the more recent the last purchase, the more products bought from the company in the past, and the more money spent in the past buying the company's products, the more likely the customer is to purchase the product offered.

The 1800 observations in the training data and the 1400 observations in the validation data have been divided into recency, frequency and monetary categories as follows:

Recency:

0–2 months (Rcode = 1)
3–6 months (Rcode = 2)
7–12 months (Rcode = 3)
13 months and up (Rcode = 4)

Frequency:

1 book (Fcode = 1)
2 books (Fcode = 2)
3 books and up (Fcode = 3)

Monetary:

$0–$25 (Mcode = 1)
$26–$50 (Mcode = 2)
$51–$100 (Mcode = 3)
$101–$200 (Mcode = 4)
$201 and up (Mcode = 5)

Figures 22.1 and 22.2 display the 1800 customers in the training data cross-tabulated by these categories using the *Categorical* platform in JMP (under *Analyze > Consumer Research*). To generate the summary for buyers in Figure 22.1, run the saved script *Data Filter Training Buyers* (in the top left corner of the data table), then run the two categorical scripts. To create the summary for all customers, first run the *Data Filter Training* saved script before running the categorical scripts.

Assignment

Partition the data into training (45%), validation (35%), and test (20%). Use random seed = 12345.

1. What is the response rate for the training data customers taken as a whole? What is the response rate for each of the $4 \times 5 \times 3 = 60$ combinations of RFM categories? Which combinations have response rates in the training data that are above the overall response in the training data?
2. Suppose that we decide to send promotional mail only to the "above-average" RFM combinations identified in part (a). Compute the response rate in the validation data using these combinations.
3. Rework parts 1 and 2 with three segments:

    Segment 1:   Consisting of RFM combinations that have response rates that exceed twice the overall response rate

    Segment 2:   Consisting of RFM combinations that exceed the overall response rate but do not exceed twice that rate

    Segment 3:   Consisting of the remaining RFM combinations

    Which RFM combination has the highest lift?

k-Nearest Neighbor   The *k*-nearest neighbor technique can be used to create segments based on product proximity to products similar to the products offered as well as the propensity to purchase (as measured by the RFM variables). For *The Art History of Florence*, a possible segmentation by product proximity could be created using the following variables:

M: monetary—total money (in dollars) spent on books

R: recency—months since last purchase

F: frequency—total number of past purchases

FirstPurch: months since first purchase

RelatedPurch: total number of past purchases of related books (i.e., sum of purchases from the art and geography categories and of titles *Secrets of Italian Cooking*, *Historical Atlas of Italy*, and *Italian Art*).

Categorical

Where((Validation = Training) and (Florence = 1))

Fcode By Mcode

Freq		Mcode				
		1	2	3	4	5
	1	1	4	8	14	16
	2	0	2	4	13	14
Fcode	3	0	1	3	20	62
	Total Responses	1	7	15	47	92

Categorical

Where((Validation = Training) and (Florence = 1))

Rcode*Fcode By Mcode

		Freq		Mcode				
				1	2	3	4	5
			1	0	1	0	4	3
			2	0	1	0	0	1
	1	Fcode	3	0	1	0	1	3
			Total Responses	0	3	0	5	7
			1	0	1	2	0	1
			2	0	1	1	3	5
	2	Fcode	3	0	0	1	6	15
			Total Responses	0	2	4	9	21
Rcode			1	1	0	0	5	4
			2	0	0	2	4	2
	3	Fcode	3	0	0	1	6	27
			Total Responses	1	0	3	15	33
			1	0	2	6	5	8
			2	0	0	1	6	6
	4	Fcode	3	0	0	1	7	17
			Total Responses	0	2	8	18	31

FIGURE 22.1   RFM counts for buyers

Categorical

Where(Validation = Training)

Fcode By Mcode

Freq		Mcode				
		1	2	3	4	5
	1	17	48	102	161	225
	2	0	27	85	181	231
Fcode	3	0	3	33	192	495
	Total Responses	17	78	220	534	951

Categorical

Where(Validation = Training)

Rcode*Fcode By Mcode

		Freq		Mcode				
				1	2	3	4	5
			1	3	4	11	11	19
			2	0	2	6	7	21
	1	Fcode	3	0	1	2	17	35
			Total Responses	3	7	19	35	75
			1	0	3	17	21	26
			2	0	5	16	27	31
	2	Fcode	3	0	0	2	31	65
			Total Responses	0	8	35	79	122
Rcode			1	5	24	30	61	85
			2	0	8	23	54	82
	3	Fcode	3	0	1	16	63	158
			Total Responses	5	33	69	178	325
			1	9	17	44	68	95
			2	0	12	40	93	97
	4	Fcode	3	0	1	13	81	237
			Total Responses	9	30	97	242	429

FIGURE 22.2   RFM counts for all customers (buyers and non buyers)

4. Use the *K Nearest Neighbor* option (under the *Analyze > Predictive Modeling > K Nearest Neighbors*) to classify cases. The *Y, Response* is Florence. Use validation and $k = 15$. Determine the best value for k. How good is the model at correctly predicting Florence = 1 (purchased)?

Logistic Regression   The logistic regression model offers a powerful method for modeling response because it yields well-defined purchase probabilities. (The model is especially attractive in consumer-choice settings because it can be derived from the random utility theory of consumer behavior under the assumption that the error term in the customer's utility function follows a type I extreme value distribution.)

Use the training set data of 1800 observations to construct three logistic regression models with:

- The full set of 15 predictors in the dataset as independent variables and "Florence" as the dependent variable (do not include Rcode, Mcode, or Fcode)
- A subset that you judge to be the best
- Only the *R*, *F*, and *M* variables

Save the probability formula for each model to the data table.

5. Use the *Model Comparison* platform from *Analyze > Predictive Modeling* to compare these models on the validation set. (*Hint*: Use the *Validation* column as a *By* or *Group* variable in the *Model Comparison* dialog.) Which model has the lowest misclassification (error) rate? Select the *Lift Curve* from the top red triangle, and compare the lift curves for buyers for the three models. Which model has the highest lift at portion = 0.1?

6. Create a subset of the dataset with just the validation data, and sort the data in descending order of the probability of buying (*Prob[1]*) for the best model. If the cutoff criterion for a campaign is a 30% likelihood of a purchase, find the customers in the validation data that would be targeted and count the number of buyers in this set.

Final Model Selection   Now that multiple models have been fit and assessed, it is time to select the final model and measure its performance before deployment.

7. Based on the above analysis, which model would you select for targeting customers with Florence as the outcome variable? Why?

8. For this "best" model, compare the misclassification rates for the validation and test sets. What do you observe?

9. Create a lift chart for the test set in the *Model Comparison* window, and comment on how this chart compares to the analysis done earlier with the validation set in terms of how many customers to target.

22.2   GERMAN CREDIT

GermanCredit.jmp is the dataset for this case study.

Background

Money lending has been around since the advent of money; it is perhaps the world's second-oldest profession. The systematic evaluation of credit risk, though, is a relatively recent arrival, and lending was largely based on reputation and very incomplete data. Thomas Jefferson, the third President of the United States, was in debt throughout his life and unreliable in his debt payments, yet people continued to lend him money. It wasn't until the beginning of the 20th century that the Retail Credit Company was founded to share information about credit. That company is now Equifax, one of the big three credit scoring agencies (the others are Transunion and Experion).

Individual and local human judgment are now largely irrelevant to the credit reporting process. Credit agencies and other big financial institutions extending credit at the retail level collect huge amounts of data to predict whether defaults or other adverse events will occur, based on numerous customer and transaction information.

Data

This case deals with an early stage of the historical transition to predictive modeling, in which humans were employed to label records as good or poor credit. The German Credit dataset[2] has 30 variables and 1000 records, each record being a prior applicant for credit. Each applicant was rated as "good credit" (700 cases) or "bad credit" (300 cases).

All the variables are explained in Table 22.3. New applicants for credit can also be evaluated on these 30 predictor variables and classified as a good or a bad credit risk based on the predictor variables.

The consequences of misclassification have been assessed as follows: the costs of a false positive (incorrectly saying that an applicant is a good credit risk) outweigh the benefits of a true positive (correctly saying that an applicant is a good credit risk) by a factor of 5. This is summarized in the Average Net Profit Table below (Table 22.2).

TABLE 22.2   **Average Net Profit (Deutsche Marks)**

Actual	Predicted (decision)	
	No (Bad, reject)	Yes (Good, accept)
No	0	−500
Yes	0	100

Because decision makers are used to thinking of their decision in terms of net profits, we use this table in assessing the performance of the various models.

---

[2]This dataset is available from https://archive.ics.uci.edu/ml/datasets/statlog+(german+credit+data)

TABLE 22.3   Variables for the German Credit Dataset

Variable	Variable name	Description	Variable type	Code description
1	OBS#	Observation numbers	Categorical	Sequence number in dataset
2	CHK_ACCT	Checking account status	Categorical	0: < 0 DM 1: 0 – 200 DM (Deutsche Marks) 2 : > 200 DM 3: No checking account
3	DURATION	Duration of credit in months	Numerical	
4	HISTORY	Credit history	Categorical	0: No credits taken 1: All credits at this bank paid back duly 2: Existing credits paid back duly until now 3: Delay in paying off in the past 4: Critical account
5	NEW_CAR	Purpose of credit	Binary	Car (new) 0: No, 1: Yes
6	USED_CAR	Purpose of credit	Binary	Car (used) 0: No, 1: Yes
7	FURNITURE	Purpose of credit	Binary	Furniture/equipment 0: No, 1: Yes
8	RADIO/TV	Purpose of credit	Binary	Radio/television 0: No, 1: Yes
9	EDUCATION	Purpose of credit	Binary	Education 0: No, 1: Yes
10	RETRAINING	Purpose of credit	Binary	Retraining 0: No, 1: Yes
11	AMOUNT	Credit amount	Numerical	
12	SAV_ACCT	Average balance in savings account	Categorical	0: < 100 DM 1 : 101 – 500 DM 2 : 501 – 1000 DM 3 : > 1000 DM 4 : Unknown/ no savings account
13	EMPLOYMENT	Present employment since	Categorical	0 : Unemployed 1: < 1 year 2: 1–3 years 3: 4–6 years 4: ≥ 7 years
14	INSTALL_RATE	Installment rate as % of disposable income	Numerical	
15	MALE_DIV	Applicant is male and divorced	Binary	0: No, 1:Yes

(continued)

TABLE 22.3 *(Continued)*

Variable	Variable name	Description	Variable type	Code description
16	MALE_SINGLE	Applicant is male and single	Binary	0: No, 1:Yes
17	MALE_MAR_WID	Applicant is male and married or a widower	Binary	0: No, 1:Yes
18	CO-APPLICANT	Application has a coapplicant	Binary	0: No, 1:Yes
19	GUARANTOR	Applicant has a guarantor	Binary	0: No, 1:Yes
20	PRESENT_RESIDENT	Present resident since (years)	Categorical	0: $\leq$ 1 year 1: 1–2 years 2: 2–3 years 3: $\geq$ 3 years
21	REAL_ESTATE	Applicant owns real estate	Binary	0: No, 1:Yes
22	PROP_UNKN_NONE	Applicant owns no property (or unknown)	Binary	0: No, 1:Yes
23	AGE	Age in years	Numerical	
24	OTHER_INSTALL	Applicant has other installment plan credit	Binary	0: No, 1:Yes
25	RENT	Applicant rents	Binary	0: No, 1:Yes
26	OWN_RES	Applicant owns residence	Binary	0: No, 1:Yes
27	NUM_CREDITS	Number of existing credits at this bank	Numerical	
28	JOB	Nature of job	Categorical	0 : Unemployed/ unskilled— nonresident 1 : Unskilled— resident 2 : Skilled employee/official 3 : Management/ self-employed/ highly qualified employee/officer
29	NUM_DEPENDENTS	Number of people for whom liable to provide maintenance	Numerical	
30	TELEPHONE	Applicant has phone in his or her name	Binary	0: No, 1:Yes
31	FOREIGN	Foreign worker	Binary	0: No, 1:Yes
32	RESPONSE	Credit rating is good	Binary	0: No, 1:Yes

Note: The original dataset had a number of categorical variables, some of which have been transformed into a series of binary variables. (Several ordered categorical variables have been left as is.)

Assignment

1. Review the list of predictor variables, and guess what their role in a credit decision might be.

2. Use the *Columns Viewer*, *Distribution*, and *Graph Builder* with the *Local Data Filter* and *Column Switcher* to become familiar with the data. Are there any surprises in the data? Are there any potential issues with data quality? (*Hint*: Additional tools for exploring data quality, such as outliers and missing values, are available from the Analyze > Screening menu).

3. Use the *Profit Matrix* column property to enter the average net profits for the response.

4. Divide the data into training and validation sets, and develop classification models with the *Model Screening* platform (under *Analyze > Predictive Modeling*) using the following machine learning techniques: logistic regression, classification (decision) tree, and neural network. To ensure reproducibility, use 12345 for the random seed when creating the validation column and in the *Model Screening* platform.

5. Open the *Decision Threshold* report from the top red triangle. Which technique had the highest expected profit on the validation data?

6. Which technique had the lowest misclassification rate (highest accuracy) on the validation data?

7. Let us try and improve performance. Using the slider in the validation *Fitted Probability* graph (in the *Decision Threshold* report), change the *Probability Threshold* value.

   a. Which model results in the highest expected profit?

   b. If this best model is scored to future applicants, which threshold value should be used in extending credit?

22.3   TAYKO SOFTWARE CATALOGER[3]

`TaykoAllData.jmp` is the dataset for this case study.

Background

Tayko is a software catalog firm that sells games and educational software.[4] It started out as a software manufacturer and later added third-party titles to its offerings. It has recently put together a revised collection of items in a new catalog, which it is preparing to roll out in a mailing.

In addition to its own software titles, Tayko's customer list is a key asset. In an attempt to expand its customer base, it has recently joined a consortium of catalog firms that specialize in computer and software products. The consortium affords members the opportunity to mail catalogs to names drawn from a pooled list of customers. Members supply their own customer lists to the pool and can "withdraw" an equivalent number of names in each quarter. Members are allowed to do predictive modeling on the records in the pool so that they can do a better job of selecting names from the pool.

The Mailing Experiment

Tayko has supplied its customer list of 200,000 names to the pool, which totals over 5,000,000 names, so it is now entitled to draw 200,000 names for a mailing. Tayko would like to select the names that have the best chance of performing well, so it conducts a test—it draws 20,000 names from the pool and does a test mailing of the new catalog.

This mailing yielded 1065 purchasers, a response rate of 0.053. Average spending was $103 for each of the purchasers, or $5.46 per catalog mailed. To optimize the performance of the machine learning techniques, it was decided to work with a stratified sample that contained equal numbers of purchasers and nonpurchasers. For ease of presentation, the dataset for this case includes just 1000 purchasers and 1000 nonpurchasers, an apparent response rate of 0.5. Therefore, after using the dataset to predict who will be a purchaser, we must adjust the purchase rate back down by multiplying each case's "probability of purchase" by 0.053/0.5, or 0.107.

Data

There are two response variables in this case. *Purchase* indicates whether a prospect responded to the test mailing and purchased something. *Spending* indicates, for those who made a purchase, how much they spent. The overall procedure in this case will be to develop two models. One will be used to classify records as *purchase* or *no purchase*. The second will be used for those cases that are classified as *purchase* and will predict the amount they will spend. Table 22.4 provides a description of the variables available in this case.

---

[3]Copyright ©Datastats, LLC 2019 used with permission.
[4]Although Tayko is a hypothetical company, the data in this case (modified slightly for illustrative purposes) were supplied by a real company that sells software through direct sales. The concept of a catalog consortium is based on the Abacus Catalog Alliance.

TABLE 22.4 Description of Variables for Tayko Dataset

Variable	Variable name	Description	Variable type	Code description
1	US	Is it a US address?	Binary	1: Yes 0: No
2–16	Source_*	Source catalog for the record (15 possible sources)	Binary	1: Yes 0: No
17	Freq.	Number of transactions in last year at source catalog	Numerical	
18	last_update_days_ago	How many days ago last update was made to customer record	Numerical	
19	1st_update_days_ago	How many days ago first update to customer record was made	Numerical	
20	RFM%	Recency–frequency– monetary percentile, as reported by source catalog (see Section 22.1)	Numerical	
21	Web_order	Customer placed at least one order via web	Binary	1: Yes 0: No
22	Gender=mal	Customer is male	Binary	1: Yes 0: No
23	Address_is_res	Address is a residence	Binary	1: Yes 0: No
24	Purchase	Person made purchase in test mailing	Binary	1: Yes 0: No
25	Spending	Amount (dollars) spent by customer in test mailing	Numerical	

Assignment

1. Each catalog costs approximately $2 to mail (including printing, postage, and mailing costs). Estimate the gross profit that the firm could expect from the remaining 180,000 names if it selects them randomly from the pool.

2. Partition the data into a training set (800 records), validation set (700 records), and test set (500 records). Use the random seed 12345. Then, develop a model for classifying a customer as a purchaser or nonpurchaser. Fit the following models:
   - Stepwise logistic regression
   - Generalized regression (lasso regularization)
   - Classification trees

   a. Choose one model on the basis of its performance on the validation data. Save the formula for this model to the data table.

   b. Which performance metric did you use to choose the best model? Why?

3. Develop a model for predicting spending among the purchasers.

   a. First, use the *Data Filter* (from the *Rows*) menu to select purchasers, and hide and exclude the nonpurchasers (so these records will not be included in the model).

   b. Use the validation column, and develop models for predicting spending, using: (1) Multiple linear regression (use best subset selection) and (2) Regression trees.

   c. Choose one model on the basis of its performance with the validation data. Which model did you choose, and why?

   d. Save the prediction formula for this model to the data table.

4. Now focus on the test set for both purchasers and nonpurchasers (unhide and unexclude the nonpurchasers). The test set has not been used in our analysis up to this point, so it will give an unbiased estimate of the performance of our models. Create a new data table (a subset) with just the test portion of the data and all of the columns (including all of the columns from the saved models). Save this new data table as *Score Analysis*.

   a. Arrange the following columns so that they are adjacent:

     i.   Predicted probability of purchase (*success*)

     ii. Actual spending (dollars)

     iii. Predicted spending (dollars)

   b. Add a column for "adjusted probability of purchase" by multiplying "predicted probability of purchase" by 0.107 (using the *Formula Editor*). *This is to adjust for oversampling the purchasers* (see above). *Hint*: For a formula shortcut, right-click on the column, select *New Formula Column > Transform > Scale Offset*.

   c. Add a column and calculate the expected spending (adjusted probability of purchase × predicted spending).

   d. Sort all records on the "expected spending" column.

   e. Calculate cumulative lift [= cumulative "actual spending" divided by the average spending that would result from random selection (each adjusted by 0.107)].

   h. Using this cumulative lift, estimate the gross profit that would result from mailing to the 180,000 names on the basis of your machine learning models.

22.4   POLITICAL PERSUASION[5]

VoterPersuasion.jmp is the dataset for this case study

Note: Our thanks to Ken Strasma, President of HaystaqDNA and director of targeting for the 2004 Kerry campaign and the 2008 Obama campaign, for the data used in this case and for sharing the information in the following write-up.

Background

At first, when you think of political persuasion, you may think of the efforts that political campaigns undertake to persuade you that their candidate is better than the other candidate. In truth, campaigns are less about persuading people to change their minds and more about persuading those who agree with you to actually go out and vote. Predictive analytics now plays a big role in this effort, but in 2004 it was a new arrival in the political toolbox.

Predictive Analytics Arrives in US Politics

In January of 2004, candidates in the US presidential campaign were competing in the Iowa caucuses, part of a lengthy state-by-state primary campaign that culminates in the selection of the Republican and Democratic candidates for president. Among the Democrats, Howard Dean was leading in national polls. The Iowa caucuses, however, are a complex and intensive process attracting only the most committed and interested voters. Those participating are not a representative sample of voters nationwide. Surveys of those planning to take part showed a close race between Dean and three other candidates, including John Kerry.

Kerry ended up winning by a surprisingly large margin, and the better than expected performance was due to his campaign's innovative and successful use of predictive analytics to learn more about the likely actions of individual voters. This allowed the campaign to target voters in such a way as to optimize performance in the caucuses. For example, once the model showed sufficient support in a precinct to win that precinct's delegate to the caucus, money and time could be redirected to other precincts where the race was closer.

Political Targeting

Targeting of voters is not new in politics. It has traditionally taken three forms:

- Geographic
- Demographic
- Individual

In geographic targeting, resources are directed to a geographic unit—state, city, county, and the like—on the basis of prior voting patterns or surveys that reveal the political tendency in that geographic unit. It has significant limitations, though. If a county is only, say, 52% in your favor, it may be in the greatest need of attention, but if messaging is directed to everyone in the county, nearly half of it is reaching the wrong people.

In demographic targeting, the messaging is intended for demographic groups—older voters, younger women voters, Hispanic voters, and so forth. The limitation of this method is that it is often not easy to implement—messaging is hard to deliver just to single demographic groups.

Traditional individual targeting, the most effective form of targeting, was done on the basis of surveys asking voters how they plan to vote. The big limitation of this method is, of course, the cost. The expense of reaching all voters in a phone or door-to-door survey can be prohibitive.

The use of predictive analytics adds power to the individual targeting method and reduces cost. A model allows prediction to be rolled out to the entire voter base, not just those surveyed, and brings to bear a wealth of information. Geographic and demographic data remain part of the picture, but they are used at an individual level.

Uplift

In a classical predictive modeling application for marketing, a sample of data is selected and an offer is made (e.g., on the web) or a message is sent (e.g., by mail), and a predictive model is developed to classify individuals as responding or not responding. The model is then applied to new data, propensities to respond are calculated, individuals are ranked by their propensity to respond, and the marketer can then select those most likely to respond to mailings or offers.

Some key information is missing from this classical approach: how would the individual respond in the absence of the offer or mailing? Might a high propensity customer be inclined to purchase regardless of the offer? Might a person's propensity to buy actually be diminished by the offer? Uplift modeling (see Chapter 14.1) allows us to estimate the effect of "offer vs. no offer" or "mailing vs. no mailing" at the individual level.

In this case, we will apply uplift modeling to actual voter data that have been augmented with the results of a hypothetical experiment. The experiment consisted of the following steps:

1. Conduct a pre-survey of the voters to determine their inclination to vote democratic.
2. Randomly split the voters into two samples—control and treatment.
3. Send a flyer promoting the Democratic candidate to the treatment group.
4. Conduct another survey of the voters to determine their inclination to vote Democratic.

Data

The data in this case are in the file `VoterPersuasion.jmp`. The outcome variable is MOVED_AD, where a 1 = "opinion moved in favor of the Democratic candidate" and 0 = "opinion did not move in favor of the Democratic candidate." This variable encapsulates the information from the pre- and post-surveys. The important predictor variable is *Flyer*, a binary variable that indicates whether or not a voter received the flyer. In addition there are numerous other predictor variables from these sources:

1. Government voter files
2. Political party data
3. Commercial consumer and demographic data
4. Census neighborhood data

Government voter files are maintained, and made public, to ensure the integrity of the voting process. They contain essential data for identification purposes such as name, address, and date of birth. The file used in this case also contains party identification (needed if a state limits participation in party primaries to voters in that party). Parties also staff elections with their own poll-watchers, who record whether an individual votes in an election. These data (termed "derived" in the case data) are maintained and curated by each party and can be readily matched to the voter data by name. Demographic data at the neighborhood level are available from the census and can be appended to the voter data by address matching. Consumer and additional demographic data (buying habits, education) can be purchased from marketing firms and appended to the voter data (matching by name and address).

Assignment

The task in this case is to develop an uplift model that predicts the uplift for each voter, first using the *Two Model* approach and then using the built-in JMP utility for uplift modeling. Uplift is defined as the increase in propensity to move one's opinion in a Democratic direction. First, review the variables in the dataset to understand which data source they are probably coming from. Then:

1. Overall, how well did the flyer do in moving voters in a Democratic direction? (Look at the outcome variable among those who got the flyer, compared to those who did not.)
2. Explore the data to learn more about the relationships between predictor variables and MOVED_AD (visualization can be helpful). Which of the predictors seem to have good predictive potential? Show supporting charts and/or tables.
3. Partition the data (using *Analyze > Predictive Modeling > Make Validation Column*, with the random seed 12345) and make decisions about predictor inclusion, and fit three predictive models accordingly. For each model, give sufficient detail about the method used, its parameters, and the predictors used, so that your results can be replicated, and save the scripts for your models to the data table. Save the prediction formulas to the data table.
4. Among your three models, choose the best one in terms of predictive power. Which one is it? Why did you choose it?
5. Using your chosen model, report the propensities for the first three records in the validation set.
6. Create a derived variable that is the opposite of *Flyer*. Call it *Flyer-reversed*. Using your chosen model, refit the same model using the *Flyer-reversed* variable as a predictor, instead of *Flyer*. Report the propensities for the first three records in the validation set.
7. For each record, uplift is computed based on the following difference:

$$P(\text{success}|\text{Flyer} = 1) - P(\text{success}|\text{Flyer} = 0)$$

   Compute the uplift for each of the voters in the validation set, and report the uplift for the first three records.
8. If a campaign has the resources to mail the flyer only to 10% of the voters, what uplift cutoff should be used?

9. Use the Uplift platform in JMP Pro (under *Analyze > Consumer Research*) to fit an uplift model for MOVED_AD. Recall that the Uplift platform in JMP is based on the *Partition* platform.

   • What are the overall rates (propensities) for Flyer = 0 and Flyer = 1 (before splitting)?

   • What is the uplift before splitting? Interpret this value.

   • Click *Go* to automatically build and prune the model. Using the leaf report and the uplift graph, determine which node has the highest uplift for the training set? What is the uplift for this node? Provide an interpretation of this uplift value.

   • Save the difference formula to the data table. Create a subset of this table containing only the validation set, and sort the differences (the uplift) in descending order. What is the highest uplift for the validation set?

   • Based on this model, if a campaign has the resources to mail the flyer only to 10% of the voters, what uplift cutoff should be used? Characterize these voters in terms of the predictors in your model.

22.5 TAXI CANCELLATIONS[6]

TaxiCancellation.jmp is the dataset for this case study.

Business Situation

In late 2013, the taxi company Yourcabs.com in Bangalore, India, was facing a problem with the drivers using their platform—not all drivers were showing up for their scheduled calls. Drivers would cancel their acceptance of a call, and, if the cancellation did not occur with adequate notice, the customer would be delayed or even left high and dry.

Bangalore is a key tech center in India, and technology was transforming the taxi industry. Yourcabs.com featured an online booking system (though customers could phone in as well) and presented itself as a taxi booking portal. The Uber ride sharing service would start its Bangalore operations in mid-2014.

Yourcabs.com had collected data on its bookings from 2011 to through 2013 and posted a contest on Kaggle, in coordination with the Indian School of Business, to see what it could learn about the problem of cab cancellations.

The data presented for this case are a randomly selected subset of the original data, with 10,000 rows, one row for each booking. There are 17 input variables, including user (customer) ID, vehicle model, whether the booking was made online or via a mobile app, type of travel, type of booking package, geographic information, and the date and time of the scheduled trip. The outcome variable of interest is the binary indicator of whether a ride was canceled. The overall cancellation rate is between 7% and 8%. The data are in an Excel file.

Assignment

1. How can a predictive model based on these data be used by Yourcabs.com?
2. How can a profiling model (identifying predictors that distinguish canceled/ uncanceled trips) be used by Yourcabs.com?
3. Explore, prepare, and transform the data to facilitate predictive modeling. Here are some hints:
   - In exploratory modeling, it is useful to move fairly soon to at least an initial model without solving *all* data preparation issues. One example is the GPS information— other geographic information is available, so you could defer the challenge of how to interpret/use the GPS information.
   - How will you deal with missing data, such as cases where "NULL" is indicated? (Consider using utilities under the *Analyze > Screening* menu, a column property, or a column utility in JMP.)
   - Think about what useful information might be held within the date and time fields (the booking timestamp and the trip timestamp). The Excel file has a tab with hints on how to transform the date variables. Recall that the JMP formula editor can be used to extract information from variables and derive new variables.
   - Think also about the categorical variables and how to deal with them. Should we recode some of them? Use only some of the variables?

---

4. Fit several predictive models of your choice; you can build various models individually or leverage *Model Screening*. Do they provide information on how the predictor variables relate to cancellations? Compare the models, and select one best-performing model.

5. Report the predictive performance of your model in terms of error rates (the confusion matrix). How well does the model perform? Can the model be used in practice?

6. Examine the predictive performance of your model in terms of ranking (lift). How well does the model perform? Can the model be used in practice?

22.6 SEGMENTING CONSUMERS OF BATH SOAP[7]

BathSoap.jmp is the dataset for this case study.

Business Situation

CRISA is an Asian market research agency that specializes in tracking consumer purchase behavior in consumer goods (both durable and nondurable). In one major research project, CRISA tracks numerous consumer product categories (e.g., "detergents") and, within each category, perhaps dozens of brands. To track purchase behavior, CRISA constituted household panels in over 100 cities and towns in India, covering most of the Indian urban market. The households were carefully selected using stratified sampling to ensure a representative sample; a subset of 600 records is analyzed here. The strata were defined on the basis of socioeconomic status and the market (a collection of cities).

CRISA has both transaction data (each row is a transaction) and household data (each row is a household), and for the household data, CRISA maintains the following information:

- Demographics of the households (updated annually)
- Possession of durable goods (car, washing machine, etc., updated annually; an "affluence index" is computed from this information)
- Purchase data of product categories and brands (updated monthly)

CRISA has two categories of clients: (1) advertising agencies that subscribe to the database services, obtain updated data every month, and use the data to advise their clients on advertising and promotion strategies, and (2) consumer goods manufacturers that monitor their market share using the CRISA database.

Key Problems

CRISA has traditionally segmented markets on the basis of purchaser demographics. They would now like to segment the market based on two key sets of variables more directly related to the purchase process and to brand loyalty:

1. Purchase behavior (volume, frequency, susceptibility to discounts, and brand loyalty)
2. Basis of purchase (price, selling proposition)

Doing so would allow CRISA to gain information about what demographic variables are associated with different purchase behaviors and degrees of brand loyalty and thus deploy promotion budgets more effectively. More effective market segmentation would enable CRISA's clients (in this case, a firm called IMRB) to design more cost-effective promotions targeted at appropriate segments. Thus multiple promotions could be launched, each targeted at different market segments at different times of the year. This would result in a more cost-effective allocation of the promotion budget to different market segments. It would also enable IMRB to design more effective customer reward systems and thereby increase brand loyalty.

[7]Copyright © 2019 Cytel, Inc. and Datastats, LLC. Used by Permission.

Data

The data in BathSoap.jmp profile each household, so that each row contains data for one household. The variable descriptions are provided in Table 22.5.

TABLE 22.5   Description of variables for each household

Variable type	Variable name	Description
Member ID	Member id	Unique identifier for each household
Demographics	SEC	Socioeconomic class (1 = high, 5 = low)
	FEH	Eating habits(1 = vegetarian, 2 = vegetarian but eat eggs, 3 = nonvegetarian, 0 = not specified)
	MT	Native language (see table in worksheet)
	SEX	Gender of homemaker (1 = male, 2 = female)
	AGE	Age of homemaker
	EDU	Education of homemaker (1 = minimum, 9 = maximum)
	HS	Number of members in household
	CHILD	Presence of children in household (4 categories)
	CS	Television availability (1 = available, 2 = unavailable)
	Affluence Index	Weighted value of durables possessed
Purchase summary over the period	No. of Brands	Number of brands purchased
	Brand Runs	Number of instances of consecutive purchase of brands
	Total Volume	Sum of volume
	No. of Trans	Number of purchase transactions (multiple brands purchased in a month are counted as separate transactions
	Value	Sum of value
	Trans/ Brand Runs	Average transactions per brand run
	Vol/Trans	Average volume per transaction
	Avg. Price	Average price of purchase
Purchase within promotion	Pur Vol	Percent of volume purchased
	No Promo - %	Percent of volume purchased under no promotion
	Pur Vol Promo 6%	Percent of volume purchased under promotion code 6
	Pur Vol Other Promo %	Percent of volume purchased under other promotions
Brandwise purchase	Br. Cd. (57, 144), 55, 272, 286, 24, 481, 352, 5, and 999 (others)	Percent of volume purchased of the brand
Price categorywise purchase	Price Cat 1 to 4	Percent of volume purchased under the price category
Selling propositionwise purchase	Proposition Cat 5 to 15	Percent of volume purchased under the product proposition category

Measuring Brand Loyalty

Several variables in this case measure aspects of brand loyalty. The number of different brands purchased by the customer is one measure. However, a consumer who purchases one or two brands in quick succession, then settles on a third for a long streak, is different from a consumer who constantly switches back and forth among three brands. How often customers switch from one brand to another is another measure of loyalty. Yet a third perspective on the same issue is the proportion of purchases that go to different brands—a consumer who spends 90% of his or her purchase money on one brand is more loyal than a consumer who spends more equally among several brands.

All three of these components can be measured with the data in the purchase summary worksheet.

Assignment

1. Use k-means clustering to identify clusters of households based on the following:
   a. Variables that describe purchase behavior (including brand loyalty)
   b. Variables that describe the basis for purchase
   c. Variables that describe both purchase behavior and basis of purchase

   Note 1: How should k be chosen? Think about how the clusters would be used. It is likely that the marketing efforts would support two to five different promotional approaches.
   Note 2: How should the percentages of total purchases comprised by various brands be treated? Isn't a customer who buys all brand A just as loyal as a customer who buys all brand B? What will be the effect on any distance measure of using the brand share variables as is? Consider using a single derived variable.

2. Select what you think is the best segmentation, and comment on the characteristics (demographic, brand loyalty, and basis for purchase) of these clusters. (This information would be used to guide the development of advertising and promotional campaigns.)

3. Develop a model that classifies the data into these segments. Since this information would most likely be used in targeting direct mail promotions, it would be useful to select a market segment that would be defined as a *success* in the classification model.

22.7 CATALOG CROSS-SELLING[8]

CatalogCrossSell.jmp is the dataset for this case study.

Background

Exeter, Inc., is a catalog firm that sells products in a number of different catalogs that it owns. The catalogs number in the dozens but fall into nine basic categories:

1. Clothing
2. Housewares
3. Health
4. Automotive
5. Personal electronics
6. Computers
7. Garden
8. Novelty gift
9. Jewelry

The costs of printing and distributing catalogs are high. By far, the biggest cost of operation is the cost of promoting products to people who buy nothing. Having invested so much in the production of artwork and printing of catalogs, Exeter wants to take every opportunity to use them effectively. One such opportunity is in cross-selling—once a customer has "taken the bait" and purchases one product, try to sell them another while you have their attention.

Such cross-promotion might take the form of enclosing a catalog in the shipment of a purchased product, together with a discount coupon to induce a purchase from that catalog. Or, it might take the form of a similar coupon sent by email, with a link to the web version of that catalog.

But which catalog should be enclosed in the box or included as a link in the email with the discount coupon? Exeter would like it to be an informed choice—a catalog that has a higher probability of inducing a purchase than simply choosing a catalog at random.

Assignment

Using the dataset CatalogCrossSell.jmp, perform an association rules analysis, and comment on the results. Your discussion should provide interpretations in plain, non-technical language of the meanings of the various output statistics (lift ratio, confidence, support) and include a very rough estimate (precise calculations are not necessary) of the extent to which this will help Exeter make an informed choice about which catalog to cross-promote to a purchaser.

---

[8]Copyright ©Datastats, LLC 2019; used with permission.

Acknowledgment

The data for this case have been adapted from the data in a set of cases provided for educational purposes by the Direct Marketing Education Foundation ("DMEF Academic Data Set Two, Multi Division Catalog Company, Code: 02DMEF"), used with permission.

22.8 DIRECT-MAIL FUNDRAISING

Fundraising.jmp and FutureFundraising.jmp are the datasets used for this case study.

Background

Note: Be sure to read the information about oversampling and adjustment in Chapter 5 before starting to work on this case.

A national veterans' organization wishes to develop a predictive model to improve the cost-effectiveness of their direct marketing campaign. The organization, with its in-house database of over 13 million donors, is one of the largest direct-mail fundraisers in the United States. According to their recent mailing records, the overall response rate is 5.1%. Out of those who responded (donated), the average donation is $13.00. Each mailing, which includes a gift of personalized address labels and assortments of cards and envelopes, costs $0.68 to produce and send. Using these facts, we take a sample of this dataset to develop a classification model that can effectively capture donors so that the expected net profit is maximized. Weighted sampling is used, underrepresenting the nonresponders so that the sample has equal numbers of donors and nondonors.

Data

The file Fundraising.jmp contains 3120 records with 50% donors (TARGET_$B = 1$) and 50% nondonors (TARGET_$B = 0$). The amount of donation (TARGET_D) is also included but is not used in this case. The descriptions for the 22 variables (including two outcome variables) are listed in Table 22.6.

Assignment

Step 1: Prepare the data: Partition the dataset into 60% training and 40% validation (set the seed to 12345). Specify a profit matrix for TARGET_B. (Again, the expected donation, given that they are donors, is $13.00, and the total cost of each mailing is $0.68.)

Step 2: Build the model: Follow the next series of steps to build, evaluate, and choose a model.

1. *Select classification tool and parameters:* Run at least two classification models of your choosing. Be sure NOT to use TARGET_D in your analysis. Describe the two models that you chose, with sufficient detail (method, parameters, variables, etc.) so that they can be replicated. Save the scripts for the models to the data table, and save the probability formulas to the data table.

2. *Classification under asymmetric response and cost:* What is the reasoning behind using weighted sampling to produce a training set with equal numbers of donors and nondonors? Why not use a simple random sample from the original dataset?

TABLE 22.6   Description of Variables for the Fundraising Dataset

Variable	Description
ZIP	Zip code group (Zip codes were grouped into five groups: 1 = the potential donor belongs to this zip group.) 00000–19999 ← zipconvert_1 20000–39999 ← zipconvert_2 40000–59999 ← zipconvert_3 60000–79999 ← zipconvert_4 80000–99999 ← zipconvert_5
HOMEOWNER	1 = homeowner, 0 = not a homeowner
NUMCHLD	Number of children
INCOME	Household income
GENDER	0 = male, 1 = female
WEALTH	Wealth rating uses median family income and population statistics from each area to index relative wealth within each state. The segments are denoted 0 to 9, with 9 being the highest wealth group and zero the lowest. Each rating has a different meaning within each state.
HV	Average home value in potential donor's neighborhood in hundreds of dollars
ICmed	Median family income in potential donor's neighborhood in hundreds of dollars
ICavg	Average family income in potential donor's neighborhood in hundreds
IC15	Percent earning less than $15K in potential donor's neighborhood
NUMPROM	Lifetime number of promotions received to date
RAMNTALL	Dollar amount of lifetime gifts to date
MAXRAMNT	Dollar amount of largest gift to date
LASTGIFT	Dollar amount of most recent gift
TOTALMONTHS	Number of months from last donation to July 1998 (the last time the case was updated)
TIMELAG	Number of months between first and second gift
AVGGIFT	Average dollar amount of gifts to date
TARGET_B	Outcome variable: binary indicator for response 1 = donor, 0 = nondonor
TARGET_D	Outcome variable: donation amount (in dollars). We will NOT be using this variable for this case

3. *Calculate net profit:* For each method, calculate the net profit for both the training and validation set based on the actual response rate (5.1%). [*Hint*: To calculate estimated net profit, we will need to undo the effects of the weighted sampling and calculate the net profit that would reflect the actual response distribution of 5.1% donors and 94.9% nondonors. To do this, divide each row's net (expected) profit by the oversampling weights applicable to the actual status of that row (create a weight formula column). The oversampling weight for actual donors is 50%/5.1% = 9.8. The oversampling weight for actual nondonors is 50%/94.9% = 0.53.]

4. *Draw lift curves:* Draw each model's net profit lift curve for the validation set onto a single graph (net profit on the *y*-axis, proportion of list or number mailed on the *x*-axis). Is there a model that dominates?

5. *Select best model:* From your answer in 2, what do you think is the "best" model?

Step 3: Testing: The file FutureFundraising.jmp contains the attributes for future mailing candidates.

6. Using your "best" model from step 2 (number 5), which of these candidates do you predict as donors and nondonors? (*Hint*: Copy and paste the formula(s) for your final model into new columns in this table.) Sort them in descending order of the probability of being a donor. Starting at the top of this sorted list, roughly how far down would you go in a mailing campaign?

22.9   TIME SERIES CASE: FORECASTING PUBLIC TRANSPORTATION DEMAND

`Bicup2006.xlsx` is the dataset for this case study.

Background

Forecasting transportation demand is important for multiple purposes such as staffing, planning, and inventory control. The public transportation system in Santiago de Chile has gone through a major effort of reconstruction. In this context, a business intelligence competition took place in October 2006 that focused on forecasting demand for public transportation. This case is based on the competition, with some modifications.

Problem Description

A public transportation company is expecting an increase demand for its services and is planning to acquire new buses and to extend its terminals. These investments require a reliable forecast of future demand. To create such forecasts, one can use data on historic demand. The company's data warehouse has data for each 15-minute interval between 6:30 and 22:00, on the number of passengers arriving at the terminal. As a forecasting consultant you have been asked to create a forecasting method that can generate forecasts for the number of passengers arriving at the terminal.

Available Data

Part of the historic information is available in the file `Bicup2006.xlsx`. The file contains the worksheet "Historic Information" with known demand for a three-week period, separated into 15-minute intervals. The second worksheet ("Future") contains dates and times for a future three-day period, for which forecasts should be generated (as part of the 2006 competition).

Import the data from the two worksheets into one JMP data table using the JMP *Excel Import Wizard.* (*Note:* Since the imported data will have two columns, Date and Time, you have a few options regarding how to handle the time variable when you fit your models. You can leave the *X, Time ID* variable field blank (which will use the row order for the time series), you can use a formula to combine the columns to create a single (date+time) column, or you can insert a new "row number" column, and use this as your time ID.)

Assignment Goal

Your goal is to create a model/method that produces accurate forecasts. To evaluate your accuracy, partition the given historic data into two periods: a training period (the first two weeks) and a validation period (the last week). Models should be fitted only to the training data and evaluated on the validation data.

Although the competition winning criterion was the lowest mean absolute error (MAE) on the future three-day data, this is *not* the goal for this assignment. Instead, if we consider a more realistic business context, our goal is to create a model that generates reasonably good forecasts on any time/day of the week. Consider not only predictive metrics such as

MAE and mean absolute percentage error (MAPE), but also look at actual and forecasted values, overlaid on a time plot, as well as a time plot of the forecast errors.

Assignment

For your final model, present the following summary:

1. Name of the method/combination of methods.
2. A brief description of the method/combination.
3. All estimated equations associated with constructing forecasts from this method.
4. Two of the following statistics: MAPE, MAE, average error, and RMSE (or RASE) for the training period and the validation (or holdback) period.
5. Forecasts for the future period (March 22–24), in 15-minute bins.
6. A single chart showing the fit of the final version of the model to the entire period, including training, validation (holdback), and future. Note that this model should be fit using the combined training and validation (holdback) data.

Tips and Suggested Steps

1. Use exploratory analysis to identify the components of this time series. Is there a trend? Is there seasonality? If so, how many "seasons" are there? Are there any other visible patterns? Are the patterns global (the same throughout the series) or local?
2. Consider the frequency of the data from a practical and technical point of view. What are some options?
3. Compare the weekdays and weekends. How do they differ? Consider how these differences can be captured by different methods.
4. Examine the series for missing values or unusual values. Think of solutions.
5. Based on the patterns that you found in the data, which models or methods should be considered?

22.10   LOAN APPROVAL[9]

`UniversalBankCase.jmp` is the dataset for this case study.

Background

You are a data scientist recently hired by Universal Bank, a mid-sized bank in the southern United States. Most of your work to this point has involved pulling reports from databases, but now you have been given a more interesting task. The bank is facing competition from online lenders that can offer rapid automated loan approvals, and it wants to develop its own predictive model so that it can do likewise. Before building the web infrastructure, which could be costly, the bank wants to pilot a prototype loan approval model, developed by you.

The bank wants to launch its model with the approval process for personal loans extended to existing customers. The bank has only been offering these loans for a relatively short time, so has little data on default rates. It does have data on prior loan applications and whether they were approved or disapproved. You have done some preliminary data prep and feature selection work, resulting in the dataset of 5000 records for this project. Each record is for a customer and consists of feature values for that customer along with a record of the human decision on their loan application. Your end goal is to develop a model to predict that human decision and an accompanying report to the bank's chief lending officer.

Regulatory Requirements

You are somewhat familiar with regulatory requirements with respect to discrimination (do a web search for the US Department of Justice Equal Credit Opportunity Act or ECOA). Bank attorneys have told you that the ECOA requirements pertain to the basis for credit decisions and do not mean that the proportion of approved loans must be the same for all groups.

Getting Started

You have the RDS framework in hand (see Chapter 21). You recognize that some aspects of the framework may take more time (e.g., consulting with other stakeholders and learning more about how the data were produced), but you want to get started with at least a draft so that you can discuss next steps with your superiors.

Assignment

1. Identify any features that might need to be excluded from the modeling task, per the ECOA.
2. Should you simply eliminate these features from the data?
3. Explore the data, with a focus on loan approval rates for different groups.

[9]Copyright Datastats, LLC, and Galit Shmueli. Used by permission.

4. Split the data into training and validation data, and fit several models of your choice to predict whether a personal loan should be approved, using only permitted features. There are a number of possible performance measures; evaluate the model performance by the metric(s) you consider useful.

5. Assess the usefulness of the model from a pure model performance standpoint.

6. Considering the "protected" categories per the ECOA, evaluate whether the model is fair.

7. Describe steps that might be taken to improve the model fairness. Implement any measures that can be taken without going beyond the dataset at hand. (*Hint*: If you did not calculate a correlation matrix earlier, you should do one now.)

8. Write a very short report summarizing your findings.

REFERENCES

Aggarwal, C. C. (2016). *Recommender Systems* (Vol. 1). Springer Cham.

Agrawal, R., Imielinski, T., and Swami, A. (1993). Mining associations between sets of items in massive databases. In: *Proceedings of the 1993 ACM-SIGMOD International Conference on Management of Data*, pp. 207–216. New York: ACM Press.

Berry, M. J. A., and Linoff, G. S. (1997). *Data Mining Techniques*. New York: Wiley.

Berry, M. J. A., and Linoff, G. S. (2000). *Mastering Data Mining*. New York: Wiley.

Breiman, L., Friedman, J., Olshen, R., and Stone, C. (1984). *Classification and Regression Trees*. Boca Raton, FL: Chapman & Hall/CRC (orig. published by Wadsworth).

Brin, S., Motwani, R., Ullman, J. D., and Tsur, S. (1997). Dynamic itemset counting and implication rules for market basket data. In: *Proceedings of the 1997 ACM SIGMOD international conference on Management of data*, pp. 255–264. New York: ACM Press.

Chatfield, C. (2003). *The Analysis of Time Series: An Introduction*, 6th ed. Boca Raton, FL: Chapman & Hall/CRC.

Delmaster, R., and Hancock, M. (2001). *Data Mining Explained*. Boston, MA: Digital Press.

Efron, B. (1975). The Efficiency of Logistic Regression Compared to Normal Discriminant Analysis. *Journal of the American Statistical Association*, vol. 70, number 352, pp. 892–898.

Few, S. (2012). *Show Me the Numbers*, 2nd ed. Oakland, CA: Analytics Press.

Few, S. (2021). *Now You See It*. 2nd ed. Oakland, CA: Analytics Press.

Fleming, G., and Bruce, P. C. (2021). *Responsible Data Science*. Hoboken, NJ: Wiley.

Golbeck, J. (2013). *Analyzing the Social Web*. Waltham, MA: Morgan Kaufmann.

Hand, D. J. (2009). Measuring classifier performance: a coherent alternative to the area under the ROC curve. *Machine Learning*, vol. 77, number 1, pp. 103–123.

Harris, H., Murphy, S., and Vaisman, M. (2013). *Analyzing the Analyzers: An Introspective Survey of Data Scientists and Their Work*. Cambridge, MA: O'Reilly Media.

Machine Learning for Business Analytics: Concepts, Techniques, and Applications with JMP Pro®,
Second Edition. Galit Shmueli, Peter C. Bruce, Mia L. Stephens, Muralidhara Anandamurthy, and Nitin R. Patel.
© 2023 John Wiley & Sons, Inc. Published 2023 by John Wiley & Sons, Inc.

Hastie, T., Tibshirani, R., and Friedman, J. (2001). *The Elements of Statistical Learning*. New York: Springer.

Hosmer, D. W., and Lemeshow, S. (2000). *Applied Logistic Regression*, 2nd ed. New York: Wiley-Interscience.

Hyndman, R., and Yang, Y. Z. (2018). tsdl: Time Series Data Library. v0.1.0. `https://pkg.yangzhuoranyang.com/tsdl/`.

Jank, W., and Yahav, I. (2010). E-Loyalty networks in online auctions. *The Annals of Applied Statistics*, vol. 4, number 1, pp. 151–178.

JMP Statistical Discovery LLC 2022–2023. *Discovering JMP® 17*. Cary, NC: JMP Statistical Discovery LLC

Johnson, W., and Wichern, D. (2002). *Applied Multivariate Statistics*. Upper Saddle River, NJ: Prentice Hall.

Kohavi, R., Tang, D., and Xu, Y. (2020). *Trustworthy Online Controlled Experiments: A Practical Guide to A/B Testing*. Cambridge University Press. NY: Cambridge University Press.

Larsen, K. (2005). Generalized naive Bayes classifiers. *SIGKDD Explorations*, vol. 7, number 1, pp. 76–81.

Le, Q. V., Ranzato, M. A., Monga, R., Devin, M., Chen, K., Corrado, G. S., Dean, J., and Ng, A.Y. (2012). Building high-level features using large scale unsupervised learning. In: *Proceedings of the Twenty-Ninth International Conference on Machine Learning*. Editors: J. Langford and J. Pineau. Edinburgh: Omnipress.

Lim, L., Acito, F., and Rusetski, A. (2006). Development of archetypes of international marketing strategy. *Journal of International Business Studies*, vol. 37, pp. 499–524. DOI: 10.1057/palgrave.jibs.8400206.

Lobo, J. M., Jiménez-Valverde, A., and Real, R. (2008). AUC: a misleading measure of the performance of predictive distribution models. *Global Ecology and Biogeography*, vol. 17, number 2, pp. 145–151.

Loh, W.-Y., and Shih, Y.-S. (1997). Split selection methods for classification trees. *Statistica Sinica*, vol. 7, pp. 815–840.

Maas, A. L., Daly, R. E., Pham, P. T., Huang, D., Ng, A. Y., and Potts, C. (2011). Learning word vectors for sentiment analysis. In: *Proceedings of the 49th Annual Meeting of the Association for Computational Linguistics: Human Language Technologies - vol. 1*, pp. 142–150. DOI: 10.5555/2002472.2002491.

Matz, S. C., Kosinski, S. C., Nave, G., and Stillwell, D. J. (2017). Psychological targeting in digital mass persuasion. In: *Proceedings of the National Academy of Sciences*, vol. 114, number 48, pp. 12714–12719.

McCullugh, C. E., Paal, B., and Ashdown, S. P. (1998). An optimisation approach to apparel sizing. *Journal of the Operational Research Society*, vol. 49, number 5, pp. 492–499.

Molnar, C. (2022). *Interpretable Machine Learning: A Guide for Making Black Box Models Explainable*, 2nd ed. Lulu.com. `https://christophm.github.io/interpretable-ml-book/`.

O'Neil, C. (2016). *Weapons of Math Destruction*. New York: Crown Publishers.

Pregibon, D. (1999). 2001: a statistical odyssey. Invited talk at *The Fifth ACM SIGKDD International Conference on Knowledge Discovery and Data Mining*. New York: ACM Press. p. 4.

Ribeiro, M. T., Singh, S., and Guestrin, C. (2016). Why should I trust you?: Explaining the predictions of any classifier. In: *Proceedings of the 22nd ACM SIGKDD International Conference on Knowledge Discovery and Data Mining*, pp. 1135–1144. DOI: 10.1145/2939672.2939778.

Russell, S. (2019). *Human Compatible: Artificial Intelligence and the Problem of Control*. New York: Penguin Books.

Saddiqi, N. (2017). *Intelligent Credit Scoring: Building and Implementing Better Credit Risk Scorecards*, 2nd ed. Hoboken, NJ: Wiley.

Sall, J. (2002). Monte Carlo calibration of distributions of partition statistics. *White Paper.* http://www.jmp.com/content/dam/jmp/documents/en/white-papers/montecarlocal.pdf.

Sall, J., Lehman, A., Stephens, M., and Loring, S. (2017). *JMP® Start Statistics: A Guide to Statistics and Data Analysis Using JMP®*, 6th ed. Cary, NC: SAS Institute Inc.

Sarle Warren S. (1983). *Cubic Clustering Criterion.* SAS Institute.

Shmueli, G., and Lichtendahl, K. C. (2016). *Practical Time Series Forecasting with R: A Hands- On Guide*, 2nd ed. Green Cove Springs, FL: Axelrod-Schnall Publishers.

Siegel, E. (2013). *Predictive Analytics.* New York: Wiley.

Sutton, R. S., and Barto, A. G. (2018). *Reinforcement Learning: An Introduction*, 2nd ed. Cambridge, MA: Bradford Books.

DATA FILES USED IN THE BOOK

1. Accidents.jmp
2. Accidents NN.jmp
3. Accidents1000DA.jmp
4. Airfares.jmp
5. Amtrak.jmp
6. AmtrakTS.jmp
7. ApplianceShipments.jmp
8. AustralianWines.jmp
9. AutoAndElectronics.zip
10. autos.jmp
11. BankBiasData.jmp
12. Bankruptcy.jmp
13. Banks.jmp
14. BareggDaily.jmp
15. BathSoap.jmp
16. Bicup2006.xlsx
17. BostonHousing.jmp
18. CanadianWorkHours.jmp
19. CatalogCrossSell.jmp
20. Cereals.jmp
21. CharlesBookClub.jmp
22. Colleges.jmp
23. Combined.jmp
24. Compas-clean.csv
25. COMPAS.jmp
26. Cosmetics.jmp
27. CourseTopics.jmp
28. DepartmentStoreSales.jmp
29. EastWestAirlinesCluster.jmp
30. EastWestAirlinesNN.jmp
31. eBayAuctions.jmp
32. electronics.jmp
33. emailABtest.jmp
34. FarmAds.jmp
35. FinancialReporting10Companies.jmp
36. FinancialReportingRaw.jmp
37. FlightDelays.jmp
38. FlightDelaysLR.jmp
39. FlightDelaysNB.jmp
40. FlightDelaysPrepared.jmp
41. Fundraising.jmp
42. FutureFundraising.jmp
43. GermanCredit.jmp
44. HairCareProduct.jmp
    (Also provided in the JMP Sample Data
    folder as Hair Care Product.jmp)
45. IMDB-Dataset-10k.jmp
46. LaptopSales.txt

Machine Learning for Business Analytics: Concepts, Techniques, and Applications with JMP Pro®,
Second Edition. Galit Shmueli, Peter C. Bruce, Mia L. Stephens, Muralidhara Anandamurthy, and Nitin R. Patel.
© 2023 John Wiley & Sons, Inc. Published 2023 by John Wiley & Sons, Inc.

INDEX